applied electronics

CARL B. WEICK

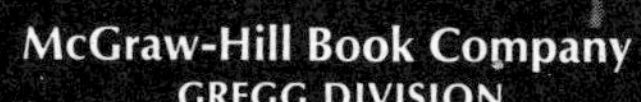

McGraw-Hill Book Company
GREGG DIVISION

York • St. Louis • Dallas • San Francisco • Auckland • Düsseldorf • Johannesburg • Kuala Lumpur • London
Mexico • Montreal • New Delhi • Panama • Paris • São Paulo • Singapore • Sydney • Tokyo • Toronto

Special adaptation of APPLIED ELECTRONIC CIRCUITS by Carl B. Weick for distribution in the United States only.

Library of Congress Cataloging in Publication Data

Weick, Carl B
Applied electronics.

Includes index.
1. Electronic circuits. 2. Electronics.
I. Title
TK7867.W37 621.381 75-35886
ISBN 0-07-069012-X

Illustrations by Bruce A. Renton
Photography by Tracy A. Glasner

Applied Electronics

Especially adapted for United States distribution.

1 2 3 4 5 6 7 8 9 0 KPKP 7 8 3 2 1 0 9 8 7 6

The editors for this book were *Gordon Rockmaker* and *Zivile K. Khoury*, The designer was *Tracy A. Glasner*, and the production supervisor was *Rena Shindelman*. It was set in Optima by Typographic Sales, Inc. Printed and bound by Kingsport Press, Inc.

contents

preface

This book is intended as a text for the study of circuit fundamentals. Although not designed as a laboratory manual, most of the circuits it covers could be inexpensively breadboarded for experiments.

Basic concepts of circuit operation can be explained in nonmathematical terms. Once they understand these concepts, serious readers can embark on a quantitative study using more advanced texts. In spite of the minimal mathematical content in this book, a serious effort was made to maintain rigor. The concepts gained here should not be in conflict with the mathematical analyses encountered in more advanced treatments of the subject.

The circuits chosen for discussion form the building blocks for electronic systems. Although the list is far from exhaustive, many circuits encountered in common appliances such as radio and television receivers, musical instruments, and industrial controls are included. Large numbers of such circuits are now available as monolithic integrated circuits, and their operation often bears little resemblance to the discrete circuits which they have replaced. Even so, it is common practice to explain their operation in comparative terms.

Some trade jargon is evident in the text and the purist will note expressions such as "frequencies are shunted" instead of "signals are shunted." Also, abbreviations are often used as adjectives; as, for example, in the phrases "an ac current" and "an FM signal."

Finally, a short but sincere note of thanks must be included. A ready source of suggestions, reference material, and equipment was always available from my colleagues and users of the previous edition of this text. Many students served unwittingly as "guinea pigs" in the testing of circuits. My special thanks to the illustrator, Mr. Bruce Renton, who made order out of my chaotic sketches.

B. Weick

1-1 PERSPECTIVE

Applied electronics has evolved from the union of wireless and electrical technology. Physics has been a prime catalyst and has become the dominant force in the expansion of electronic knowledge.

Every aspect of electronics must be visualized in its proper perspective. A large installation designed for a given task is known as a *system*. A system is all-encompassing. It is composed of lesser units, possibly subsystems, none of which alone could perform the task as specified for the system.

The hardware portion of a system is composed of a variety of *circuits*. Circuits are much more stylized than systems, and similar circuits are found in diverse systems. For this reason circuits lend themselves to cataloging, and designers rarely "design" a new circuit. Rather, they tend to trim an existing design to meet their needs.

Circuits are assembled from *components*. Many circuit forms are now produced as integrated circuits. In these all components are etched on a silicon substrate or "chip." This has led to further classification of circuits into those constructed of *discrete* components, and *integrated circuits*.

Some systems, such as digital computers, require more than electronic hardware to perform their functions. In addition to the assembly of physical components, a series of written instructions is needed. Such instructions are then coded into electrical pulses. The complete assemblage of instructions is known as the *software* for the system. Software is composed of a variety of programs, each designed to perform a specified task. The relationships among various aspects of an electronic installation are depicted in Fig. 1-1.

Even the system shown in Fig. 1-1 may be considered a subsystem of a larger system, such as accounting. Thus a system is usually also a subsystem. Only by definition can we distinguish between a system and a subsystem.

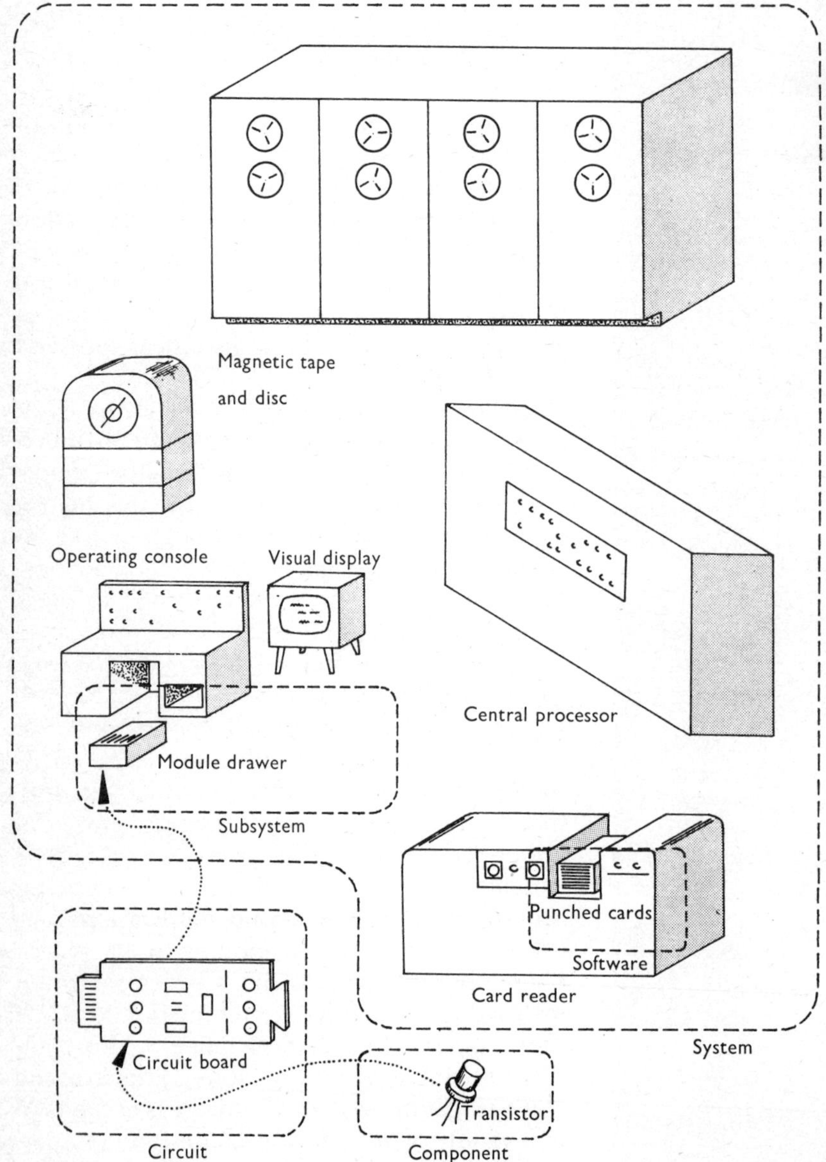

Fig. 1-1 Parts of an electronic system, consisting of various hardware subsystems and circuits, as well as the software required to make it operate.

1-2 DESCRIPTION AND DRAFTING PRACTICES

A large and complex system is described by drawings and schematic diagrams of various levels of detail. In a set of maintenance manuals you are likely to find most of the following types of diagrams:

FUNCTIONAL BLOCK DIAGRAMS give an overview of the system

SCHEMATIC DIAGRAMS show actual electrical connections

PICTORIAL DIAGRAMS show component placement

TIMING DIAGRAMS depict the chronological sequence of operations

TABLES OF TYPICAL MEASUREMENTS a technician's measured quantities can be compared with these values when troubleshooting

No universally accepted drafting practices are used, although strong convention can be found in all areas. Figure 1-2 shows examples of drawings found in maintenance literature.

In the smaller, less complex systems such as television receivers, "stereo" sets, and other domestic appliances, block diagrams are often omitted from maintenance publications. In larger systems, such as radar or computer installations, block diagrams are essential.

Schematic diagrams show in detail the electrical connections between components. It is good practice to limit them to the size of a page, or at most a foldout sheet. Larger schematics tend to obscure rather than clarify detail. Schematic diagrams can be compared with road maps. A road map of the world could be prepared, but it would be inconvenient to use on a trip. A typical size limit for a schematic diagram is that of a television receiver. In nearly every instance a complete television receiver schematic can be placed on a size-A_2 drawing sheet (420 x 594 mm). Larger systems require sets of schematic diagrams, with appropriate cross referencing between individual sheets.

Pictorial diagrams convey physical characteristics such as dimensions, location of assemblies, and grouping and placing of components. There must be complete correspondence between component identifying numbers on pictorial, schematic, and block diagrams.

When it is necessary to show the chronological sequence of events in a circuit, a timing diagram is prepared. A timing diagram is a variant of a rectangular graph. It has a common horizontal axis graduated in units of elapsed time. Several signal levels, that is, information expressed in electrical form, each with its own units of measurement in the vertical axis are then drawn on the common horizontal axis. In this way it is possible to see at each instant of time a given signal activity relative to any other signal in the group.

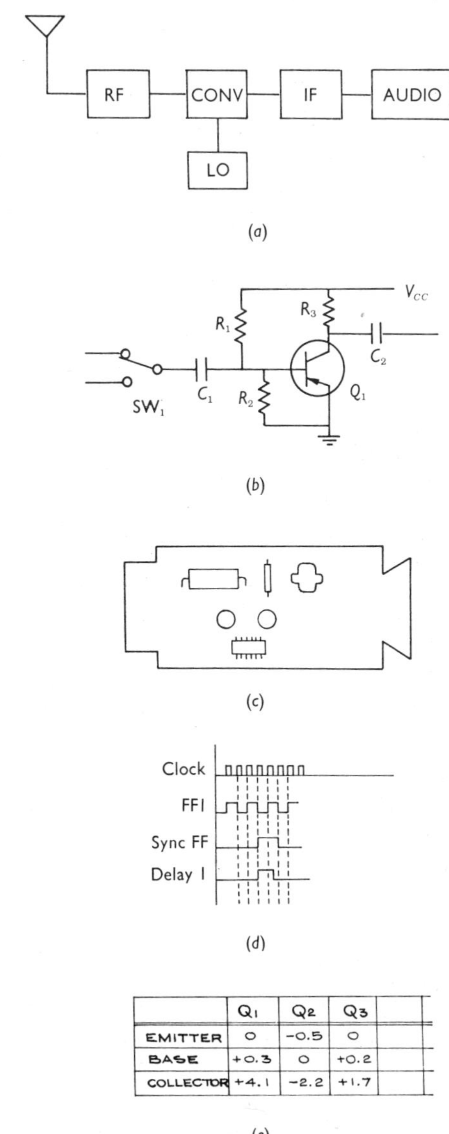

	Q1	Q2	Q3	
EMITTER	0	−0.5	0	
BASE	+0.3	0	+0.2	
COLLECTOR	+4.1	−2.2	+1.7	

(e)

Fig. 1-2 Types of diagrams found in maintenance manuals for electronic systems: (*a*) Functional, or block, diagram; (*b*) Schematic diagram; (*c*) Pictorial diagram; (*d*) Timing diagram; (*e*) Table of measurements.

Tables of measurements are listings of average measurements made on a system under specified conditions. If a malfunction occurs, the technician attempts to duplicate the specified conditions, and then to perform measurements and compare them to those of the manufacturer. It is essential to realize that published measurements are only a guide, as no two operating systems are likely to yield identical readings.

1-3 ANALYSIS TECHNIQUES

Circuit analysis is the exercise of studying a circuit to gain a given level of comprehension. Probably the two most powerful means of studying circuits are by analogy and the use of timing diagrams. Analogy implies reference to and comparison with previous knowledge and includes the preparation and study of *equivalent circuits.* Equivalent circuits are models of the real circuit constructed in such a way that a mathematical analysis is possible.

1-4 SPECIFICATIONS AND DESIGN CONSIDERATIONS

Electronic circuit designers must be aware of the environment in which equipment will operate. A given circuit may be electronically optimized, yet may fail to perform in other than laboratory conditions. It is often more difficult to design the package than the circuit.

A major source of trouble in electronic equipment arises from the use of connectors. For maintenance, equipment must be constructed in modular form. Even though each module is operating perfectly, plugs, pins, and connectors can cause faulty or intermittent system performance. In all electronic design, a great deal of effort is expended in keeping chemically unbonded metal-to-metal contacts at an absolute minimum.

Reliability is elusive and difficult to achieve. A measure of reliability is conveyed by a figure known as the *mean time between failure* or MTBF. For example, if a diary of failures is kept on a given equipment and the times between failures are averaged, an MTBF figure is obtained. This figure is dependent on the reliability of the individual components which comprise the equipment. The MTBF drops very significantly if the number of components is increased, regardless of how reliable each individual component is.

1-5 CLASSIFICATION OF CIRCUITS

Circuits are most frequently classified by function and often by the power levels they are capable of dissipating. Some common circuit functions are switching, gating, amplifying, oscillating, and comparing. Each of these classes contains a large variety of special types. For example, note some of the combinations possible with the term amplifier:

low-power (small-signal) high-power (large-signal)	audio servo ac dc wideband video IF RF differential	{amplifier}

1-6 FABRICATION TECHNIQUES

Circuits composed of large discrete components are constructed by mounting the components on terminal boards fastened to a metal chassis. Wire is used to interconnect the components. If power levels are low enough to warrant the use of small discrete components, they may be placed on an insulating board which plugs into a connector. Generally the board is copper-clad and the interconnection pattern is etched into the copper, in which case the board is known as a *printed circuit.*

Many circuits are available in integrated-circuit form. They are packaged in one of several standard forms, some of which are shown in Fig. 1-3. Integrated circuits are most commonly mounted on printed-circuit boards to form modules of subsystems. The major limitations of this construction are that (1) a large number of leads is required to connect to the module and (2) the power dissipation per module must be low because of the high packing density. Figure 1-4 illustrates one of the common mounting techniques in use.

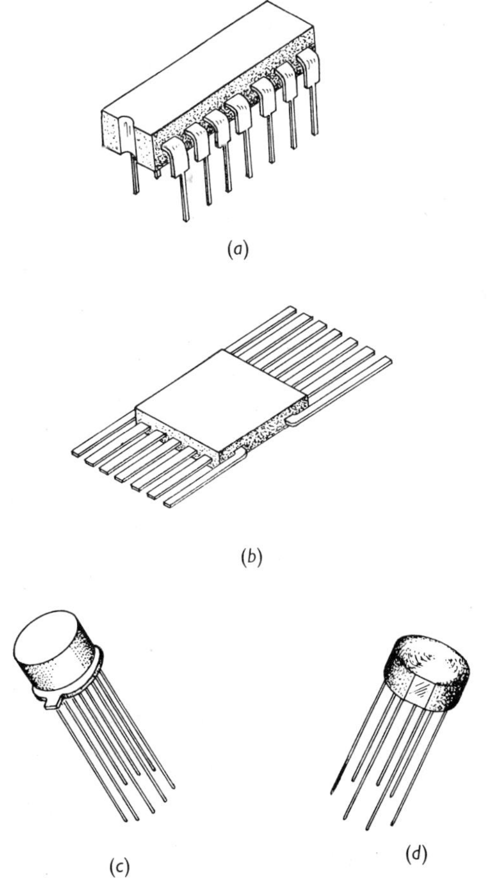

Fig. 1-3 Typical integrated-circuit packaging: (*a*) Plastic DIP (Dual Inline Pack); (*b*) Ceramic flat pack; (*c*) Metal can T pack; (*d*) Plastic T pack.

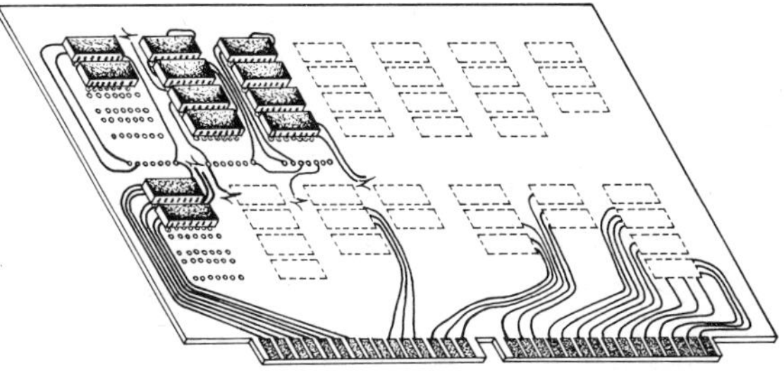

Fig. 1-4 Integrated circuits connected to form a subsystem on a plug-in type, printed-circuit board.

1-7 INTEGRATED-CIRCUIT TECHNOLOGY

A circuit is considered to be an integrated circuit if both active components (for example, transistors) and passive components (for example, resistors and capacitors) are interconnected and sealed in a common package. In most instances the circuit is hermetically sealed and only the signal and power-supply leads extend beyond the seal.

Integrated circuits fall within four classifications: (1) monolithic, (2) thin film, (3) thick film, and (4) hybrid.

In monolithic integrated circuits, both the active and passive components are fabricated as part of a single (mono) chip of semiconductor material, usually silicon. To conserve materials and obtain high yields, the circuits are made as small as the state of the art will allow. If the number of devices (transistors, diodes, FETs) located on a single chip approaches a density of the order of 50 000 per square inch, the circuit is termed Large Scale Integration (LSI).

A major problem is that of bonding leads to points on a microscopic-sized semiconductor chip. One type of lead connection is made by bonding fine gold wires with mechanical pressure to metalized contact areas. Monolithic circuits are made in a series of steps in temperature-controlled environments by injecting impurities into pure semiconductor material.

Thin-film circuits are made by depositing metallic material such as tantalum onto the surface of a glass or ceramic substrate. Note that deposition (and not injection) onto an insulating surface characterizes the thin-film process. Transistors cannot be made by deposition, so thin-film technology can be used only for fabricating the passive components such as resistors, capacitors, and inductors. The chief value of thin-film structures is that they can be combined with monolithic active elements to form "hybrid" integrated circuits.

Thick-film circuits are similar to thin film. The layer of deposited metal is much thicker than that used in thin-film circuits and as a result has little resistance. Thick films are used mainly for making interconnections on a substrate.

Hybrid circuits, combinations of monolithic and film circuits, have features not available in a single type. For example, it is costly in terms of chip space to fabricate resistors or capacitors into monolithic circuits. On the other hand, it is impossible to make transistors with the film process. A practical compromise is to manufacture as much as possible of the circuit in monolithic form and then to mount it on a thin-film circuit to make a finished hybrid product.

Summary

Electronics is an outgrowth of physics and wireless and electrical technology.

In an electronic installation it should be possible to visualize the *system* as a whole, one or more *subsystems,* numerous *circuits,* and finally large numbers of *components.*

Electronic circuits are of two types: *integrated* and *discrete.*

Electronic systems are often comprised of *software* as well as *hardware.*

Just as hardware can be subdivided into subsystems, circuits, and components, so software can be divided into software systems, programs, subroutines, and instructions.

Maintenance manuals contain block diagrams, schematic diagrams, pictorial diagrams, timing diagrams, and measurement tables.

Circuits are most commonly analyzed by analogy by using *equivalent circuits* and by studying timing diagrams.

Circuit packaging is no less important than the electrical performance.

The *mean time between failure* decreases as the number of components is increased, regardless of the reliability of the components.

Circuits are classified by function and by power dissipation.

Integrated circuits are classified as monolithic, thin film, thick film, and hybrid.

The term Large Scale Integration (LSI) is used to denote monolithic circuits with a high device density per chip.

Questions and Exercises

1. Figure 1-1 illustrates a large electronic digital-computer installation. Identify the following: **(*a*)** system, **(*b*)** subsystems, **(c)** circuit, **(*d*)** component, **(e)** software.
2. Select a communication system with which you are familiar. Draw a block diagram of major parts.
3. Describe briefly the difference between an integrated circuit and a discrete component circuit.
4. What is the difference between schematic and pictorial diagrams?
5. A technician discovers that a certain indicator lamp fails to come on in a specified sequence, even though all circuits appear to be performing normally. What type of diagram might be consulted?
6. What purpose does an equivalent circuit serve?
7. Figure 1-5 is a timing diagram of a counting circuit.

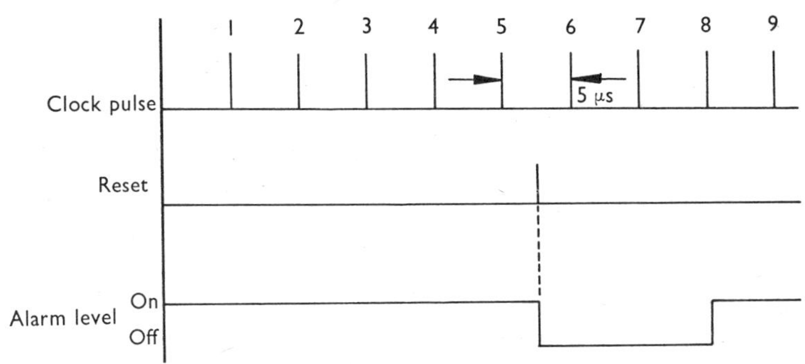

Fig. 1-5 Timing diagram.

a. What is the elapsed time between the reset pulse and the following clock pulse?

b. How many clock pulses occur between the reset pulse and the beginning of the alarm level?

c. How long does the alarm level remain off?

8. Suggest a way to reduce the number of connecting plugs in a given electronic system. What disadvantage would result if your suggestion were followed?

9. Why is it desirable to keep schematic diagrams on page-size drawing sheets?

10. What is meant by MTBF?

11. How are circuits classified?

12. State the difference between a printed circuit and a wired circuit.

13. What factors limit the size of a printed-circuit module?

14. State the meaning of the following terms: (*a*) monolithic integrated circuit, (*b*) thin-film integrated circuit, (*c*) thick-film integrated circuit, (*d*) hybrid integrated circuit.

15. State the procedure involved in the manufacture of a monolithic integrated circuit.

2-1 PRIMARY POWER SUPPLY—FUNDAMENTAL REQUIREMENTS

The primary electric power system is nearly always alternating current supplied by a public utility. It may range in voltage from a nominal 115 V, as obtainable at any wall outlet, to several thousand volts. Regardless of voltage, the frequency of the alternating current is accurately maintained. Voltage does vary and is largely beyond the control of the utility. It depends greatly on the instantaneous demand for power by the users. As loads are being switched on and off in a community, the *IR* drops in the power lines change. The varying *IR* drops cause voltage variations for every customer. Some users require primary power voltage regulators, but in most applications some voltage fluctuation is tolerated. Voltage regulation is usually performed as required at other points in the system.

Many electronic systems must operate continuously, such as process computers in automated manufacturing or electronic instruments in hospitals. These must be designed to operate not only with primary voltage fluctuations, but frequency variations as well. For instance, if a failure occurs with the utility, a local diesel electric generating system starts automatically. Invariably there is speed variation in the engine as the unit stabilizes. There is also a momentary drop in voltage when the load is switched from one source to the other.

A third commonly used primary source is an electrochemical cell or a battery. A battery supplies direct current, which is stable and free from variations, but subject to gradual drop as the source is consumed. It is suitable only for low-power portable equipment or for emergency standby. Certain types of cells and batteries are capable of having their voltage restored by recharging. Even when unused, a battery's voltage will tend to drop considerably below its rated value over a period of time.

Various primary power sources and their stability characteristics are illustrated in Fig. 2-1.

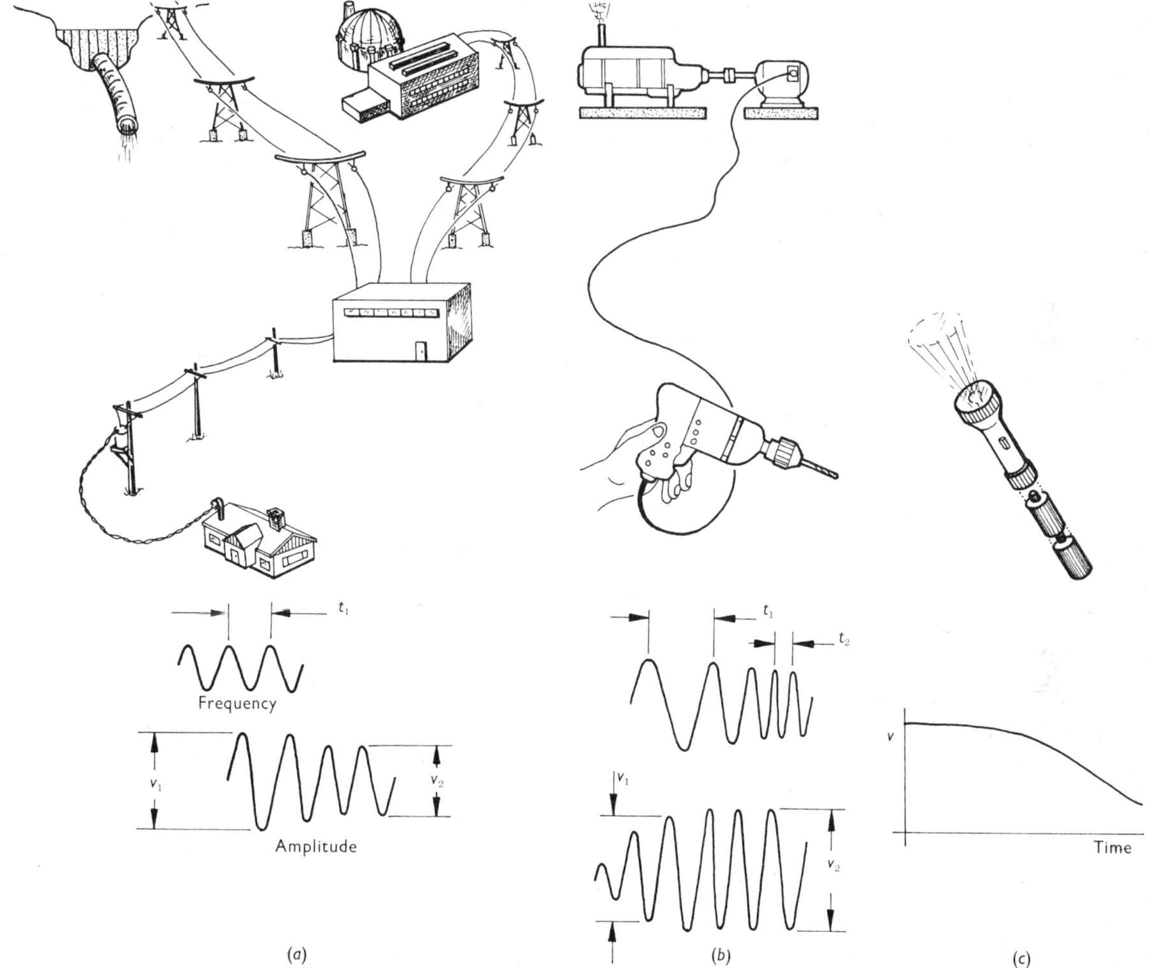

Fig. 2-1 Examples of primary electric power sources: (*a*) Public utility—excellent frequency stability, with reasonably good voltage stability; (*b*) Locally generated primary power—unpredictable frequency and voltage stability; (*c*) Battery—excellent short-term voltage stability, with gradual drop to exhaustion.

2-2 AC POWER DISTRIBUTION IN CIRCUITS AND SYSTEMS

In nearly every electronic system, input power is applied to the primary winding of some type of power transformer. This is done for two reasons: (1) The primary voltage must be stepped up or down, and (2) electrical isolation is desired between the equipment and the primary power wires. Usually, stepped-up voltage is processed immediately after it leaves the secondary winding of the transformer. The stepped-down low voltage may be further distributed to various modules or stages where it may be required to actuate other parts of the circuit.

When alternating current is distributed to various points in a chassis there is a risk that the alternating magnetic field surrounding the conductors will induce voltages in nearby components. Such voltages are referred to as noise, because in sensitive equipment they mix with the signal and degrade operation. For instance in a high-gain high-fidelity amplifier, voltages induced by stray power-supply fields are amplified and reproduced by the speaker as a hum or buzz. One way to minimize the production of the noise-inducing fields is to twist the ac distribution wires. Because current flow in each wire is of equal amplitude but opposite direction, the magnetic fields produced cancel each other's effects. Figure 2-2 illustrates some methods of minimizing noise due to power distribution.

2-3 DC POWER REQUIREMENTS

Nearly all electronic circuits contain either vacuum tubes or solid-state devices. These require direct current for their operation.

In early radio receivers only batteries were available to supply the required dc. Three distinct sources were needed: a low-voltage high-current cell for heating tube filaments, called the *A battery;* a high-voltage low-current battery for supplying plate current, known as the *B battery;* and a medium-voltage low-current battery for grid biasing, known as a *C battery.* Even after electronic power supplies replaced batteries, the labels *A*, *B*, and *C* remained. It is still common to find the labels *B*+ and *C*− on many schematic diagrams. The label *A* has disappeared from use. Most vacuum tubes used

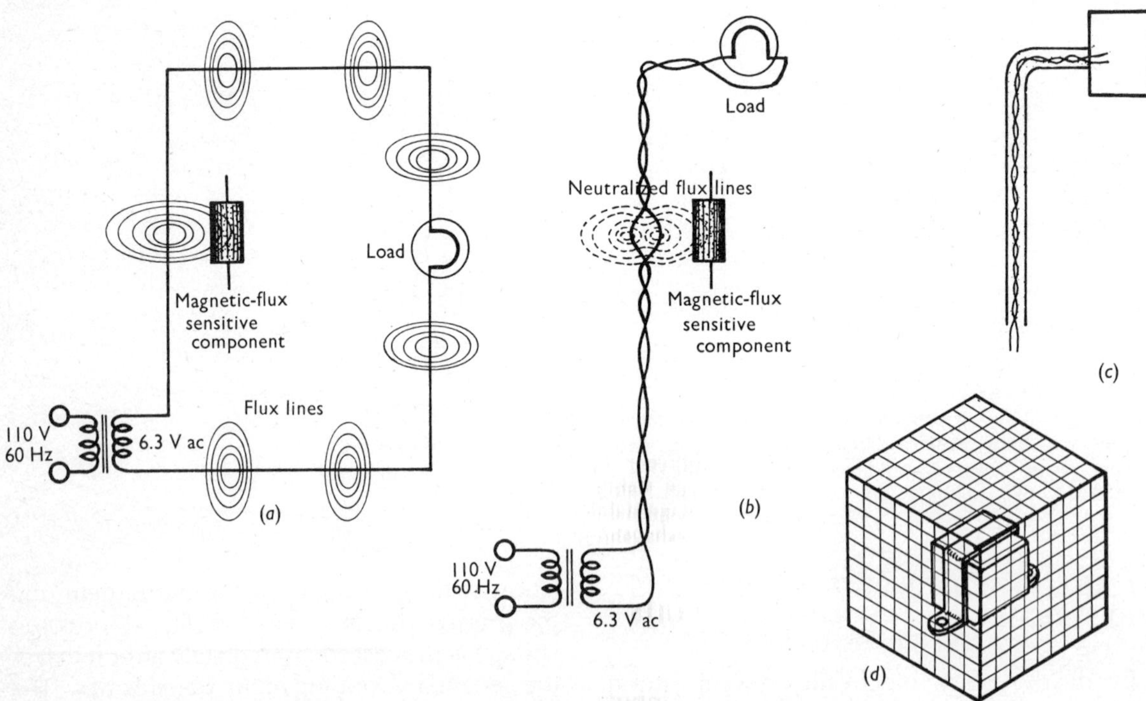

Fig. 2-2 Some methods of minimizing noise induced by power distribution: (*a*) Poor way to distribute ac—strong unopposed magnetic-flux lines induce voltages in nearby components; (*b*) Preferred way is use of twisted pair to distribute ac—opposing flux is produced, and the result is cancellation; (*c*) Twisted pair enclosed in ferromagnetic conduit; (*d*) Flux-producing component enclosed in ferromagnetic mesh cage.

Table 2-1 Listing of Typical Continuous dc Voltages Found on Electrodes of Transistors and Tubes in Electronic Equipment. All Values and Polarities Are with Reference to Ground or Common Neutral Potential. When the Applied Voltage Is in the Form of Pulses of Short Duration, the Maximum Voltages Can Be Increased

Electrode	*Small Transistor*	*Low-power Receiving-type Tubes*	*High-power Transmitter Tubes*
Emitter	0	—	—
Base	±0.5 V to ±1 V	—	—
Collector	±2 V to ±20 V	—	—
Cathode	—	0	0
Control grid	—	0 to −10 V	0 to −300 V
Screen grid	—	+50 V to +250 V	+0.5 kV to +5 kV
Suppressor grid	—	−10 V to 0	0 to +100 V
Plate	—	+50 V to +350 V	+0.75 kV to +15 kV

in low-power applications have indirectly heated cathodes wherein the filaments are heated with low-voltage ac. The dc power requirements for typical low-power electronic circuits are summarized in Table 2-1.

2-4 RECTIFICATION—DIODE AS A FAST SWITCH

The primary power supplied by public utilities is in the form of alternating current, yet most electronic circuits require direct current. It is necessary to use an *ac-to-dc converter.*

Perhaps one of the easiest forms of converter to comprehend (though not practical to construct and use) would be a switch driven by a synchronous motor. It is illustrated together with its schematic and timing diagrams in Fig. 2-3. The switch is a commutator which completes one revolution each time the alternating-current source completes one cycle. Note that the switch contacts are closed only during half of an alternation, allowing current to flow in one direction through the load. During the next half alternation the load is disconnected from the source so that reverse current cannot flow. The timing diagram shows a plot of applied voltage and the resultant voltage across the load. Note that *the load voltage always has the same polarity,* i.e., *ac-to-dc conversion has indeed taken place.* The process of ac-to-dc conversion is known as *rectification,* and the switching device is known as the *rectifier.*

There are many disadvantages associated with the use of a mechanical switch or commutator. It is slow and subject to wear and dirt accumulation. In practical applications *diodes* are used as rectifiers. Solid-state diodes block current flow because of their high resistance in the reverse-bias connection (when the *P-type* electrode is more negative than the *N-type*). They are good conductors in the forward-bias connection. Vacuum diodes block reverse current because thermionic emission of electrons can take place only at the cathode and not at the plate. The mechanical rectifier shown in Fig. 2-3 can be modified by inserting solid-state or vacuum-diode rectifiers as shown in Fig. 2-4. Diodes can switch at the rate of billions of times per second, compared with several hundred for the mechanical commutator. Diodes are silent and are not subject to wear. Solid-state diodes are extremely efficient, can switch hundreds of amperes of current, and have an indefinite life.

2-5 TYPES OF RECTIFIER CIRCUITS

The circuit shown in Fig. 2-4 is known as a *half-wave rectifier,* because only half of each alternation reaches the load. By using a second

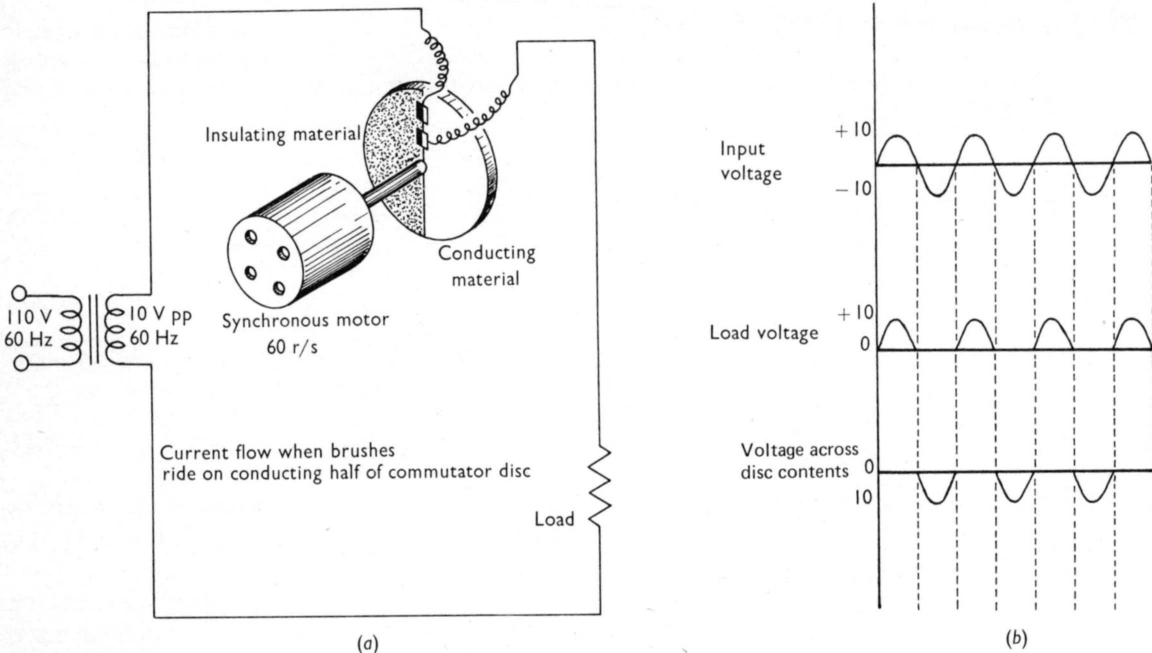

Fig. 2-3 Mechanical rectifier circuit, using a synchronous motor and a disc commutator with brushes: (*a*) Mixed pictorial and schematic diagram; (*b*) Timing diagram.

transformer and diode, both alternations can be routed through the load as shown in Fig. 2-5a. Note that D_1 conducts on the positive alternation and rests on the negative, whereas D_2 conducts on the negative alternation and rests on the positive. The load receives current flow in the same direction regardless of which diode is conducting. Such a circuit is known as a *full-wave rectifier* circuit. It is further refined in Fig. 2-5b by using a single transformer with a split phase (center-tapped) secondary. Note that with reference to the center tap, the outer ends of the secondary coil are always of opposite polarity, so that one or the other of the diodes D_1 or D_2 is always conducting and supplying current to the load.

A full-wave rectifier circuit can be constructed without the use of a center-tapped transformer by using four diodes instead of two. The circuit is commonly known as a *bridge rectifier* because the arrangement of source, diode, and load in the schematic diagram resembles a Wheatstone bridge. In Fig. 2-6a one set of diodes D_1 and D_4 conduct during one-half of the ac cycle, D_2 and D_3 on the other half.

The primary coil input voltage and the rectifier output voltage across a load are shown by the upper and lower traces, respectively, for both full- and half-wave rectifiers in the scope photos in Fig. 2-7.

Schematic diagrams of practical power supplies using circuits discussed throughout this chapter are shown in Appendix A.

2-6 RIPPLE FILTERS

In each of the rectifier circuits discussed, the rectifier output voltage applied to the load is *pulsating dc*. Pulsating dc is adequate for some applications, such as charging storage batteries. In devices which process signals (for example, amplifiers), pulsating dc would mix with the

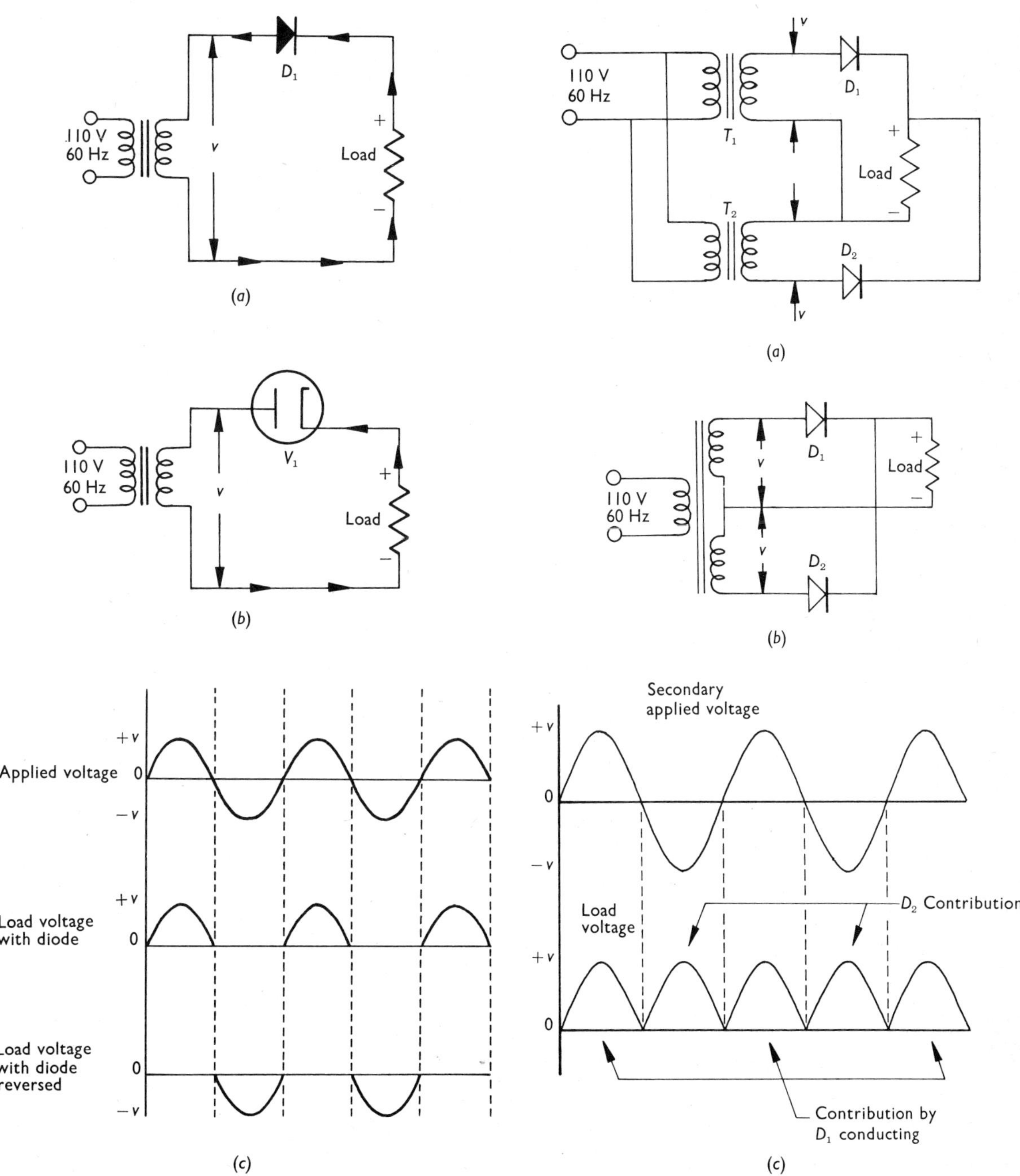

Fig. 2-4 Electronic rectifier circuits, using diodes as a form of switch: (*a*) Solid-state diode; (*b*) Vacuum-tube diode; (*c*) Timing diagram.

Fig. 2-5 Full-wave rectifier circuit: (*a*) Two transformers, each supplying current for one alternation; (*b*) Single, center-tapped, secondary transformer; (*c*) Timing diagram.

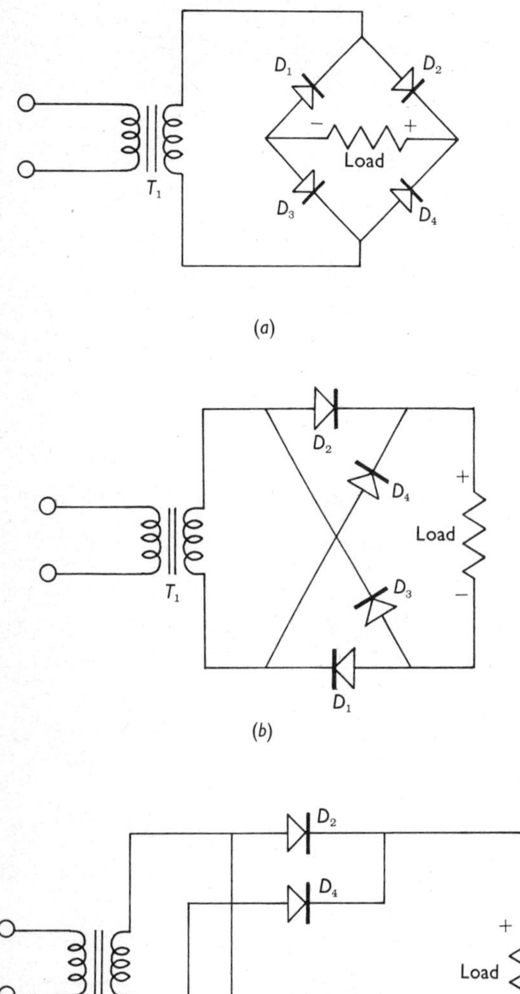

Fig. 2-6 Schematic diagrams of a bridge rectifier circuit, showing three variations in the drawing of the schematic.

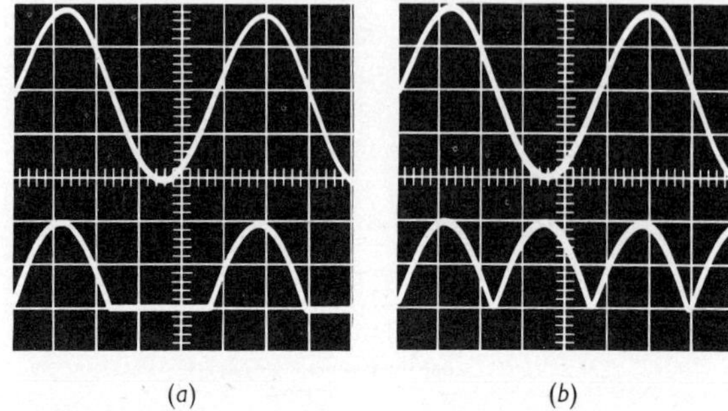

Fig. 2-7 Rectifier-output oscilloscope photographs: (*a*) 60-Hz input and pulsating dc output from half-wave rectifier; (*b*) 60-Hz input and pulsating dc output from full-wave rectifier.

signal and would appear as noise. In fact it would appear to the amplifier as though someone were switching the power on and off many times per second. (It would be 60 interruptions per second if a half-wave rectifier were supplying power, and 120 interruptions each second if a full-wave rectifier were used.) Pulsating dc output from a rectifier is shown in the scope photos in Fig. 2-7 and in the timing diagram in Fig. 2-8.

One possible though impractical means of eliminating the effects of pulsating dc would be to connect a storage battery to the rectifier output. The battery could supply the load whenever pulsation voltage dropped lower than the battery voltage. Some of the energy discharged from the battery would be replenished each time a pulsation appeared, and over a period of time the battery would remain charged. To ensure maintenance of full charge it would be necessary to adjust the pulsation voltage peaks somewhat higher than the battery voltage. The load would still be receiving pulsating dc, although the amplitude of the pulsations would be greatly reduced, as shown in Fig. 2-8c. The pulsations remaining after the initial reduction are known as *ripple*.

A more practical method of removing pulsations, and most of the ripple, is to replace the storage battery with a capacitor, as shown in Fig. 2-8d. A capacitor stores charge, but it has no ability to generate electricity as does a storage battery. Whatever energy loss occurs be-

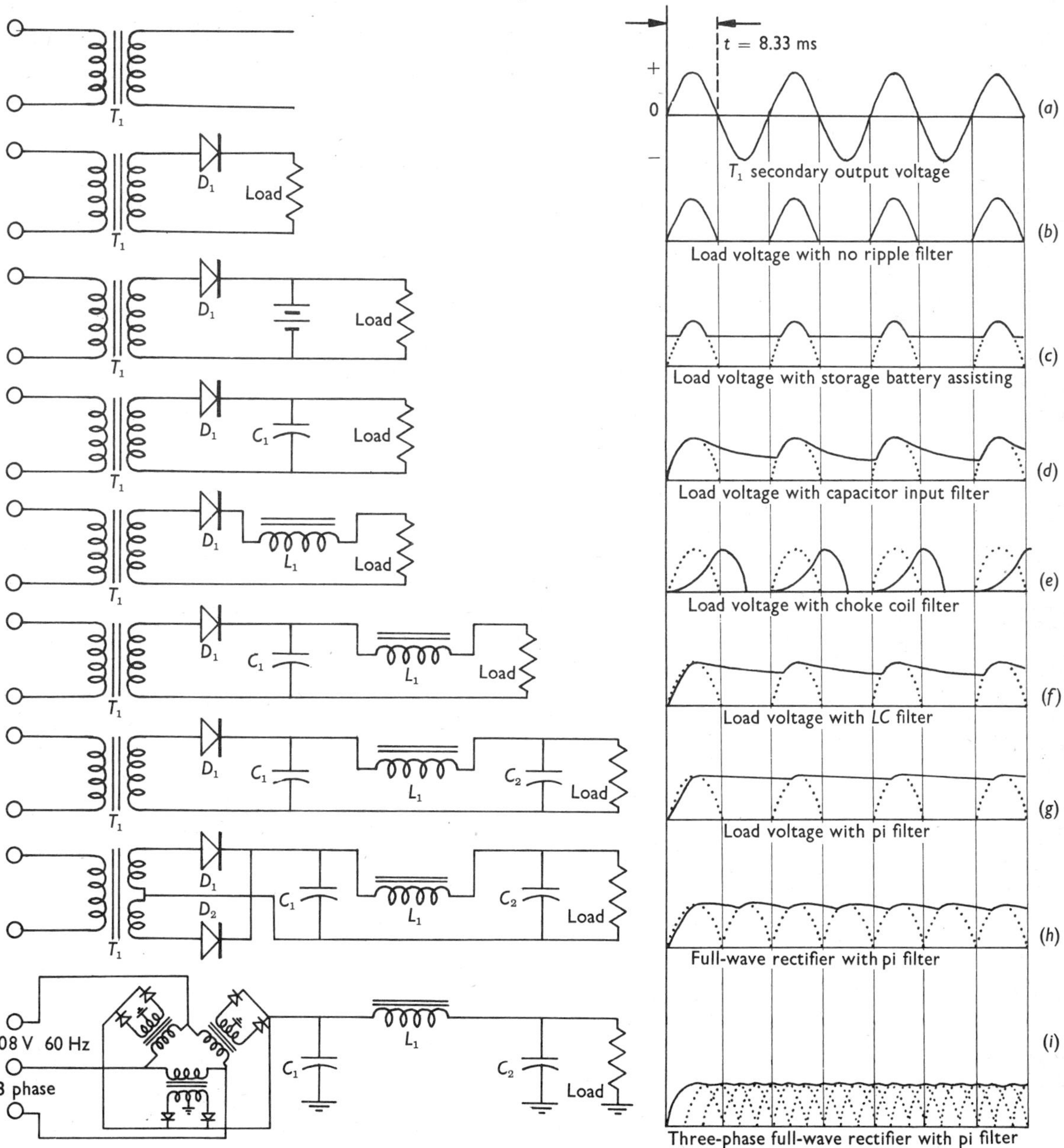

Fig. 2-8 Removal of ripple from rectifier output: (*a*) Secondary voltage before rectification; (*b*) Rectifier output, with no ripple filtering; (*c*) Use of a storage battery to reduce ripple amplitude; (*d*) Capacitor filter; (*e*) Choke filter; (*f*) Capacitor and choke filter; (*g*) Pi filter; (*h*) Full-wave rectifier with pi filter; (*i*) Three-phase, full-wave rectifier with pi filter.

tween pulsations can only come from the next pulsation. How low will capacitor voltage drop between pulsations? The answer depends on three factors: (1) The maximum charging current the rectifier can supply, (2) the capacitance of the capacitor, and (3) the current drain by the load. You can easily visualize that a rectifier capable of supplying current pulses of infinite magnitude to the largest known capacitor, connected to a load which draws no current, would result in no ripple. In practical applications the amount of ripple measurable in a circuit varies widely. Some circuits, such as battery chargers, can tolerate 100 percent ripple; others, such as sensitive amplifiers, can tolerate virtually none.

Ripple voltage alone is not indicative of power-supply performance. For example, a 6-V dc power supply with 0.6 V of ripple is not performing as well as a 100-V supply with 5 V of ripple. Ripple voltage must be compared to the dc output voltage. It is generally expressed as a percentage of the dc output voltage by the following expression:

$$\%\ \text{ripple} = \frac{\text{rms ripple voltage}}{\text{average dc output voltage}} \times 100$$

Rms and average values are used because these are measurable with portable VOM instruments.

EXAMPLE 2-1 What is the percent ripple in a 200-V dc power supply when the ripple voltage measures 5 V rms? Substituting numerical values in the expression gives

$$\%\ \text{ripple} = \frac{\text{rms ripple volts}}{\text{average dc volts}} \times 100$$

$$= \frac{5}{200} \times 100$$

ANSWER $= 2.5\%$

A further reduction in ripple can be achieved by using an inductance in series with the load, as shown in Fig. 2-8e. You will recall that an inductance opposes sudden changes in current. By Lenz's law if the source current wishes to rise, the magnetic-flux lines induce a counter current to oppose the increase. Also any level of current once established tends to be maintained, because falling source current is bolstered by the regenerating action due to collapsing magnetic-flux lines. The larger the inductance, the greater the counter current involved. In Fig. 2-8e you will note that as a pulsation is rising in amplitude, the full voltage is not transmitted to the load. Neither is the sudden drop, because the energy borrowed as represented by the IX_L drop during the rise is returned by the inductance during the fall. Ripple is therefore reduced by the addition of an inductance, which in this application is often called a *choke coil* or simply a *choke*.

Ripple filtering is further improved if both a capacitor and coil are used. For example in the circuit of Fig. 2-8d, assume that the capacitor leaves only 10 percent ripple and in the circuit of Fig. 2-8e the choke leaves 20 percent ripple. By constructing the filter of Fig. 2-8f the choke leaves 20 percent of the 10 percent left by the capacitor, resulting in only 2 percent ripple ($10/100 \times 20/100 = 200/10\,000 = 2/100 = 2$ percent). Even better ripple filtering is achieved by adding a third capacitor to remove ripple left by the choke. Such a filter network, consisting of two capacitors and a choke, is known as a *pi filter* and is shown in Fig. 2-8g. It is a form of *low-pass filter* which can limit ripple to a fraction of a percent.

2-7 VOLTAGE-DOUBLING AND -MULTIPLYING CIRCUITS

Whenever possible, transformers are used to step up voltage to required levels before rectification takes place. Yet there are many applications where the weight, bulk, and expense of a transformer are not justified. One example, a portable tube-type television receiver, requires voltage levels higher than those obtainable by rectifying domestic line voltage. A *voltage-doubler* circuit can usually be designed to supply the required power.

All voltage doublers, triplers, quadruplers, or multipliers operate on the same principle. Their principle of operation is this: Switching diodes and capacitors are so arranged that (1) the capacitors receive a charge and (2) the capacitors present the algebraic sum of their voltages to the load.

The simplest voltage doubler, known as the *half-wave doubler,* is shown in Fig. 2-9. In this circuit, a capacitor is switched across the power line during the negative alternation by a diode (D_1 in Fig. 2-9b). In Fig. 2-9b the load is shown disconnected during the charge process. Although it remains connected, it receives no voltage because of the low resistance of diode D_1 shunting it. A second diode D_2 (Fig. 2-9d) is added to ensure that only voltage of the correct polarity reaches the load. When the line volt-

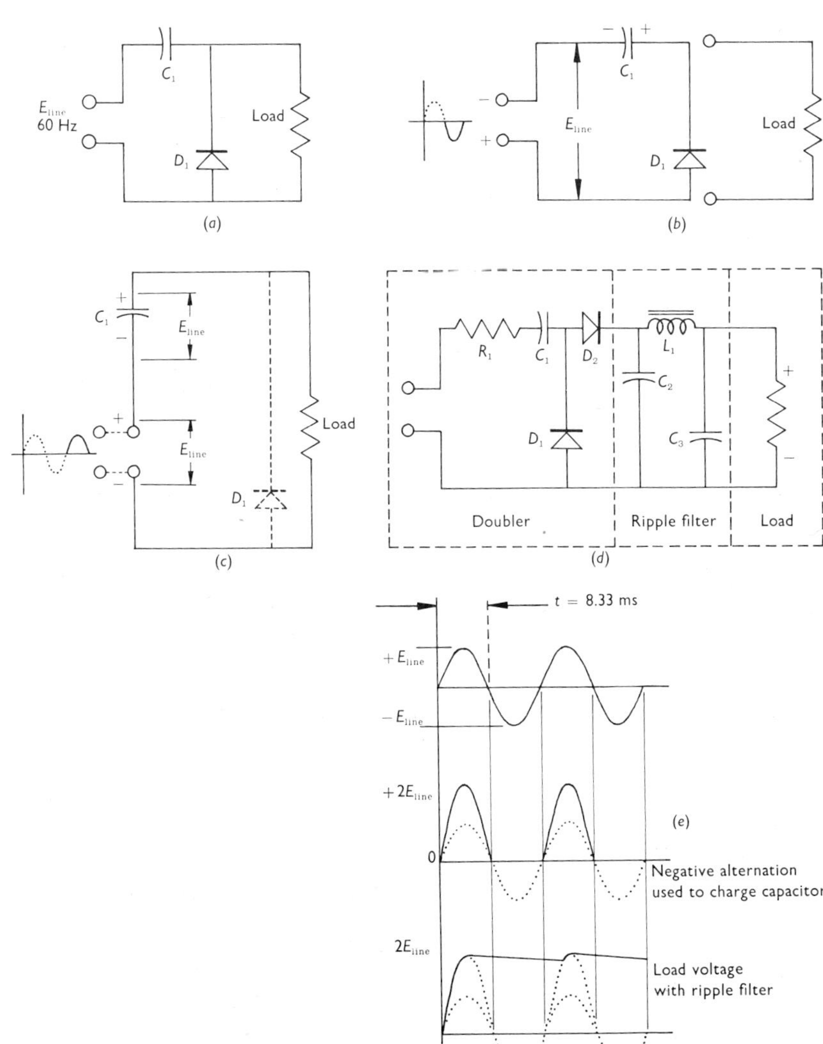

Fig. 2-9 Half-wave voltage-doubler circuit: (*a*) Doubler circuit; (*b*) Equivalent circuit during negative ac alternation, with capacitor charging; (*c*) Equivalent circuit during positive ac alternation, with capacitor and line voltage adding and the diode reverse-biased; (*d*) Doubler circuit with ripple filter isolating diode D_2 and surge current-limiting resistor R_1; (*e*) Timing diagram for half-wave doubler operation.

age reverses polarity during the next half cycle, diode D_1 becomes reverse-biased, stops conducting, and in effect is removed from the circuit as shown in Fig. 2-9c. Also, the full positive alternation, together with the accumulated capacitor voltage (connected in series) is presented to the load. It is as though the load were receiving power from two sources connected in series, the capacitor and the line. At its peak, this voltage will be twice the peak voltage of the line. In practice it is a little lower, because capacitor C_1 never quite reaches full line voltage during its charge cycle. Also it begins discharging into the load at the start of the positive alternation and as a result has lost some of its charge by the time that line voltage peaks. A study of the timing diagram in Fig. 2-9e will help you visualize the sequence of events during half-wave doubler operation.

The designation "half wave" may appear puzzling, especially since current is drawn from the line on both alternations. Nevertheless, the voltage pulses applied to the ripple filter and load are similar in waveform (see Fig. 2-9e) to those from a half-wave rectifier (see Fig. 2-8b).

Figure 2-10 illustrates a *full-wave voltage-doubler* circuit. Note that in this circuit a capacitor is being charged during each alternation, not just every second alternation as in the half-wave doubler. Both capacitors C_1 and C_2 are always connected in series, and the algebraic sum of their charge potentials is at all times applied to the load as shown in Fig. 2-10d. A study of the timing diagram (Fig. 2-10f) shows that the voltage waveform applied to the load has the same ripple frequency as the output of a full-wave rectifier (Fig. 2-5). The circuit is therefore known as a full-wave voltage doubler.

2-8 OUTPUT-VOLTAGE REGULATION

All the power-supply circuits discussed in Secs. 2-5 to 2-7 deliver maximum output voltage with minimum ripple under no-load conditions. When loads are connected, the load currents discharge the filtering capacitors to some extent. The greater the load current, the lower the output voltage becomes. A direct short across the output (loading carried to the extreme) results in no output voltage and probable damage to diodes and filter choke.

The difference in output voltage between load and no-load conditions is an indication of power-supply performance. A ratio, known as the *regulation* of the circuit, is expressed most often as a percentage by the following expression:

$$\% \text{ regulation} = \frac{E_{\text{no load}} - E_{\text{load}}}{E_{\text{load}}} \times 100$$

EXAMPLE 2-2 What is the percent regulation in a power supply which drops from 220 to 200 V when the load is connected?

$$\% \text{ regulation} = \frac{E_{\text{no load}} - E_{\text{load}}}{E_{\text{load}}} \times 100$$

$$= \frac{220 - 200}{200} \times 100$$

$$= \frac{20}{200} \times 100$$

ANSWER $= 10\%$

EXAMPLE 2-3 A power supply with 5 percent voltage regulation delivers 300 V with no load. What will be the output voltage when the rated load is connected? The expression

$$\% \text{ regulation} = \frac{E_{\text{no load}} - E_{\text{load}}}{E_{\text{load}}} \times 100$$

must first be rearranged by solving for E_{load}. Then substituting numerical values gives

$$E_{\text{load}} = \frac{E_{\text{no load}}}{\left(\frac{\% \text{ reg}}{100} + 1\right)}$$

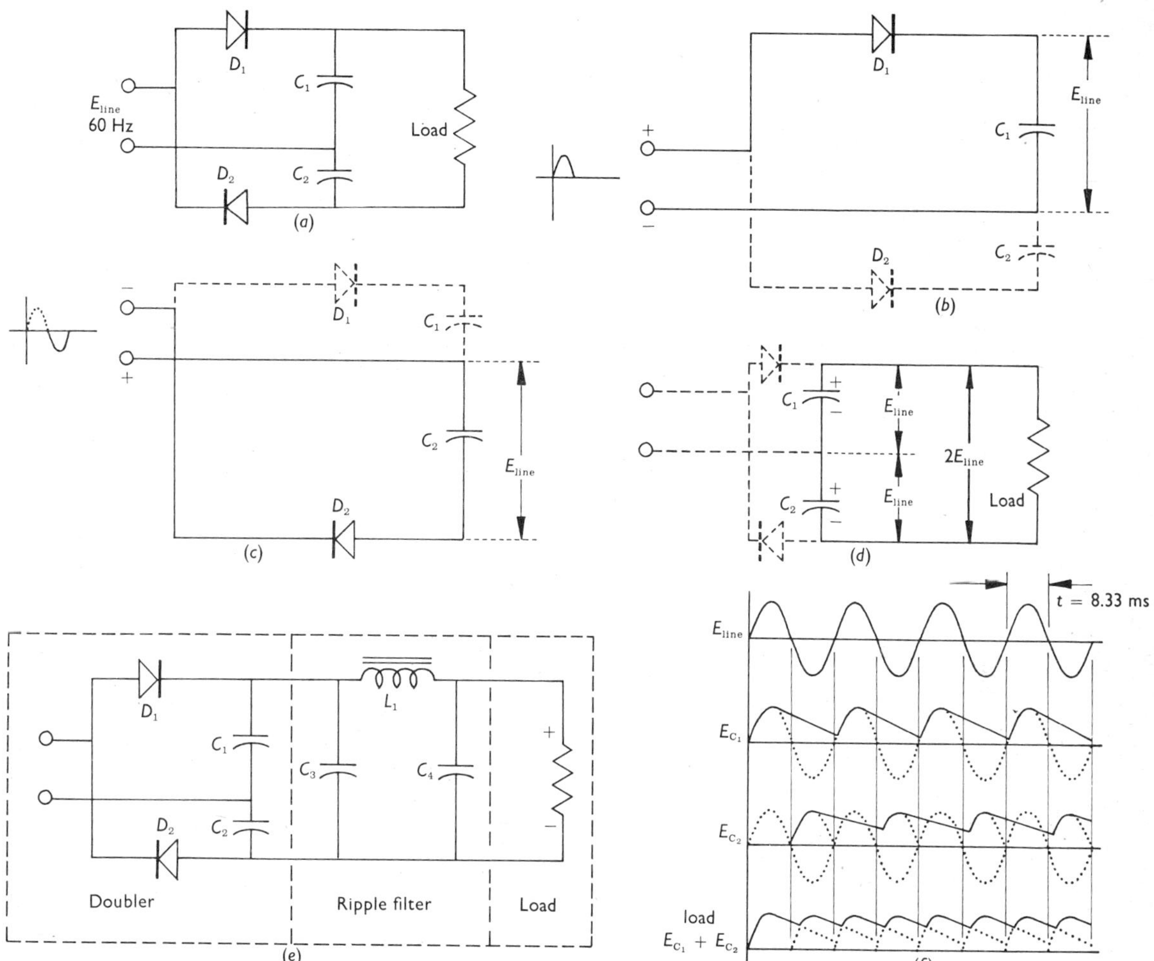

Fig. 2-10 Full-wave voltage-doubler circuit: (a) Doubler circuit; (b) C_1 charging on positive ac alternation, with C_2 disconnected by reverse-biased diode D_2; (c) C_2 charging on negative ac alternation, with C_1 disconnected by reverse-biased diode D_1; (d) Resultant voltage applied to load; (e) Full-wave doubler with ripple filter; (f) Timing diagram for full-wave doubler operation.

$$E_{\text{load}} = \frac{300}{\left(\frac{5}{100} + 1\right)}$$

$$= \frac{300}{(0.05 + 1)}$$

$$= \frac{300}{1.05}$$

ANSWER $= 286$ V

EXAMPLE 2-4 A power supply has 10 percent voltage regulation under rated load. If the load suddenly becomes disconnected, how high will the output voltage rise from its normal operating voltage of 400 V? The expression

$$\%\ \text{regulation} = \frac{E_{\text{no load}} - E_{\text{load}}}{E_{\text{load}}} \times 100$$

must first be rearranged by solving for $E_{\text{no load}}$. Substituting numerical values gives

$$E_{\text{no load}} = \left(\frac{\%\ \text{reg}}{100} + 1\right) E_{\text{load}}$$

$$= \left(\frac{10}{100} + 1\right) 400$$

$$= (0.1 + 1)\, 400$$

$$= 1.1 \times 400$$

ANSWER $= 440\ \text{V}$

What factors influence regulation? A complete answer is beyond the scope of this book. Mainly regulation depends on the internal resistance, real or equivalent, of the power supply. A power supply with low internal resistance can respond to varying current demands without appreciable variation in terminal voltage.

One way to improve regulation is to preload the power supply with a resistor. Then when the intended load is connected, the percentage drop in output voltage will not be as large. A resistor used for preloading is known as a *bleeder resistor.* Typically it is chosen such that it draws in the neighborhood of one-tenth as much current as the load. A bleeder resistor also discharges the filter capacitors when the equipment is turned off.

2-9 ZENER DIODE

Regulation can be accomplished electronically with devices such as *zener diodes* or *gas-filled tubes*. Solid-state diodes are very poor conductors in the reverse-bias connection. The reverse current is known as *leakage* current and is due to *minority carriers* (small numbers of electrons in P-type material and holes in N-type material). Leakage current is dependent mainly on temperature. Changing the reverse-bias voltage has little effect on the leakage current, unless it is raised to a certain critical level known as the *breakdown* or *avalanche potential*. Do not let the term "breakdown" lead you to believe that destruction of the diode follows. Breakdown is a technical term and is not at all synonymous with catastrophic failure. At breakdown voltage, the minority carriers are accelerated to sufficient speeds to dislodge on impact valence electrons in the semiconductor crystal. Each such impact generates two new carriers, a hole and an electron. Dislodged electrons are in turn accelerated by the strong electric field to impact velocity, causing further generation of electron-hole pairs. An "electronic landslide" develops, hence the term avalanche.

Very large current flow is possible in the breakdown condition, and if the power dissipated in the process is too large, the diode will be destroyed. Figure 2-11 is a graph of diode current in the reverse-bias connection. All values of current for less than the breakdown voltage are small, but increase many orders of magnitude when breakdown is reached. Certain diodes are specially designed to operate in the breakdown region and are known as *zener diodes* (after C. Zener for his research into the electrical breakdown of solid dielectric materials). In schematic diagrams zener diodes are indicated as shown in Fig. 2-11b. The small hooks on the N-type material or cathode in Fig. 2-11b should remind you of the zener-diode characteristic curve.

2-10 ZENER-DIODE VOLTAGE REGULATOR

Note particularly that at breakdown, the slightest voltage variation results in large current variation. Conversely, a zener diode can sink a very wide range of breakdown currents without appreciable voltage variation. It is this latter quality which makes the *zener diode useful as a voltage regulator*. In essence, a zener acts as a variable resistance bleeder, which adjusts its current to maintain breakdown potential.

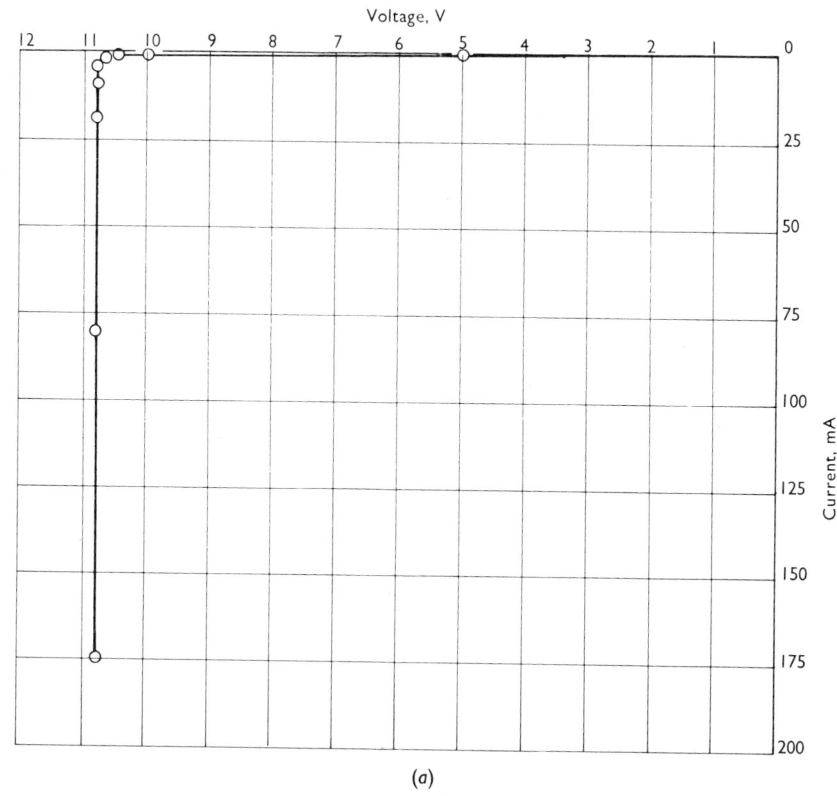

(a)

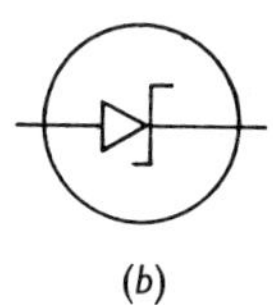

(b)

Fig. 2-11 (a) Volt-ampere curve of a typical 10-W zener diode in reverse-biased connection; (b) Schematic diagram symbol for a zener diode.

A zener-diode voltage-regulator circuit is shown in Fig. 2-12a. It is a shunt regulator, because the diode is connected in parallel with the load. By applying Ohm's and Kirchhoff's laws to the equivalent circuit of Fig. 2-12b some useful equations can be deduced. There are two *IR* drops, namely E_{R_s}, the voltage across the series resistor, and E_z, which is also the load voltage E_{R_L}. Applying Kirchhoff's law, which states that E applied = sum of *IR drops*, gives

$$E_{in} = E_{R_s} + E_z$$

(Assume E_{in} to be the lowest value to which power-supply voltage falls.)

By Ohm's law, $E_{R_s} = (I_z + I_L) R_s$. Substituting this value for E_{R_s} gives

$$E_{in} = (I_z + I_L) R_s + E_z$$

In practice, I_z, the zener current, is made approximately $\frac{1}{10}$ of the magnitude of load current, or $I_z = 0.1\, I_L$. Substituting $I_z = 0.1\, I_L$ in the equation gives

$$E_{in} = (0.1\,I_L + I_L)R_s + E_z$$
$$= (0.1 + 1)I_L R_s + E_z$$
$$= 1.1\,I_L R_s + E_z$$

Solving for R_s gives

$$R_s = \frac{E_{in} - E_z}{1.1\,I_L}$$

If R_s is too large, regulation is poor; if too small, zener current is too large and may damage the diode. A practical compromise which gives adequate regulation for most purposes is to set zener current at approximately one-tenth the load current. Then, depending on load current and voltage, a zener diode with sufficient power or wattage rating is chosen.

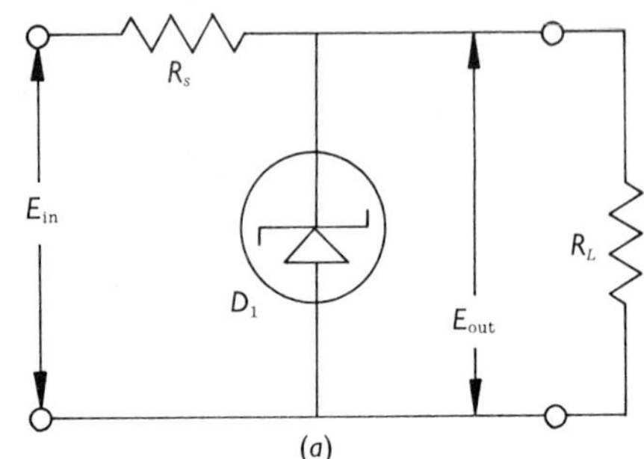

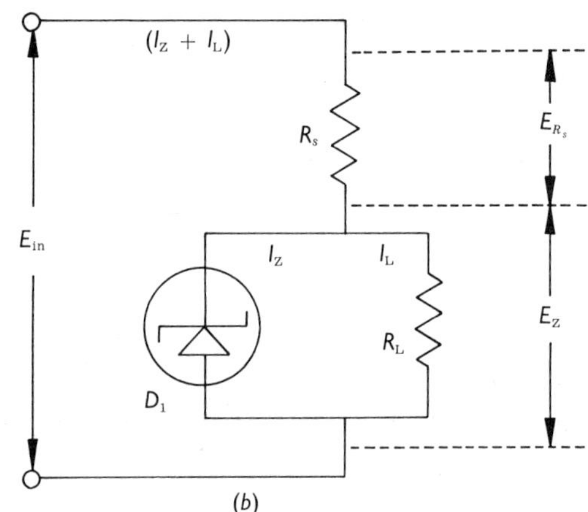

Fig. 2-12 Zener-diode voltage regulator: (*a*) Typical schematic diagram; (*b*) Equivalent circuit, with circuit currents and *IR* drops.

2-11 ZENER-DIODE POWER DISSIPATION

Power dissipated by the zener diode is the product of maximum zener current by the zener voltage. Maximum zener current can be deduced by Kirchhoff's and Ohm's laws. Reference to Fig. 2-12b shows that voltage across the series resistor is the difference $E_{in(max)} - E_z$. Dividing this voltage by the resistance R_s gives the total current

$$I_{R_s} = \frac{E_{in(max)} - E_z}{R_s}$$

It is the sum of zener and load currents. Therefore, subtracting the load current gives the maximum zener current as

$$I_{z(max)} = \left(\frac{E_{in(max)} - E_z}{R_s}\right) - I_L$$

Finally, the maximum zener-diode power dissipation is the product $I_{z(max)}E_z$. Substituting for $I_{z(max)}$ gives

$$P = \left\{\left(\frac{E_{in(max)} - E_z}{R_s}\right) - I_L\right\} E_z$$

A zener diode must be operated within its rated power dissipation, and with large diodes heat sinks are required to radiate the heat produced.

EXAMPLE 2-5 A 14-V power supply is to be connected to a load requiring 10 V at 300 mA. A 1N3020 zener diode is to be used as the shunt regulator as shown in Fig. 2-13. Calculate the value of series resistor R_s.

Applying the equation and substituting given numerical values gives

$$R_s = \frac{E_{in} - E_z}{1.1\,I_L}$$

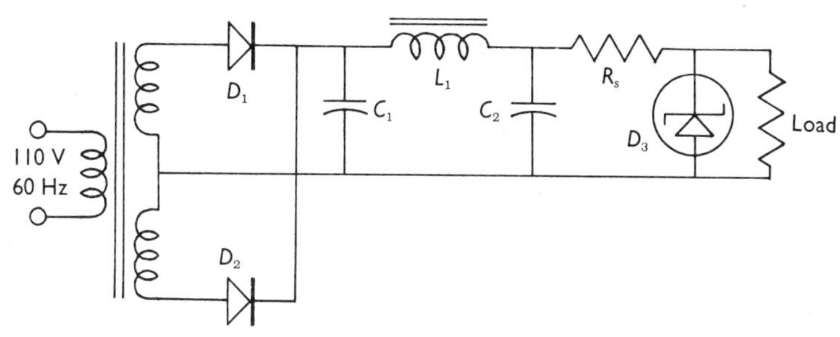

Fig. 2-13 Power supply with zener-diode voltage regulator.

$$R_s = \frac{14 - 10}{1.1 \times 300 \times 10^{-3}}$$

$$= \frac{4}{1.1 \times 0.3}$$

$$= \frac{4}{0.33}$$

ANSWER $= 12.1\ \Omega$

EXAMPLE 2-6 What power will be dissipated if the line voltage is increased and the power-supply output jumps to 14.75 V?

Applying the equation for power dissipation and substituting given numerical values gives

$$P = \left\{\left(\frac{(E_{\text{in(max)}} - E_Z)}{R_S}\right) - I_L\right\} E_Z$$

$$= \left\{\left(\frac{14.75 - 10}{12.1}\right) - 0.3\right\} \times 10$$

$$= \left(\frac{4.75}{12.1} - 0.3\right) \times 10$$

$$= \frac{47.5}{12.1} - 3$$

$$= 3.92 - 3$$

ANSWER $= 0.92$ W

NOTE The 1N3020 is rated at 1 W

2-12 GAS-FILLED GLOW-DISCHARGE TUBE

Before zener diodes came into general use, gas-filled diode tubes were used as shunt regulators. The construction of a gas-filled glow-discharge regulator tube is shown in Fig. 2-14.

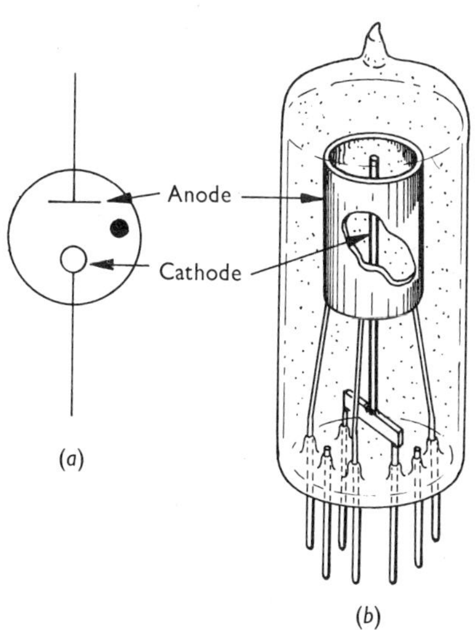

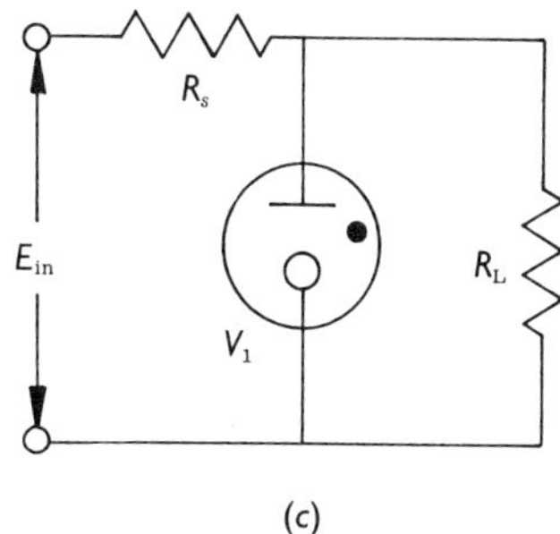

Fig. 2-14 Gas-filled glow-discharge tube: (*a*) Schematic symbol; (*b*) Construction; (*c*) VR tube voltage-regulator circuit.

There are certain similarities in the principle of operation of a gas diode and zener diode. In the gas diode, when the positive potential on the anode is raised sufficiently, free electrons in the gas—usually argon, helium, or neon—are accelerated to a speed where impact with a gas molecule dislodges a valence electron. This results in the generation of an additional electron and a positive gas ion. The process begins at "breakdown" potential and avalanches if further increase in voltage is attempted. Unlike the zener, the gas-filled diode will usually survive momentary overloads, and arcing occurs if the current is too great. Figure 2-14c illustrates a shunt-regulator circuit using a VR tube. These tubes are most commonly available for regulating at 75, 90, 105, and 150 V.

It should be noted that shunt regulators not only improve regulation, but help remove ripple as well. Ripple is regarded by a regulator as just another voltage variation. In Chap. 14, you will find further discussion of electronic regulators.

2-13 COMPONENT RATING—SAFE OPERATING CONDITIONS

Although power-supply circuits tend to be reliable, failures occur occasionally. Electrolytic filter capacitors usually give the most trouble. If moisture enters the capacitor, or if chemical impurities are present, the unit slowly degenerates and may become short-circuited. If a short occurs, the diodes may not withstand the current drain and may either open or short-circuit internally. If they open, the primary power is disconnected and no further harm occurs. If they short-circuit, rectification stops and alternating current is applied to the filter. This may further damage other capacitors. When cost permits it is preferable to use capacitors with working dc voltages of 150 percent or even double the required voltages.

Rectifier diodes are selected not only on the basis of current rating but also on *peak inverse voltage* or *PIV* (sometimes designated as *PRV* for peak reverse voltage). Figure 2-15 shows

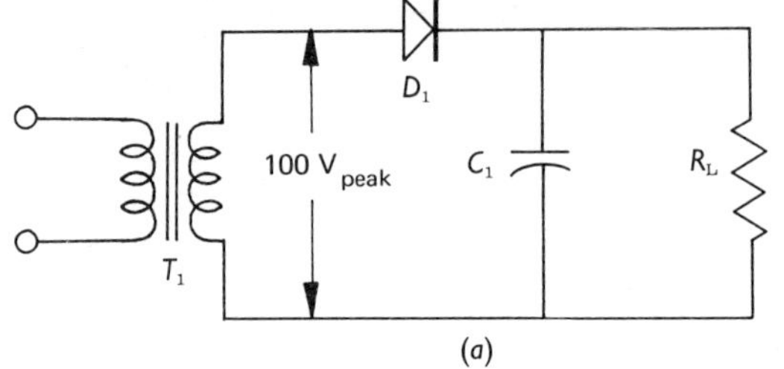

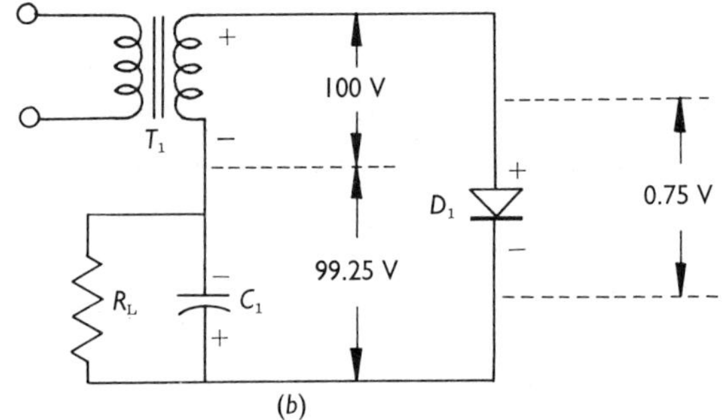

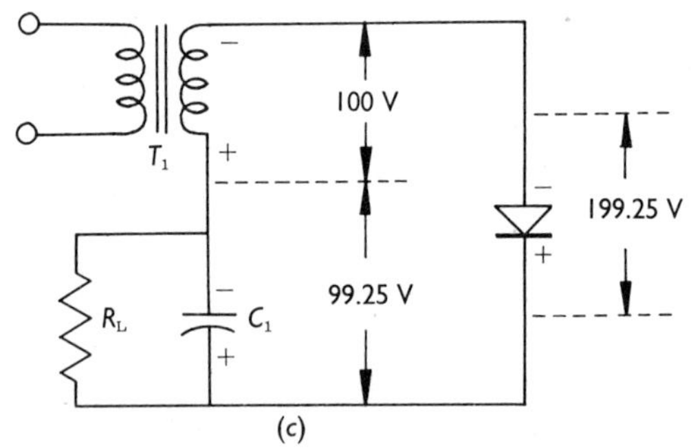

Fig. 2-15 Illustration of peak inverse voltage on a diode: (*a*) Half-wave rectifier circuit; (*b*) Equivalent circuit and voltages across components when the diode conducts; (*c*) Equivalent circuit when the diode is not conducting. Note the reverse bias on the diode is nearly twice the secondary peak voltage.

voltage across a diode during both alternations in a rectifier. Note that when the diode is not conducting, it must withstand both the secondary voltage and the accumulated capacitor voltage without breakdown. Manufacturers specify the PIV for their products, and wherever cost permits diodes should be selected with an extra margin of safety.

Summary

Nearly all electronic circuits have a power source, a power load, a signal source, and a signal load.

Signal and power currents may flow in the same conductors.

Signal and power currents may be separated by electrical filters.

Primary power voltage variation is inevitable because of the fluctuating power demands in a community.

Power transformers are used for electrical isolation and for transforming voltages to required levels.

Noise radiation from wires carrying ac can be minimized by using a twisted pair.

Diodes can switch currents of hundreds of amperes and at rates of billions of times per second.

The three fundamental forms of rectifier circuit are the *half wave, full wave,* and *bridge.*

All rectifiers produce *pulsating dc.* The pulsations which remain after filtering are known as *ripple.* Ripple is also referred to as the ac component of the output voltage.

One power-supply performance factor is *percent ripple.* A second factor is *percent regulation.*

An advantage of voltage-doubler circuits is that they require no costly transformers.

A *bleeder* resistor preloads a power supply and improves regulation.

Zener diodes and gas-filled glow-discharge tubes are simple and convenient shunt regulators.

Rectifier diodes must have a sufficiently high PIV. They must never enter the breakdown point on the reverse alternation.

Questions and Exercises

1. How can signal and power currents be separated if they are present in the same circuit?
2. Why is exact line voltage not guaranteed by public utility companies?
3. Give an example of a primary power source with excellent short-term voltage stability.
4. Give two reasons for using power transformers in electronic power supplies.
5. What is the purpose of twisting filament distribution wires?
6. What do the labels $B+$ and $C-$ signify on schematic diagrams? How did these labels originate?
7. List some advantages a diode has compared with a mechanical switch.
8. What is meant by rectification?
9. If the input to a half-wave rectifier is a 60-Hz sine wave, what is the ripple frequency?
10. Redraw the schematic in Fig. 2-4 such that the load-voltage polarity is reversed.
11. What is one advantage of a bridge rectifier as compared with a split secondary full-wave circuit?
12. What is one advantage of a split secondary full-wave circuit as compared with a bridge rectifier?
13. Redraw the circuits of Fig. 2-5 to give load voltage of opposite polarity to that shown.
14. State one application of a rectifier where ripple filtering is not required.
15. Explain in your own words how a capacitor removes some ripple in the output of a rectifier.
16. Explain in your own words how a choke coil removes some ripple in the output of a rectifier.
17. Which power supply has the lowest percent ripple? **(*a*)** 200 V dc with 0.3 V rms ripple; **(*b*)** 50 V dc with 0.075 V rms ripple; **(*c*)** 90 V dc with 0.1 V rms ripple. Show all calculations.
18. State the principle of operation of all

voltage-doubler circuits.

19. Why are voltage-doubler circuits classified as half wave and full wave?

20. Redraw the circuit of Fig. 2-9d so that the load-voltage polarity will be reversed.

21. Redraw the circuit of Fig. 2-10e so that the load-voltage polarity will be reversed.

22. What is meant by the term regulation?

23. The output of a 100-V power supply rises to 105 V when the load is removed. What is the percent regulation of the power supply?

24. How high will the output voltage become if a 200-V power supply with 20 percent regulation is disconnected from its load?

25. What will be the output voltage of an unloaded power supply with 10 percent regulation if its rated load is connected to its 40-V unloaded output?

26. Give two functions performed by a bleeder resistor.

27. A power supply delivers 200 V at 50 mA to a load. What is a typical value of bleeder resistor on this power supply?

28. What is meant by breakdown potential in a reverse-biased solid-state diode?

29. In your own words explain how a zener diode performs the function of a shunt-type voltage regulator.

30. The circuit shown in Fig. 2-13 is to be used for testing 12-V car radios in an appliance store. The voltage across C_2 is expected to be 14 V, and car radio drain is typically 500 mA. D_3 is a 12-V zener diode. Calculate the value of R_S.

31. What power will be dissipated by the zener diode in Question 30 if the voltage during normal operation jumps from 14 to 14.5 V across C_2?

32. What is one advantage of a gas-filled glow-discharge tube over a zener diode?

33. Obtain a current electronic components catalog. Make a table showing zener-diode and rectifier-diode minimum and maximum voltage, power dissipation, operating current, PIV, and price ranges. List only those entries which are applicable to the device you are working on.

34. Figure 2-16 shows a schematic diagram of a voltage tripler.

a. Draw the equivalent circuit for the charge path for C_1.

b. Draw the equivalent circuit for the charge path of C_2.

c. Draw the equivalent circuit for the charge path of C_3.

d. What will be the ripple frequency as seen by the load?

35. Figure 2-17 shows a schematic diagram of a voltage quadrupler.

a. Draw the equivalent circuits for the charge paths for each of C_1, C_2, C_3, and C_4.

b. What will be the ripple frequency as seen by the load?

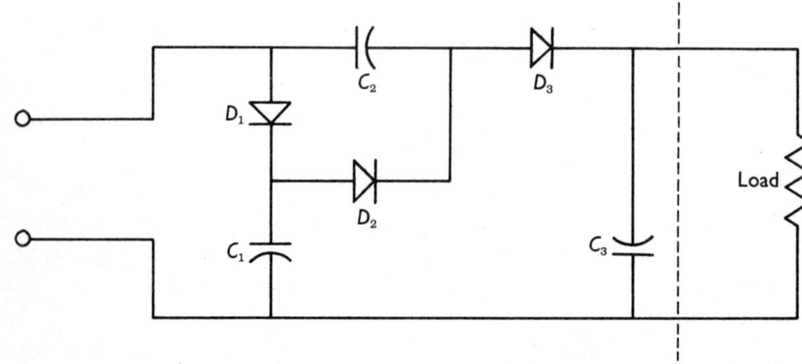

Fig. 2-16 Voltage tripler. Ripple frequency equals line frequency, because voltage is taken across one capacitor, receiving a pulse once every alternation.

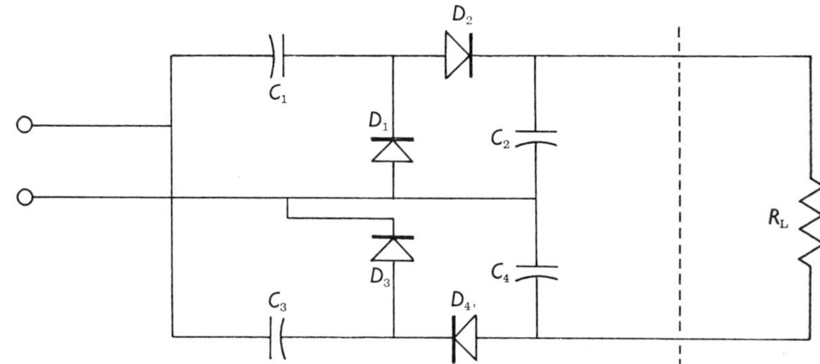

Fig. 2-17 Voltage quadrupler.

36. Redraw the high-voltage secondary circuit in Fig. A-2 (Appendix A). With a pen make small arrow marks to show electron flow leaving the red lead, completing the circuit, and returning to the green-red lead of secondary coil. Next assume the alternation is reversed and, with a different color ink, show electron flow leaving the green-red lead, completing the circuit, and returning to the red lead of the secondary coil.
37. What would be the result if any one of the diodes SR_{201}, SR_{202}, SR_{203}, or SR_{204} in Fig. A-2 (Appendix A) became open (ceased conducting in either direction)? Would the ripple frequency remain the same or would it change, and to what value?
38. Which of the types of circuits discussed in this chapter is the + 135-V supply circuit in Fig. A-3 (Appendix A)?
39. If diode Y_{402} in Fig. A-3 (Appendix A) became short-circuited, which other component in the circuit would likely be destroyed? Explain your answer.
40. If capacitor C_{32D} in Fig. A-4 (Appendix A) became short-circuited, and assuming that the 350-V output continued to function, what current would flow through R_{104}? Would R_{104} survive, or would it likely be destroyed by excess heat? (Hint: Calculate the power dissipated in R_{104} and compare with the power rating.)

3-1 CONCEPT OF GAIN

Devices which have the capability of raising signal-power level are known as amplifiers. The ratio

$$\frac{\text{output-signal power}}{\text{input-signal power}}$$

is known as the *gain* of a device or a circuit. Unless further qualified, gain means power gain, although it is also meaningful to speak of voltage gain or current gain.

3-2 TRANSISTOR AMPLIFIER

Figure 3-1 shows a set of collector characteristic curves for a 2N3565 transistor. Reference to the curves will show the various collector currents which result when collector voltage is 10 V (represented by dotted vertical line), and when the base current is varied between 5 and 25 μA. In particular, note that a base current of 10 μA gives a collector current of 3 mA. Raising or lowering base current results in correspondingly larger or smaller collector current.

Suppose that the circuit in Fig. 3-2a were constructed. There are actually two complete circuits in the diagram, the base circuit and the collector circuit. To understand circuit operation you must be able to trace electron flow in each of the two. First, consider the base circuit shown in Fig. 3-2b. There is negligible resistance in the conductors, the moving-coil-type microphone, and the emitter-base junction (because the emitter-base junction is forward-biased). If resistor R_1 were removed, there would be sufficient base current to destroy the transistor instantly. The resistor R_1 is therefore placed in the circuit to limit the base current to a desired level. It is known as a *biasing resistor,* because it determines the base current which will flow in the absence of microphone current.

How is the value of a biasing resistor chosen? For dc bias current the base circuit resolves

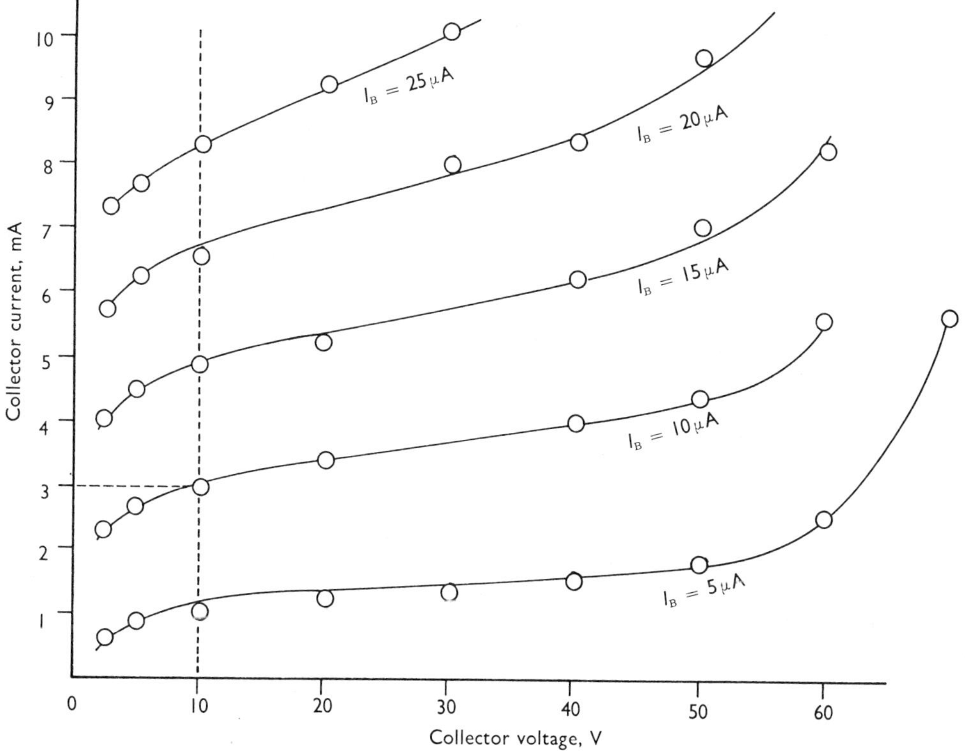

Fig. 3-1 Collector characteristic curves for a 2N3565 silicon transistor.

into two effective components, the source battery and the resistor R_1 (see Fig. 3-2c). If a base current of 10 μA is desired, Ohm's law gives the resistance as

$$R_1 = \frac{E}{I}$$

$$= \frac{10}{10 \times 10^{-6}}$$

$$= 1 \times 10^6 \Omega$$

$$= 1 \text{ M}\Omega$$

Next examine the collector circuit shown in Fig. 3-2d. Headphone current is controlled by transistor collector current. In a sense the transistor can be regarded as a variable resistor the value of which at any instant is determined by base current. If a collector current varies, then the varying current through the headphones produces sound waves. When someone speaks into the microphone, sound waves strike the diaphragm causing it and the attached voice coil to vibrate in the permanent magnetic field. Magnetic-flux lines are cut and a small signal current is induced in the wires of the voice coil. It will be an alternating current, because the voice coil moves in both directions. Since the voice coil is connected in series with the base circuit, its induced currents will alternately add to and subtract from the base-bias current.

Figure 3-3 shows the resulting base current when the microphone current is assumed to be sinusoidal with 5-μA peak amplitude. On one signal alternation, base current reaches a peak of 15 μA (sum of 5-μA signal and 10-μA bias).

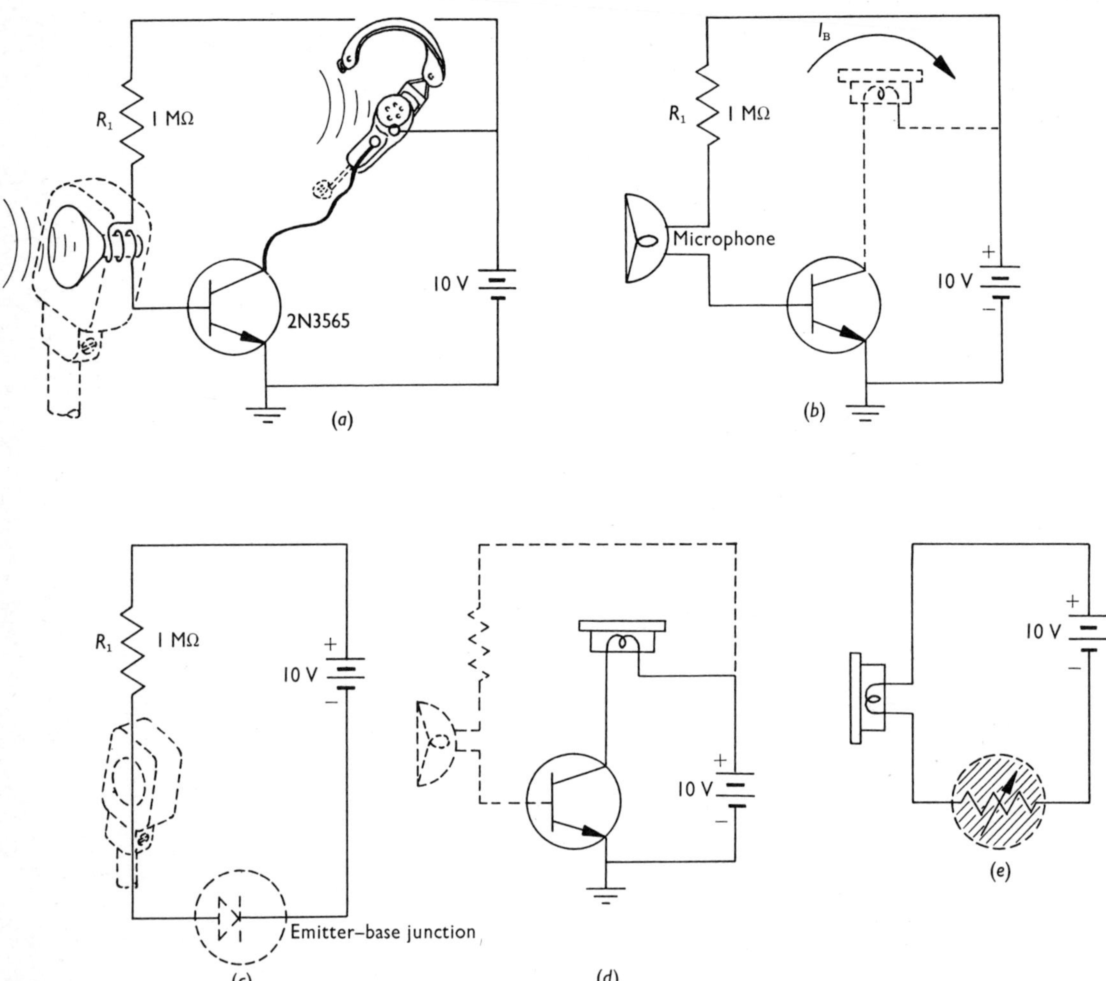

Fig. 3-2 Elementary transistor audio amplifier: (*a*) Schematic diagram; (*b*) Base circuit; (*c*) Simplified base equivalent circuit; (*d*) Collector circuit; (*e*) Simplified collector equivalent circuit.

On the other alternation it drops to a low of 5 μA (difference of 5-μA signal and 10-μA bias). There is a corresponding variation in collector current but the values involved are far larger, which means that current gain has taken place.

The current-gain factor is obtained by a simple calculation. First, determine the maximum collector-current change, and divide this by the base-current change which caused it. In this instance collector-current swing is

$$\Delta I_C = 5.5 \text{ mA} - 1.4 \text{ mA} = 4.1 \text{ mA} = 4.1 \times 10^{-3} \text{ A}$$

Base-current swing producing this change is

$$\Delta I_B = 15\ \mu\text{A} - 5\ \mu\text{A} = 10\ \mu\text{A} = 10 \times 10^{-6} \text{ A}$$

The current-gain factor then simplifies to

$$\text{gain} = \frac{\Delta I_C}{\Delta I_B}$$

$$= \frac{4.1 \times 10^{-3}}{10 \times 10^{-6}}$$

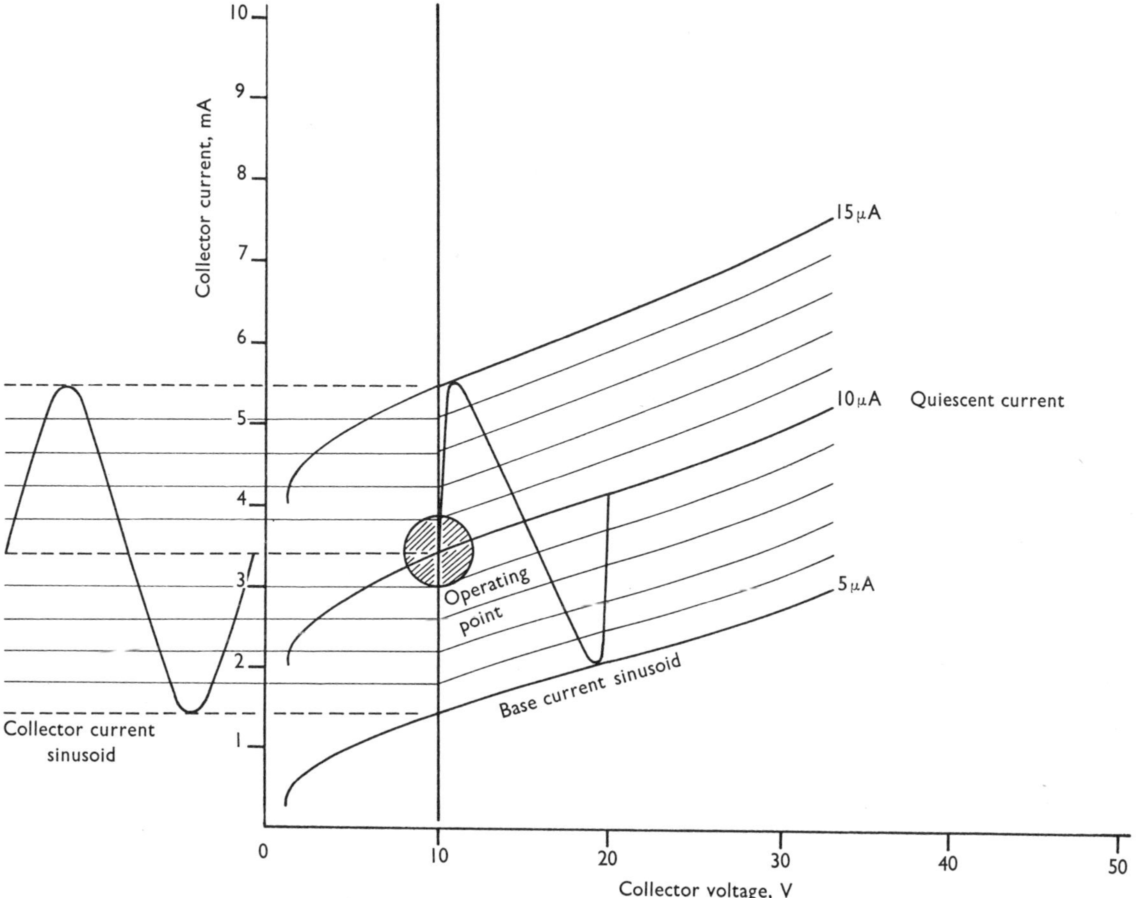

Fig. 3-3 Variation of collector current, resulting from base-current variation, assuming the load has negligible resistance.

$$\begin{aligned} \text{gain} &= 4.1 \times 10^2 \\ &= 410 \end{aligned}$$

What this indicates is that the microphone signal current is boosted by a factor of 390 times.

All transistor amplifiers operate on a similar principle. A signal current is introduced into the circuit in such a way that it causes the base current to vary. The varying base current then causes the much larger collector current to follow these variations. In this way a relatively weak current can control a much larger current, and amplification takes place.

3-3 OPERATING POINT

In the simple amplifier circuit described in the previous section, a given value of base current always flows, even when the microphone picks up no sound. (In the circuit of Fig. 3-2 this value is 10 μA as established by the biasing resistor.)

The value of base current when no signal is present is known as the base *quiescent* current or rest current. When the quiescent current, together with quiescent voltage, is plotted on a graph such as in Fig. 3-3, it is known as the operating or quiescent point.

3-4 CLASS OF OPERATION

Reference to Fig. 3-3 shows that base and collector current flows at all times during both positive and negative alternations of the signal. Another way of stating this is to say that base and collector current flows for 360° of the signal waveform. Your intuition may tell you that this is to be expected, because if collector current is stopped for any reason during a portion of the signal cycle, there would be a "chunk" missing from the output-signal waveform. As is so often the case in electronics you must not rely on intuition alone, because amplifiers which reproduce only a fraction of a signal cycle are very useful in many applications. The term which identifies how much of a waveform is amplified is known as the *class of operation*. Letters are used to designate the various classes. Class A amplifiers are biased so that 360° of a signal waveform is amplified. Class B amplifers amplify only 180° or half the waveform, whereas class C amplifiers amplify less than half the waveform.

Class of operation is determined by the bias current or voltage because it is the bias value which determines the quiescent or operating point. Figure 3-4 illustrates how changing the bias shifts the operating point and therefore determines the class of operation of an amplifier. Table 3-1 summarizes some characteristics of each of these three classes of operation. Note that even though the class C amplifier is the most efficient (because current drain occurs in short bursts only at signal peaks), it is entirely unsuitable for high-fidelity audio applications because of its severe waveform distortion. On the other hand, in applications such as transmitters, the intense signal

Table 3-1 Comparison of Class A, B, and C Amplifier Characteristics

Amplifier Class of Operation	*Relative Advantages*	*Relative Disadvantages*	*Applications*
A	Lowest possible waveform distortion	Lowest efficiency because quiescent current is high Poor efficiency limits use to small signals	Preamplifiers Low-power stages Amplifiers in which fidelity of waveform reproduction is more important than power output
B	Fairly good efficiency since no power is used unless signal is present	Only half the waveform is reproduced, must be used in pairs to amplify both halves of waveform More distortion than class A	Power amplifiers for audio applications (output stages to drive speakers) Output stages in some transmitters
C	Highest efficiency because current flows in bursts during part of each signal alternation	Severe distortion, only 30–50% of each waveform is amplified	High-power amplifiers in transmitters, pulse amplifiers, high-voltage power supplies

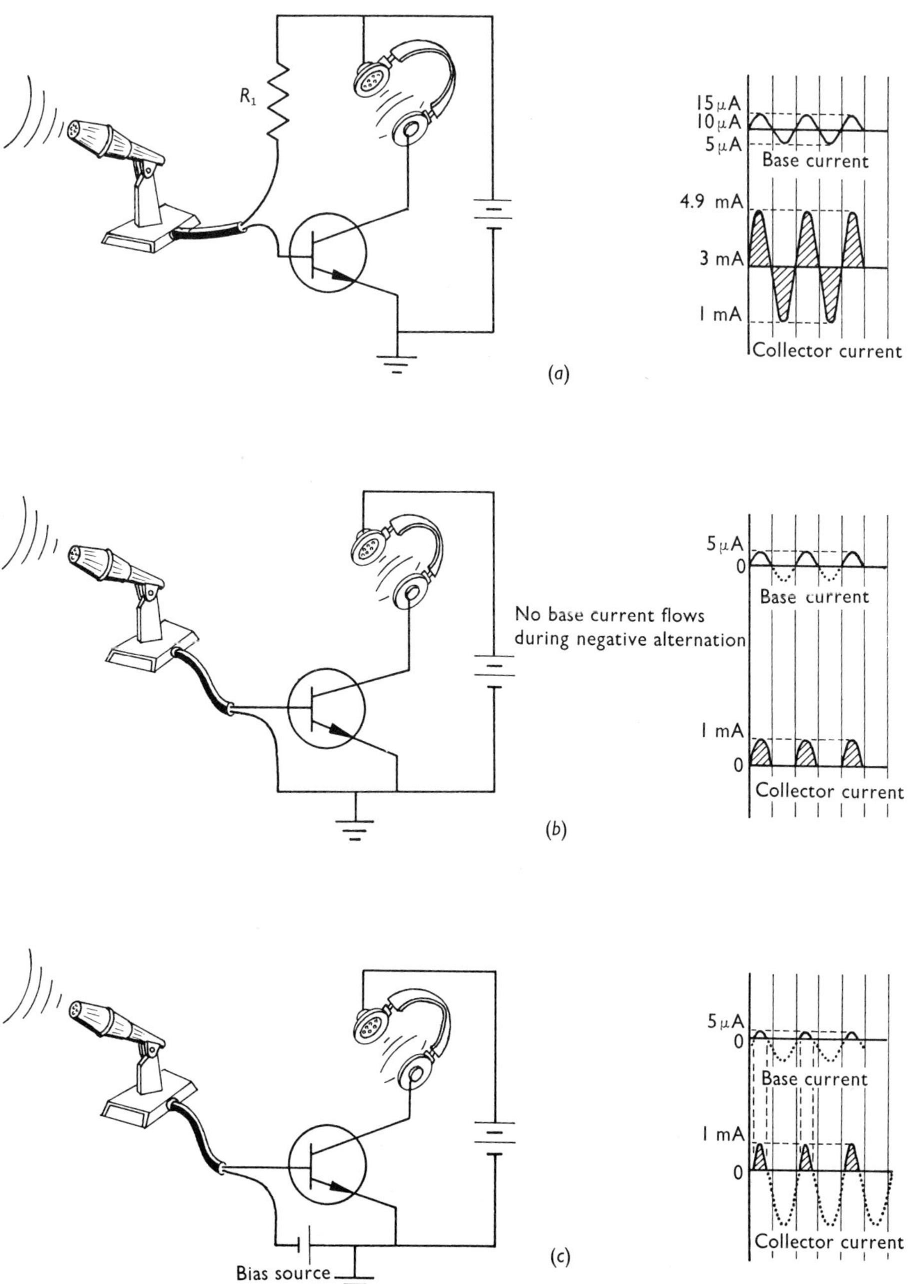

Fig. 3-4 Effect of bias level on amplifier operation: (*a*) Class A operation—full wave, or 360°, of waveform receives amplification; (*b*) Class B operation—half wave, or 180°, of waveform receives amplification; (*c*) Class C operation—less than half the waveform receives amplification.

bursts can be used to energize an *LC*-tuned circuit, which restores the missing parts of the waveform.

3-5 LOAD LINE

In the amplifier circuit shown in Fig. 3-2, it was assumed for simplicity that the headphones offered little resistance to collector current. In practical circuits, load resistance cannot be neglected. What effect does load resistance have on amplifier operation? A good way to determine the effect is to replace the headphones with a resistor so that the load is independent of frequency. Then examine the amplifier circuit under two sets of extremes. First consider the transistor to be at cutoff (no collector current). At the second extreme assume that the transistor is fully conducting and offers negligible or zero resistance.

Figure 3-5a shows the amplifier circuit of Fig. 3-2 redrawn with the headphones replaced by a load resistor of 8000 Ω. With a load resistor of this magnitude collector supply voltage must be increased to obtain sufficient collector current. A value of 50 V has been chosen. It is now possible to analyze the circuit under the two sets of ex-

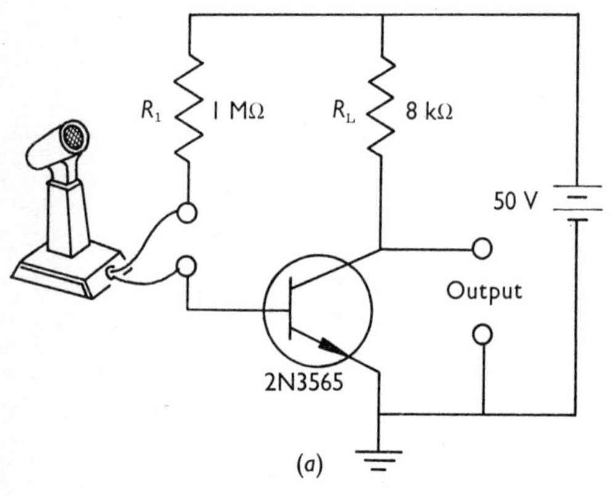

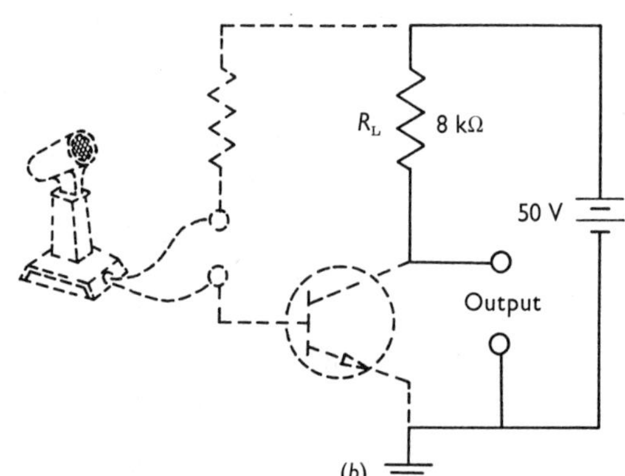

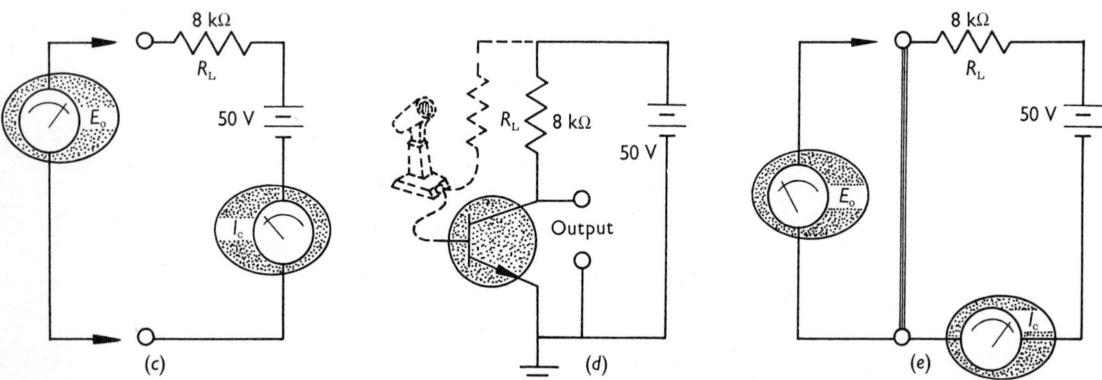

Fig. 3-5 Developing a load line: (*a*) Actual circuit biased at operating point; Equivalent circuits for extreme conditions, when the transistor is cut off [(*b*), (*c*)] and when it is fully conducting [(*d*), (*e*)].

tremes, i.e., transistor cut off and transistor fully conducting. Collector current and output voltage will be calculated at each of the two extremes.

When the transistor is cut off, no collector current flows and the collector-to-emitter connection appears as an open circuit as shown in Fig. 3-5b. To focus attention on the important characteristics, the equivalent circuit of Fig. 3-5c shows even more clearly what is taking place. Note, $I_c = 0$, and because no current flows, there is no *IR* drop across R_L. For these reasons, full supply voltage (50 V) appears at the output. These data are entered in Table 3-2. Now what happens at the other extreme, i.e., the transistor fully conducting?

A fully conducting transistor as shown in Fig. 3-5d has a low collector-to-emitter resistance. It is so low in comparison with the 8-kΩ load resistor R_L that it may be of wire. Now note that output voltage is zero because there is negligible *IR* drop across the conducting transistor. Only the load resistor R_L limits current flow, and the current is found by applying Ohm's law. Current through an 8-kΩ resistor when 50 V is applied is given by

$$I = \frac{E}{R}$$

$$= \frac{50}{8 \times 10^3}$$

$$= 6.25 \times 10^{-3}\ \text{A}$$

$$= 6.25\ \text{mA}$$

These data are also entered into Table 3-2.

Table 3-2 Collector Current and Output Voltage at Extremes of Transistor Cutoff and Full Conduction

Transistor Status	*Output Voltage*	*Collector Current*
Cut off	50 V	0 A
Fully conducting	0 V	6.25 mA

The data accumulated in the form of coordinate pairs in Table 3-2 are now plotted as two points on the graph of the family of characteristic curves in Fig. 3-6. Furthermore, the two points (50,0) and (0,6.25) are joined by a straight line as shown. The name given to this line is the *load line* for the circuit. A load line has one major feature. It gives the value of output voltage and load current at any value of collector current from cutoff to full conduction. The load line gives dynamic or operating characteristics for the amplifier.

3-6 TRANSFER CURVES

A load line is an example of a *transfer characteristic curve*. Transfer characteristic curves are widely used in electronics. They display graphically the effect a circuit has upon a signal passing through it. A transfer curve can be visualized somewhat like a "mirror" which reflects the input signal. As you know, only flat or perfectly straight mirrors give undistorted output. Similarly, only devices with a straight or linear transfer curve can transfer electrical signals from one circuit point to another without distortion. As your experience in electronics grows you will soon realize that linear transfer curves are the exception rather than the rule.

3-7 APPLICATION OF THE LOAD LINE

Every amplifier has a *load line*. Loads with very low resistance or impedance give a steep, nearly vertical load line (see Fig. 3-6, the dotted line). Large-value load resistors or impedances give a shallow sloped, nearly horizontal load line, as shown by the dashed line in Fig. 3-6.

When a load line is constructed it is possible to predict all values of amplifier output voltage and current under any given input signal. Load current is the same as collector current, because all current leaving or enter-

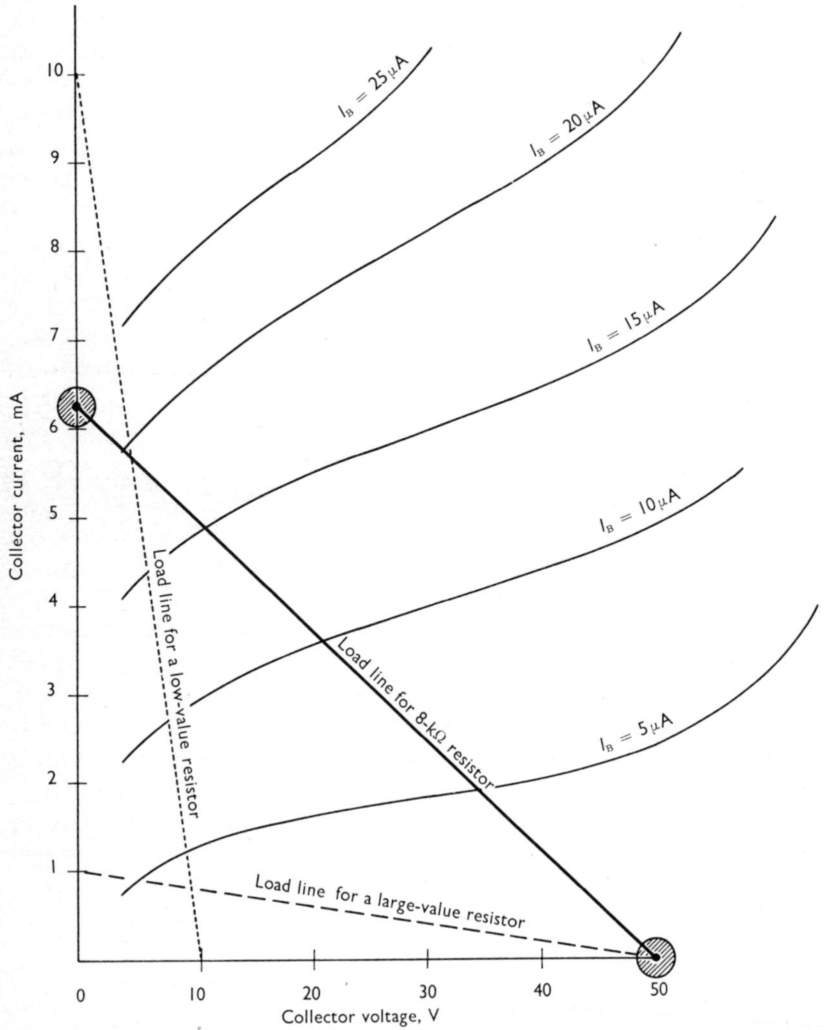

Fig. 3-6 Load line for an 8-kΩ resistor and a 2N3565 transistor.

ing the collector does so via the load. It is not always obvious that load and output voltage are the same.

In Fig. 3-7, the original amplifier schematic is progressively reduced into an equivalent circuit. *The essential concept to be gained here is that for signals a battery or power supply behaves as a very low resistance path.* In its simplest form the transistor can be regarded as a generator of signals connected to the signal load (see Fig. 3-7e and f).

Output-signal voltage and current can be found when the magnitude of base-current variation is known. First the output *quiescent* or rest values must be determined. Refer to Fig. 3-8. A base-quiescent current of 10 μA gives a collector (and load) current of approximately 3.5 mA. Output voltage is then approximately 22 V. These values are obtained by running to each axis a straight line from the intersection of the load line and the 10-μA base-current graph (see Fig. 3-8). The

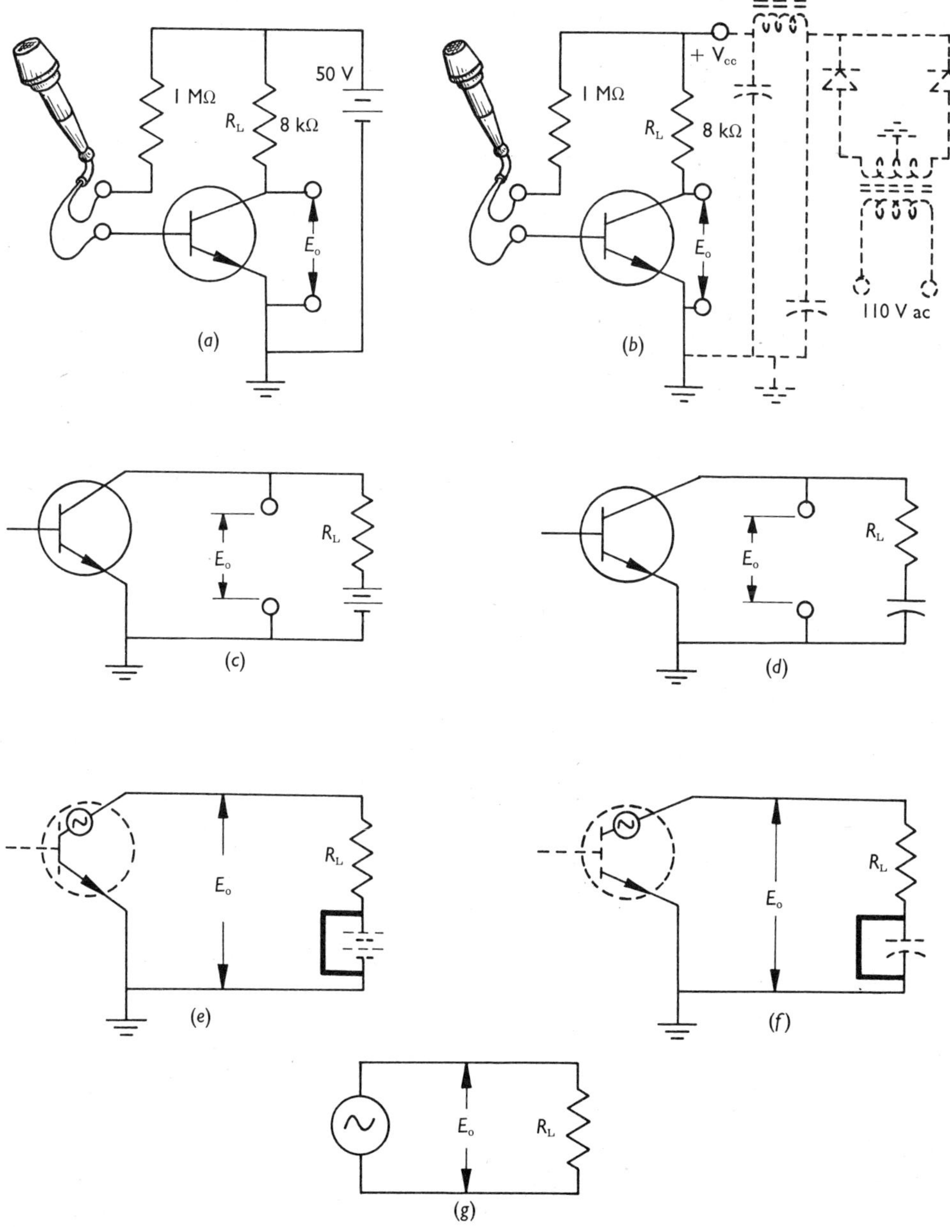

Fig. 3-7 For signals, the load is considered to be connected in parallel with the transistor: (*a*) Amplifier schematic—battery power supply; (*b*) Amplifier schematic—electronic power supply; (*c*), (*d*) Collector circuit; (*e*), (*f*) Battery and power-supply ripple-filter capacitors have low impedance, and are shown shunted for signal current; (*g*) Final equivalent circuit—collector is considered to be the signal generator driving the load.

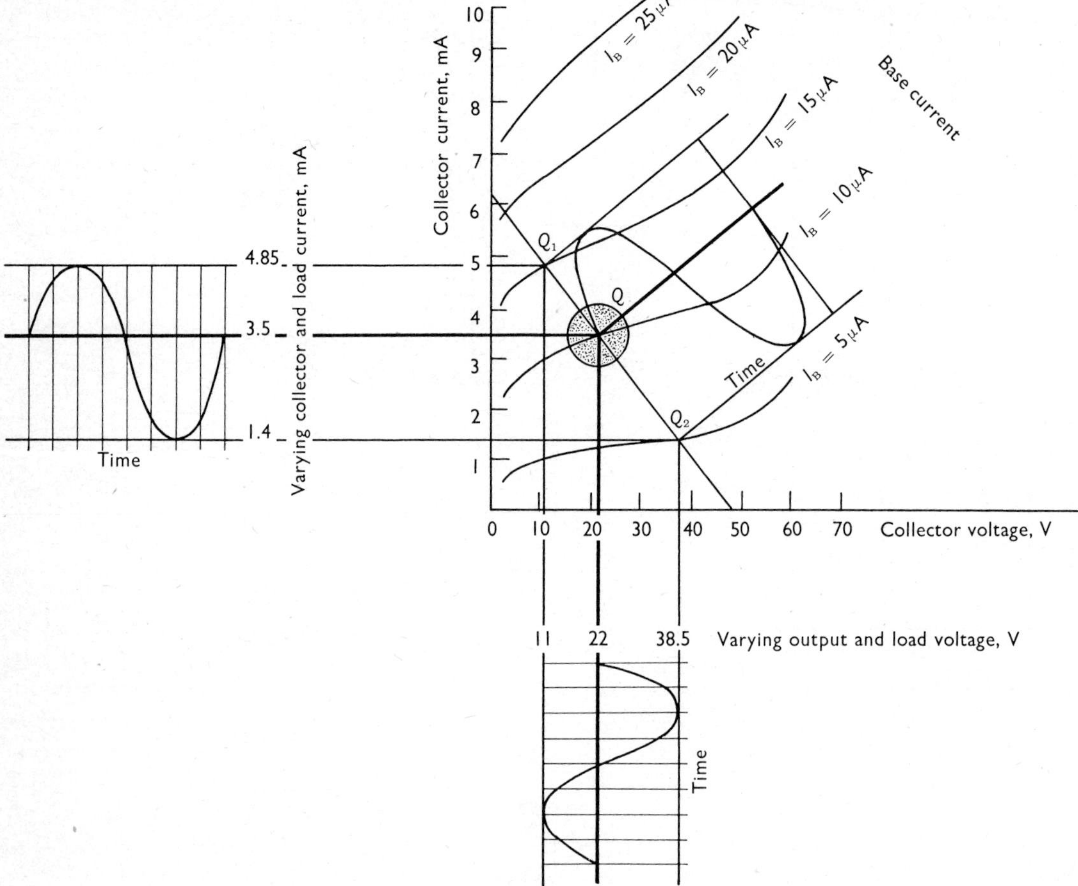

Fig. 3-8 Variation in collector current and output voltage with base signal current.

point Q on the load line in Fig. 3-8 is the operating point for the amplifier when base-bias current is set at 10 μA and an 8-kΩ load resistor is used.

Signal current varies the base current above and below its rest value of 10 μA. As a result the operating point moves back and forth along the load line. As it does so, its horizontal and vertical projections on the x and y axes also move back and forth and up and down. The output-voltage variations are obtained from the x axis, and the load-current variations from the y axis. In Fig. 3-8 the input signal causes the base current to vary between 15 and 5 μA. The net effect is to shift the operating point. First the operating point shifts to Q_1, which gives a 4.85-mA collector-current shift and an 11-V output-voltage excursion. On the negative signal peak, the operating point shifts to Q_2, giving a collector current of 1.4 mA and output voltage of 38.5 V. In the example shown in Fig. 3-8 input and output current and voltage waveforms for a 60-Hz signal are drawn with reference to the load line. Examine this drawing and note the following items:

1. A given base-current change causes greater collector-current change at low values of collector current than at high

values. For this reason, the negative alternation receives slightly more amplification than the positive. Similar distortion occurs to some extent in all transistor amplifiers because transistors are not perfectly linear devices.

2. The input- and output-current waveforms are in phase, i.e., each reaches its maximum and minimum at the same time.
3. The output voltage and current are 180° out of phase, i.e., when the current is greatest, the voltage is smallest, and vice versa.

3-8 BIAS SUPPLY AND STABILIZATION

In all class A transistor amplifiers a certain value of quiescent base current is required to fix the operating point. The quiescent current is known as *bias* current. It can be established with a bias battery or bias power supply, but usually it comes from the collector supply through a current-limiting resistor. The simplest biasing circuit has already been introduced in Figs. 3-2, 3-5, and 3-7.

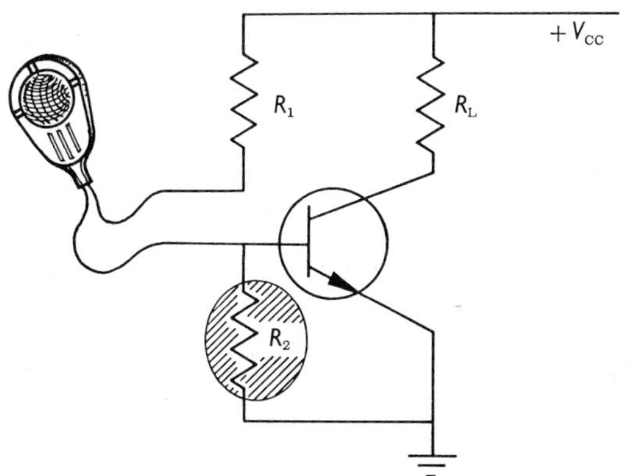

Fig. 3-9 Addition of a bias-stabilizing resistor.

If the emitter-base conduction were to vary over a wide range because of temperature effects, the bias current (and the operating point) could shift over a period of time. A second resistor known as a bias-stabilizing resistor is usually connected across the emitter-base junction as shown in Fig. 3-9 (R_2). Its behavior is similar to that of a bleeder resistor (see Sec. 2-8). Typically, the current through the stabilizing resistor (R_2) is five times the magnitude of the base current. Note that both the base current and the stabilizing-resistor current must pass through the bias resistor R_1. Because the base current is only a percentage of the total current through R_1, minor base-current variations cause less *IR* drop variation across R_1. In the amplifier circuit of Fig. 3-9, if the base-quiescent current were 10 μA, the total current through R_1 would then be the sum of the base current (10 μA) and the current through R_2 (50 μA), or 60 μA.

3-9 THERMAL-RUNAWAY PROTECTION

Germanium transistors are more temperature-sensitive than silicon types. When germanium transistors are used, a third resistor R_3 is connected in the emitter as shown in Fig. 3-10a. The emitter resistor guards against thermal runaway, a condition in which excessive collector current heats the transistor. This further increases collector current, culminating in the destruction of the transistor. Reference to Fig. 3-10b and c shows the polarities of the *IR* drops developed across R_2 and R_3 for both NPN and PNP transistors. Note that the polarity across R_2 is always such that it forward-biases the base and enhances collector-current flow. On the other hand, R_3 develops an *IR* drop which opposes the voltage across R_2. The effective base-to-emitter voltage is the difference or algebraic sum of the two *IR* drops as a consequence of Kirchhoff's law. Stated as an equation, this is,

$$\text{(emitter-to-base voltage)} = (IR \text{ drop on } R_2) - (IR \text{ drop on } R_3)$$

$$E_B = E_{R_2} - E_{R_3}$$

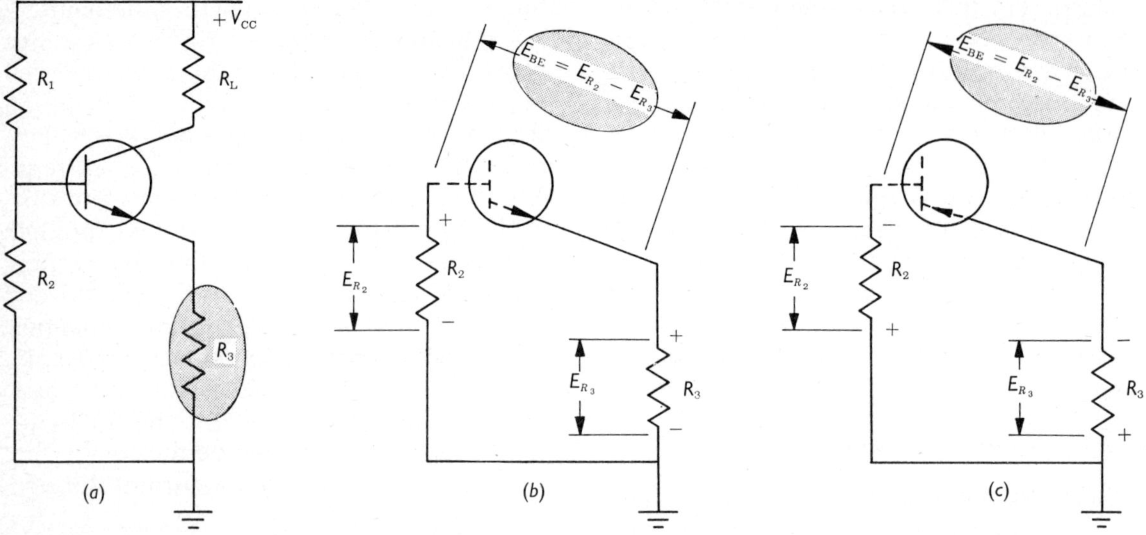

Fig. 3-10 Addition of an emitter resistor to guard against thermal runaway. The polarity of the *IR* drop on R_3 always opposes the forward bias on the base: (*a*) Schematic diagram; (*b*) NPN transistor; (*c*) PNP transistor.

In practice, the values of R_2 and R_3 are calculated such that for a given operating point, E_B is sufficient to overcome the barrier potential of the emitter-base PN junction. For silicon the barrier potential is approximately 0.6 V, whereas for germanium it is 0.3 V.

3-10 FIELD-EFFECT-TRANSISTOR (FET) AMPLIFIER

One disadvantage of bipolar transistors is that base current is required to cause a flow of collector current. This means that the signal source must be capable of supplying both signal current and signal voltage, i.e., signal power. Often a signal source cannot provide the required power to drive the base of a transistor yet it may possess considerable voltage amplitude. One example of such a signal source is a high-impedance crystal microphone. The instrument has an internal resistance of the order of 500 kΩ. You can easily visualize that signal current would be severely limited by an internal resistance of this magnitude.

Field-effect transistors (FETs) and the metal-oxide silicon field-effect transistors (MOSFETs) have no base electrode and require no base current. The drain electrode performs the same function as the collector in a bipolar transistor. Drain current is varied by changing the voltage on the gate electrode. Figure 3-11a shows drain characteristics for a 2N3819 N-channel FET, and an experimental circuit for acquiring data to produce the curves (Fig. 3-11b). Note the essential practical difference between bipolar and field-effect transistors. In bipolar transistors load current is varied by *signal current* applied to the base. In field-effect transistors load current is varied by *signal voltage* applied to the gate.

Figure 3-12 shows the schematic diagram of a FET amplifier. The quiescent gate-to-source voltage measures − 1 V, since no bias voltage is lost across the input resistor. (No gate current flows in a FET, therefore no *IR* drop takes place across the large input resistor R_1 which is typically 1 to 10 MΩ.) A glance at the characteristic curves in Fig. 3-11 shows that a − 1-V gate-bias operating point

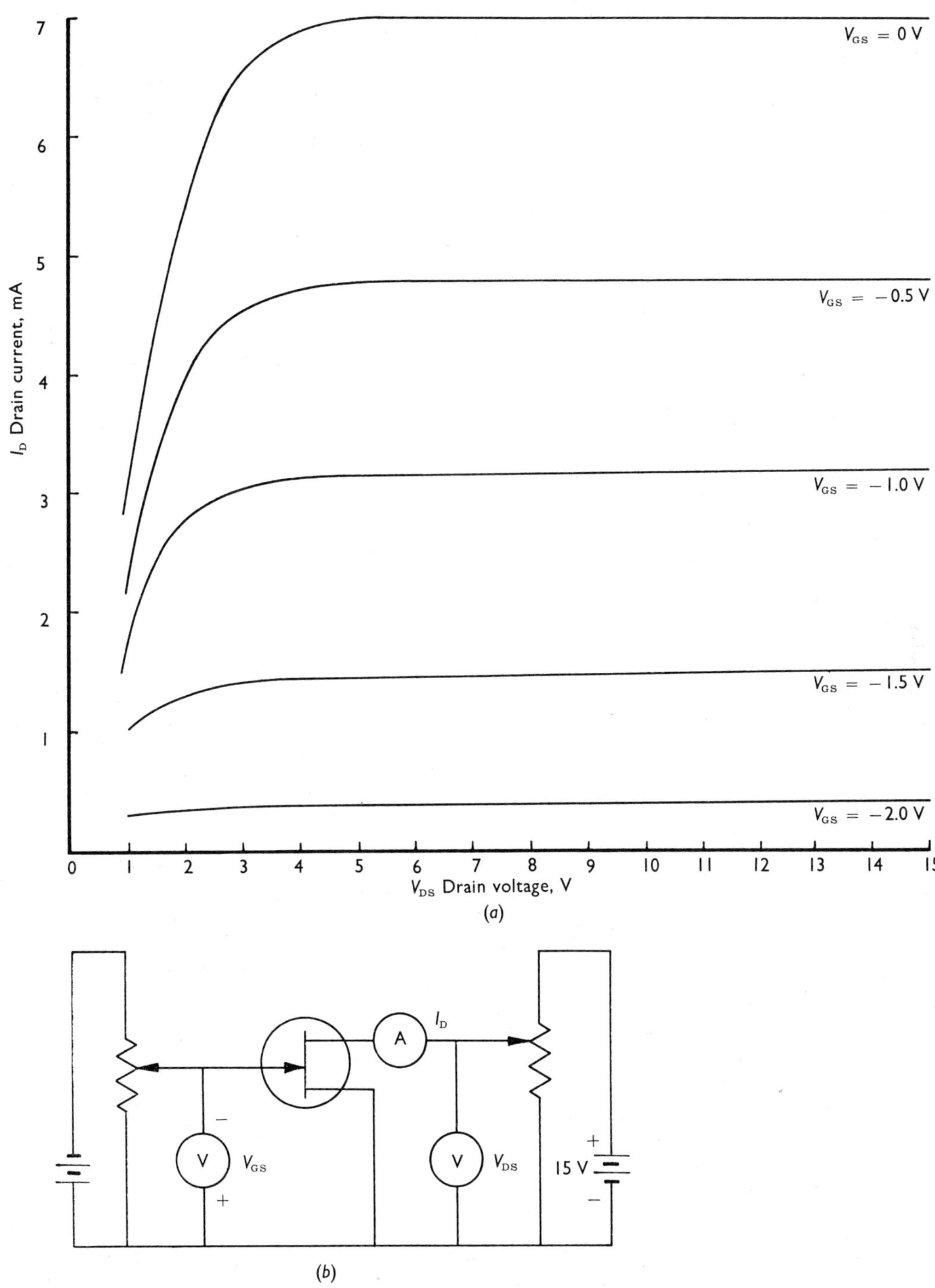

Fig. 3-11 Field-effect transistor: (*a*) Drain characteristic curves; (*b*) Circuit for collecting data to prepare drain characteristic curves.

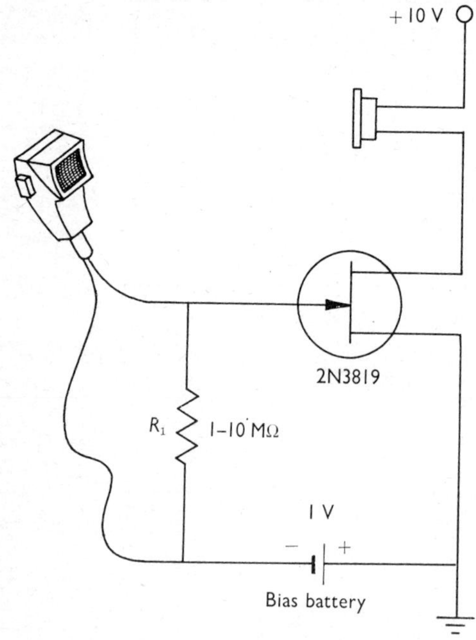

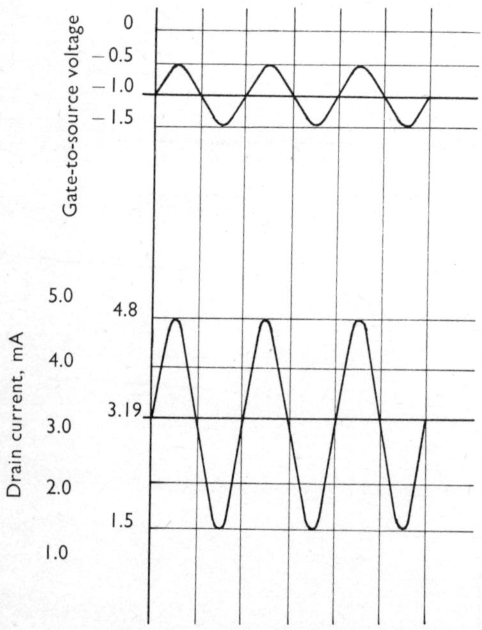

Fig. 3-12 FET amplifier with timing diagram showing output current vs. input voltage.

causes a quiescent drain current of 3.19 mA through the headphones. Now visualize a 1-V peak-to-peak alternating signal voltage arriving from the microphone. This signal voltage is impressed across the input resistor. The input resistor is of high ohmic value and will therefore not allow much microphone current to flow even if it were available. Since the input resistor and the bias battery are in series, the effective gate-to-source voltage is the algebraic sum of the bias battery voltage and the signal voltage. It will be − 1.5 V on the negative alternation and − 0.5 V on the positive, as shown in the timing diagram.

It is not always necessary to use a battery to provide bias voltage for a FET. In Fig. 3-13 a resistor R_2 connected between the source electrode and ground develops an *IR* drop caused by the source-to-drain current flowing through it. Effectively, this *IR* drop is connected through R_1 (which itself has no *IR* drop) *between the source and the gate electrodes* (Fig. 3-13b). The voltage is of correct polarity to bias the FET. Ohm's law is used for calculating the value of R_2, since we know the current through the resistor (drain current) and the bias voltage required. In the present instance, bias voltage is 1 V, and quiescent drain current is 3.19 mA. Applying Ohm's law gives the value of R_2 as

$$R_2 = \frac{E}{I}$$

$$= \frac{1}{3.19 \times 10^{-3}}$$

$$= \frac{1000}{3.19}$$

$$= 314\ \Omega$$

In practice a 330-Ω resistor would be adequate.

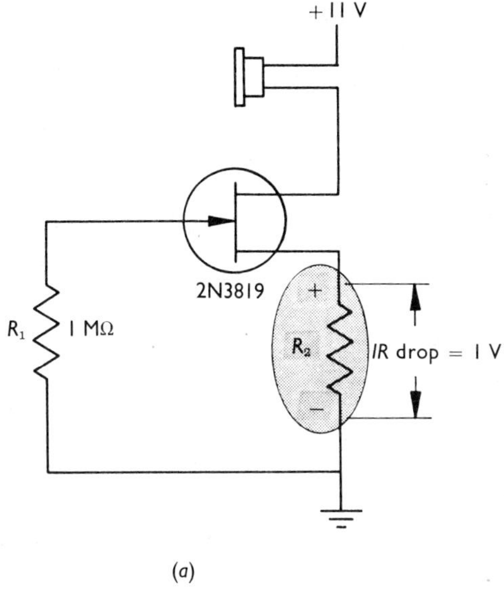

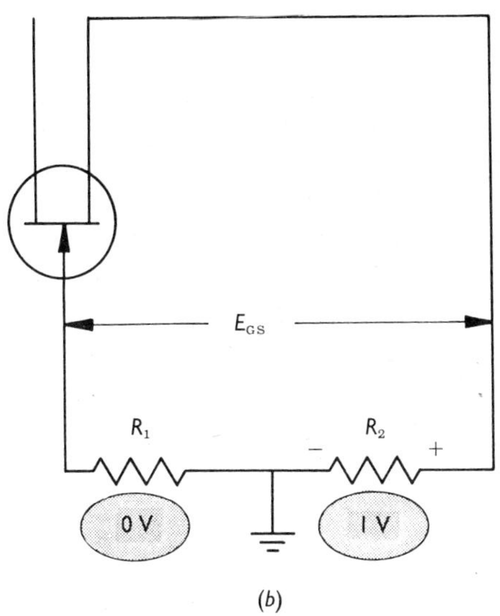

Fig. 3-13 FET amplifier circuit with biasing resistor: (*a*) Schematic diagram; (*b*) Gate-to-source circuit, showing gate-to-source voltage developed by R_2.

3-11 FET AMPLIFIER LOAD LINE

To appreciate how FET devices operate under actual conditions we must construct a load line for the value of load resistor used. The procedure is similar to the construction of a bipolar transistor load line (Sec. 3-5).

Power-supply voltage and zero drain current determine one end of the load line on the horizontal axis of the characteristic curves. The other end intersects the vertical axis at a current which depends on the value of the resistor used. The following example shows how a load line is constructed from given data.

EXAMPLE 3-1 Construct a load line for a 2N3819 FET with a 3-kΩ load and a drain power supply of 15 V. With the drain at cutoff, the output is 15 V. With the FET fully conducting, the maximum drain current possible is obtained by assuming 15 V across the load. Ohm's law then gives load and drain current as

$$I_D = \frac{E_{\text{supply}}}{R_L}$$

$$= \frac{15}{3000}$$

$$= 5 \times 10^{-3}\ \text{A}$$

ANSWER $\quad = 5\ \text{mA}$

These facts are summarized in tabular form.

Status of FET	*Output Volts*	*Output Current*
Cut off	15	0
Fully conducting	0	5 mA

The two points (15,0) and (0,5) are plotted on the characteristic curves and joined to form the load line, as shown in Fig. 3-14.

The next example shows how to calculate the value of the bias resistor inserted in the source lead.

EXAMPLE 3-2 Compute the value of source-biasing resistor required to give a gate-source bias of 1.25 V.

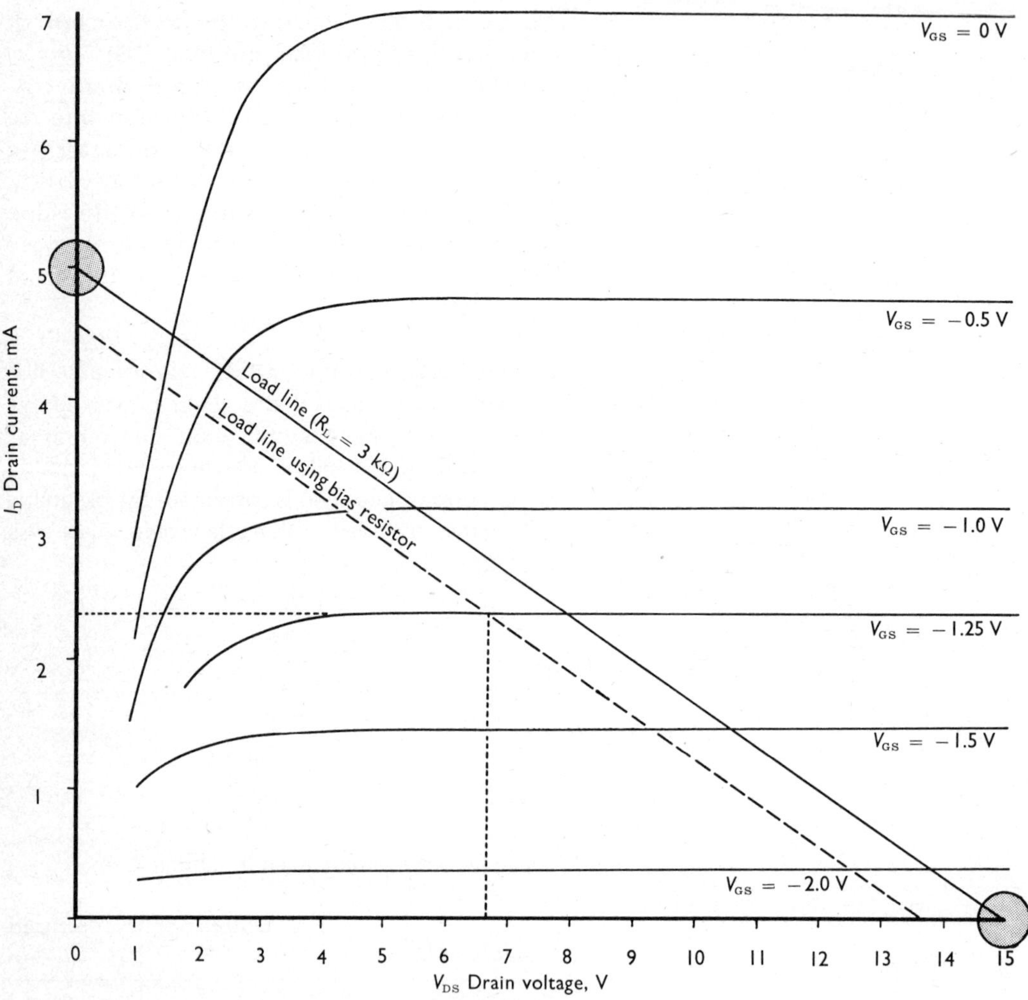

Fig. 3-14 Family of drain characteristic curves for a 2N3819 N-channel FET, showing the load line for a 3-kΩ load.

To be strictly correct, the *IR* drop developed across the bias resistor must be subtracted from the available power-supply voltage. The remaining voltage (15 − 1.25 = 13.75 V) is applied across the FET and load (Fig. 3-15). A new load line is calculated and constructed using these revised data (dashed load line in Fig. 3-14). A horizontal projection of the intersection point of the 1.25-V_{GS} curve and the dashed load line gives a drain current of 2.35 mA.

Applying Ohm's law gives

$$R_s = \frac{E_{bias}}{I_D}$$

$$= \frac{1.25}{2.35 \times 10^{-3}}$$

$$= \frac{1250}{2.35}$$

ANSWER $= 532\ \Omega$

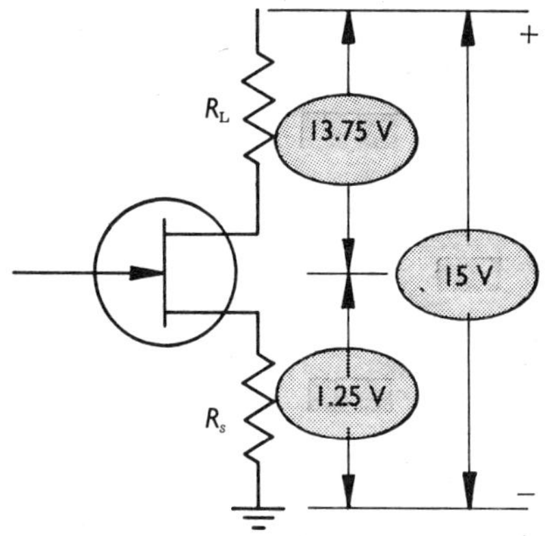

Fig. 3-15 Voltage distribution in a typical FET, source-biased amplifier circuit.

The complete FET amplifier circuit schematic diagram with component values is shown in Fig. 3-16.

3-12 VACUUM-TUBE AMPLIFIER

A vacuum tube can perform the same functions as a FET or MOSFET in that it is also a voltage-amplifying device. For a discussion of the important features of a vacuum-tube amplifier circuit refer to Appendix B.

3-13 FUNDAMENTAL AMPLIFIER CIRCUIT CONFIGURATIONS

Each of the amplifying devices, the bipolar transistor, the field-effect transistor, and the vacuum tube, has three fundamental electrodes which are required by all amplifiers. These are an electrode which emits current, an electrode which controls the amount of current flow, and an electrode which receives the emitted current. Table 3-3 lists each of the devices and the names of the electrodes that perform these three essential functions.

Table 3-3 Electrodes and Their Functions

Device	*Emitting Electrode*	*Controlling Electrode*	*Receiving Electrode*
Bipolar transistor	Emitter	Base	Collector
Field-effect transistor	Source	Gate	Drain
Vacuum tube	Cathode	Control grid	Plate or anode

Every electrical source and load has *two terminals,* one for current to enter, the other for current to leave. An amplifier circuit consists of a signal source and signal load connected to an amplifying device which as noted from Table 3-3 has only three terminals. This means that four terminals in all are connected to a three-terminal device. In every instance one of the terminals on the amplifying device must be shared by the source and load. The term "common" is applied to the shared electrode.

In the circuit of Fig. 3-2 the signal source (a microphone) is connected to the base and *emitter* through resistor R_1 and the battery. The signal load is connected to the collector and *emitter* (through the battery). Since both source and load each have one connection to the *emitter,* the circuit is known as a *common-emitter* circuit. Other possible arrangements for transistor and FET amplifiers are shown in Fig. 3-17. Equivalent vacuum-tube connections are shown in Fig. B-2 (Appendix B). Essential characteristics of each type of circuit are listed in Table 3-4.

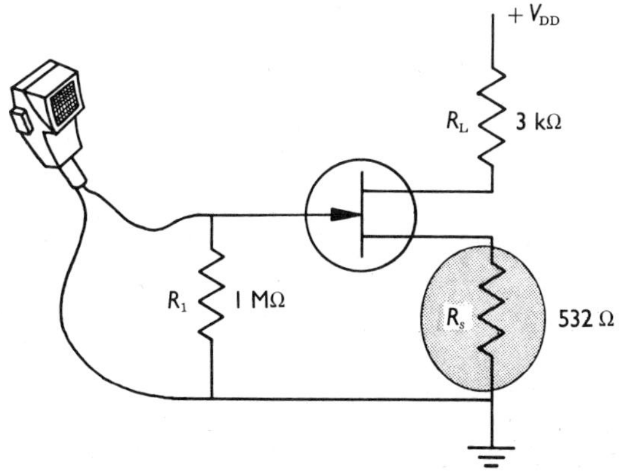

Fig. 3-16 FET amplifier circuit with source-biasing resistor.

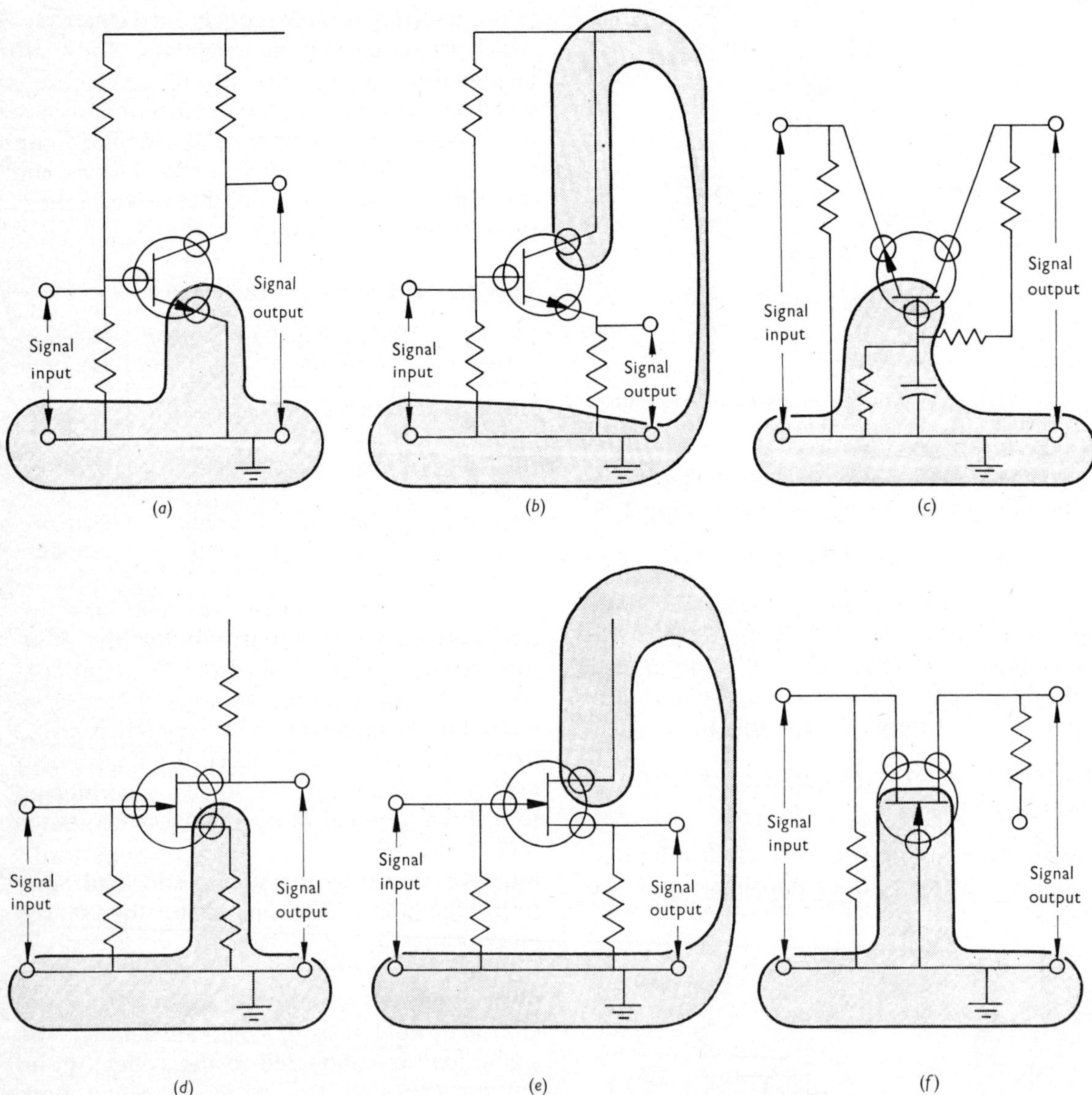

Fig. 3-17 Basic types of amplifier configurations: (*a*) Common emitter; (*b*) Common collector (emitter follower); (*c*) Common base; (*d*) Common source; (*e*) Common drain (source follower); (*f*) Common gate.

When classifying a circuit as to type, always remember that power supplies and batteries offer no effective opposition to signal currents. For signal sources and loads they can be mentally removed and replaced by a *jumper*. (A jumper is a length of wire or conductor placed across two terminals, usually temporarily, to make a direct connection between them.)

Table 3-4 Essential Characteristics of Various Amplifier Configurations in Fig. 3-17

Type	*Input Impedance*	*Output Impedance*	*Gain Characteristics*	*Applications*
Common {emitter, source, cathode} Fig. 3-17a, d Appendix B, Fig. B-2a	Very high for FET and tube, determined mainly by R_1 which may be several megohms. In transistor circuits R_2 and R_1 are in parallel with base-emitter resistance, therefore resistance ranges from several hundred to several thousand ohms	Several thousand ohms for triodes. Several hundred thousand for pentodes, transistors, and FETs	Highest overall power gain of the three types of configurations, good voltage, and current gain	Majority of amplifier requirements can be implemented by this type
Common {collector, drain, plate} or {emitter, source, cathode} follower Fig. 3-17b, e Appendix B, Fig. B-2b	Extremely high. Can be several megohms, because some of the *IR* drop developed across the load is fed back to the input such that signal-source current is reduced. To the signal source this appears as an extremely high resistance	Very low—several ohms to several hundred ohms	Good current gain and moderate power gain. Voltage gain is always less than 1	Used in applications with minimum signal-source loading. Follower circuit requires nearly no signal current to operate, leaving the signal source lightly loaded. The output can be heavily loaded because it can supply considerable output current. Followers are used to drive signals through cables, because the available signal current can quickly charge the capacitance between the cable conductors

Table 3-4 continued

Type	*Input Impedance*	*Output Impedance*	*Gain Characteristics*	*Applications*
Common {base, gate, grid} Fig. 3-17c, f Appendix B, Fig. B-2c	Low (input current is higher than output current)	Very high	High voltage gain. Current is less than 1. Provides good gain at high frequencies	Used in high-frequency preamplifiers in TV and FM tuners

Summary

Gain is the ratio

$$\frac{\textit{output-signal level}}{\textit{input-signal level}}$$

The signal level can be expressed in volts, amperes, or watts, but unless specifically stated it should be assumed to be power gain, i.e., signal level expressed in watts.

In a bipolar transistor amplifier circuit gain occurs because input-signal base current controls a much larger collector current. This produces an enlarged replica of the signal current.

In FET amplifier circuits gain occurs because signal voltage applied to the gate electrode controls drain current. The operation of vacuum-tube amplifiers is similar except that grid voltage controls plate current.

Quiescent current or voltage is the steady-state current and voltage when no signal is present. Only bias current flows under quiescent conditions.

The *operating point* is a point located on a load line. It represents quiescent conditions. Bias, quiescent conditions, and operating point are closely related and convey essentially the same information.

Class of operation refers to the proportion of a signal waveform which receives amplification. Class of operation is established by the bias voltage and/or currents.

A *load line* shows graphically how load current and voltage vary under actual operating conditions. It is a graph of load currents versus load voltage for every possible combination of voltage and current which the amplifying device can assume.

A *transfer curve* shows graphically the effect a given circuit has on a given signal. A load line is one example of a transfer curve.

An emitter resistor protects against *thermal runaway*.

The three fundamental amplifier circuits are summarized in Table 3-4.

Questions and Exercises

1. What factor is expressed by the mathematical quantity (output-signal power)/(input-signal power)?
2. A certain amplifier boosts the signal from a phonograph cartridge from 750 mV to 3.75 V. What is its voltage gain?
3. A television booster amplifier operating in a fringe area picks up a weak signal of 10 μV. It has a rated voltage gain of ten thousand. What will be the voltage amplitude of the boosted signal at its output?
4. A team of medical specialists wishes to record certain electrical signals originating in an area of the brain. These are to be recorded on an instrument which requires

a 5-V input signal. The brain signals are known to be only of 2.5×10^{-2} mV maximum amplitude from previous experience. How much voltage gain will be required in the amplifier chosen to boost the signals?

5. Plot a graph of collector current versus base current (i.e., base current on the horizontal axis and collector current on the vertical) for a 2N3565 transistor operating at a fixed collector voltage of 10 V. Obtain your data from Fig. 3-1.
6. Calculate the collector current in the circuit of Fig. 3-2a if the 1-MΩ resistor were replaced by a 0.66-MΩ resistor.
7. In your own words explain how the circuit of Fig. 3-2a amplifies a signal.
8. What is meant by the term quiescent current?
9. Plot a graph of collector current if a 2-μA peak sine-wave signal is applied to the base of the circuit in the circuit of Fig. 3-2a. (Hint: Refer to Fig. 3-3.)
10. Is the amplifier circuit in Fig. 3-2a operating class A, B, or C?
11. What determines the class of operation in any given amplifier?
12. Which class of operation is most efficient? Explain your answer.
13. Construct a load line for the circuit in Fig. 3-5a using a 40-V collector supply voltage and a 10-kΩ load resistor.
14. From your load line constructed in Question 13, determine collector current and voltage for the following base currents: $I_B = 5\ \mu A$ $I_B = 10\ \mu A$ $I_B = 15\ \mu A$.
15. Explain why load voltage (measured across the load) and output voltage (measured from collector to ground) are the same for signal voltages.
16. Why is output-signal voltage and output-signal current 180° out of phase in the circuit of Fig. 3-7a?
17. What is the purpose of resistor R_1 in Fig. 3-9?
18. What is the purpose of resistor R_2 in Fig. 3-9?
19. Explain the operation of a bias-stabilizing resistor.
20. What general rule is used to determine the amount of current flowing in the bias-stabilizing resistor?
21. If the base current in the circuit of Fig. 3-9 is 14 μA and the voltage across the emitter-base junction is 0.72 V, what is the value of R_2?
22. Explain the purpose of R_3 in Fig. 3-10a.
23. What is one advantage of using a FET rather than a bipolar transistor when attempting to amplify a weak signal?
24. Draw schematic diagrams to illustrate two methods of providing bias voltage for a FET amplifier stage.
25. A FET amplifier circuit requires 1.5 V of gate bias and operates with a drain quiescent current of 3 mA. What value of source resistor is required to produce the bias voltage?
26. Construct a load line using the drain characteristic curves in Fig. 3-14. A 2-kΩ resistor is to be used with a 10-V drain supply voltage.
27. Under the conditions of Question 26, plot a new load line if a source resistor is added for biasing such that 1 V of bias voltage is developed across it.
28. Calculate the value of cathode resistor required in Fig. B-1a (Appendix B) to supply 3 V of grid bias if plate current is 10 mA.
29. A screen voltage-dropping resistor is needed to reduce the power-supply voltage from 300 to 200 V for a tube which draws 2 mA screen current. Find the value of the resistor required. (Refer to Appendix B.)

4-1 CASCADING AND INTERSTAGE COUPLING

A single transistor, FET, or vacuum tube, and its associated circuit components, is known as a *stage*. Each of the amplifier circuits referred to in Chap. 3 are *single-stage* circuits. In most applications one stage of amplification is insufficient to raise the signal to the required level. Additional gain is obtained by *cascading* stages, which means that the output of the first stage is used as input for the second, the output of the second serves as input for the third, etc.

One of the most important considerations in cascading stages is to ensure that the signal connections between stages do not disturb the biasing arrangements of each stage. For instance, if you simply connect a wire jumper from the collector of the first stage to the base of the second, the dc voltage levels in both stages would be drastically altered, and at least the second, if not both stages, would cease functioning.

4-2 AC AND DC AMPLIFIERS

One way to prevent dc levels in one stage from interfering with those in another is to pass only the ac component of a signal from one stage to the next. Amplifiers which pass only the ac component of a signal are known as ac amplifiers. In many electronic systems including radio, television, and home-entertainment equipment ac amplifiers are adequate because the ac signal component is restricted to a given frequency range.

There are many applications where ac amplifiers cannot be used. One example would be the amplification of the voltage change from a thermocouple. The output of a thermocouple is either a level or a slowly varying voltage, and we do not normally regard it as having an ac component (although strictly speaking, any varying voltage, regardless of rate of change, has an ac component). Amplifiers which respond to dc signal levels are known as *dc amplifiers*. With dc amplifiers there is a unique output for each voltage-level input (within limits), but with ac amplifiers there is an output only at those instants when the input is changing in amplitude.

4-3 TRANSFORMER COUPLING

The earliest form of interstage coupling was accomplished with transformers because they pass only the ac component of a signal, leaving the bias levels unaltered. Current is induced in the secondary coil only when a magnetic-flux change occurs.

A two-stage amplifier with transformer coupling is shown in Fig. 4-1. An equivalent vacuum-tube circuit is shown in Fig. C-1 (Appendix C). The primary coil of T_1 causes a varying magnetic flux in the transformer core as a result of the varying collector current in transistor Q_1. The varying magnetic-flux amplitude is an exact replica of the current which created it, and it induces an exact replica of the primary current in the secondary coil. The relative magnitude of the two currents depends on the turns ratio.

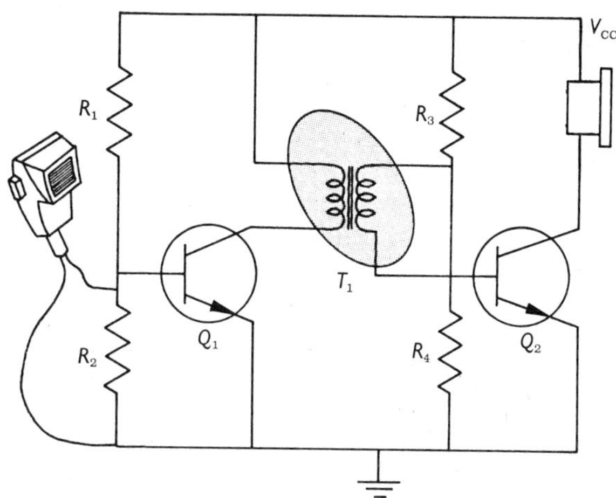

Fig. 4-1 Two-stage, transformer-coupled, transistor audio amplifier.

Induced secondary current is then used to vary the base current of transistor Q_2. Note that because there is no direct electrical contact between the primary and secondary coils the dc collector voltage on Q_1 does not upset the dc bias voltage on the base of Q_2. The bias on transistor Q_1 is set by resistors R_1 and R_2 and on transistor Q_2 by resistors R_3 and R_4.

If silicon transistors are used, it is possible to operate the amplifier without thermal runaway resistors in the emitter leads, but in most instances they are included. They are omitted here to simplify the diagram.

4-4 ADVANTAGES AND DISADVANTAGES OF TRANSFORMER COUPLING

Several advantages occur with transformer coupling. In addition to the electrical isolation between stages, impedance matching can be obtained between the output of the driving stage and the input of the driven stage. A primary-to-secondary turns ratio is chosen to ensure that the *E/I* ratio in both primary and secondary coils is ideal for source and load, respectively. With transistor amplifiers you will commonly find voltage stepdown transformers in use. The output impedance at the collector circuit can be higher than in the base circuit of the input of the next stage. This is not the case with FET or vacuum-tube stages, and as a rule, there will be more turns in the secondary to obtain voltage stepup.

The major disadvantages of transformers, especially at low frequencies, are that they are bulky, heavy, and expensive. Also, unless they are well designed, and therefore expensive, not all frequencies in a given band are passed at the same amplitude. Designers avoid using transformers whenever other methods are equally acceptable. Perhaps the predominant use of audio transformers is to match the output stage of an amplifier to its load, such as the power-amplifier stage to the speaker.

4-5 TUNED-TRANSFORMER COUPLING

As well as interstage isolation and good matching, a third and possibly most important advantage of transformers is that both transformer coils may be tuned. For example, if a capacitor is placed across the secondary coil as shown in Fig. 4-2a, a series *LC* circuit is formed (Fig. 4-2e). Such a circuit resonates at the frequency

$$f = \frac{1}{2\pi\sqrt{LC}}$$

and at this resonant frequency the output voltage will be maximum. At frequencies above resonance, circulating induced current is curtailed by the coils' increased inductive reactance. At frequencies below resonance, the capacitor provides high reactance and again circulating signal current is reduced. Signal-output voltage developed across the capacitor is maximum only at resonant frequency. In practical applications, the capacitor or coil or both may be variable, allowing the user to select the desired resonant frequency. This is done whenever transistor radio receivers are tuned to select different stations.

Some examples of single-tuned and double-tuned transformer-coupled amplifier circuits are shown in Fig. 4-3. Note that in Fig. 4-3b the signal is delivered to and taken from taps on the primary and secondary coils. This is done to match the low impedance of the transistor to the high impedance of the coil at resonant frequency. Connecting a transistor directly across either the coil or capacitor would excessively load it. To the tuned circuit it would seem as though a wire jumper or a low resistance had been connected across it, and its function would be impaired. Enough signal current can be supplied to the base of a transistor from a tap on the secondary, and yet the tuned circuit is not loaded sufficiently to seriously interfere with its operation. Similarly, enough primary current can be injected by the collector to a tap on the primary coil. A vacuum-tube circuit using tuned-coupling transformers is shown in Fig. C-2 (Appendix C).

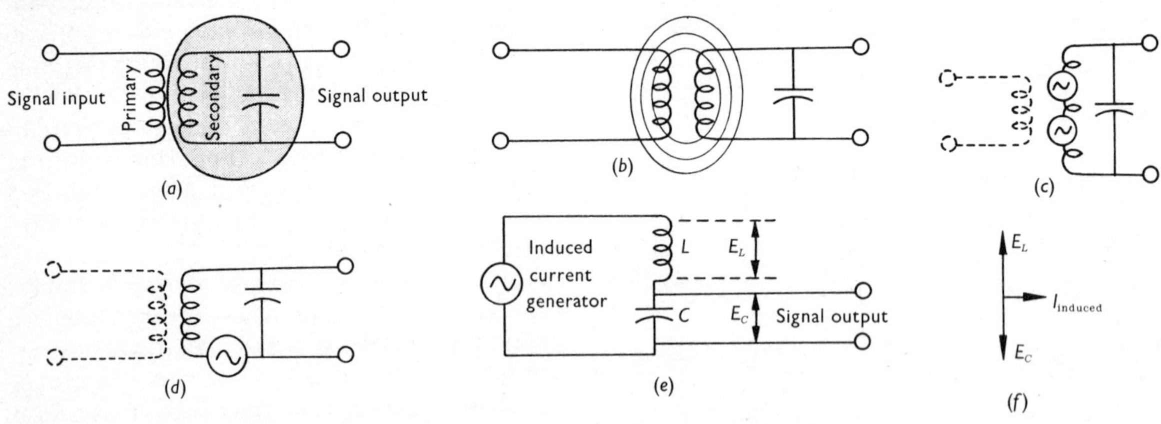

Fig. 4-2 Tuned transformer for interstage coupling: (*a*) Schematic diagram; (*b*) Magnetic-flux coupling between primary and secondary; (*c*) Partial schematic diagram, showing location of imaginary current generators in the windings of the secondary; (*d*) Equivalent circuit, showing all induced current generators lumped together; (*e*) Tuned secondary equivalent circuit; (*f*) Phasor diagram of current and voltage in the secondary circuit. At resonance, $E_L = E_C$ and both are maximum.

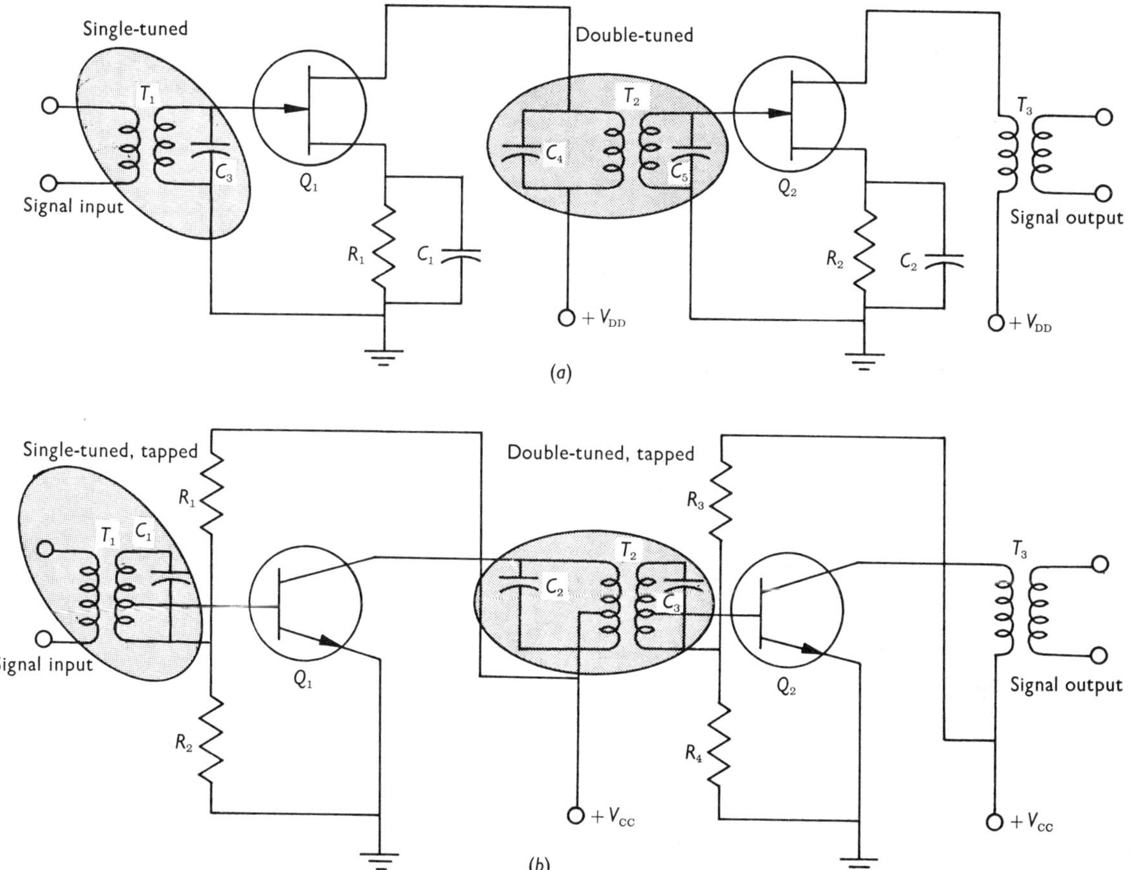

Fig. 4-3 Single-tuned, double-tuned, and untuned coupling transformers: (*a*) FET amplifier; (*b*) Bipolar transistor amplifier.

4-6 CAPACITOR COUPLING

A second method of transferring only the ac signal component can be accomplished by *capacitor coupling*. To help grasp the essential concept involved, refer to Fig. 4-4.

The signal voltage, which is a time-varying voltage, has an ac component which can be represented by an ac generator as shown in Fig. 4-4b. The generator's output is applied to the series-connected resistor and capacitor, as in the equivalent circuit of Fig. 4-4c. By Kirchhoff's voltage law, there will be voltage division across the capacitor and resistor. Since the resistor voltage is also the signal voltage which arrives at the gate of the FET, the aim is to keep the capacitor voltage low and the resistor voltage as high as possible. How can this be done?

Capacitor voltage depends on two things, (1) the generator current and (2) the capacitive reactance. If the capacitive reactance is large in comparison with the resistance of R_1, consider-

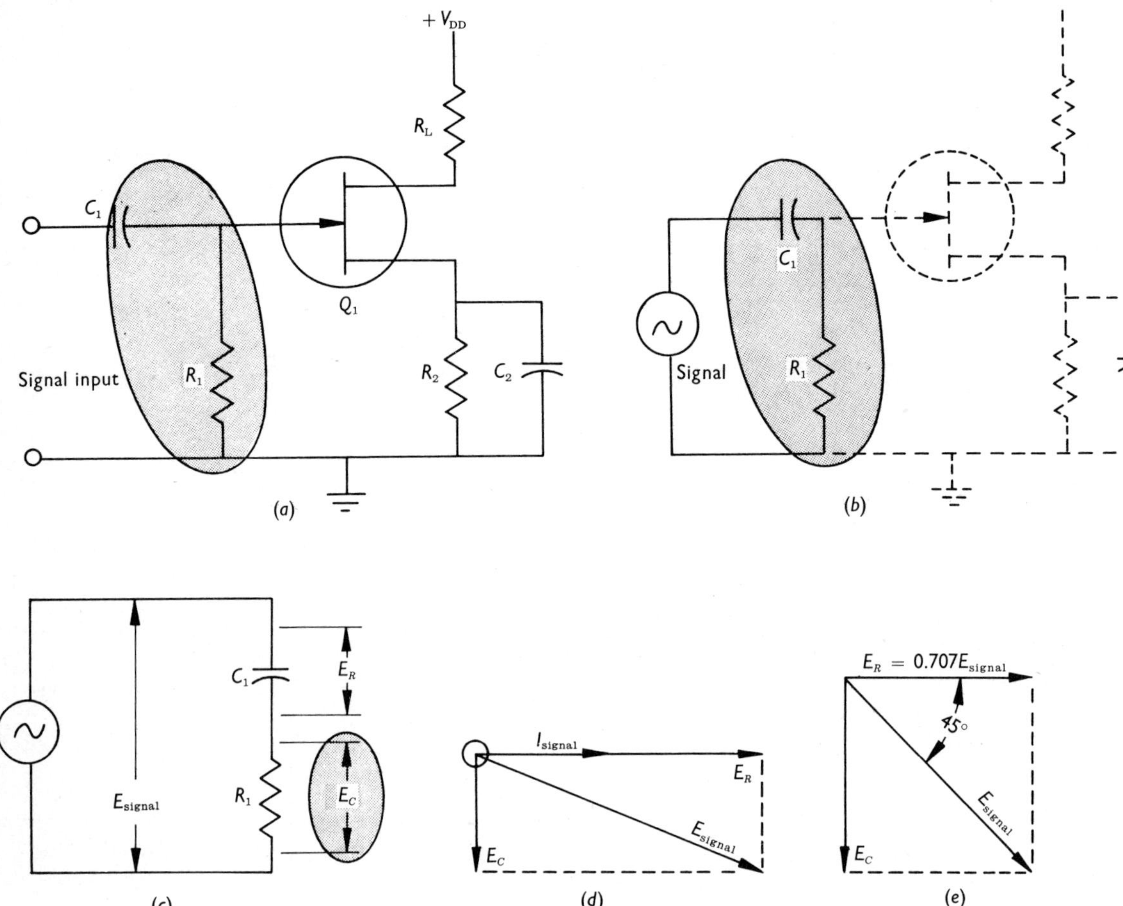

Fig. 4-4 Principles of capacitor coupling: (*a*) FET amplifier; (*b*) Signal source and coupling circuit only; (*c*) Equivalent circuit of signal source and coupling network; (*d*) Phasor diagram at frequencies for which $X_C < R$; (*e*) Phasor diagram at the frequency at which $X_C = R$—the half-power point.

able voltage drop will occur across the capacitor. This means that proportionately less voltage drop will be available across the resistor.

Capacitive reactance depends on the frequency of the signal. For a given value of capacitance, the reactance drops as the frequency increases. This means that the higher the frequency, the less voltage drop across the capacitor and the greater the signal voltage which reaches the input of the amplifier. Note that dc levels do not get transferred from one side of the capacitor to the other, only voltage variations do. Bias voltages and currents in each stage are therefore undisturbed when capacitor coupling is used.

4-7 HALF-POWER POINTS

Every amplifier is designed to amplify a band of frequencies. The center frequency in this band is known as the *mid-frequency,* and the lowest and highest frequencies to be included are known as the *low cutoff frequency* and *high cutoff frequency,* respectively. The value of coupling capacitance is selected on the basis of the low cutoff frequency.

In terms of amplifier-power output the low cutoff frequency is referred to as the *half-power point.* It is the frequency at which signal power falls to one-half its mid-frequency level. A drop in voltage level from full signal to 0.707 (70.7 percent) causes power to drop to one-half (50 percent). The reason for this fact is that power is the product of volts times amperes. By Ohm's law, if a given voltage E falls to $0.707E$, the current I will also fall to $0.707I$. The net power in the circuit then changes from IE W to $(0.707)I \times (0.707)E = (0.707)^2IE = 0.5IE$ W.

The following example illustrates how to calculate the minimum value of coupling capacitance required when the input resistance and cutoff frequency are known.

EXAMPLE 4-1 Calculate the value of C_1 in Fig. 4-4a when R_1 is 1 MΩ and the amplifier is to operate over the audio band 20 to 20 000 Hz. In this instance R_1 represents the true input resistance, because the gate-to-source impedance is extremely high and does not shunt R_1 appreciably.

At the half-power point the voltage across R_1 and therefore the current through R_1 drops to 0.707 of the maximum or mid-frequency value. As shown in Fig. 4-4e, this occurs when capacitive reactance equals the resistance. Stated mathematically, this is

$$X_{C_1} = R_1$$

Substituting $1/2\pi fC_1$ for X_{C_1} gives

$$\frac{1}{2\pi fC_1} = R_1$$

Solving for C_1 gives the expression

$$C_1 = \frac{1}{2\pi fR_1}$$

Next, substitute known values for resistance and the lowest operating frequency:

$$C_1 = \frac{1}{2\pi(20)(1 \times 10^6)}$$

$$= \frac{0.159}{2 \times 10^7}$$

$$= \frac{0.079}{10^7}$$

$$= 0.079 \times 10^{-7} \text{ F}$$

$$= (0.079 \times 10^{-7}) \times 10^6\ \mu\text{F}$$

ANSWER $= 0.0079\ \mu\text{F}$

NOTE In practice a 0.01-μF capacitor would more than adequately fulfill the requirement.

4-8 CHOKE COUPLING

Although *CR* coupling is popular and economical, the resistor wastes signal energy. In high-power stages the signal-power loss may be considerable. In those instances the resistor may be replaced by a choke coil, as shown in Fig. 4-5. The coil, known as a *radio frequency choke (RFC),* has a high reactance at the operating frequency and as a result a large signal is developed across it. There is almost no signal-

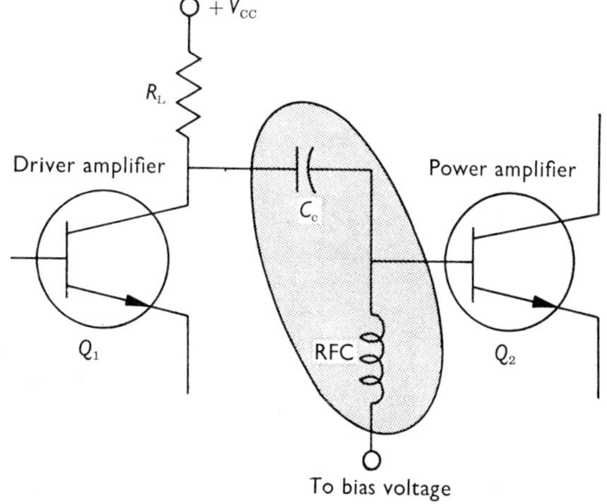

Fig. 4-5 Capacitor-choke coupling.

power loss because the coil exhibits reactance rather than resistance. Capacitor-choke coupling can be used effectively at audio frequencies, but the expense and the added bulk and weight of audio chokes tend to discourage their use. At radio frequencies choke coils are very effective and are widely used in transmitter circuits.

4-9 DIRECT COUPLING

When carefully designed, two or more stages can be directly connected without the use of capacitors or transformers. Such amplifiers are known as *dc amplifiers* or dc coupled amplifiers. Voltage *levels* as well as voltage *variations* are amplified. Component values and power-supply voltages must be adjusted so that each of the stages is properly biased. You can easily visualize that the design is exacting, because a change in any one component generally dictates changes in many others to compensate for the disturbance. Figure 4-6 shows examples of direct-coupled stages using FETs and bipolar transistors. A vacuum-tube circuit is shown in Fig. C-3 (Appendix C).

If for any reason quiescent voltage levels change in a dc amplifier stage, the change becomes amplified by succeeding stages. Even a small change can become very noticeable after several boosts in amplitude. The condition is known as *drift*. In solid-state circuits drift is caused by changes in temperature. The problem has been greatly reduced in integrated circuits, because all transistors and FETs reside on the same chip of semiconductor. It is possible to incorporate compensating circuits which are subjected to the same temperature environment as the amplifying devices.

In the dc amplifier circuits shown in Fig. 4-6a the zener diode is used to maintain a fixed bias on the second stage. A bypass capacitor is unsuitable for this purpose because the signal variations have no low-frequency limit. The signal could vary in amplitude slowly enough so that even the largest bypass capacitor has sufficient time to charge and discharge. Capacitor voltage would then follow signal voltage, and bypass action would not occur.

In Fig. 4-6b the value of R_3 is chosen to provide correct base-quiescent current for Q_2. It also acts as the load resistor for Q_1, which receives its collector current through it. Note that a base-to-emitter resistor is not shown for transistor Q_2. It is not required here as the collector circuit of the preceding stage (Q_1) serves the same purpose.

4-10 COMPOUND OR DARLINGTON CONNECTION

Another form of direct coupling, noted for its high gain, is shown in Fig. 4-7. It is known as the *compound connection* or more commonly as the Darlington connection.

Reference to Fig. 4-7a shows that all the base current for Q_2 must come from Q_1. Assuming that the gain of Q_1 is β_1, if a signal is applied to the base of Q_1, a signal β_1 times as great is applied to the base of Q_2. If the gain of Q_2 is β_2, then the total gain is $\beta_1 \times \beta_2$. A Darlington connection is noted for its high-gain and high-input resistance (because very little signal-input current is required).

In practical circuits the collector-quiescent current of Q_1 is small (because otherwise the base current of Q_2 would be too large). Because of the low collector current the gain of Q_1 could be unnecessarily low. Additional collector current can be provided by a "long-tail" resistor as shown in Fig. 4-7b. Although this resistor "robs" some of the current otherwise destined for the base of Q_2, the additional gain provided by Q_1 more than compensates for the loss.

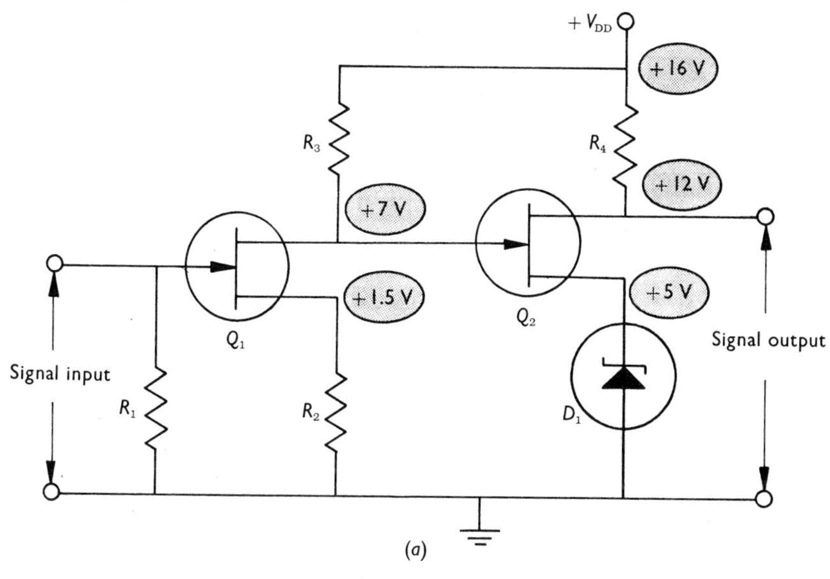

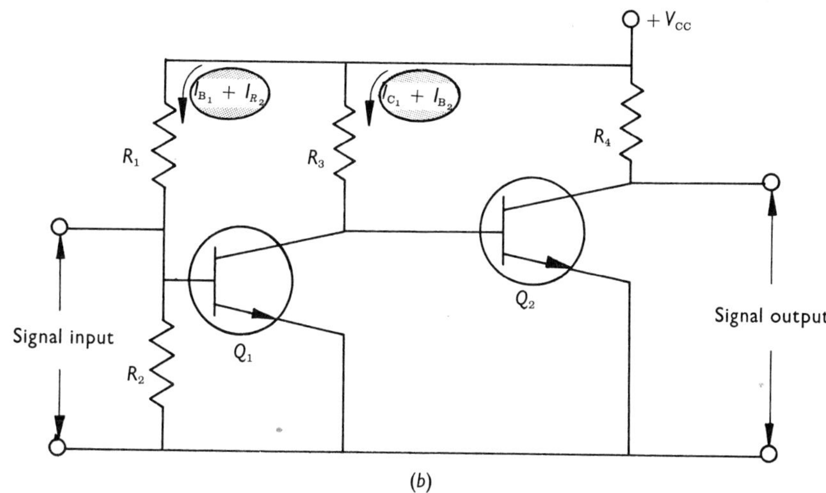

Fig. 4-6 Direct-coupled amplifier stages: (*a*) FET amplifier; (*b*) Transistor amplifier.

Many manufacturers offer the Darlington circuit in a single encapsulated package. In this form it is especially useful when combined internally with a power transistor. The result is a "power transistor" with very small signal-drive requirements. The Darlington connection is also used as an integral part of many integrated circuits.

4-11 SOURCE- AND CATHODE-BYPASS CAPACITORS

Bias voltage can be developed across a source resistor in a FET amplifier stage. The same principle can be used to develop grid bias in vacuum-tube circuits, if cathode resistors (R_2 and R_3) are used (see Fig. C-1, Appendix C).

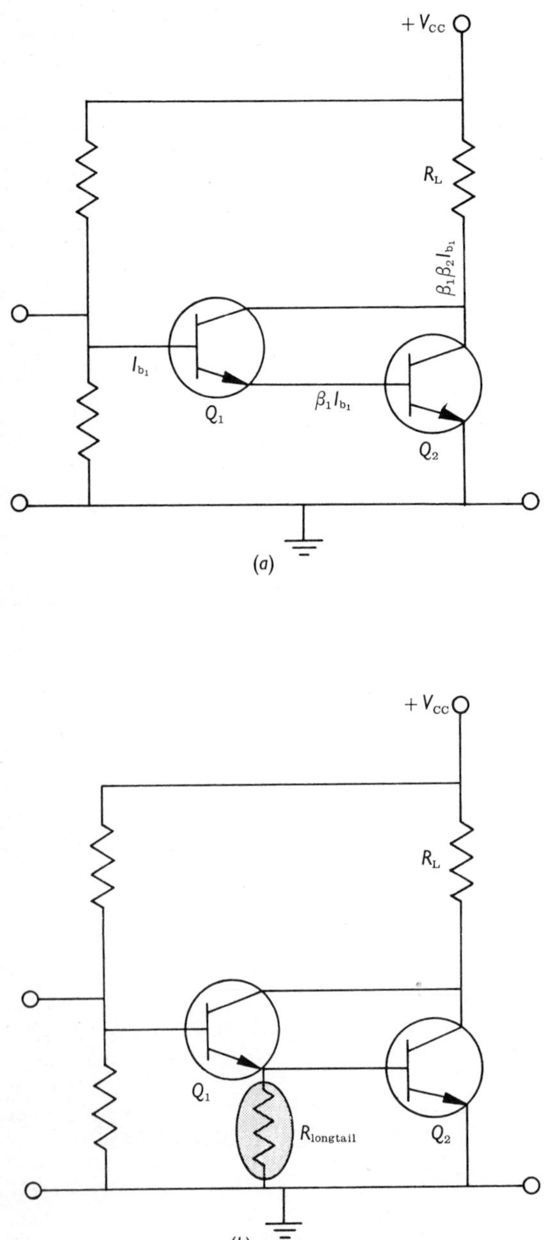

Fig. 4-7 Compound or Darlington connection: (*a*) Darlington circuit showing how gain multiplication takes place; (*b*) Addition of a "long-tail" resistor to increase Q_1 quiescent current and improve its operating characteristics.

A two-stage FET amplifier using source-resistor biasing is shown in Fig. 4-8. Note that in both FET and vacuum circuits capacitors are connected across the source and cathode resistors. These are known as bypass capacitors. The capacitors are used to ensure that the *IR* drop across each of the resistors remains constant irrespective of momentary current fluctuations through the resistors.

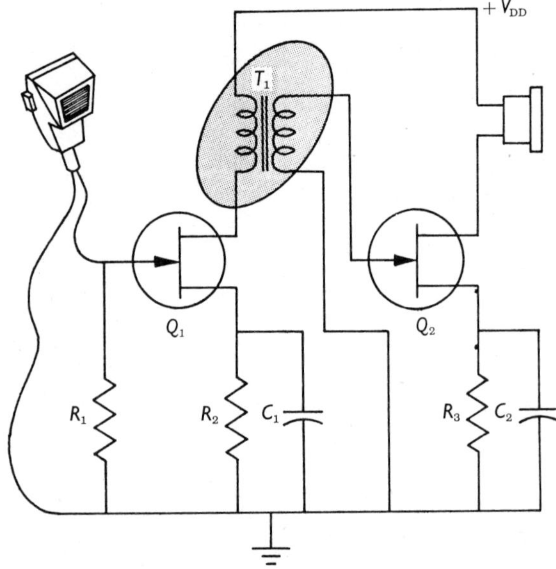

Fig. 4-8 Transformer-coupled FET audio amplifier.

There are two ways to understand the operation of bypass capacitors. The first is to visualize them in the same role as ripple filters in power supplies. If the *IR* drop across a given resistor tends to rise, the capacitor cannot charge instantly, and there is a delay before the *IR* drop can increase. In a similar fashion if the *IR* drop tends to fall, the charge on the capacitor maintains the average or quiescent voltage level. Signal fluctuations across the biasing resistor are virtually eliminated. Stage gain is reduced without sufficiently large bypass capacitors to remove fluctuations because their polarity is always such that they cancel part of the input signal.

The second explanation of the operation of bypass capacitors is perhaps more pleasing mathematically. It relies on the concept that capacitive reactance is low for high frequencies but tends toward infinity when frequency nears zero. This means that bypass capacitors act as an electrical jumper for audio and higher frequencies, and detour the ac component around the resistor, which otherwise would attenuate. The dc *IR* drop across the resistor remains undisturbed, so in effect it is the dc current through the resistor which is causing the *IR* drop, whereas the ac component is bypassed directly to the source or cathode by the capacitor.

Bypass capacitors in practical applications are chosen so that their reactance at the lowest frequency is $\frac{1}{2}$ to $\frac{1}{10}$ that of the resistance of the resistor they are bypassing. At higher frequencies the capacitors will automatically be adequate because their reactance drops even more, and bypass action is even better than at low frequencies.

The following example shows how to calculate the value of a bypass capacitor.

EXAMPLE 4-2 What value of bypass capacitor is required to bypass a 300-Ω source resistor in a high-fidelity audio amplifier designed to operate from 20 to 20 000 Hz? The capacitor must be chosen to bypass adequately at 20 Hz. If it does so, then it will also perform well at higher frequencies. The resistance to bypass is 300 Ω. Assume that the capacitive reactance must be $\frac{1}{10}$ of this value or 30 Ω.

The expression for capacitive reactance is

$$X_C = \frac{1}{2\pi fC}$$

or

$$X_C = \frac{0.159}{fC}$$

Substituting 20 Hz for f and 30 Ω for X_C gives

$$30 = \frac{0.159}{20C}$$

or

$$C = \frac{0.159}{20 \times 30}$$

$$= \frac{0.159}{600}\ \text{F}$$

$$= \frac{0.159}{600} \times 10^6\ \mu\text{F}$$

$$= \frac{159{,}000}{600}$$

ANSWER $= 265\ \mu\text{F}$

NOTE In practice, a 250- or 300-μF electrolytic capacitor would serve the purpose. Caution is required to ensure that correct polarity is observed during installation.

Summary

Signal-coupling circuits between cascaded stages must not disturb bias voltage and current.

Ac amplifiers respond only to time-varying signals. Dc amplifiers respond both to constant levels and time-varying signals.

With transformer coupling only the ac component of a signal is transferred.

Transformer-coupled stages have electrical isolation between stages.

In addition to interstage coupling, transformers can provide impedance matching.

Coupling transformers can be tuned to provide frequency selection.

Capacitor-coupling circuits pass only the ac component of a signal and block the dc.

Amplifiers designed to amplify a band of frequencies provide maximum gain at the *center frequency*. At the *upper* and *lower* cutoff frequencies the power gain drops to one-half the gain at center frequency. On a response curve these two frequencies are known as the half-power points.

Normally resistors are used with capacitor coupling, but choke coils can be used as well. Because they are reactive at the frequencies used, choke coils waste no signal power, whereas resistors do.

Dc amplifiers require direct coupling.

A compound or Darlington connection provides high-gain and high-input resistance and thereby requires very little signal current.

Bypass capacitors are a form of "jumper" for ac signals. Bypass capacitors do not disturb dc voltage levels.

Questions and Exercises

1. What is meant by the term cascading as applied to amplifier stages?
2. What important precaution is required when one stage is coupled to another?
3. Distinguish between ac and dc amplifiers.
4. Is a transformer-coupled amplifier an example of an ac or dc amplifier?
5. What are some advantages and disadvantages of transformer coupling?
6. Redraw the circuits in Fig. 4-6 as transformer-coupled stages.
7. What is the purpose of capacitor C_3 in the circuit in Fig. 4-3a?
8. Assume that the secondary coil of T_1 in Fig. 4-3b has an effective inductance of 256 μH, and capacitor C_1 has a value of 100 pF. What frequency will be selected for amplification by transistor Q_1?
9. Assume that the amplifier in Fig. 4-3a is required to operate in the FM broadcast band (88 to 108 MHz), tuned to a frequency of 100 MHz. Transformer T_1 consists of only a few turns of wire wound on $\frac{1}{2}$-inch polystyrene tubing, and the secondary coil has an effective inductance of 0.126 mH. What value of C_3 is required to tune it to resonance?
10. When capacitive coupling is employed, does the amount of signal voltage transferred from one stage to the next increase or decrease if the frequency is reduced? Explain your answer.
11. Redraw the circuits in Fig. 4-6 as *CR* type capacitive-coupled circuits.
    a. How many additional components are required in Fig. 4-6a?
    b. How many additional components are required in Fig. 4-6b?
12. Define the following terms: (***a***) half-power point, (***b***) low cutoff frequency, (**c**) high cutoff frequency, and (***d***) mid-frequency.
13. The amplifier in Fig. 4-4a has the following component values: $C_1 = 0.01\ \mu$F, $R_1 = 0.5\ \text{M}\Omega$, $R_2 = 250\ \Omega$, $C_2 = 250\ \mu$F, $R_L = 5\ \text{k}\Omega$. What is the low cutoff frequency for this amplifier?
14. In certain communication circuits the audio amplifiers are used only for voice frequencies, and a frequency range of 300 to 3000 Hz is adequate. What is the maximum value for C_1 in Question 13 if a low cutoff frequency of 300 Hz is required?
15. What would be the effect on the low cutoff frequency if, under the conditions stated in Question 13, the resistor R_1 were changed to a value of 100 kΩ?
16. Why is a choke coil sometimes used instead of a resistor with capacitive coupling?
17. Why must dc amplifiers necessarily use direct coupling?
18. Why is the design of dc amplifiers more painstaking and critical than that of ac amplifiers?
19. Define drift as it applies to dc amplifiers.
20. Why are bypass capacitors useless in direct-coupled amplifier circuits?
21. What are the outstanding features of the compound or Darlington connection?
22. In the circuit in Fig. 4-7a assume that transistor Q_1 has a gain of $\beta_1 = 100$, and transistor Q_2 a gain of $\beta_2 = 50$. If the input-signal current at the base of Q_1 changed from 1.5 to 2 μA, what would be the change in output current?

23. What is the purpose of the "long-tail" resistor in Fig. 4-7b?

24. What function do source-bypass capacitors perform?

25. What practical value of source-bypass capacitor is required to bypass a 500-Ω source resistor if the amplifier is to operate in the voice frequency range 300 to 3000 Hz?

5-1 AMPLIFIER FREQUENCY RESPONSE

Ideally, amplifiers should perform equally well at all frequencies. Ideal ac amplifiers should amplify equally all frequencies in a given band. In practice neither of these desirable qualities is achieved.

First consider some of the practical limitations of dc amplifiers. Although they have no low-frequency limitations, dc amplifiers produce extraneous output signals because of drift. In addition they have a high-frequency cutoff. The cutoff frequency depends mainly on the amplifying device (tube, transistor, FET) and on the placement of components.

In solid-state devices the amplifying action is achieved by forcing the output current to follow the signal pattern. Even if the signal pattern contains rapid variations (high frequencies) the collector (or drain) currents must follow these. You will recall that current is the movement of electrical charges, and varying current implies acceleration of electric charge. This takes time; time to accumulate charges, and time to disperse them. In germanium semiconductors charge mobility is greater than in silicon, so that germanium devices are capable of operating at higher frequencies. Even so, there is an upper frequency limit at which a device will function effectively.

In vacuum tubes amplifying action is achieved by varying the electron flow through vacuum in accordance with the signal pattern. It takes time for the electrons to travel between cathode and plate. The time interval is known as transit time. It is of short duration because of the high speed of electrons. Yet if signal amplitude varies quickly enough, it will change significantly while electrons are in transit, and electron control is less effective. An upper frequency limitation occurs in practice at around 1 GHz.

In most amplifiers, neither charge mobility nor transit time imposes the upper cutoff frequency. The upper half-power point is reached because a combination of components and their relative placement attenuates high frequencies. Attenuation is due to either capacitive or inductive reactance. Figure 5-1

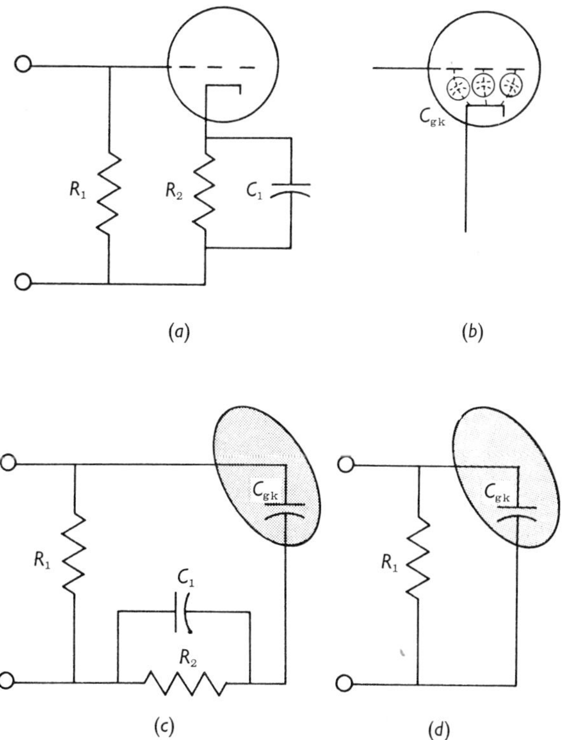

Fig. 5-1 Effect of grid-to-cathode capacitance—shunting action increases at higher frequencies: (*a*) Schematic of input circuit; (*b*) Capacitance between the grid and cathode electrodes; (*c*) Equivalent input circuit; (*d*) Equivalent circuit at very high frequencies, when X_{C_1} is negligible.

shows one type of shunt capacitance found in vacuum-tube amplifiers. Transistor and FET devices also exhibit a shunt capacitance, both at the input and output. Any capacitance which shunts the signal has negligible effect at low frequencies because the capacitive reactance is great. As the frequency is raised, the capacitive reactance decreases. If the frequency is high enough, a shunt capacitor behaves like a short circuit; just as though a wire jumper were connected across the terminals. Another way to describe the effect of shunt capacitance is to say that signal current is bypassed through the shunt to ground and therefore never reaches its intended destination. High-frequency signal flow is also impeded by the connecting leads, because of their appreciable inductive reactance at high frequencies.

Low-frequency cutoff in ac amplifiers takes place because the coupling circuit fails to transfer enough signal to the next stage. One such instance, already noted (Sec. 4-6), is due to an insufficiently large coupling capacitor. Transformer coupling also results in low-frequency cutoff; if the frequency is too low, magnetic flux is not changing rapidly enough to induce sufficient current in the secondary. Also, if choke coils are used as coupling network components, their inductive reactance drops at low frequency, and they tend to shunt low-frequency signals.

In summary, you will appreciate that the causes of high- and low-frequency cutoff are complex. A truly wideband amplifier is both very difficult to design and construct and full of compromise.

5-2 AMPLIFIER FREQUENCY-RESPONSE CURVE

Amplifier behavior over a band of frequencies is best portrayed with a *frequency-response curve*. A frequency-response curve is a graph; a plot of the *amplifier gain versus signal frequency*. The gain may be expressed either as a ratio of signal output to signal input, or in decibels (dB). (In audio amplifier measurements 1 dB is the smallest variation in output level discernible by the human ear.)

It is inconvenient to calculate power gain in amplifier response measurements. Voltage gain is more easily found, because nearly all common measuring instruments such as oscilloscopes, VTVMs, TVMs, and VOMs respond readily to voltage amplitude. Current amplitude could be measured, but it usually involves opening a circuit to insert the current-sensing device, whereas voltage measurement involves only a probe contacting a given circuit point.

Power gain is expressed in decibels by the equations

$$\text{dB} = 10 \log\left(\frac{P_{\text{out}}}{P_{\text{in}}}\right) \quad \text{when power is known}$$

$$\text{dB} = 20 \log\left(\frac{E_{\text{out}}}{E_{in}}\right) \quad \text{when voltage only is known}$$

$$\text{dB} = 20 \log\left(\frac{I_{\text{out}}}{I_{\text{in}}}\right) \quad \text{when current only is known}$$

Strictly speaking, the second and third expressions involving voltage and current are true only if input and load resistance are equal. In amplifier circuits this is rarely the case. For example, in a phono amplifier the pickup cartridge works into several thousand ohms, whereas the load may be a 4-Ω speaker. Yet, purely for convenience, a decibel figure is calculated without regard to input and load resistance, but it is not necessarily the true power gain. Rather, it is a defined figure used only when referring to amplifier voltage or current gain; it is especially useful when comparing gain at different frequencies.

When frequency-response measurements are made, a constant-amplitude oscillator is most convenient to use as the signal source. Output voltage is measured across a resistive load to ensure that the load responds equally to all frequencies. Figure 5-2 illustrates a bench layout to perform a frequency-response test. The input signal is adjusted to a convenient level, low enough to prevent overdriving of any amplifier stage. Overdriving overloads one or more stages and results in a clipped waveform. Clipping is easily detected with an oscilloscope, and for this reason it is preferable to use an oscilloscope to measure the output. If a scope is not available or if measurements are being made at frequencies of several MHz, a good-quality VTVM or TVM with an RF probe should be used. There is no reliable way to detect clipping with the latter instruments, and erroneous readings could result if the amplifier under test is being overdriven.

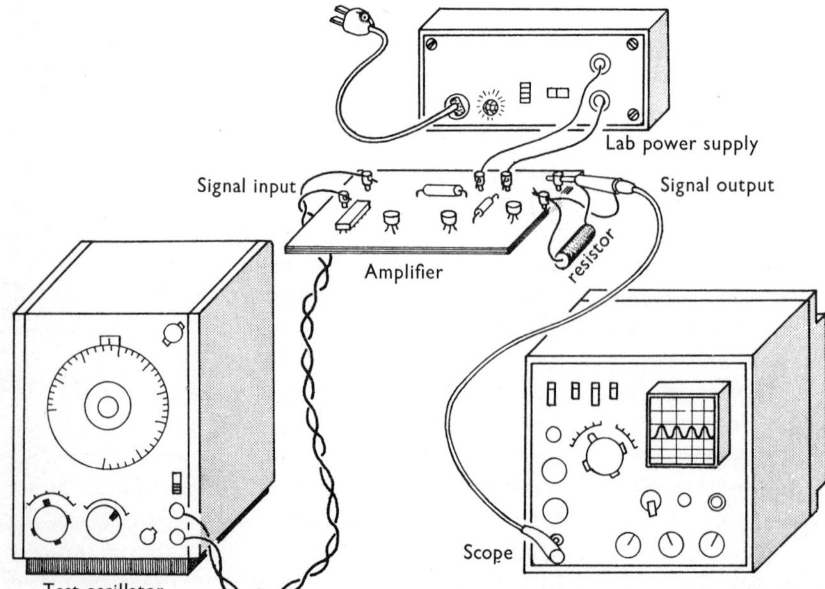

Fig. 5-2 Bench layout for measuring amplifier frequency response.

The voltage read across the load resistor is an indication of the power output. Amplifier gain in decibels can be calculated from the expression

$$\text{dB} = 20 \log\left(\frac{E_{\text{out}}}{E_{\text{in}}}\right)$$

The following example shows how gain in decibels is calculated when input and output voltage is known.

EXAMPLE 5-1 A voltage amplifier has 0.1 V applied at a given frequency. The output voltage read across the output load resistor is 5 V. What is the gain in decibels at this frequency? Applying the decibel equation and substituting numerical values gives

$$\text{dB} = 20 \log\left(\frac{E_o}{E_i}\right)$$

$$= 20 \log\left(\frac{5}{0.1}\right)$$

$$= 20 \log 50$$

$$= 20 \times 1.699$$

ANSWER $= 33.98$ dB

5-3 PLOTTING THE RESPONSE CURVE

A response curve is plotted as shown in Fig. 5-3 on semilog graph paper. On semilog paper, equal ratios occupy equal distances on one of the axes. Distances represent the logarithms of magnitude, not magnitudes themselves as on linear graph paper. The distance from 10 to 1 has ratio 10/1 as has the distance from 100 to 10 or 1000 to 100, etc. On semilog paper you will find one axis on which the numbers 1 to 10 occupy the same distance as the numbers 10 to 100, and 100 to 1000, etc. Use this axis for frequency. The second axis is linear and is used for plotting the gain.

An amplifier which has a constant gain over its rated bandwidth is termed *flat*. This term is used because the response curve lacks dips

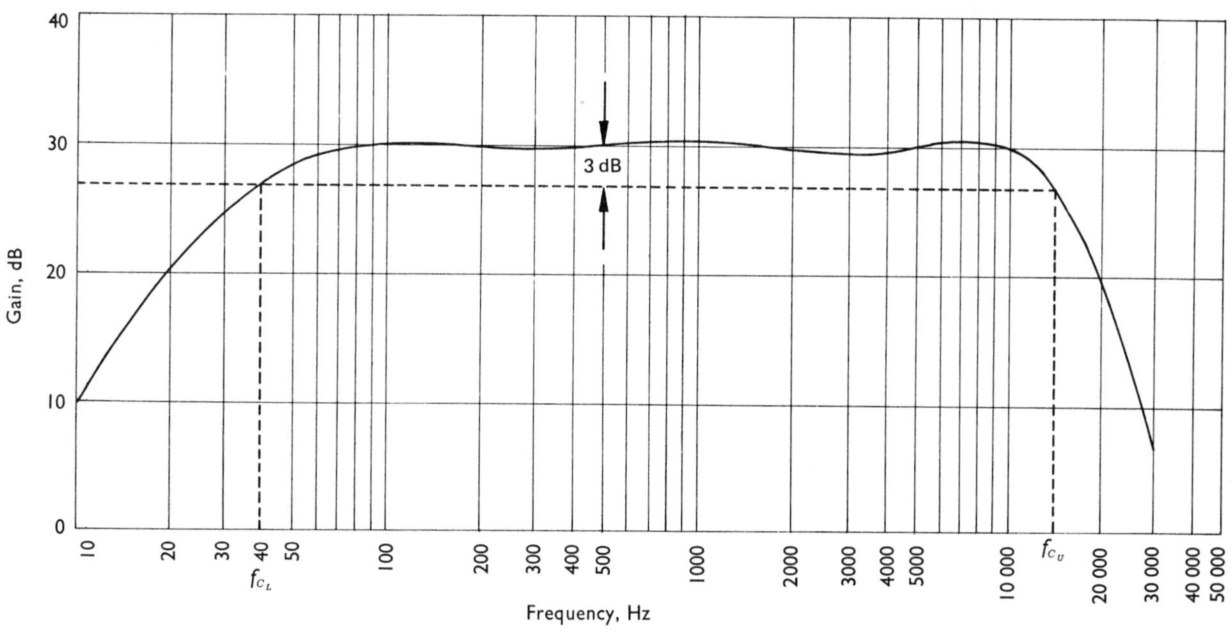

Fig. 5-3 Typical audio amplifier response curve.

or humps, i.e., the curve is flat. On some amplifiers flatness is a desirable feature, as it is in a high-fidelity audio amplifier. All frequencies produced by an orchestra would receive equal gain, and the reproduced sound would be natural. In other applications, amplifier-response curves are deliberately peaked or dipped, perhaps to compensate for a weakness in the signal source or signal load.

5-4 HALF-POWER POINTS EXPRESSED IN DECIBELS

On the response curve in Fig. 5-3 note the upper and lower cutoff frequencies. These occur at the half-power points. Note that although absolute power gain cannot be calculated from voltage amplitude measurements unless input and output resistance is equal, relative power-gain changes can be. When the voltage output falls by a certain percentage, output current drops proportionally because it is assumed that load resistance remains the same. For instance, if there is a 3-dB drop indicated by voltage measurement, there is in fact a 3-dB drop in power output as well.

Expressed in decibel notation, a 3-dB drop is equivalent to a drop to half power. This can easily be shown by substituting numerical values in the decibel equation

$$\text{dB} = 10 \log\left(\frac{P_{out}}{P_{in}}\right)$$

In the equation, P_{out} and P_{in} are two power levels, and the result gives the difference in the levels in decibels. Assume one level to be half the other, or $P_{in} = \frac{1}{2}P_{out}$. Substituting into the equation gives

$$\begin{aligned}\text{dB} &= 10 \log\left(\frac{P_{out}}{P_{in}}\right) \\ &= 10 \log\left(\frac{P_{out}}{\frac{1}{2}P_{out}}\right)\end{aligned}$$

$$\begin{aligned}\text{dB} &= 10 \log\left(\frac{1}{\frac{1}{2}}\right) \\ &= 10 \log 2 \\ &= 10(0.3) \\ &= 3\end{aligned}$$

5-5 DISTORTION

An amplifier's function is to raise the level of a signal with a minimum of distortion. A distortionless amplifier would be capable of reproducing an exact replica of any signal waveform or amplitude fed to the input. In practice it is not possible to construct a distortionless amplifier. No practical amplifier can boost all frequencies by the same amount, nor can it withstand unlimited input amplitude without clipping the waveform peaks. Distortion can be attributed to three causes as indicated in the following sections.

5-6 FREQUENCY DISTORTION

Amplifier gain is not the same for all frequencies. Since complex waveforms are composed of the algebraic sum of many simple waveforms (sine waves) of varying frequencies and amplitudes, their relative amplitudes must be preserved. If one or a group of frequencies in the set is amplified more than others, the original character of the complex waveform is changed. This is illustrated in Fig. 5-4. Distortion originating from this cause is known as *frequency distortion*. When the response curve of an amplifier is not flat, frequency distortion occurs.

5-7 AMPLITUDE DISTORTION

Often there is clipping or disproportionate amplification at different amplitude points of a waveform. For example, if a sinusoidal signal of 1 V is applied to an amplifier as shown in Fig. 5-5 and the amplifier has more gain at low signal levels than at higher levels, the portion

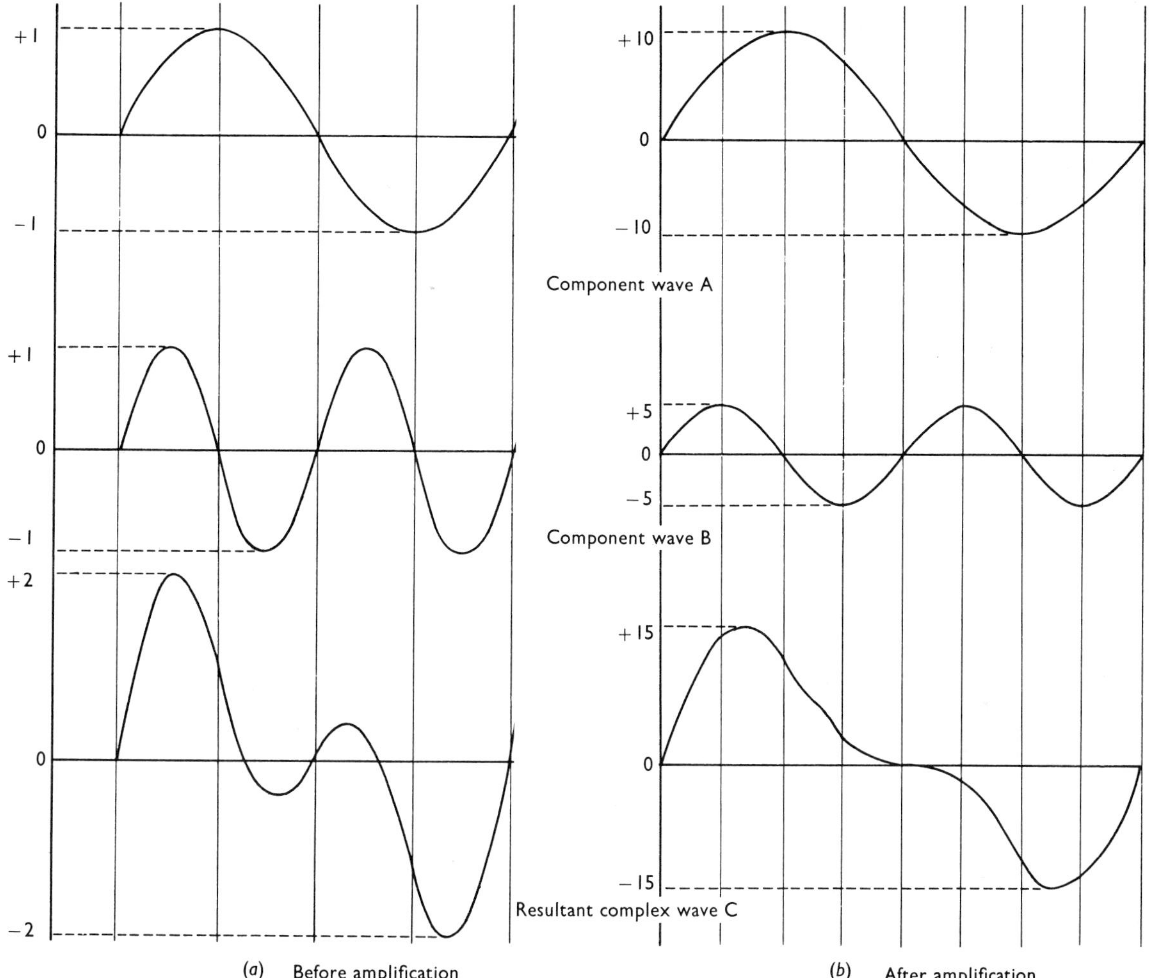

Fig. 5-4 Frequency distortion: (*a*) Complex waveform applied to the input of the amplifier; (*b*) Resultant waves at the output of amplifier with poor high-frequency response. The high-frequency waveform receives less amplification than the low-frequency waveform.

of the wave near its horizontal axis will get more amplification than the portion near the peak of the wave. In severe cases, the peak may get no amplification at all or may even be attenuated, in which case the output waveform bears little resemblance to the input. This type of distortion is known as *amplitude distortion* or also as *nonlinear distortion* (meaning that gain is not uniform at all amplitudes).

5-8 HARMONIC DISTORTION

Amplitude distortion is also known as *harmonic distortion,* because the new components of the resultant waveform created by amplitude distortion are *harmonics* of the original signal. This means that their frequencies will be *integral multiples* of the frequency of the original signal. In audio

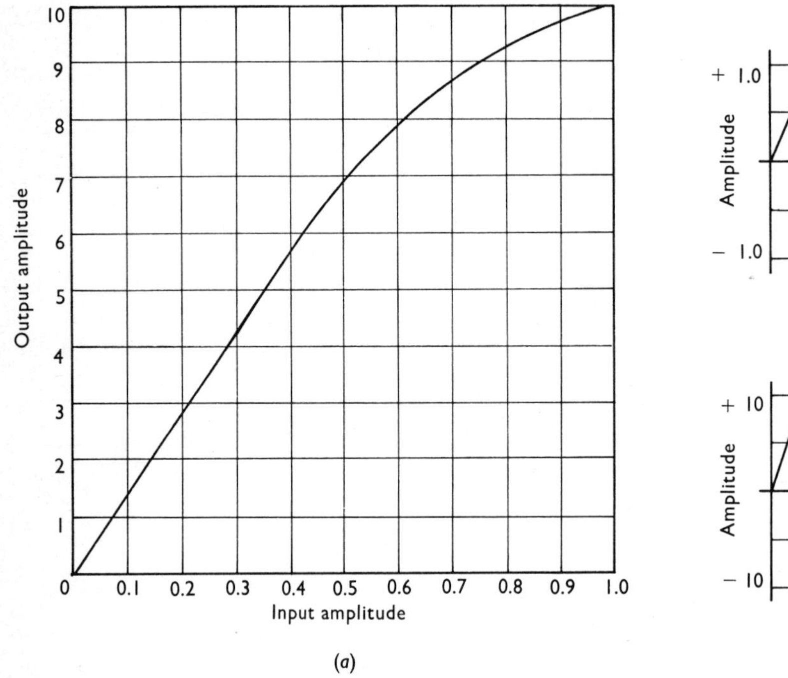

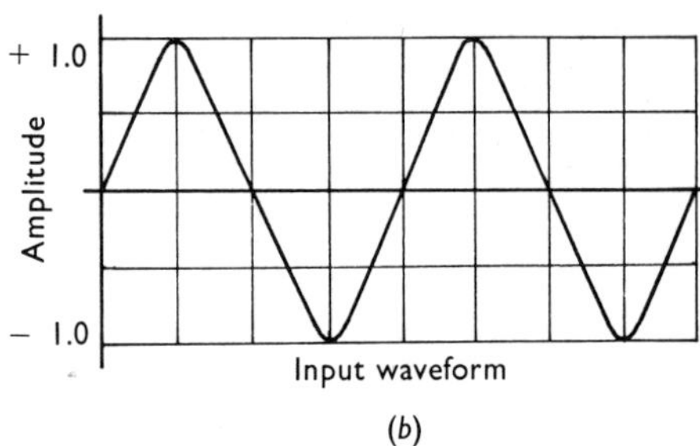

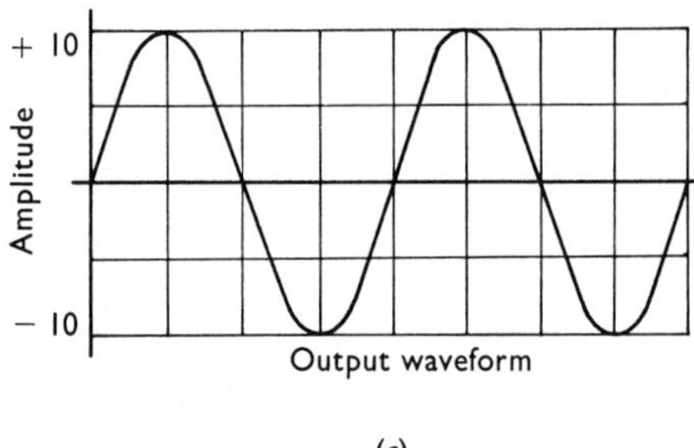

Fig. 5-5 Amplitude distortion. The signal does not receive constant amplification at all amplitudes: (*a*) Amplifier gain curve; (*b*) Input waveform; (*c*) Resultant output waveform.

amplifiers harmonic distortion is very objectionable because it produces harsh unpleasant sounds. It can be produced on almost every radio or recordplayer by advancing the volume control sufficiently. At such volume settings the output stage is being driven out of its linear operating range.

Harmonic distortion is expressed as a percentage of the fundamental frequency amplitude. The equation for percent harmonic distortion is

$$\% \text{ harmonic distortion} = \left(\frac{\sqrt{A_2^2 + A_3^2 + A_4^2 + \cdots + A_n^2}}{A_1} \right) \times 100$$

where A_1 is the amplitude of the fundamental frequency, and A_2, A_3, etc., are the relative amplitudes of the second, third, etc., harmonics. The following example shows how percent harmonic distortion is calculated if the amplitudes of the fundamental and its generated harmonics are known.

EXAMPLE 5-2 What is the percent harmonic distortion at 400 Hz if the amplitude of the fundamental is 2 V, the second harmonic (800 Hz) has amplitude 0.05 V, and the third harmonic (1200 Hz) has amplitude 0.02 V. Ignoring the amplitude of higher-order harmonics, we have

$$A_1 = 2 \text{ V}$$

$$A_2 = 0.05 \text{ V}$$

$$A_3 = 0.02 \text{ V}$$

$$A_4 \cdots A_n = 0 \text{ V}$$

Substituting these values into the equation gives,

% harmonic distortion

$$= \frac{\sqrt{(0.05)^2 + (0.02)^2}}{2} \times 100$$

$$= \frac{\sqrt{(5 \times 10^{-2})^2 + (2 \times 10^{-2})^2}}{2} \times 100$$

$$= \frac{\sqrt{25 \times 10^{-4} + 4 \times 10^{-4}}}{2} \times 100$$

$$= \frac{\sqrt{29 \times 10^{-4}}}{2} \times 100$$

$$= \frac{\sqrt{29} \times \sqrt{10^{-4}}}{2} \times 100$$

$$= \frac{5.4 \times 10^{-2}}{2} \times 100$$

$$= \frac{5.4}{2}$$

ANSWER $= 2.7\%$

NOTE In practice the amplitudes A_1, A_2, A_3, etc., are measured with a distortion analyzer. A distortion analyzer is an instrument which contains a fixed-frequency audio generator and a tunable voltmeter. By tuning the voltmeter to the fundamental and the various harmonic frequencies, their relative amplitudes may be measured.

A special type of distortion occurs when the harmonics created by amplitude distortion mix together producing sum and difference frequencies which were not present in the original signal. The generation of new frequencies from harmonics in a nonlinear amplifier is known as intermodulation distortion. *Intermodulation distortion is usually expressed as a percentage. For instance, if an amplifier has less than 1 percent intermodulation distortion, the maximum amplitude of intermodulation products will be less than $\frac{1}{100}$ of the amplitude of the signal.*

5-9 PHASE DISTORTION

Not all signal components experience identical phase shifts as they progress through the amplifier. Varying phase differences among the component frequencies cause resultant amplitude variations from one instant to another. Since it is the resultant amplitude of all components which make up the signal, the original pattern of the waveform is altered. Distortion caused by phase change is known as *phase distortion* or *delay distortion*. It is illustrated in Fig. 5-6 but is greatly exaggerated to convey the concept.

5-10 EXTENDING FREQUENCY RESPONSE

Amplifier frequency response can be improved by using larger coupling capacitors (Fig. 5-7a). At frequencies approaching the low-frequency cutoff, the capacitor-voltage phasor has nearly the same magnitude as the resistor voltage (Fig. 5-7b). The larger the value of the coupling capacitor, the less its reactance and the smaller its voltage phasor. Consequently, more applied voltage will appear across the resistor (Fig. 5-7c).

High-frequency response is limited by shunt capacitances in circuit components and circuit wiring. It can be extended by the use of *peaking coils,* inductances placed in series with load resistors as shown in Fig. 5-8. At low frequencies the inductive reactance of a peaking coil is so low that it is not "seen" by the signal. At high frequencies the coil reactance becomes significant, and the load is no longer composed only of resistor R_L. Because the resistor and coil are connected in series, they have a total impedance Z_L. The value of this impedance is obtained from the series *LR* circuit formula: $Z_L = \sqrt{R_L^2 + X_L^2}$. At high frequencies more signal-voltage drop is developed across the combination than would occur with the resistor alone.

Amplifier gain will not rise beyond certain limits, even with the use of peaking coils. Beyond a given frequency, the capacitance between the turns in the coil bypasses signal, and the coil loses its effectiveness. Peaking coils are found in oscilloscope and television video amplifier circuits.

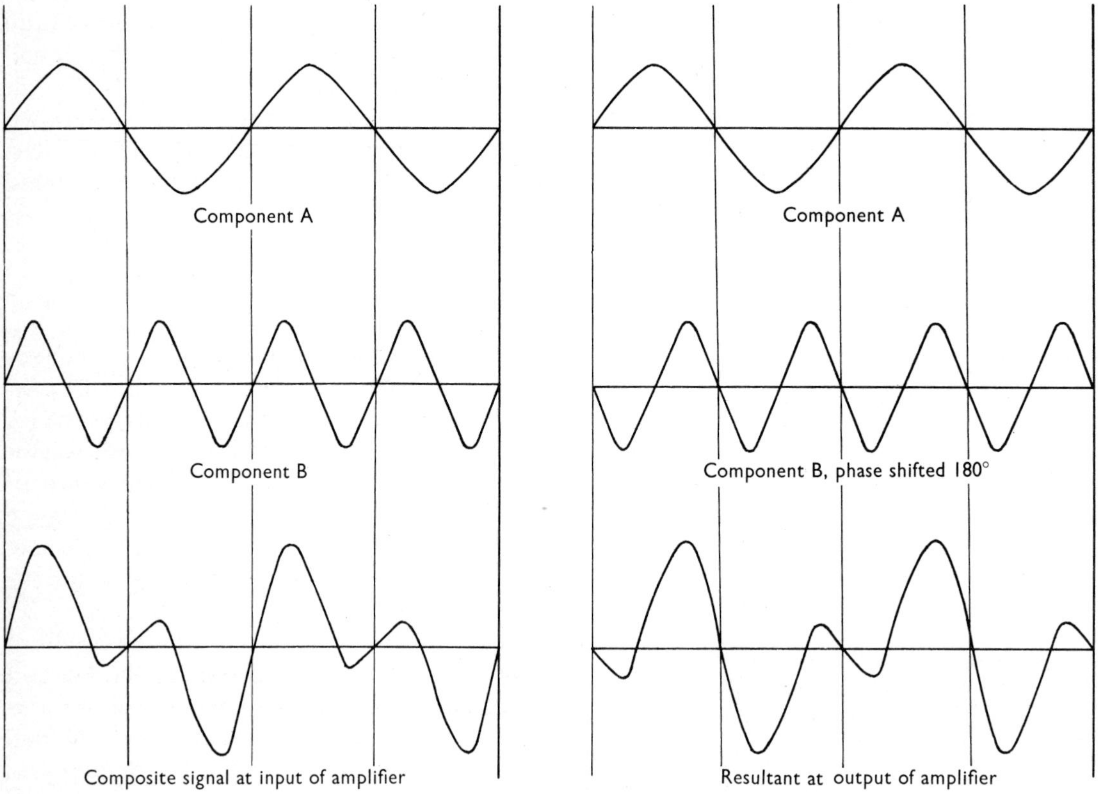

Component A

Component A

Component B

Component B, phase shifted 180°

Composite signal at input of amplifier

Resultant at output of amplifier

Fig. 5-6 Phase or delay distortion. The component frequencies of a complex waveform experience unequal delay in an amplifier.

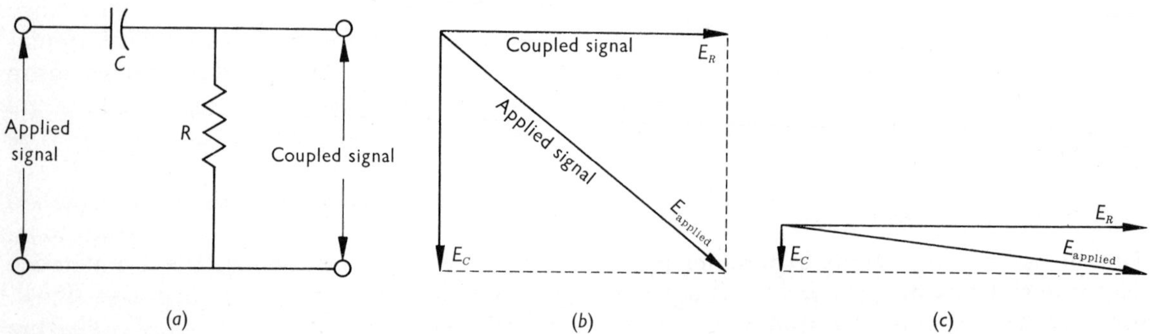

Fig. 5-7 Phase shift in a *CR* coupling circuit: (*a*) *CR* coupling circuit; (*b*) Phasor diagram for the low cutoff frequency—at the half-power point; (*c*) Phasor diagram near the mid-band frequency.

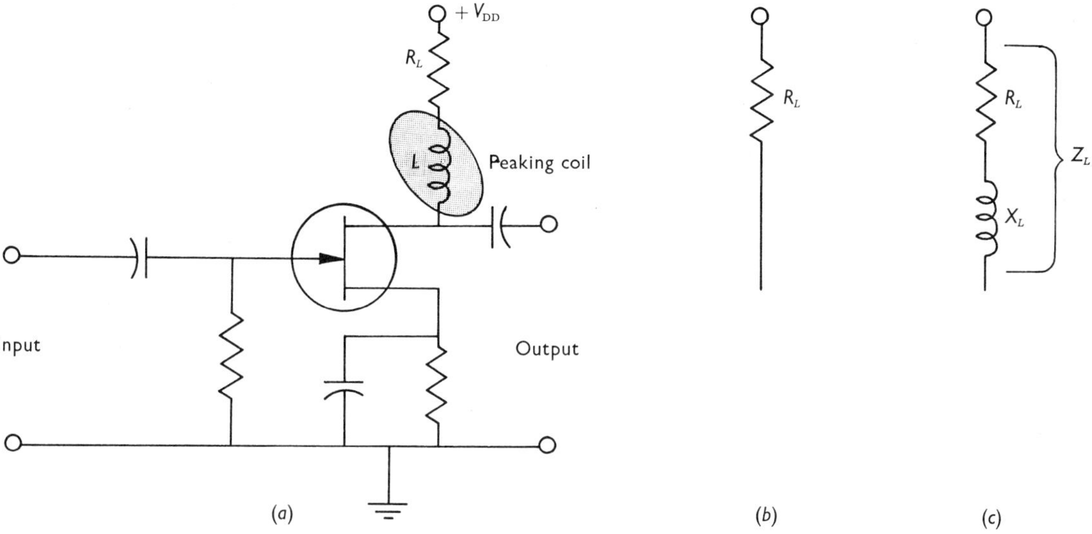

Fig. 5-8 Use of a peaking coil to extend an amplifier's high-frequency response: (*a*) Schematic for a one-stage amplifier; (*b*) Load as it appears at low frequencies; (*c*) Load as it appears at high frequencies.

5-11 REDUCING PHASE DISTORTION

Phase distortion results when some frequencies reach the output before others, i.e., when some frequencies are delayed more than others. Phase delay is caused by reactance exhibited by certain components. It is a painstaking task to design an amplifier which is free from phase distortion. It usually involves the addition of capacitors, coils, or both at circuit points such that the natural circuit reactances are balanced out. In audio systems phase delay may go unnoticed because the ear is insensitive to phase shift. In video amplifiers found in television systems, phase shift degrades picture quality and cannot be tolerated.

5-12 MINIMIZING NONLINEAR DISTORTION

Nonlinear distortion occurs when the gain is not the same at all amplitudes. The most effective way to deal with nonlinear distortion is to introduce negative feedback into the circuit. In essence this means feeding a sample of the output signal into an electronic comparator which compares the input signal with the output. If there is a difference in the waveform, the amplifier gain is instantaneously adjusted to minimize the difference. The techniques used to incorporate negative feedback into amplifiers are discussed in the next chapter.

Summary

An amplifier frequency-response curve is a graph showing gain versus frequency.

Amplifier gain may be expressed in decibels (dB).

Gain variation with frequency is usually stated in dB.

An amplifier with a uniform frequency response over a given frequency range is said to have a "*flat*" response.

A signal drop in 3 dB is equivalent to a drop to *half power*.

Distortion is present in an amplifier if the output waveform is not an exact replica of the input.

When gain is not equal at all frequencies, *frequency distortion* is present.

When gain is not equal at all amplitudes of the input waveform, *amplitude* or *nonlinear distortion* is present.

Amplitude distortion results in *harmonic distortion,* because frequency components are generated which have a harmonic relationship to the input signal.

If the harmonics produced by amplitude distortion mix together to produce sum and difference frequencies, *intermodulation distortion* is present.

If some component frequencies of a signal waveform experience more delay than others, *phase* or *delay distortion* is present.

Upper frequency response can be extended through the use of *peaking coils.*

Lower frequency response can be extended by (1) larger coupling capacitors, (2) larger transformers, and (3) direct coupling.

Phase or delay distortion can be reduced by changing the component layout and by introducing reactive components to cancel the reactive effects of existing components.

Nonlinear or amplitude distortion can be minimized with negative feedback.

Questions and Exercises

1. List three causes of high-frequency cutoff in practical amplifier circuits.
2. Give two causes of low-frequency cutoff in practical amplifier circuits.
3. What is meant by frequency response?
4. State one method of portraying amplifier frequency response.
5. It is usually more convenient to plot a graph of the logarithm of gain rather than the gain itself. Write a formula and state the units used to express gain when this is done.
6. Why is it more convenient to measure voltage gain rather than current or power gain?
7. The true power gain in decibels is given by the equation

$$\text{dB} = 10 \log\left(\frac{P_{out}}{P_{in}}\right)$$

only if a certain precaution is observed. What is this precaution?
8. Prove by using the decibel equation (Sec. 5-2) that doubling the power output from a given value is equivalent to a rise in 3 dB (show your calculations). To get an increase of 3 dB, voltage output need only increase by 1.414 times the initial voltage. Prove this and show your calculations. Why does it take less voltage increase than power increase to get the same increase in decibels?
9. Why is it preferable to use an oscilloscope rather than a meter when making amplifier-output measurements?
10. Express in decibels the voltage gain of an amplifier which gives 2 V output when 2 mV of input signal is applied. Show your calculations.
11. Suggest an advantage of using semilog graph paper for plotting frequency-response curves.
12. What is meant by the term "flat" when dealing with amplifier response?
13. Sketch freehand a response curve for an ideal amplifier used for amplifying only voice frequencies in the range of 300 to 3000 Hz. With a dotted line superimpose on your sketch the response curve for a high-fidelity audio amplifier.
14. What is the technical meaning of distortion?
15. Define frequency distortion and state how it is produced.
16. **a.** On a sheet of graph paper plot to scale the equations $y = 2 \sin x$ and $y = \sin 3x$ one above the other as is done in Fig. 5-4a. Use a common x axis and advance x at 10° intervals so that one complete cycle of

the equation $y = 2 \sin x$ is plotted (360°). In the same interval the equation $y = \sin 3x$ should have three complete cycles at half the amplitude.

b. Next, add graphically the amplitudes at 10° intervals and use these to obtain the points of the resultant waveform which is plotted in the same time relationship below the waveforms obtained in (***a***). The resultant waveform is the sum of a fundamental wave ($y = 2 \sin x$) and its third harmonic with half the amplitude ($y = \sin 3x$).

c. Now, replot on another sheet of graph paper the two waves after they have passed through an amplifier which boosts the amplitude of the first wave by 10 and the second wave by only 5. Your equations at the output will then be $y = 20 \sin x$ and $y = 5 \sin 3x$. Again produce the resultant of the sum of these two waves.

d. Compare the waveshape of the resultant with the resultant obtained in (***b***). Is it the same? If need be, replot the resultant obtained in (**c**) at $\frac{1}{10}$ amplitude and then compare.

e. What does this exercise infer about amplifiers which boost some frequencies more than others?

17. Define carefully the following types of distortion and state briefly how each is produced: (***a***) amplitude distortion, (***b***) harmonic distortion, (**c**) intermodulation distortion, and (***d***) phase distortion.

18. How many distinctly unique kinds of distortion are mentioned in this chapter?

19. The output of an amplifier is measured and found to be 10 V at 1000 Hz. Also present at the output is a second harmonic of 0.6 V, a third harmonic of 0.3 V, and a fourth harmonic of 0.1 V. Calculate the percent harmonic distortion.

20. What is the purpose of a peaking coil?

21. An amplifier has a 5-kΩ load resistor in series with a 0.1-H peaking coil. What is the effective load impedance at 10 kHz? (Note: $X_L = 2\pi fL$.)

22. How can phase distortion be reduced?

23. Obtain a copy of a recent industrial electronic components catalog and find the listings for public address amplifiers. Prepare a table listing at least five amplifiers under the headings: Amplifier make and model, power output, frequency response, percent harmonic distortion at rated output, and price.

6-1 THE CONCEPT OF FEEDBACK

You may have noticed that as you move your hand there is an awareness of its position. The brain issues not only the "movement" order but also the necessary "continue" signals. A corrected continue signal can be given only if the brain is receiving positional information from instant to instant, i.e., only if information feedback is taking place.

This is an example of a closed-loop system with *negative feedback*. Feedback is negative when the information dampens the action and positive when it enhances it. Examples of both kinds, positive and negative feedback, abound in nature, and both kinds can be introduced into electronic circuits. Most body movements are stabilized by negative feedback.

Positive feedback makes a mechanism unstable and, if uncontrolled, can end in destruction of the mechanism.

In electronic circuits negative feedback is used to achieve stability. More specifically, in amplifier circuits negative feedback causes the output to faithfully follow the input. Conversely, positive feedback causes instability, and in the extreme it causes amplifiers to become independent of external input signals. Amplifiers then become oscillators and generate output signals whether an input is present or not.

The basic principle of negative feedback in amplifiers is this: *A portion of the output-signal waveform is compared with the original input and if they are not identical, the circuit attempts to minimize the differences.*

6-2 TECHNIQUES OF ELECTRICAL COMPARISON

How are two signals compared electrically and automatically? First, what is meant by "identical" signals, and secondly, how can differences be detected?

Two signals are defined as identical if their waveshapes (harmonic content), frequencies, and amplitudes are identical. Not only must the amplitude of the fundamental frequency be the same, but the relative amplitudes of all harmonic components must be the same, as this is what determines the shape of the wave.

One device for detecting differences in signals is a differential amplifier (see Sec. 12-15). A differential amplifier has zero output unless there is a difference in the two input signals. The output from a differential amplifier is often referred to as the *error signal,* because its amplitude depends on how greatly two signals differ.

6-3 ELECTRICAL SUMMING NETWORK

An alternate way of comparing signals is to use a resistive summing network. By connecting three resistors as shown in Fig. 6-1,

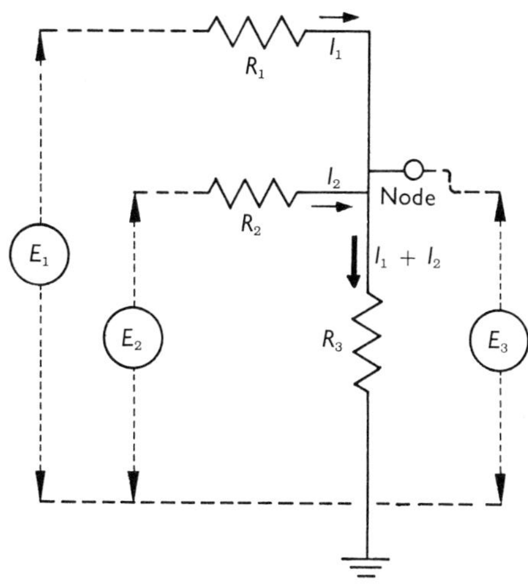

Fig. 6-1 Resistance-type, voltage-summing network.

the voltage E_3 developed at the node point (the electrical junction) depends on the *IR* drop across R_3. This *IR* drop in turn depends on the applied voltages and the resistance values.

It is easy to show mathematically that summation of voltages does indeed take place. The resultant voltage E_3, the *IR* drop across R_3, is given by Ohm's law. It is $E_3 = (I_1 + I_2)R_3$. In this equation it is now necessary to substitute terms which contain E_1 and E_2 to see how these voltages are mathematically related to the output voltage E_3. The potential difference across R_1 is $E_1 - E_3$, so that by Ohm's law the current through R_1 is given by $I_1 = (E_1 - E_3)/R_1$. Similarly, $I_2 = (E_2 - E_3)/R_2$. Substituting these values for I_1 and I_2 in the equation $E_3 = (I_1 + I_2)R_3$ gives

$$E_3 = \left[\left(\frac{E_1 - E_3}{R_1}\right) + \left(\frac{E_2 - E_3}{R_2}\right)\right] R_3$$

$$= \left(\frac{E_1}{R_1} - \frac{E_3}{R_1} + \frac{E_2}{R_2} - \frac{E_3}{R_2}\right) R_3$$

$$= E_1\frac{R_3}{R_1} - E_3\frac{R_3}{R_1} + E_2\frac{R_3}{R_2} - E_3\frac{R_3}{R_2}$$

Collecting terms by transposing all terms with E_3 to the left side of the equation gives

$$E_3 + E_3\frac{R_3}{R_1} + E_3\frac{R_3}{R_2} = E_1\frac{R_3}{R_1} + E_2\frac{R_3}{R_2}$$

Factoring out E_3 gives

$$E_3\left(1 + \frac{R_3}{R_1} + \frac{R_3}{R_2}\right) = \left(E_1\frac{R_3}{R_1} + E_2\frac{R_3}{R_2}\right)$$

The term

$$1 + \frac{R_3}{R_1} + \frac{R_3}{R_2}$$

can also be written as

$$\frac{R_1R_2 + R_2R_3 + R_3R_1}{R_1R_2}$$

Substituting, the equation becomes

$$E_3\left(\frac{R_1R_2 + R_2R_3 + R_3R_1}{R_1R_2}\right) = E_1\frac{R_3}{R_1} + E_2\frac{R_3}{R_2}$$

Finally, solving for E_3 gives

$$E_3 = \left(\frac{R_1R_2}{R_1R_2 + R_2R_3 + R_3R_1}\right)\left(E_1\frac{R_3}{R_1} + E_2\frac{R_3}{R_2}\right)$$

The term

$$\frac{R_1R_2}{R_1R_2 + R_2R_3 + R_3R_1}$$

is a constant, no matter what voltages are applied to the circuit. The term is therefore a constant multiplier in the equation. Stated in words the equation says that the resultant voltage E_3 depends on the sum of the applied voltages E_1 and E_2, each multiplied by a resistance ratio. The simplest test to illustrate summation is to make all three resistors 1 Ω. Then the equation becomes

$$E_3 = \frac{1 \times 1}{(1 \times 1) + (1 \times 1) + (1 \times 1)}\left(E_1\frac{1}{1} + E_2\frac{1}{1}\right)$$

or simply

$$E_3 = \tfrac{1}{3}(E_1 + E_2)$$

The last expression clearly shows that for 1-Ω resistors, or indeed for any equal-valued resistors, the output will always be one-third of the voltages E_1 and E_2 added together.

Note that by varying the choice of resistors it is possible to reduce the contribution made by either E_1 or E_2. In the equation the term $(E_1R_3/R_1 + E_2R_3/R_2)$ shows that if a small value of R_3 is used and if R_1 is made large, then the term R_3/R_1 becomes a small fraction. Since E_1 is multiplied by this fraction, the term E_1R_3/R_1 represents only a small fraction of the applied voltage E_1. Similarly, E_2R_3/R_2 represents only a fraction of E_2. In this way it is possible to regulate the contribution from each source.

6-4 APPLICATION OF THE SUMMING NETWORK

It is possible to use the summing network for detecting differences in two signals. If we wish to test two signals for similarity in waveform they may be applied to a summing network 180° out of phase with each other. Resistance ratios are chosen such that the signals arrive at the node point with approximately the same amplitude. If they are indeed identical, the positive amplitudes of one wave will cancel the negative amplitudes of the other. The resultant will be 0 V. Complete cancellation takes place only if the waveforms are identical. This cancellation process is illustrated in Fig. 6-2.

It must be emphasized that summing networks can be constructed using capacitors, inductors, resistors, or a mixture of all three. If reactance is present in the summing network, it cannot perform effectively at all frequencies. It will have different characteristics for high-order harmonics than it does for the fundamental frequency.

To ensure that the amplitude of each signal is the same at the node point, it is necessary to make the expressions E_1R_3/R_1 and E_2R_3/R_2 in the summation term equal.

If
$$E_1\frac{R_3}{R_1} = E_2\frac{R_3}{R_2}$$

then
$$\frac{E_1}{E_2} = \frac{R_1}{R_3}\frac{R_3}{R_2}$$

and finally
$$\frac{E_1}{E_2} = \frac{R_1}{R_2}$$

This means that the contribution from each of the input voltages is proportional to the inverse of the input resistances. The next example shows how the resistance ratio can be calculated when two input voltages to be summed are given.

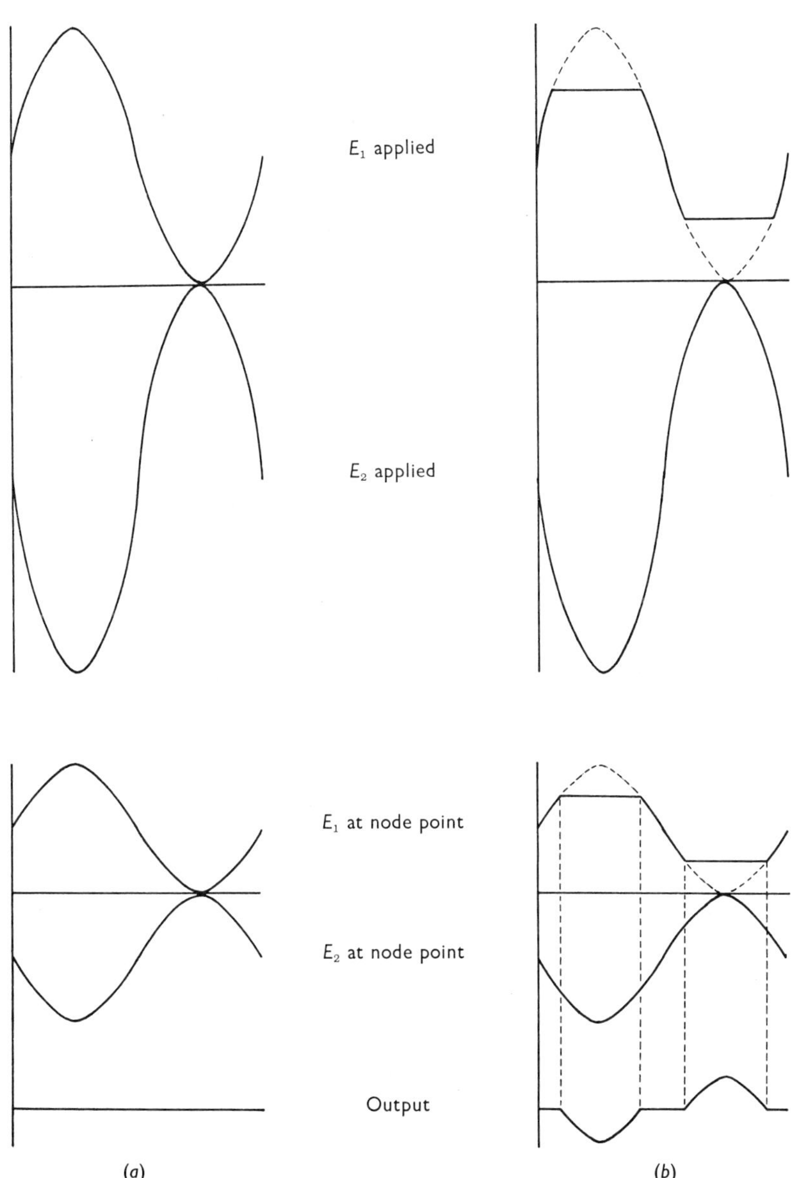

Fig. 6-2 Summation of two signals (opposite phase) with summing network adjusted to compensate for different initial amplitudes: (a) Identical waveforms; (b) Differing waveforms.

EXAMPLE 6-1 Two signals of equal frequency and opposite phase are to be compared. One has an rms amplitude of 30 V, the other 10 V. What resistance ratio must be used to get equivalence at the node point? Assuming that R_3 is small in comparison with R_1 and R_2 and using the equation $E_1/E_2 = R_1/R_2$ substituting numerical values gives

$$\frac{30}{10} = \frac{R_1}{R_2} = 3$$

This means $R_1 = 3R_2$ or $R_2 = \frac{1}{3}R_1$, i.e., the resistor used for inputting the higher amplitude signal must be three times the value of resistor in the other input. The actual values chosen depend on other considerations. If the signal sources are low-impedance, low-value resistances of the order of several hundred ohms would suffice. In high-impedance sources it might be necessary to use resistances of several megohms. A low value may load the sources excessively, whereas a high value (say 10 and 30 MΩ) can result in excessive noise pickup induced by stray electromagnetic fields.

6-5 AMPLIFIERS WITH NEGATIVE FEEDBACK

Negative feedback when used in amplifiers has two major features: (1) *The gain is reduced* and (2) *the fidelity is improved* (i.e., all three types of distortion—frequency, amplitude, and phase—are reduced).

To show how gain is reduced, consider first the amplifier shown symbolically in Fig. 6-3a. Voltage gain is defined as output-voltage amplitude divided by input, which gives the expression

$$A = \frac{e_o}{e_s}$$

Next examine the resultant amplifier when a negative-feedback loop is added as shown in Fig. 6-3b. The feedback loop could consist of an electrical summing network as discussed in the previous section. Although the gain of the amplifier module itself is A, system gain is not equal to A because of the effect of the feedback. The system is shown as a new composite amplifier enclosed by the broken line. The initial amplifier now receives input from an electrical summimg network. Inputs to this summing network are the input signal and a fraction (β) of the output signal e'_o, shown as $\beta e'_o$. The term β is known as the *feedback fraction* or *feedback factor.* It is important that the signal $\beta e'_o$ be 180° out of phase with the input signal to achieve negative feedback.

How does the gain of the original amplifier (A) compare with the resultant gain (A') when negative feedback is added? Using the definition of gain, output amplitude divided by input, gives

$$A' = \frac{e'_o}{e_s}$$

Nevertheless, the actual amplifier module as a component of the system still functions intrinsically with gain A. Its output is e'_o and its input is $e_s + \beta e'_o$ (Fig. 6-3b), so that the gain expression is

$$A = \frac{e'_o}{e_s + \beta e'_o}$$

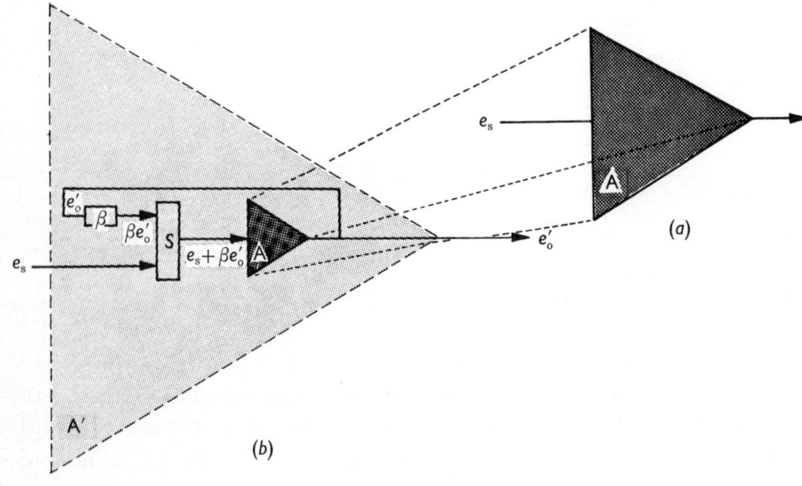

Fig. 6-3 Amplifiers shown symbolically: (*a*) Amplifier with no feedback and gain A; (*b*) Amplifier with negative-feedback loop added results in amplifier with different characteristics and gain A'.

Solving for e_s gives

$$A(e_s + \beta e'_o) = e'_o$$
$$Ae_s + A\beta e'_o = e'_o$$
$$Ae_s = e'_o - A\beta e'_o$$
$$e_s = \frac{e'_o - A\beta e'_o}{A}$$

or

$$e_s = \frac{e'_o}{A} - \beta e'_o$$

Substituting this expression for e_s in the overall gain expression

$$A' = \frac{e'_o}{e_s}$$

gives

$$A' = \frac{e'_o}{\frac{e'_o}{A} - \beta e'_o}$$
$$= \frac{e'_o}{\left(\frac{1}{A} - \beta\right) e'_o}$$

or*

$$A' = \frac{1}{\frac{1}{A} - \beta}$$

**NOTE Most textbooks show the expression for A′ as $A' = A/(1 - A\beta)$, which is obtained by multiplying the numerator and denominator of $1/[(1/A) - \beta]$ by A. Gain A′ is referred to as the closed-loop gain (i.e., the feedback path or loop is completed). Gain A is referred to as the open-loop gain (i.e., gain before feedback is applied). Open-loop gain is always greater than closed-loop gain when negative feedback is present.*

6-6 NEGATIVE FEEDBACK AND SYSTEM GAIN

One feature of the expression $A' = 1/[(1/A) - \beta]$ is that if A is a large number, the term $1/A$ will be small. (For example, if $A = 2000$, then $1/A = 1/2000 = 0.0005$.) By using a high-gain amplifier $1/A$ can be made very small. Then the gain of the amplifier system with feedback is more nearly given by the expression $A' \cong 1/-\beta$. This shows that it is the amount of negative feedback which really determines the gain of the system. The implications are that even if power-supply voltages, tubes, transistors, FETs, etc., are varied over wide margins, the gain of an amplifier with negative feedback will not change appreciably, because it depends mostly on β.

6-7 NEGATIVE FEEDBACK AND SYSTEM FIDELITY

To show how fidelity is improved let us examine more critically the composition of the signal entering the amplifier module in Fig. 6-3b. This signal $(e_s + \beta e'_s)$ is a waveform which results from the summation of two signals, the input to be amplified, e_s, and the finished product, i.e., a fraction of the output signal $\beta e'_o$. Let the term e_{in} designate this composite input signal. Then $e_{in} = e_s + \beta e'_o$, and by transposing, we obtain the equation $\beta e'_o = e_{in} - e_s$. Notice that in the latter equation $\beta e'_o$ would be equal to $-e_s$ if the term e_{in} were not present. What this means is that the smaller the amplitude of e_{in} the more nearly alike are the input and output signal. The signal e_{in} is a mixture of the input signal and a waveform which tries to cancel the noise and distortion signals produced inside the amplifier (see Fig. 6-4). In a system with a high-gain amplifier and substantial negative feedback, e_{in} will be very small, meaning that the output signal is nearly an exact replica of the input.

6-8 EXPRESSING THE AMOUNT OF NEGATIVE FEEDBACK

The gain of an amplifier is conveniently expressed in decibels by the equation

$$\text{dB} = 10 \log\left(\frac{P_{out}}{P_{in}}\right)$$

The amount of negative feedback used in a system is also conveniently expressed in decibels and is given by the following equation:

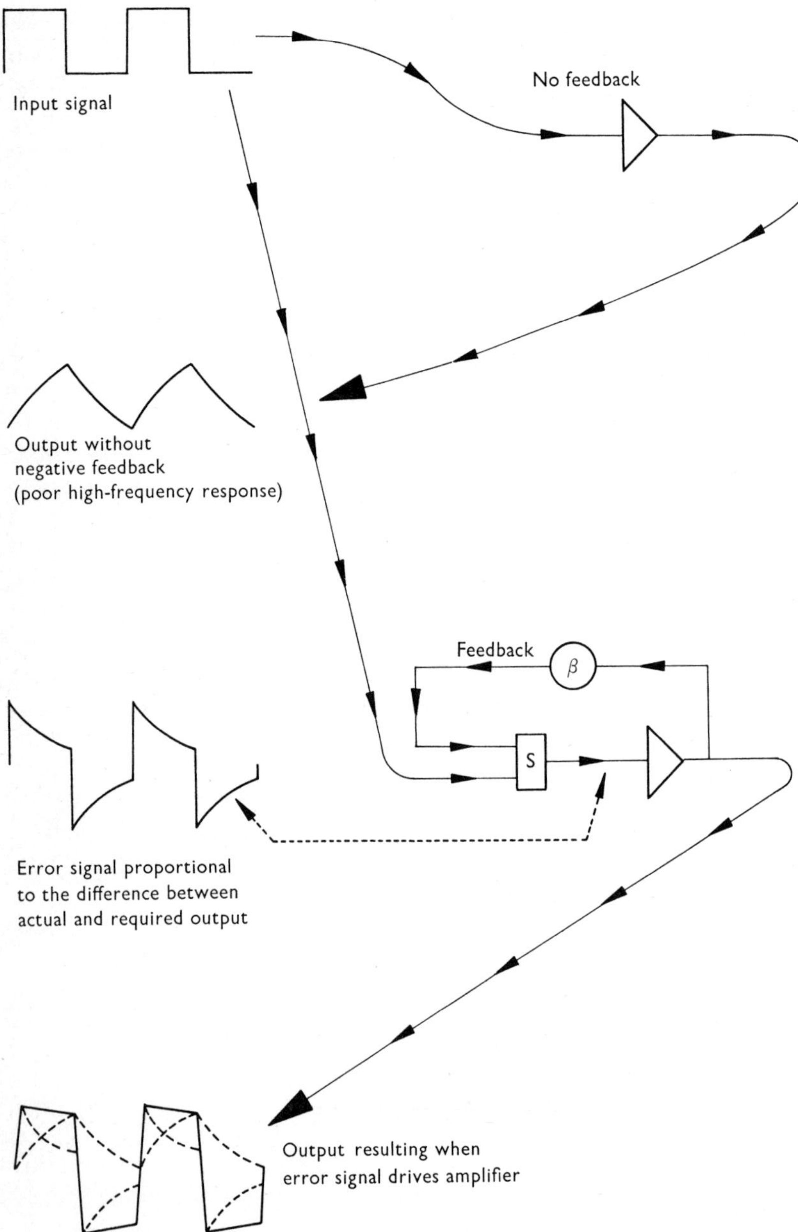

Fig. 6-4 How negative feedback corrects distortion. The exaggerated waveforms show an amplifier with poor high-frequency response without feedback, the corrected signal to combat the amplifier's deficiencies, and the improved output.

(dB negative feedback) = (gain in dB without negative feedback)–(gain in dB with feedback).

The next example shows how this equation is applied to calculate the amount of negative feedback in a given amplifier.

EXAMPLE 6-2 How many decibels of negative feedback are introduced into an amplifier which has 10 W output for a 1-mW input signal without feedback, and only 2 W output when a negative-feedback loop is added? Gain in decibels without feedback is

$$\text{gain} = 10 \log\left(\frac{P_{out}}{P_{in}}\right)$$
$$= 10 \log\left(\frac{10}{1 \times 10^{-3}}\right)$$
$$= 10 \log (1 \times 10^4)$$
$$= 10 \times 4$$
$$= 40 \text{ dB}$$

Gain in decibels with negative feedback is

$$\text{gain} = 10 \log\left(\frac{P_{out}}{P_{in}}\right)$$
$$= 10 \log\left(\frac{2}{1 \times 10^{-3}}\right)$$
$$= 10 \log 2 \times 10^3$$
$$= 10(3.3)$$
$$= 33 \text{ dB}$$

$$(\text{dB neg feedback}) = (\text{gain without feedback}) - (\text{gain with neg feedback})$$
$$= 40 - 33$$

ANSWER $= 7$ dB

6-9 FEEDBACK CIRCUITS

Negative feedback may be introduced into amplifier circuits in a variety of ways. Whatever form the circuit takes, two criteria must be met. First and foremost the input and feedback signals must be combined out of phase so that cancellation takes place. Secondly, the feedback signal must have suitable amplitude.

A direct means of providing negative feedback is shown in Fig. 6-5a. Assume momentarily that the input-signal amplitude is increasing in a positive direction. Collector current will also be increasing as a result. The rising collector current will cause a rising IR drop across R_L, causing collector potential to fall. The falling collector voltage is transferred back through R_f, the feedback resistor, to oppose the rising signal amplitude applied to the input.

In Fig. 6-5b the same circuit as Fig. 6-5a has been redrawn to focus attention on the components of the electrical summing network. The amount of feedback can be regulated by the size of resistor R_f. Whenever a pure resistor is used in the circuit, feedback will always be negative for all frequencies.

6-10 REACTIVE COMPONENTS IN THE FEEDBACK LOOP

In some applications it is desirable to introduce more feedback at certain frequencies than at others. For instance, if an amplifier with bass boost is desired, the feedback path can contain a capacitor in series with the feedback resistor. At low frequencies the capacitive reactance inhibits negative feedback and the amplifier has more gain (but also more distortion). High-frequency output is reduced because at sufficiently high frequencies only the resistor determines the amount of feedback. The capacitor behaves as a jumper wire. Similarly, high-frequency boost can be obtained by using an inductance in the feedback path because the inductive reactance inhibits negative feedback at high frequencies. A band of frequencies can be boosted or attenuated using a resonant circuit. An elementary circuit showing the various combinations of R, L, and C in the feedback path is illustrated in Fig. 6-5c.

6-11 FEEDBACK WITH UNBYPASSED EMITTER RESISTOR

One widely applied means of introducing negative feedback is the use of an unbypassed emitter resistor, as shown in the circuits in Fig. 6-6a, Fig. 6-6b, and Fig. D-1 (Appendix D). Refer again to Sec. 3-9, which explains the use of an emitter resistor to guard against thermal runaway. Note that in effect the input-signal voltage and a portion of the output-signal voltage developed across the emitter resistor are always summed 180° out of phase. (The signal voltage

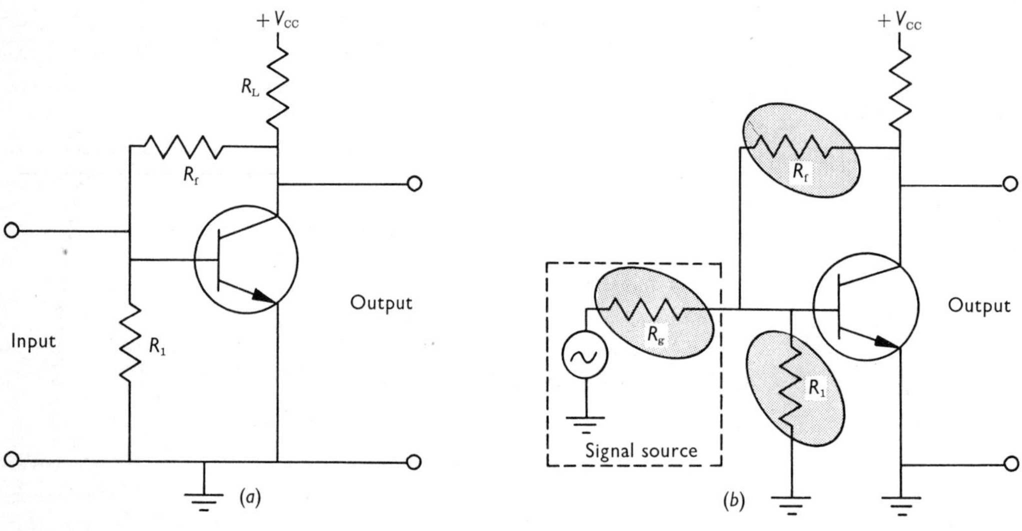

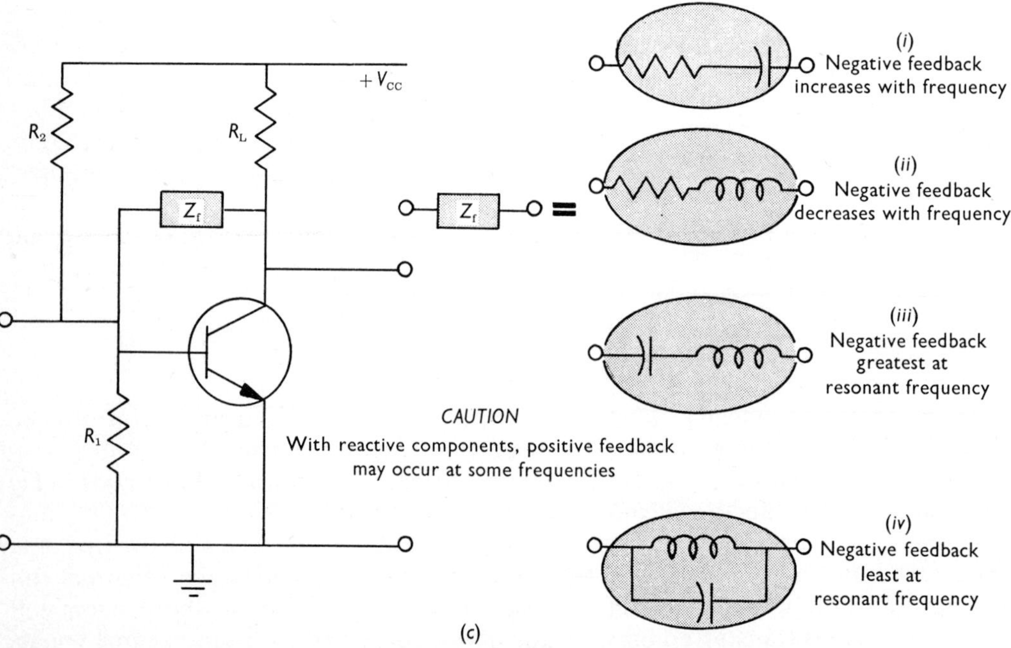

Fig. 6-5 Direct feedback path from output to input: (*a*) Resistive path; (*b*) Equivalent circuit illustrating summing action at input; (c) Combinations of *R*, *L*, and *C* in the feedback path.

developed across the emitter resistor always subtracts from the applied input-signal voltage, i.e., negative feedback is achieved.)

6-12 EMITTER-FOLLOWER CIRCUIT

If the collector resistor is left out entirely as in Figs. 6-6c, 6-6d, and D-2 (Appendix D), all the output-signal voltage is returned as negative feedback. Assuming a reasonably high-gain amplifying device is used (say, gain $A = 100$), the gain equation with feedback gives

$$A' = \frac{1}{\frac{1}{A} - \beta}$$

$$= \frac{1}{\frac{1}{100} - \beta}$$

$$= \frac{1}{0.01 - \beta}$$

With 100 percent feedback the feedback fraction β is equal to -1. Substituting this value gives

$$A' = \frac{1}{0.01 + 1}$$

$$\cong 1$$

The circuit has a voltage gain of nearly 1. In fact, the voltage gain can never be greater than one. (Power gain can of course be quite large, because of the possible increase in signal current.) At best the output-signal voltage follows the input-signal in amplitude, and such amplifier circuits go under the name of emitter

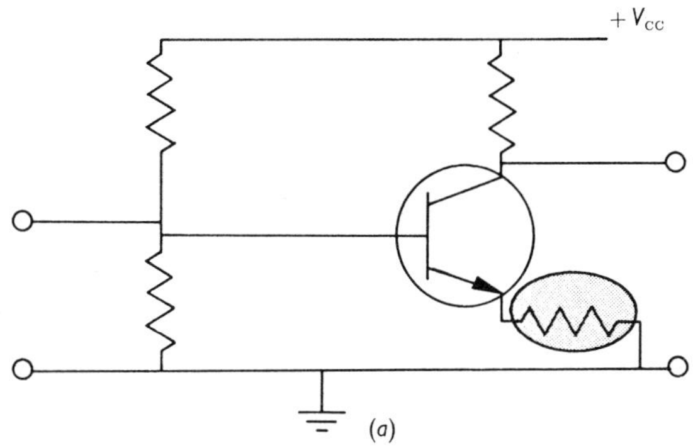

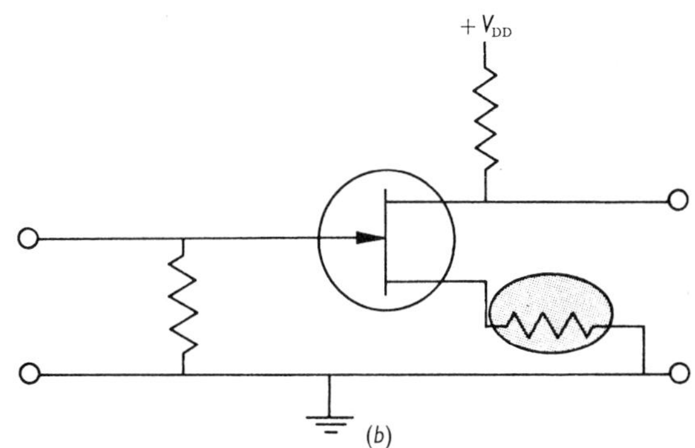

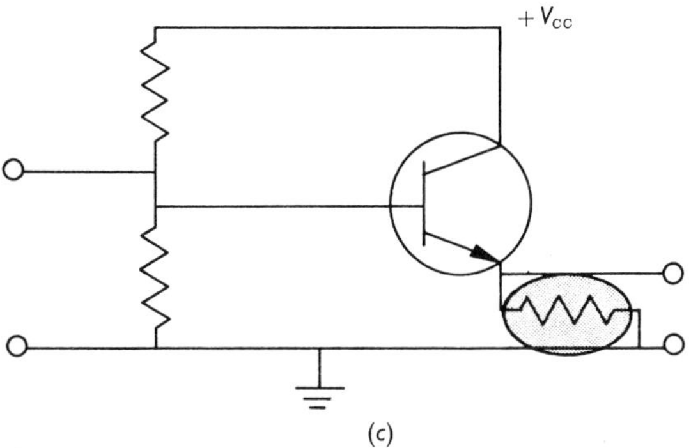

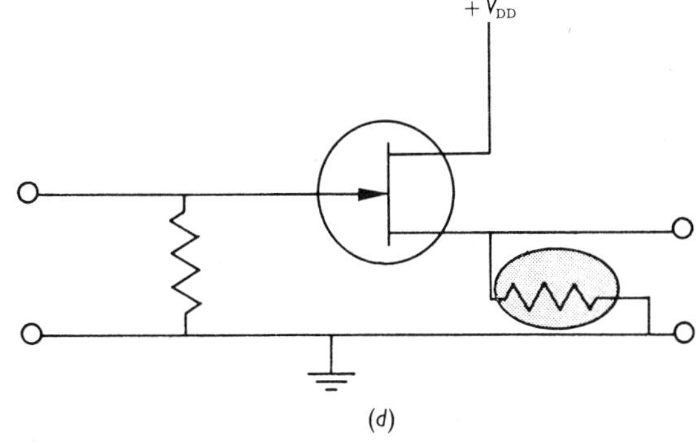

Fig. 6-6 Circuits in which negative feedback is obtained by means of an unbypassed resistor: (*a*) Unbypassed emitter resistor; (*b*) Unbypassed source resistor; (*d*) Emitter follower; (*d*) Source follower.

follower, source follower, or cathode follower, depending on the amplifying device used.

An emitter follower can accommodate a wide variety of load resistors, because regardless of load, 100 percent negative feedback is occurring. Emitter followers are therefore very useful for source-to-load matching (and they provide power gain in the process). An emitter follower has a high-input impedance (because the feedback signal cancels some of the current drain from the signal source) and therefore offers minimal load to a source. Its output can drive loads with power requirements. Widely used applications include driving signals through long cables, driving speakers, servomotors, indicator lamps, etc.

6-13 ANALYSIS OF A TYPICAL AMPLIFIER CIRCUIT

A typical amplifier circuit incorporating many of the features discussed in this chapter is dealt with in Appendix D.

Summary

Negative feedback tends to stabilize an operation.

In amplifiers a negative-feedback signal from the output attempts to compensate for deficiencies in circuit operation. This is accomplished at the expense of gain.

An electrical summing network can be used to compare two signals. If the signals are identical, there is zero output; if not, an error signal is generated.

If sufficient negative feedback is provided, amplifier characteristics become nearly independent of changes of component values and power-supply voltage variations.

The amount of negative feedback can be expressed in terms of the drop in gain resulting from the addition of negative feedback. The figure is expressed in decibels.

Feedback can be introduced via a reactive path so that the amount of feedback changes with frequency.

An emitter-follower circuit has nearly 100 percent negative feedback and is therefore highly stable and virtually free of distortion. The voltage gain of the circuit is always less than one, i.e., there is always voltage attenuation. Even so, the overall power gain may be large because of the substantial current gain.

Questions and Exercises

1. Why is negative feedback used in amplifier circuits?
2. What is the basic principle of operation of negative feedback?
3. What characteristics of signals are compared to determine whether they are identical?
4. State two types of circuits which can be used to test for differences in signals.
5. What two precautions must be observed when feeding a pair of signals into a summing network for comparison?
6. What is meant by the term error signal?
7. Two signals of 4 V and 6 V are to be compared in a summing network as shown in Fig. 6-1. If the 4-V signal is injected into a 200-Ω resistor (R_1), what must be the value of the other resistor (R_2), if equal amplitudes are to appear at the node point? (Assume R_3 is much smaller than either R_1 or R_2.)
8. ***a.*** If the 6-V signal in Question 7 were switched off, what would be the amplitude of the other if R_3 were 10 Ω?
   b. Now if the 4-V signal were switched off, and only the 6 V applied, what would be its amplitude at the node point?
   c. Are the amplitudes calculated in (***a***) and (***b***) nearly equal?
9. What are two major features of negative feedback, i.e., what effect does it have on amplifier performance?

10. What is meant by feedback factor?

11. Refer to the gain equation for an amplifier with negative feedback (end of Sec. 6-5). A' is often referred to as the "closed-loop" gain and A is the "open-loop" gain, meaning that it is the gain without the feedback path or loop completed. If an amplifier has voltage gain $A = 100$, and $\beta = -0.06$, calculate the closed-loop gain.

12. How many decibels of feedback were added in the amplifier in Question 11?

13. What determines the gain of an amplifier with negative feedback if the open-loop gain is extremely large?

14. Draw a schematic diagram of a simple transistor amplifier with direct negative feedback applied via (*a*) a resistor, (*b*) a resistor and capacitor, and (*c*) a resistor and inductance.

15. In your circuits of Question 14, which one provides (*a*) the most gain at low frequencies, (*b*) equal gain at all frequencies, and (*c*) the most gain at high frequencies?

16. Draw a schematic diagram of an emitter-follower circuit and explain how negative feedback is achieved.

17. What is the maximum theoretical voltage gain of an emitter-follower circuit?

18. What theoretical value of open-loop gain would be required to obtain a voltage gain of 1 from an emitter-follower circuit?

7-1 SMALL-SIGNAL AND POWER AMPLIFIERS

One of the many seemingly mysterious practices in the field of electronics is the subdivision of amplifiers into *small-signal* and *power* amplifiers. It is due in part to convention, although many good reasons exist for continuing it.

How do power and small-signal amplifiers differ? Both raise the signal level, but only a power amplifier can raise both signal voltage and current to a level where the signal can do useful work. A small-signal amplifier can multiply the minute signal voltages from a microphone thousands of times, yet it cannot drive a speaker. On the other hand, a power amplifier driving many speakers may be capable of filling an auditorium with sound, yet might offer no response from a microphone signal. In electronic systems, small-signal and power amplifiers are used in combination to achieve a required result. It is the job of a power amplifier to supply energy-consuming loads such as speakers, relays, deflection coils, and servomotors, with signal voltage *and* current at an E/I ratio which matches the load impedance.

7-2 SWITCHING AND LINEAR POWER AMPLIFIERS

The applications of power amplifiers are nearly endless, but they fall into two main categories, switching and linear amplifiers. Both categories are illustrated in Fig. 7-1.

A *switching* amplifier need not reproduce a faithful replica of the input signal and is therefore much easier to design in practice. Switching amplifiers may be used to extend the life of the switch contacts (e.g., the points in an auto ignition system); in this case power-handling ability is more important than switching speed. Other applications (e.g., gates in computer circuits) may call for extremely fast switching speed, but the current involved is very small. It is difficult to achieve both high-speed and high-

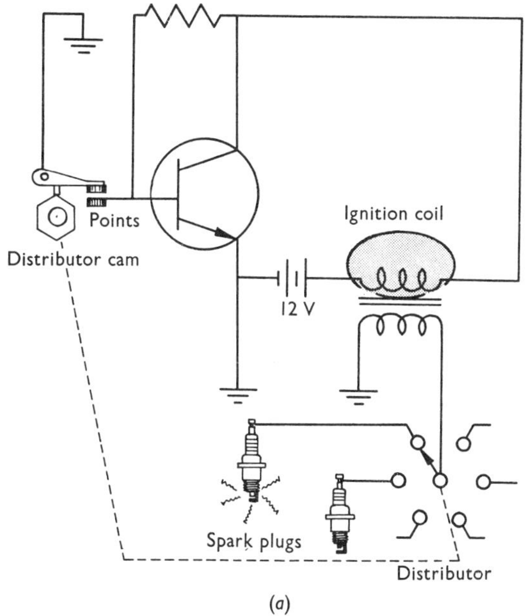

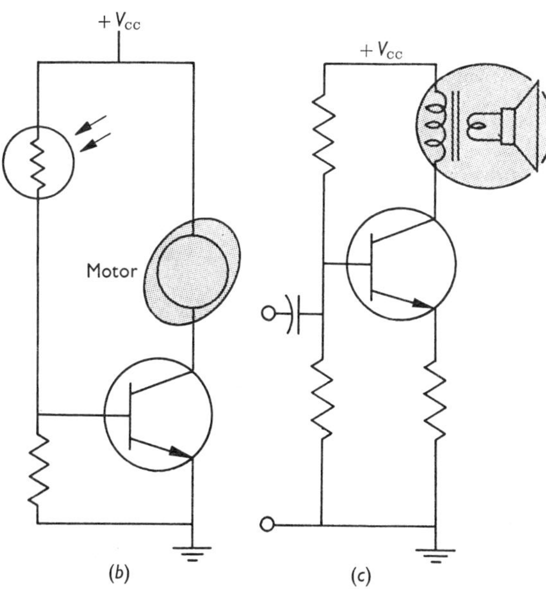

Fig. 7-1 Examples of power-amplifier circuits: (*a*) Simplified circuit of a switching amplifier as used in auto ignitions; (*b*) Motor controlled by transistor collector current, activated by the base current supplied through a photo resistor; (*c*) Linear power-amplifier circuit—for example, audio output stage in a radio receiver.

current capability as the two characteristics tend to be mutually exclusive.

Linear power amplifiers, those which must faithfully reproduce a replica of the input signal, present many design difficulties. The major problem is that amplifying devices such as transistors, tubes, and FETs are themselves linear on only a fraction of their characteristic curves. With small signals it is usually possible to limit signal variations to the linear sections of the curves. To obtain appreciable power, signal variations must be so large that amplifying devices may operate over their entire range from cutoff to full conduction. A great deal of amplitude distortion is produced by the amplifying device. This must be neutralized to acceptable levels by negative feedback.

7-3 HEAT-DISSIPATION REQUIREMENTS OF POWER-AMPLIFYING DEVICES

Tubes and semiconductor devices used in power amplifiers generate heat in proportion to the power they are supplying to the load. The heat developed must be removed or thermal runaway may destroy the device. In most applications heat may be dissipated into the surrounding air and removed by convection or forced air circulation. In the largest amplifiers, such as those found in high-powered communication transmitters, liquid cooling is required.

Every amplifying device can dissipate a given amount of power. It must be clearly understood that this dissipation rating is not the same as the signal power supplied to the load; rather it is the power wasted inside the device. For example, if a manufacturer states that a certain model of transistor is rated at 10 W collector dissipation, this means that collector voltage times collector current must not exceed 10. For example, the transistor can operate safely with 1 A of collector current provided the collector-to-emitter voltage does not exceed 10 V. Similarly, it could operate with 2 A collector current at a collector potential of 5 V, and likewise 10 A at 1 V, etc. One further note of caution: There are limits on the maximum current and voltage, even though dissipation is not exceeded. For instance, it is unlikely that the transistor could withstand 200 V at 0.05 A even though these values are within the dissipation rating. The intense electric field would probably result in damage to the device.

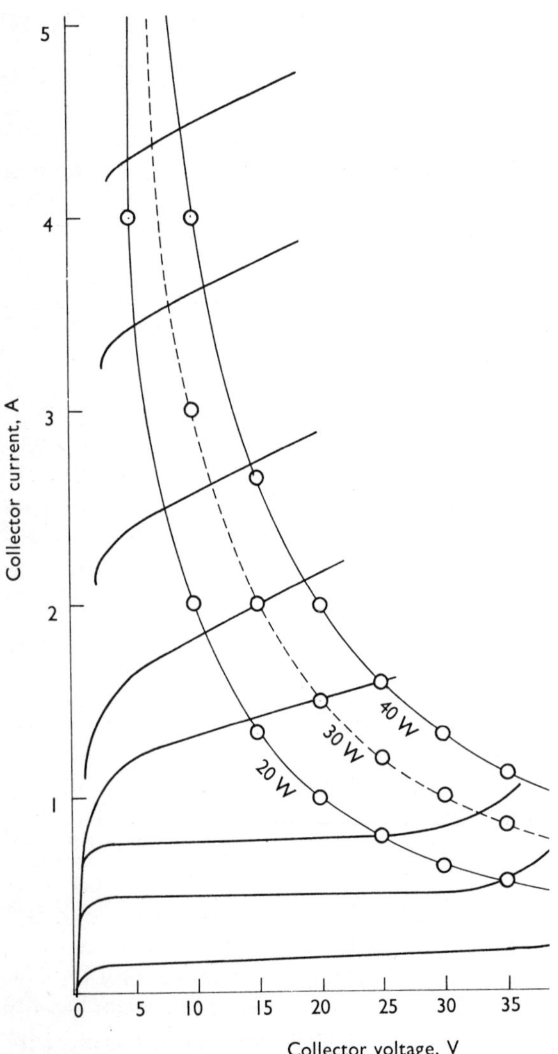

Fig. 7-2 Power dissipation curves superimposed on collector characteristic curves.

7-4 POWER DISSIPATION CURVE

When a series of constant collector-current collector-voltage product pairs are plotted and joined to produce a graph, a power dissipation curve results. Figure 7-2 shows a power dissipation curve for 20, 30, and 40 W superimposed on a family of collector curves for a typical power transistor. When designing power amplifiers it is essential that all points on the amplifier load line be located below the maximum power dissipation curve. In practice the load line should come well below the power dissipation curve, in the event that unforeseen signal peaks cause damage to the transistor.

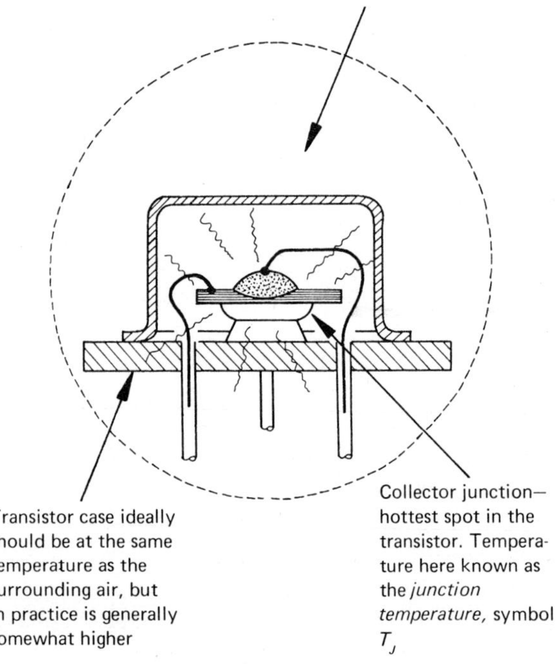

Fig. 7-3 Temperature difference $T_J - T_A$ indicates how much hotter the collector junction is than the surrounding air. The temperature difference depends on (1) the power dissipation taking place—collector voltage × collector current, (2) the thermal resistance figure θ_{JA} of the transistor and associated heat sinks.

7-5 JUNCTION AND AMBIENT TEMPERATURES

Heat accumulation in an amplifying device raises the temperature and can result in thermal runaway with subsequent destruction of the device. The problem of heat removal is far more critical in semiconductors than in tubes. Important factors such as gain, leakage current, and distortion are dependent on temperature.

In a power transistor, the highest temperature is found at the collector-base PN junction (see Fig. 7-3). Manufacturers specify the maximum junction temperature (T_J). For germanium power transistors it is typically 100°C, while for silicon devices it is typically 200°C.

Three factors determine what the junction temperature will be at any given instant. These are (1) the power being dissipated, (2) the temperature of the surrounding air, i.e., the ambient temperature, and (3) the amount of heat insulation resisting heat flow from the collector junction to the surrounding air.

7-6 THERMAL RESISTANCE

Resistance to heat flow in a transistor depends upon case construction and materials. To enable the user to calculate expected junction temperature, manufacturers specify resistance to heat flow by a number known as *thermal resistance*. Its symbol is θ, and its units are degrees Celsius/watt. Thermal resistance enables you to calculate how much greater the junction temperature is than the transistor case. If you multiply thermal resistance by power dissipation, you obtain temperature. This can easily be verified by examining the units involved [(°C/watt) × (watt) = °C]. The following example will clarify this concept.

EXAMPLE 7-1 How much higher than its case is the junction temperature of a transistor dissipating 10 W if the thermal resistance θ_{JC} (junction to case) is 0.6°C/W? Temperature difference, junction to case, is given by

$$T_{JC} = \theta_{JC} \times P_{\text{diss}}$$
$$= 0.6 \times 10$$

ANSWER
$$= 6°\text{C}$$

NOTE If the case were at room ambient temperature, say 20°C, the junction temperature would be 26°C.

7-7 HEAT SINK

Thermal resistance θ_{JC} (from junction to case) remains fixed for a given transistor. A transistor case accumulates heat, because there is thermal resistance between the case and the surrounding air. It is impossible for manufacturers to specify the case-to-ambient (θ_{CA}) thermal resistance because they cannot foretell how the user will mount a transistor. A user may simply connect wires to the transistor and rely on air to remove the heat from the case. A more effective way is to mount the case in contact with a good heat radiator known as a *heat sink*.

Heat sinks have thermal resistances ranging from 3 to 0.3°C/W. It is impossible to give a single thermal-resistance figure for a heat sink. Its sinking properties depend on the ambient temperature of the surrounding air, its rate of flow, and the power dissipation of the semiconductor device mounted on the sink. Manufacturers prefer to disclose a given heat sink's properties with graphical data as shown in Fig. 7-4a. The slope of the curve gives the exact thermal resistance at all points. To determine thermal resistance approximately at a given dissipation, proceed as in the following example.

EXAMPLE 7-2 Determine thermal resistance of the heat sink in Fig. 7-4 at 50 W dissipation. As in Example 7-1, use the expression: temperature difference = thermal resistance × power dissipation. When rearranged, and numerical data from the graph inserted, the expression gives

$$\text{Thermal resistance (sink to air } \theta_{SA}) = \frac{\text{temperature difference}}{\text{power dissipation}}$$

$$= \frac{58.8°\text{C}}{50\ \text{W}}$$

ANSWER $= 1.17°\text{C/W}$

7-8 TOTAL THERMAL RESISTANCE

The total thermal resistance between the collector junction and the outside air is obtained by adding together all the thermal resistances in the heat-flow path. As shown in Fig. 7-5, these may include θ_{JC} (junction-to-case resistance), θ_{CS} (thermal resistance of insulating washer between case and heat sink, if one is used), θ_{SA} (sink to ambient air). To determine the rise in temperature (and from this the actual temperature) of a collector junction, use the expression

Temperature difference (junction-to-ambient θ_{JA})
= (total thermal resistance in heat path) × (power dissipated)

The next example will show how junction temperature is calculated.

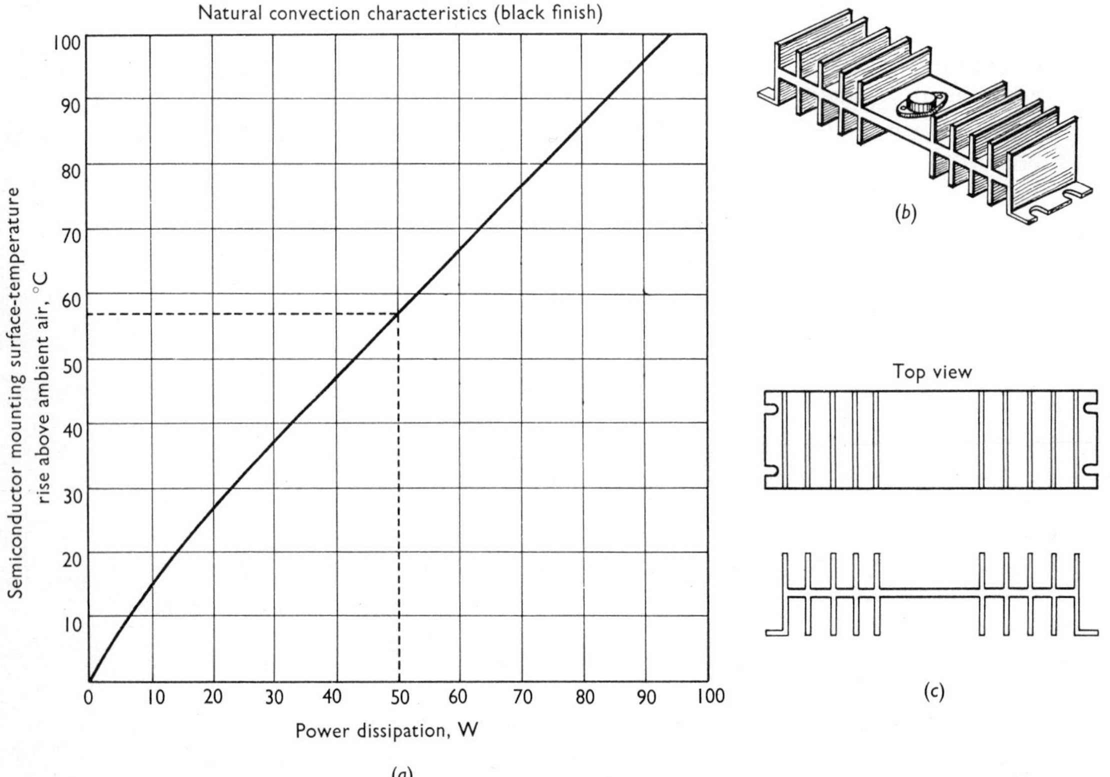

Fig. 7-4 Natural convection–thermal characteristics for a typical fin-type heat sink: (a) Dissipation—temperature curve; (b) pictorial view; (c) dimensional view.

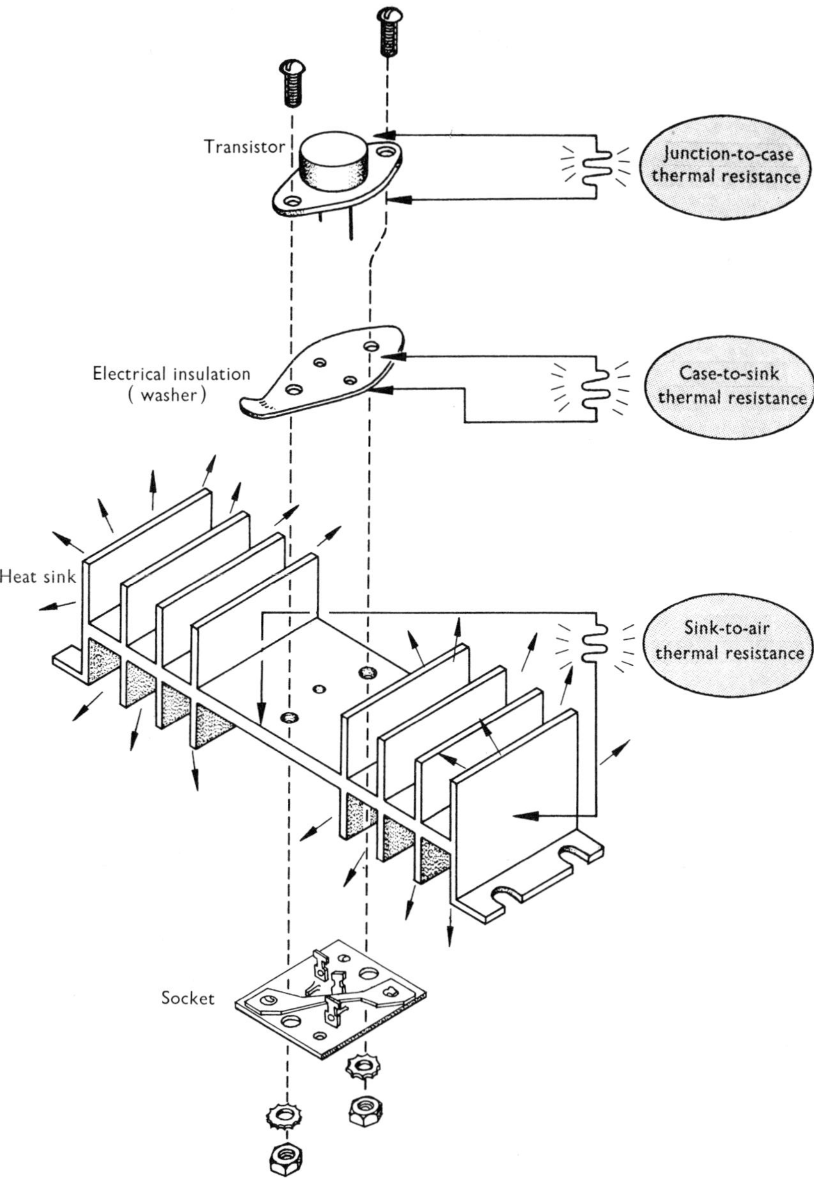

Fig. 7-5 Some thermal resistances to impede heat flow from the collector junction to ambient air.

EXAMPLE 7-3 A power transistor dissipating 20 W is mounted on a heat sink with the characteristics shown in Fig. 7-4a. Junction-to-case thermal resistance is 0.6°C/W. The transistor is insulated electrically from the heat sink with a mica washer of thermal resistance of 0.8°C/W. Find the collector junction temperature if the assembly is operating in a room at an ambient temperature of 20°C.

The following facts are given:

$$\theta_{JC} = 0.6°\text{C/W}$$

$$\theta_{CS} = 0.8°\text{C/W}$$

$\theta_{SA} = 1.32°C/W$ determined from Fig. 7-4a,

$$\frac{26.5}{20} = 1.32$$

$P_{diss} = 20$ W

Temperature difference (between junction and ambient) = (total thermal resistance) × (power dissipation)

or $T_{JA} = \theta_{JA} \times P_{diss}$, but θ_{JA} is the sum of the individual thermal resistances or $\theta_{JA} = \theta_{JC} + \theta_{CS} + \theta_{SA}$, so that the expression becomes

$$T_{JA} = (\theta_{JC} + \theta_{CS} + \theta_{SA})P_{diss}$$

Substituting numerical values gives

$$T_{JA} = (0.6 + 0.8 + 1.32)20$$
$$= 2.72 \times 20$$
$$= 55°C$$

If the junction temperature is 55°C higher than ambient, which is 20°C, the junction temperature is

ANSWER $55°C + 20°C = 75°C$

NOTE From the procedure in this example it is possible to give one general expression for junction temperature, because $T_J = T_{JA} + T_A$. *Combining gives*

$$T_J = [(\theta_{JC} + \theta_{CS} + \theta_{SA})P_{diss}] + T_A$$

7-9 DISPLAYING THE RELATIONSHIP BETWEEN QUANTITIES

It is common in electronic circuits to compare related variable quantities. Examples are voltage and current, power dissipation and temperature, and input and output signals. Timing diagrams illustrate instantaneous amplitude relationship between signals, but how are timing diagrams obtained? One way is to record the image obtained on a dual-beam oscilloscope, i.e., by direct measurement. Another is to calculate a number of amplitude points in the resultant signal, obtained from a given set of points on the generating signal. A third way is to construct a *transfer curve* and map points from a given signal through the transfer curve. This graphical method is equivalent to calculation, except that computation is avoided. A transfer curve is therefore a graph of the formula which expresses the mathematical relationship between two given quantities. Figure 7-6 illustrates three methods of displaying a mathematical relationship.

A mapping through a transfer curve shows relative amplitude variation between signals. Time dependence is preserved. If the transferring and transferred signal amplitudes vary with time, an identical scale is chosen for the time axis for both. This is illustrated in Fig. 7-7. Note that each unit of elapsed time on the input-waveform graph has its counterpart on the transferred-waveform graph. In Fig. 7-7 these time points are shown paired with lines labeled with the same number of arrowheads at each end.

Transfer curves will be used in the following sections in discussing input and output waveforms in power amplifiers.

7-10 SINGLE-ENDED AMPLIFIER CLASS OF OPERATION

Nearly every type of amplifier will give some power gain. The circuit shown in Fig. 7-1c is an example of an uncomplicated power amplifier typical of those found in portable radio or automobile receivers. Essentially it differs from previously discussed circuits on two points: (1) A transistor capable of higher power dissipation is used, and (2) a power-sinking load is used. To get undistorted operation the transistor must be operated class A (Sec. 3-4). Signal

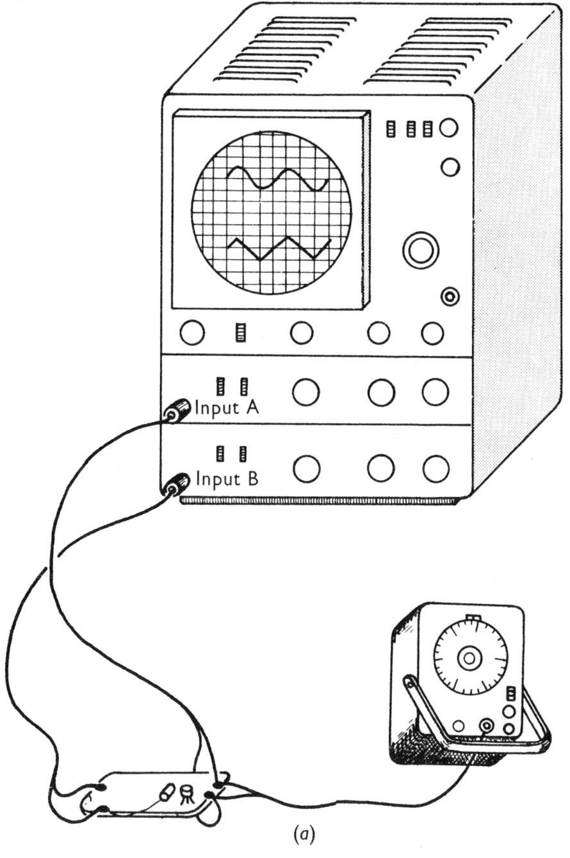

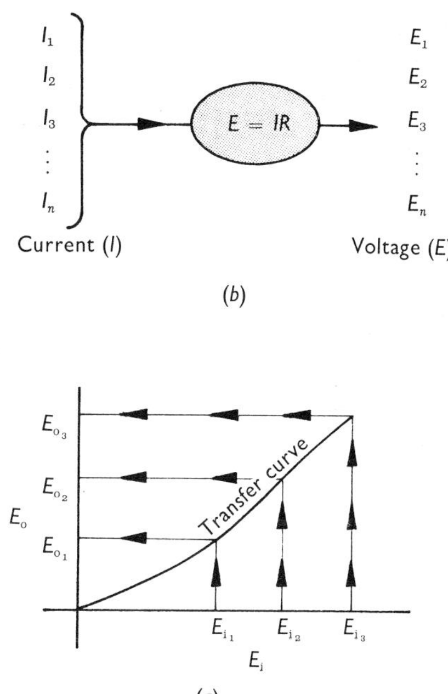

Fig. 7-6 Some methods of illustrating the relationship between signals at different points in a circuit: (*a*) By direct measurement and comparison on a dual-beam oscilloscope; (*b*) By mathematical calculation of sets of amplitude values by means of a connecting formula; (c) By graphical means, using a graph of the connecting formula, that is, *the transfer curve.*

fidelity is ensured because base and collector current remain on the linear part of the transistor's characteristic curve. The amplifier is known as *single-ended,* because only one transistor is used.

7-11 PUSH-PULL AMPLIFIER

It is usually more advantageous to use a pair of transistors in power-amplifier circuits. The pair is most often so arranged that one amplifies mainly the positive signal alternation and the other the negative. Circuit arrangements of this type are known as *push-pull* stages; one example is shown in Fig. 7-8. An equivalent vacuum-tube circuit is shown in Appendix E, Fig. E-1.

Briefly, the operation is as follows. Signal current in the primary coil L_1 of transformer T_1 induces currents in the secondary coil L_2. The secondary is divided into sections L_{2A} and L_{2B}. One end of each is common center tap and is returned to the resistor-biasing network consisting of R_1 and R_2. Since a signal voltage is developed across the entire secondary, the extreme ends will always be of opposite polarity. This means that with respect to the common center-tap point the base of Q_1, which is connected to L_{2A}, will always receive a signal voltage of opposite polarity to that being fed to

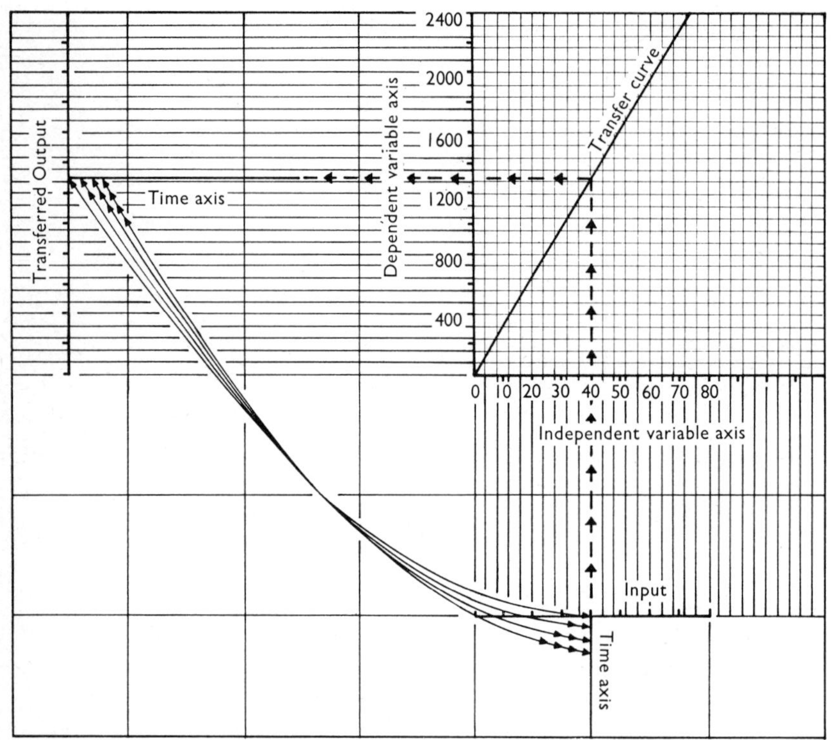

Fig. 7-7 Mapping amplitude points through a transfer curve. The time relationship between cause and effect (input and output) can be displayed by using the same time units on each time axis.

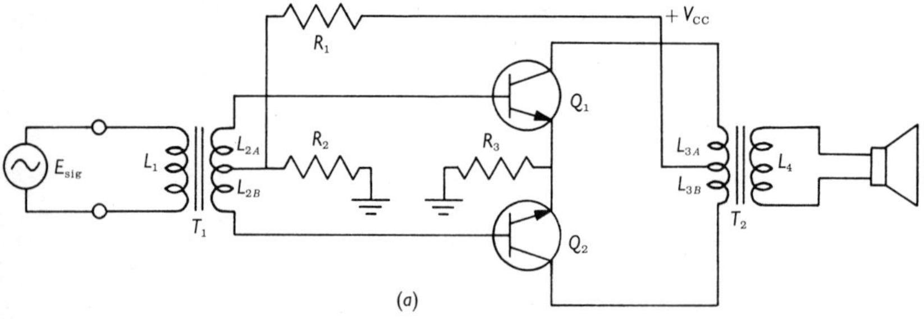

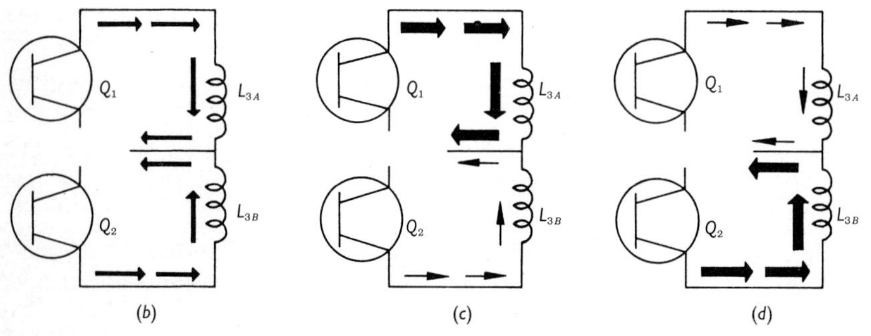

Fig. 7-8 Push-pull power-amplifier stage: (a) Schematic diagram; (b) Simplified diagram showing balanced collector current at zero signal; (c) Large collector current in Q_1, reduced current in Q_2; (d) Large collector current in Q_2, little current in Q_1—on opposite signal excursion.

the base of Q_2 via coil L_{2B}. As Q_1 is being forced into greater conduction, Q_2 is being forced toward cutoff, and vice versa.

At instants of zero signal input, collector currents at both Q_1 and Q_2 are equal as shown in Fig. 7-8b. On positive signal alternations one transistor (assume Q_1) has the greater collector current (as in Fig. 7-8c) whereas just the opposite occurs for negative alternations, as shown in Fig. 7-8d. Through the coil L_3, one transistor has the effect of driving current in one direction, the other transistor in the opposite direction. If we say Q_1 is "pulling" current through the coil, then Q_2 is "pushing." Both transistors help to move current but at opposite ends of the coil, hence the origin of the name "push-pull" circuits. Through the years the name push-pull circuit has acquired a more idiomatic meaning. It is applied freely to a class of circuit in which signal output depends on the difference in conduction between a pair of amplifying devices.

7-12 EFFICIENCY OF PUSH-PULL AMPLIFIERS

A push-pull amplifier can be more efficient (more watts of signal output per power-supply watt input) than a single-ended stage (Fig. 7-9). Ideally, a push-pull circuit need not draw any collector current when signal input is zero. Each transistor could operate class B and conduct only when the signal polarity causes base current to flow. A single-ended stage under these circumstances would amplify every half-signal cycle as shown in Fig. 7-9c, because only during half the cycle is base polarity correct to enhance conduction. In a push-pull arrangement, the addition of a second transistor ensures that the second half of the cycle is amplified by the second transistor. When no signal is present, as during a lull in program material, neither transistor draws collector current. This results not only in power economy but also in reduced power dissipation and heat generation. Essentially it means that

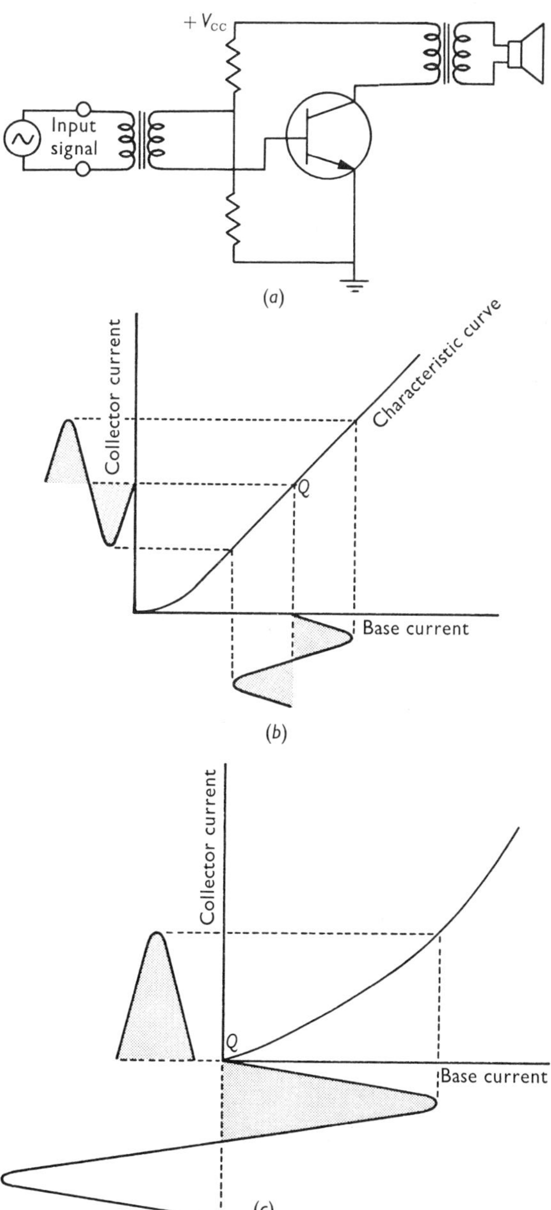

Fig. 7-9 Effect of shifting the operating point: (a) Schematic diagram; (b) Operating class A. Operating point Q and input-signal amplitude peaks are located on linear portion of the characteristic curve; (c) Operating class B. Operating point Q is located at cutoff. Only the positive or negative signal excursion—depending on whether the base is P or N type—receives amplification.

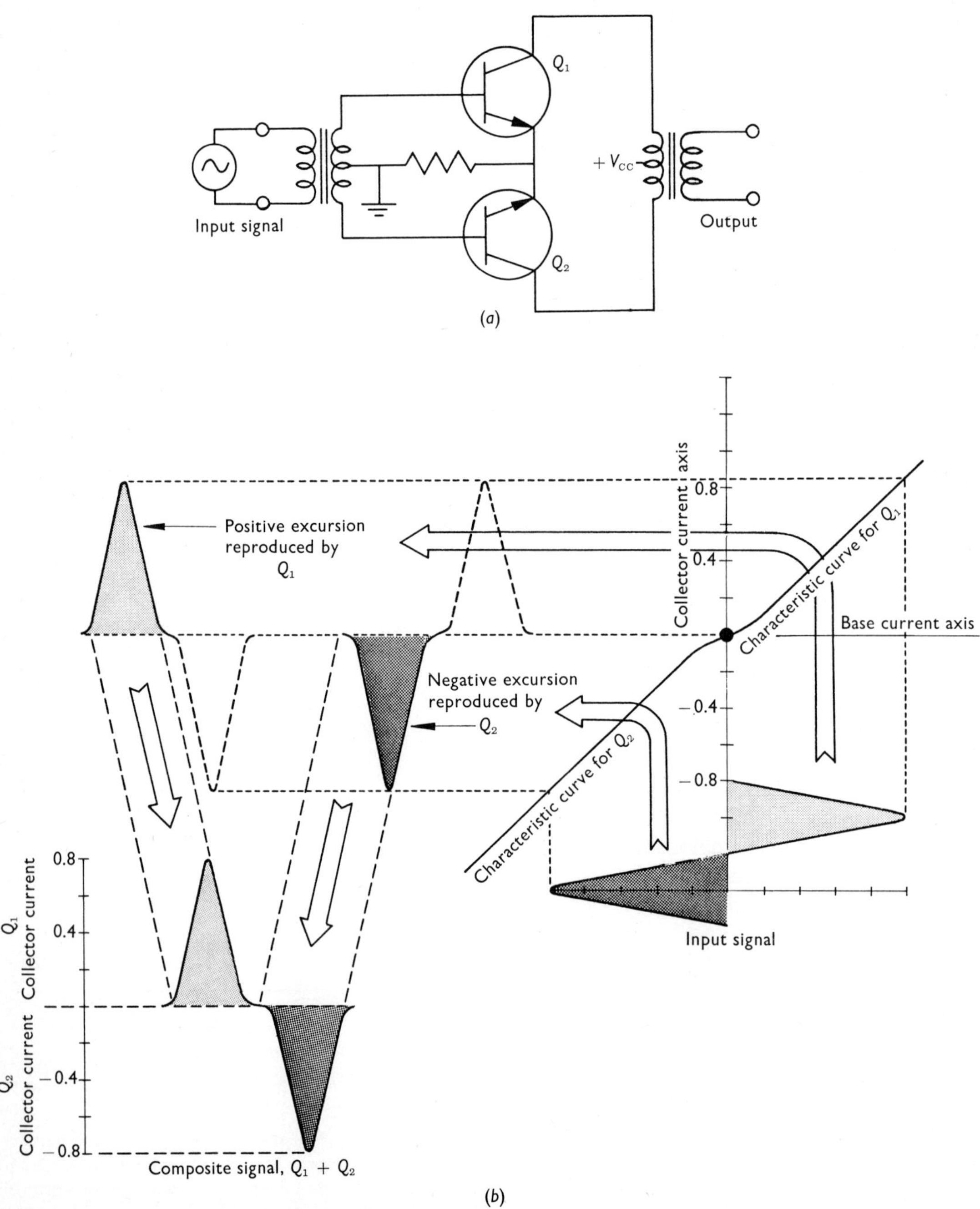

Fig. 7-10 Class B power-amplifier features: (a) Schematic diagram—zero bias; (b) Composite characteristic curve for class B push-pull power-amplifier stage, illustrating the function of each transistor.

collector current is signal current, whereas in class A stages it is a combination of signal and bias current. Figure 7-10 illustrates this concept.

7-13 CROSSOVER DISTORTION

Note that unless the composite characteristic or transfer curve is a perfectly straight line, amplitude distortion will occur. In push-pull amplifiers the nonlinearity is most pronounced near zero base and collector current. Amplitude distortion produced in this region is commonly referred to as *crossover distortion* because it occurs in the region where amplification is being transferred from one transistor to the other. Crossover distortion is better illustrated in Fig. 7-11, in which the crossover region of the composite transfer curve is enlarged. Note how the transferred wave has reduced amplitude from point *A′* to *B′*, but more nearly proportional amplitude from *C′* to *D′* where it is being transferred through a linear section of the characteristic curve.

7-14 COMPOSITE CHARACTERISTIC CURVE

Even though each transistor has a nonlinear curve, the composite curve can be made more linear overall by adding some base-bias current to the circuit. This has the effect of horizontally displacing the individual characteristic curves as shown in Fig. 7-12. Note that horizontal displacement of the characteristic curves by ad-

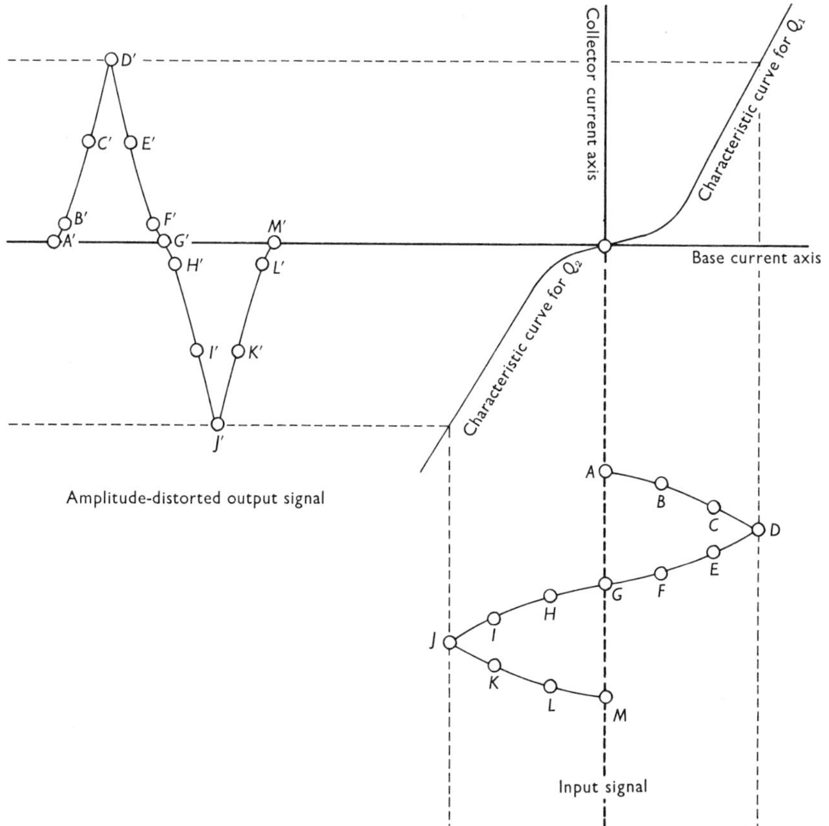

Fig. 7-11 Crossover distortion due to nonlinearity near the zero base-current region (transistor operating near cutoff).

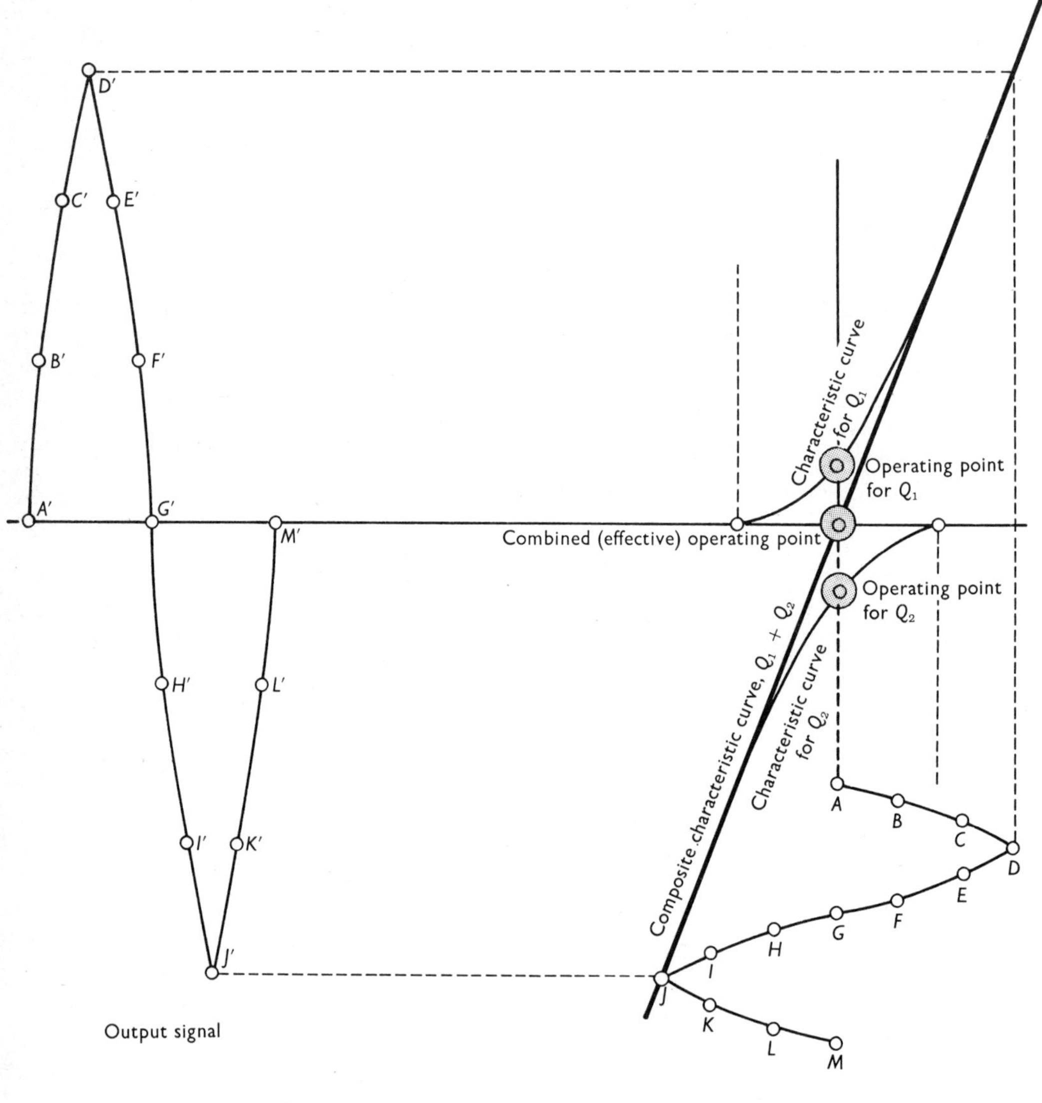

Fig. 7-12 Composite transfer curve for two transistors operating in a push-pull arrangement. Improved linearity results when some forward bias is used to shift the operating point into the more linear region of the characteristic curve.

justing bias values (operating points) changes the class of operation. Figure 7-13 illustrates some possible variations and the class of operation which results. Below each resultant transfer curve is a vector diagram showing how the points are obtained by vertically summing each point on the individual characteristic curves. Only ten points are shown for simplicity, although the transfer curve is made up of an infinite number of such points.

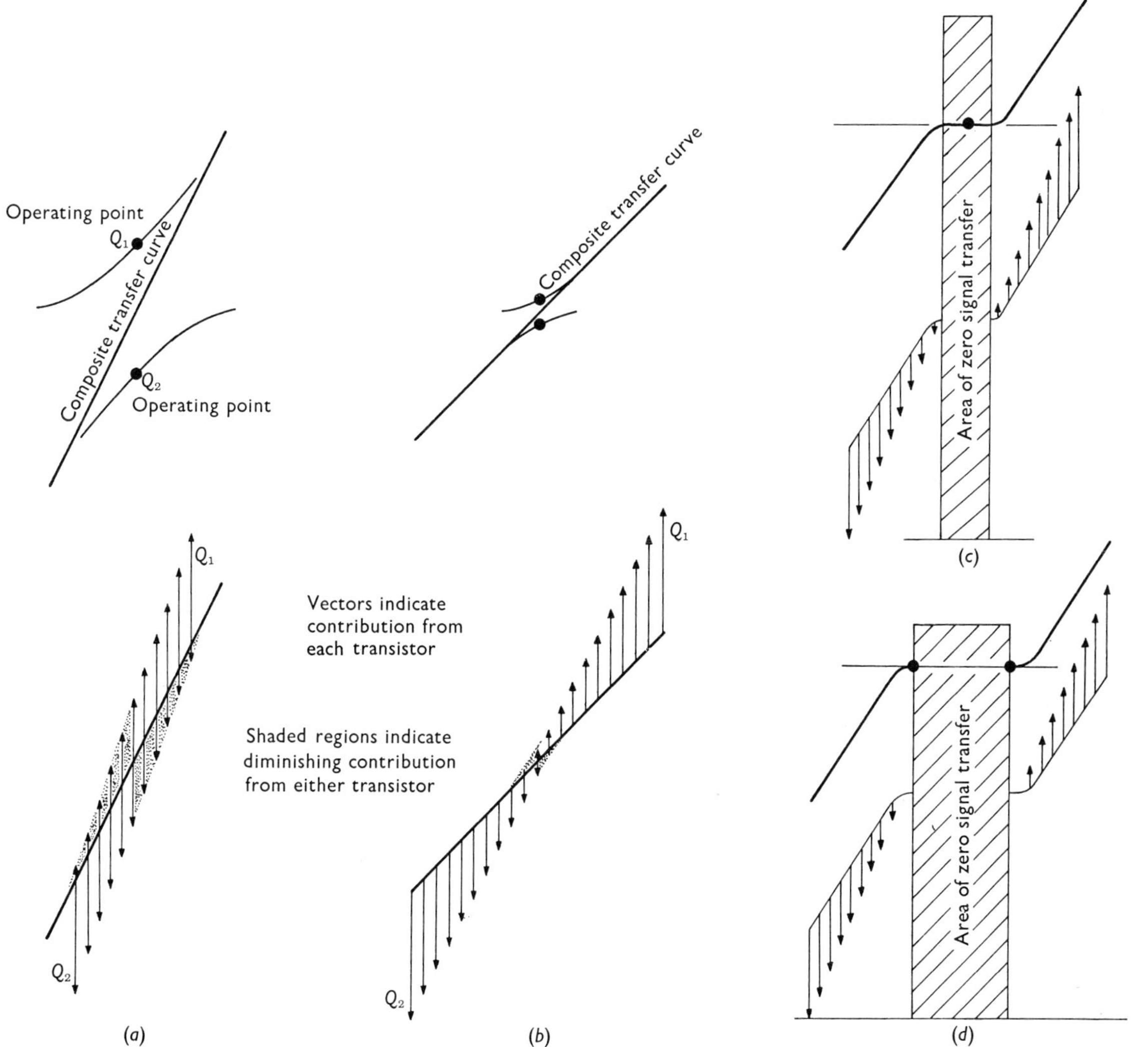

Fig. 7-13 Composite transfer curve for push-pull stage obtained by graphical summation of individual transistor characteristic curves: (*a*) Class A. Note that transistor nonlinearity lies in an area of diminishing contribution and has a minor effect on the resultant transfer curve; (*b*) Class AB; (*c*) Class B. Note the crossover nonlinearity; (*d*) Class C—unsuitable for audio reproduction.

7-15 PRACTICAL FEATURES OF PUSH-PULL AMPLIFIERS

Push-pull stages have these advantages over single-ended stages:

1. Reduced quiescent current when no signal is present and therefore greater efficiency.
2. Reduced transformer core size because quiescent-current components flow in opposing directions and neutralize each other,

thereby reducing core magnetization.

3. Smaller transistors and heat sinks can be used for a given power output, because average power dissipation is reduced with lower quiescent currents.

Some disadvantages are

1. More components are required than in single-ended stages.
2. More complex circuit because the driving signal must be split into two antiphase components to drive the "push" and the "pull" sections in opposite polarity.

Summary

Amplifiers are classed as *small-signal* and *power* amplifiers.

Power amplifiers are required for driving signals into energy-consuming loads, such as speakers, servomotors, relays, and signal lamps.

Switching power amplifiers are used for interrupting large current flow to a load by means of a small control current. *Linear* power amplifiers must reproduce an amplified but exact replica of the input signal.

In power-amplifier circuits heat is dissipated in both the amplifying device and the load.

In a transistor the highest temperature point is the collector-base PN junction.

Heat can be conducted away from the collector by means of a *heat sink*. A heat sink is a good conductor and radiator of heat.

The relative ability of a heat sink to remove and discharge heat from the collector is expressed as *thermal resistance*. The lower the thermal resistance, the better the heat conductor or sink.

Some methods of showing relationships between time-varying quantities are (1) by direct display on an oscilloscope, (2) by tabulating instantaneous amplitudes as calculated from a connecting equation, and (3) by graphical mapping with the construction of a transfer curve.

Push-pull amplifiers make use of a pair of amplifying devices. One amplifies the positive signal excursion, the other the negative.

Push-pull amplifiers can be more efficient than single-ended types because the quiescent current can be low.

Questions and Exercises

1. How do small-signal and power amplifiers differ in the functions they perform?
2. What is the important distinction between linear and switching power amplifiers?
3. Power amplifiers usually require the amplifying devices to operate over the nonlinear sections of their characteristic curves. How is distortion minimized under such operating conditions?
4. What is meant by power dissipation in an amplifying device?
5. A power transistor rated at 40 W is to be operated from a power supply of 10 V. What is the maximum permissible collector current if the power dissipation is not to exceed the rated limit?
6. What do the points on a power dissipation curve represent?
7. Construct a power dissipation curve for the transistor mentioned in Question 5.
8. What determines the upper and lower current-voltage pairs on your dissipation curve?
9. In practical circuits, why must devices be operated well within the bounds of a power dissipation curve?
10. What is the location of highest operating temperature in a transistor?
11. Which solid-state devices can operate at higher temperatures, silicon or germanium?
12. What is meant by thermal resistance?
13. If the junction-to-case thermal resistance

θ_{JC} of a power transistor is 0.6°C/W, how much warmer is its collector junction than its case? Assume the transistor is dissipating 40 W.

14. A transistor has a case temperature of 75°C but its collector-junction temperature is 111°C. If it is dissipating 60 W, what is its junction-to-case thermal resistance?

15. A transistor is operating with a junction temperature of 92.5°C and its case temperature is only 50°C. If its junction-to-case thermal resistance is 0.55°C/W, what power is being dissipated by the transistor?

16. What is meant by ambient temperature?

17. What purpose does a heat sink serve?

18. Why is it not possible to give a single figure for the thermal resistance of a heat sink?

19. Determine the thermal resistance of a heat sink of the type shown in Fig. 7-4b from the graphical data in Fig. 7-4a. Assume that the sink is dissipating 30 W.

20. What power is being dissipated by the heat sink in Question 19, if on a warm day (25°C) its temperature is 65°C?

21. If heat must flow through several materials (e.g., the transistor case, insulating washer, heat sink) and each has a different thermal resistance, how is the total thermal resistance obtained?

22. What is the collector-junction temperature of a silicon power transistor which is dissipating 30 W under the following conditions: (***a***) Junction-to-case thermal resistance 0.5°C/W, (***b***) insulating washer thermal resistance 0.7°C/W, (**c**) use of a heat sink of the type shown in Fig. 7-4, and (***d***) room temperature of 22°C?

23. Discuss three methods of showing relationship between quantities in an electronic circuit.

24. What is a single-ended stage?

25. If minimum distortion is a requirement, what class of operation is permissible in a single-ended power-amplifier stage?

26. What is meant by a push-pull stage?

27. Why is it possible to realize a higher efficiency with a push-pull stage as compared with a single-ended stage? (Assume that distortionless operation is required.)

28. What is crossover distortion?

29. Is crossover distortion more noticeable at low or high signal levels?

30. What is a composite characteristic curve?

31. List at least three advantages of push-pull amplifiers as compared with single-ended types.

32. What are some disadvantages of push-pull stages?

33. In your opinion, what is the greatest obstacle in miniaturizing power amplifiers?

8-1 HIGH-CURRENT REQUIREMENTS OF LOW-IMPEDANCE LOADS

In nearly all audio systems, moving-coil-type speakers constitute the signal load for the output power-amplifier stage. Since the coil assembly moves significant distances and reverses direction thousands of times per second, it experiences considerable mechanical acceleration. Every effort must be made to keep the mass of the system as small as possible. Only a few turns of wire can be used in the voice coil if the mass is to be kept to a minimum. A large coil current is needed to obtain the necessary magnetization to generate the accelerating force. A speaker voice coil is therefore a low-impedance signal load and must be driven by a low-impedance source to obtain acceptable power transfer.

Before the advent of transistors, vacuum-tube power amplifiers employed output impedance-matching transformers. These have a large number of turns in the primary coil, so that relatively small plate currents cause considerable core magnetization. They also have a small number of turns of large wire in the secondary coil into which large currents at low voltage levels can be induced. Ideally output transformers require a flat-frequency response over the audio-frequency range (20 Hz to 20 kHz). Good output transformers are expensive, and whenever possible their use is avoided.

8-2 HIGH-CURRENT CAPABILITIES OF POWER TRANSISTORS

Power transistors can supply the large voice-coil currents without the use of output transformers. Nevertheless, transformers are often used for improved load matching. A simple output transformerless (OTL) power-amplifier circuit is illustrated in Fig. 8-1. It is an emitter-follower-type circuit. With large power transistors the peak emitter signal current can attain several amperes. For example, if even 2 A of signal current were to flow through a 4-Ω-speaker voice coil, the signal power would be $P = I^2R = (2)^2 \times 4 = 16$ W. (Strictly speaking, this would represent true power only if the 4-Ω load were purely resistive.)

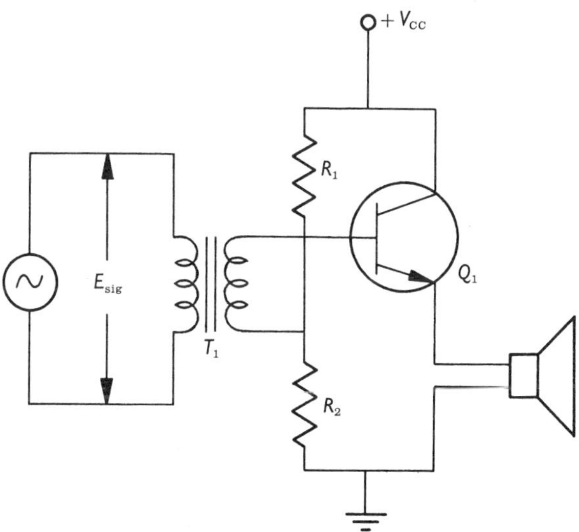

Fig. 8-1 Emitter-follower output circuit.

8-3 A PUSH-PULL EMITTER-FOLLOWER POWER AMPLIFIER—COMPLEMENTARY SYMMETRY

One disadvantage of a single-transistor emitter-follower output stage is that it operates class A. Transistor power dissipation is high because of the large quiescent current. A more efficient OTL circuit, operating class B and push-pull, consists of two emitter followers with outputs connected in parallel as shown in Fig. 8-2a. In this circuit transistor Q_1 conducts when point *A* is positive. Only then does the P-type base experience base current. Meanwhile, transistor Q_2 which has an N-type base is driven further into cutoff by a positive alternation. However, when point *A* goes negative, i.e., on the negative signal alternation, Q_2 conducts, but Q_1 is cut off. In this way Q_1 supplies the speaker voice coil with power from $+V_{CC}$ during the positive signal alternation, and Q_2 supplies power from $-V_{CC}$ during the negative alternation. At zero signal, both transistors are cut off because all base current is zero.

Crossover distortion occurs in this circuit unless both transistors have a linear characteristic curve near the cutoff region. Also, the transistors must be matched, or one alternation is amplified more than the other. It is possible to purchase PNP–NPN matched pair of transistors for constructing this type of circuit. It is often referred to as a *complementary-symmetry* circuit, because of the complementary transistors, namely, PNP–NPN.

It is not always convenient to provide a power supply with positive and negative outputs with respect to ground. The complementary-symmetry circuit can be employed with a single dc power source in the slightly modified OTL circuits shown in Fig. 8-2b and c. In each instance a capacitor C_1 is used to transfer the ac component of the signal to the speaker. The dc level at the emitters is left undisturbed by the load.

Although in theory the complementary-symmetry circuit appears attractive because of its direct approach and stark simplicity, in practice there are problems associated with it. First, the transistors must be carefully selected for compatibility, a task which is not conducive to mass production. Secondly, if the transistors are to be biased to reduce crossover distortion, separate biasing networks would be required for each because of the complementary-type bases.

8-4 "TOTEM-POLE" POWER-AMPLIFIER CIRCUIT

A most popular output transformerless circuit is the so-called totem-pole OTL circuit illustrated in Fig. 8-3a. It is a push-pull circuit in that transistor pair Q_1Q_2 take turns in supplying

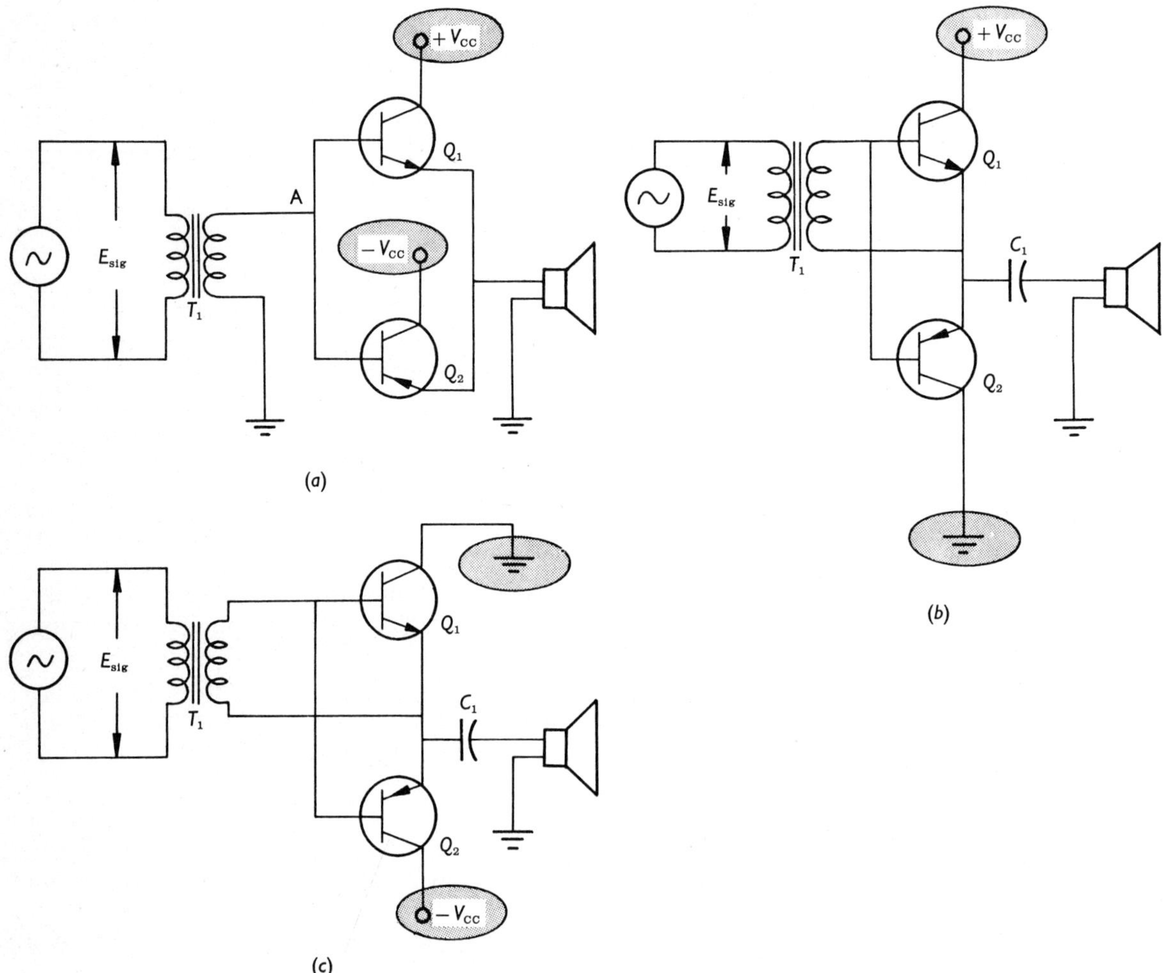

Fig. 8-2 Complementary-symmetry, output transformerless (OTL), push-pull power-amplifier stage: (*a*) Emitter followers connected to individual dc power sources; (*b*) Followers connected to single positive dc power source; (*c*) Followers connected to single negative dc source.

the load or speaker with signal current. Transformer T_1 has two secondaries. A black dot indicates the end of the coil winding which is instantaneously positive with respect to its opposite end.

Note that transistor Q_1 is connected to a positive-going end and Q_2 to a negative-going end. In this way if a positive-going signal is entering the primary, the base of Q_1 will also be going positive and collector current will be increasing, as shown in the simplified circuit in Fig. 8-3b. Simultaneously, the base of Q_2 will be receiving a negative-going signal from its driver transformer secondary. Its collector current will decrease during this alternation. On the reverse alternation, all signal polarities change. The base of Q_1 now receives the negative alternation and its collector current

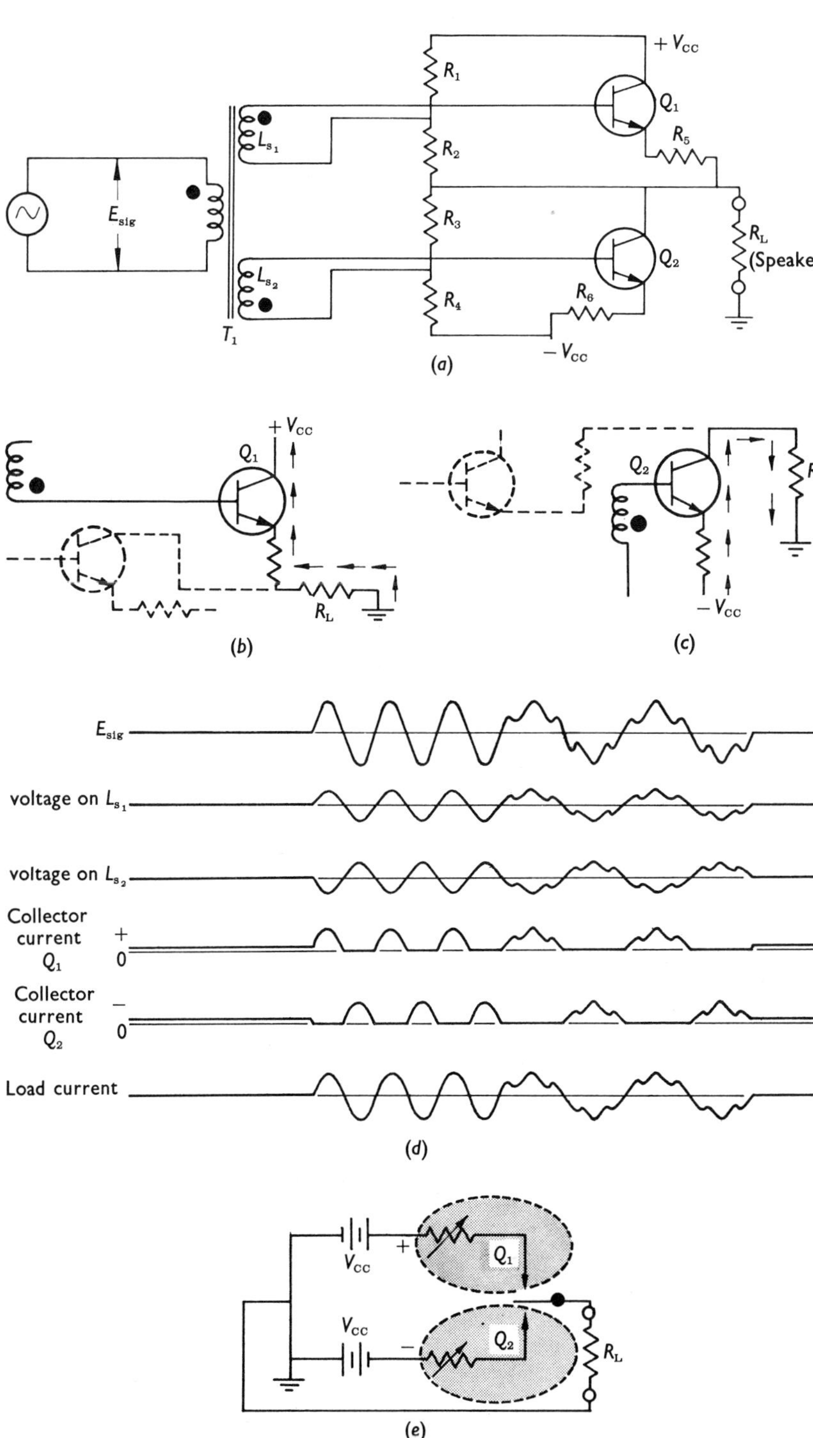

Fig. 8-3 One version of an OTL power amplifier: (*a*) Schematic diagram; (*b*) Load current during positive signal alternation; (*c*) Load current during negative signal alternation; (*d*) Timing diagram to show phase relationships at various circuit points; (*e*) Simplified equivalent circuit.

decreases. Transistor Q_2 supplies current to the load (but it flows in the opposite direction) as shown in Fig. 8-3c. A timing diagram to illustrate circuit function at various points is shown in Fig. 8-3d.

8-5 TOTEM-POLE CIRCUIT VISUALIZED AS A SWITCH

To help you to better understand its operation, think of the totem-pole circuit as a form of switch. (It is used as such in certain digital circuits, see Sec. 13-17.) During the positive and negative alternations the load is connected either to the positive ($+V_{CC}$) or negative ($-V_{CC}$) power source. When Q_1 conducts, the load is connected through the conducting transistor to $+V_{CC}$, and as a result load current flows as shown in Fig. 8-3b. All the while that the load connection to $+V_{CC}$ is made through Q_1, it is simultaneously disconnected from $-V_{CC}$ by Q_2. Similarly, when Q_2 is on, Q_1 is off, and $-V_{CC}$ is connected to the load, while $+V_{CC}$ is not (Fig. 8-3c). This concept is illustrated in schematic diagram form in Fig. 8-3e.

Resistors R_1 and R_2 provide base bias for transistor Q_1, and R_3 and R_4 provide the bias for Q_2. Although both transistors operate nearly class B, in practice a small amount of forward bias is introduced to minimize crossover distortion. This has the effect of producing a more linear composite transfer curve (see Sec. 7-14). The resistors in the emitter leads R_5 and R_6 guard against thermal runaway. Because they are unbypassed, they also provide some negative feedback to improve the fidelity of the stage.

8-6 CAPACITOR COUPLING TO LOAD

An alternate form of the OTL amplifier circuit is shown in Fig. 8-4a. This circuit has two advantages: (1) It requires only one dc power source (positive in the circuit shown) and (2) dc cannot reach the load in case of transistor failure. It also has the disadvantage that low-frequency response is limited by the size of the coupling capacitor C_1, and the initial capacitor-charge current could damage the load when power is turned on.

Circuit operation is briefly as follows. Base bias is provided by resistors R_1, R_2, R_3, and R_4 and negative feedback by R_5 and R_6. Under zero signal-input conditions, transistors Q_1, Q_2 and resistors R_5, R_6 form a voltage-divider circuit. The circuit is so designed that with zero signal input, the collector-to-ground voltage of Q_2 is exactly half of V_{CC}. Choosing a typical value of $V_{CC} = 20$ V, this is then 10 V as shown in the simplified diagram in Fig. 8-4b.

When a signal is applied, Q_1 and Q_2 alternately vary their conductance, and the output connection swings between a maximum of +20 V and ground. On the positive-signal alternation, capacitor C_1 experiences a large charging current which is limited only by the load R_L. It tries to charge to 20 V, but because of the large CR time constant (C_1 is typically several hundred μF) it cannot charge appreciably. The sudden rise in voltage on the output connection therefore appears as a 10-V increase on the load as shown in Fig. 8-4c.

On the reverse signal alternation, conduction through Q_1 decreases and in effect disconnects the 20-V V_{CC} source from the capacitor. At the same time Q_2 conducts and receives collector current from the 10-V charge accumulated on C_1, as shown in Fig. 8-4d. Again, C_1 does not have time to discharge appreciably during one signal alternation, but a heavy discharge current which constitutes the load current flows briefly as shown. An alternate way to comprehend the appearance of signal on R_L during this alternation is to visualize the simplified circuit shown in Fig. 8-4e. Here the transistor is regarded as a switch which connects the charged capacitor in parallel with the load.

8-7 REQUIREMENTS OF A DRIVER STAGE

Power transistors require substantial base currents. To supply these currents at high signal levels requires considerable power. This

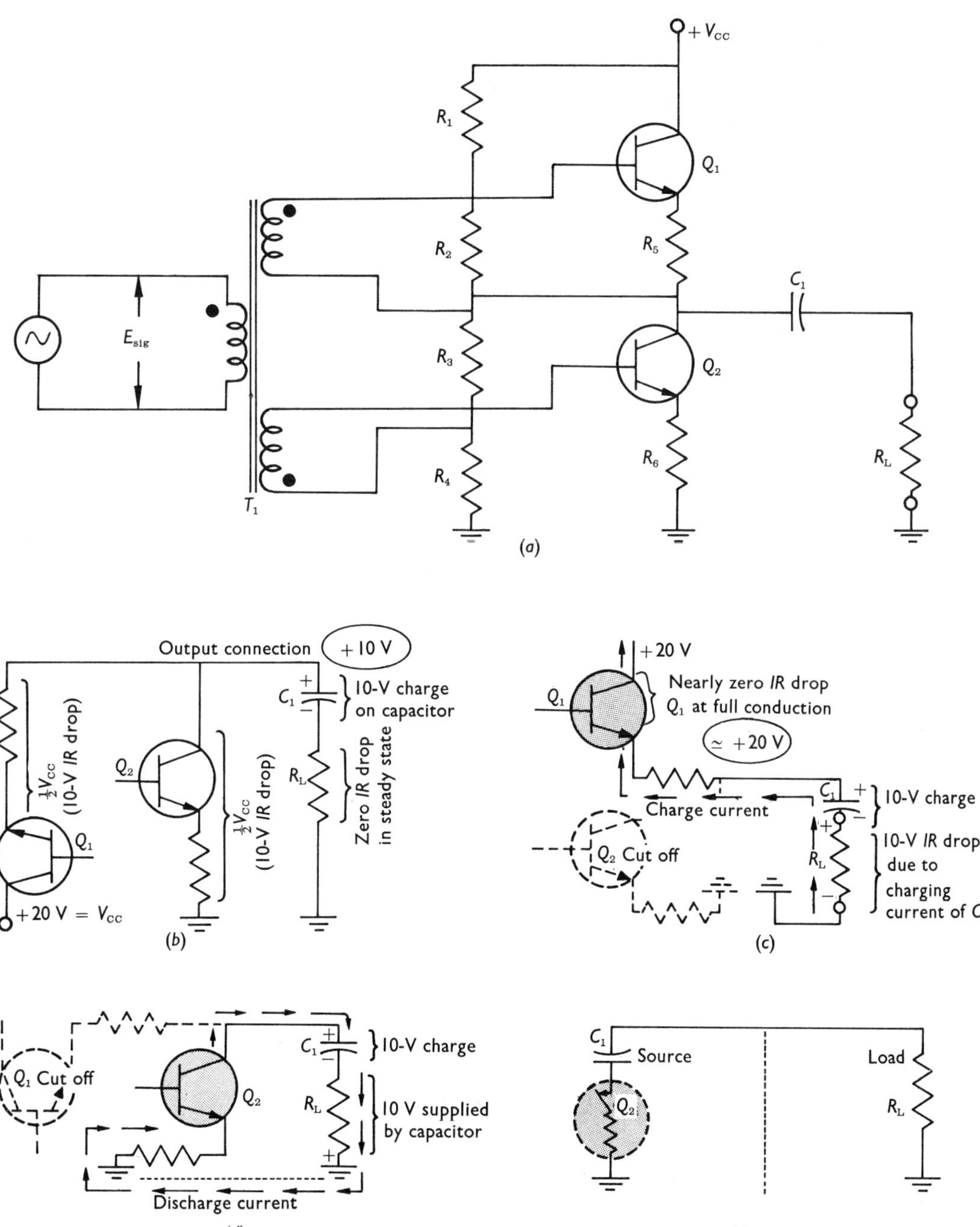

Fig. 8-4 Power amplifier with single power source: (*a*) Schematic diagram; (*b*) Simplified circuit to illustrate voltage division during zero signal conditions; (*c*) Conditions under positive signal alternation; (*d*) Conditions under negative signal alternation; (*e*) Conditions under negative signal alternation, regarding transistor Q_2 as a switch.

means that the amplifiers driving the power amplifier are also power amplifiers, though they operate at a much lower power level. A driver amplifier must fulfill two basic requirements: (1) Supply signal at a sufficiently high power level to overcome losses at the input of driven amplifier, and (2) supply signals of the correct polarity (phase) to each base of the driven push-pull stage.

The most direct means of ensuring correct phase is to use a driver transformer (such as T_1 in Figs. 8-3 and 8-4). Since transformer windings are electrically isolated from each other, each base circuit can have its own bias adjusted independently. This important feature makes balance adjustments easier (balance means that each push-pull transistor has the same quiescent current). In spite of all the desirable features of transformer coupling, the fact still remains that good transformers are heavy and bulky and cannot transfer a dc signal.

8-8 PHASE-INVERTER DRIVER

One widely used driving circuit is shown in Fig. 8-5. It provides some gain, but even more important, it is a *phase inverter,* or *phase splitter.* Figure 8-6 illustrates how the two driving signals of 180° phase difference are developed. As in all amplifier circuits, the signal applied to the base of Q_1 causes the collector current to vary at signal rate. Transistor conductance may vary between the extremes of a closed switch at full conduction (Fig. 8-6b) to open circuit (Fig. 8-6c). Under these conditions R_3 and R_4 are connected together at full conduction, and both outputs are approximately equal to $\frac{1}{2}V_{CC}$. On the reverse signal alternation R_3 and R_4 are disconnected by the slightly conducting transistor. Emitter output falls toward ground and collector output toward V_{CC}. In this way whatever voltage change takes place on the collector output occurs also at the emitter out-

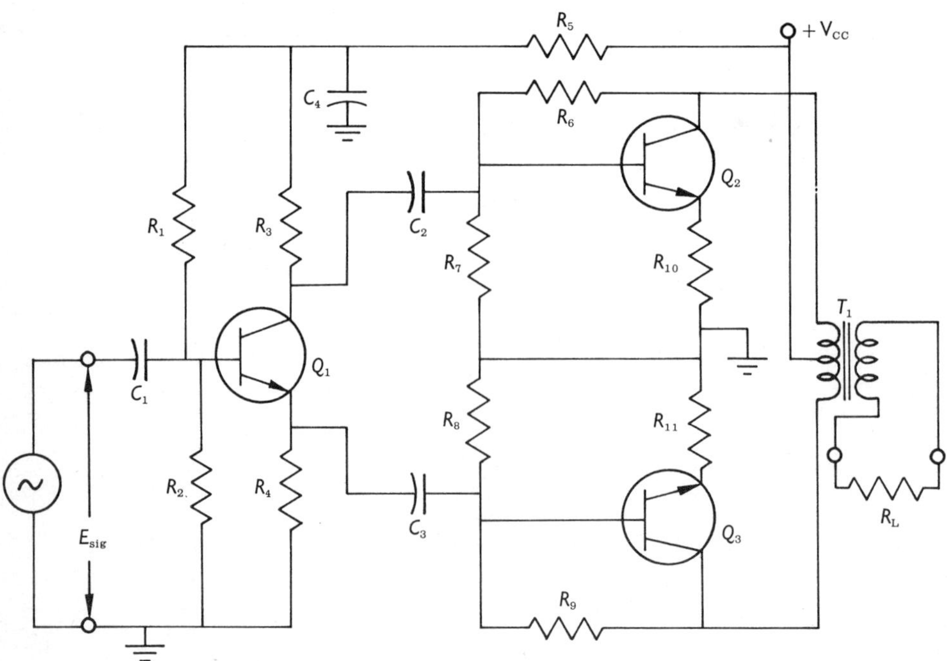

Fig. 8-5 Push-pull power amplifier with phase splitter.

put, but in antiphase as shown in the timing diagram of Fig. 8-6d.

The two outputs of a phase splitter are ideally suited for driving a push-pull amplifier. If the collector output causes the base of transistor Q_2 (Fig. 8-5) to go positive, the emitter output simultaneously causes the base of Q_3 to go negative, ensuring push-pull operation.

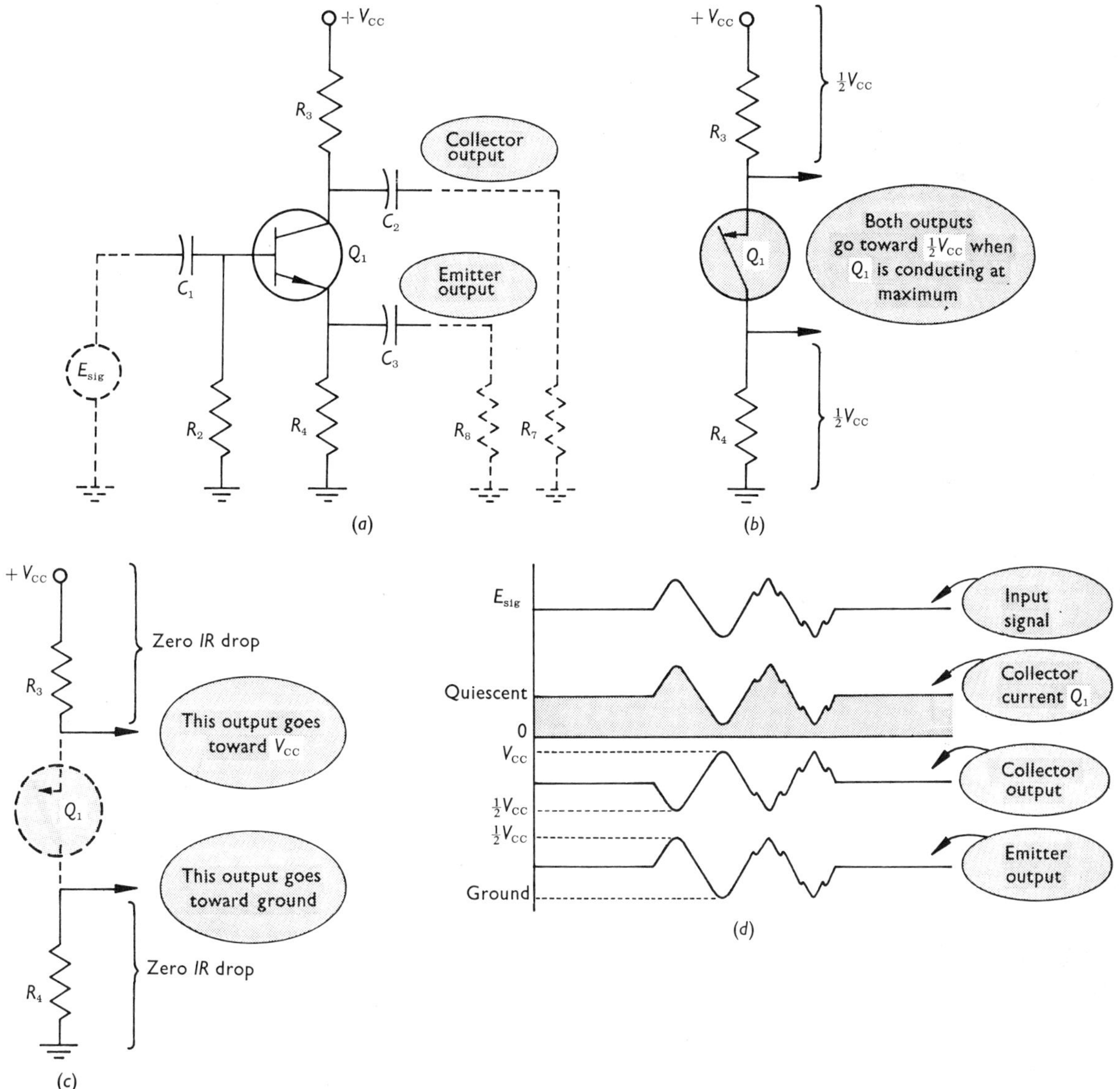

Fig. 8-6 Phase splitter: (*a*) Schematic diagram, showing input and output points; (*b*) Simplified phase-splitter output equivalent circuit, showing transistor Q_1 at maximum conduction; (*c*) Transistor Q_1 at minimum conduction; (*d*) Timing diagram (not to scale).

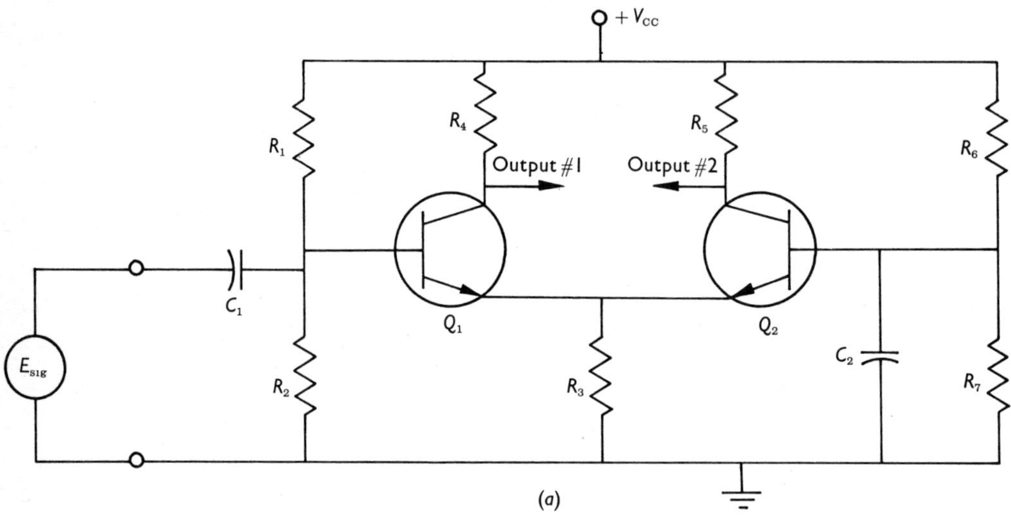

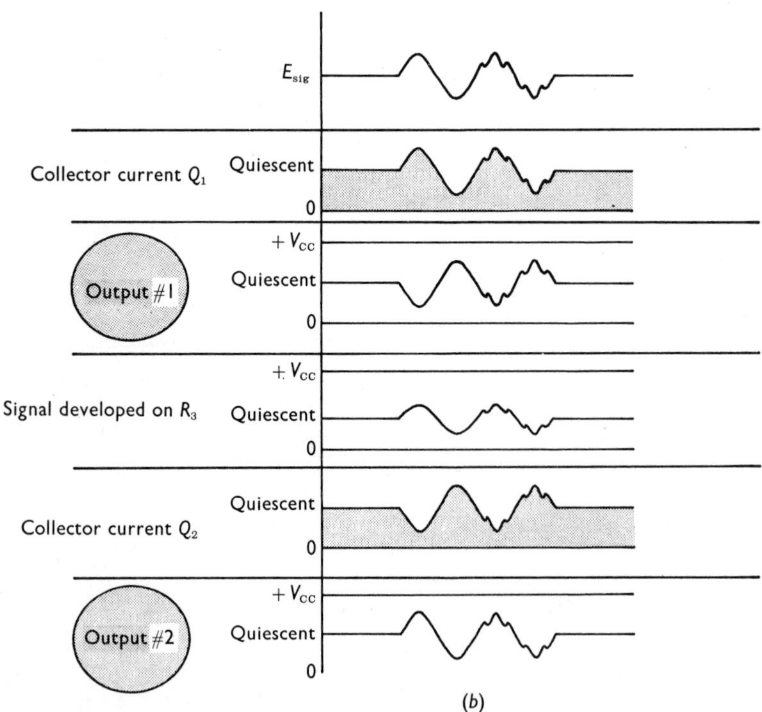

Fig. 8-7 Emitter-coupled phase splitter: (*a*) Schematic diagram; (*b*) Timing diagram (not to scale).

8-9 TWO-STAGE PHASE INVERTER

A two-transistor emitter-coupled phase splitter is shown in Fig. 8-7a. Instead of using emitter output directly to drive one of the push-pull transistors, the emitter signal is further amplified by a common-base-type amplifier circuit Q_2. (The base of Q_2 is grounded for signals by capacitor C_2 which has virtually no opposition at signal frequencies.)

Assume a positive-signal alternation arriving at the base of Q_1. Output #1 is negative-going because of phase inversion in the collector circuit. At the same time the emitter of Q_1 (and therefore Q_2 which is connected) is positive-going. Remember that the base of Q_2 is anchored at ground by C_2. The effect of the emitter of Q_2 going positive is the same as if the emitter were anchored and the base were going negative. A negative-going base-to-emitter signal on Q_2 causes its collector to rise, and a positive-going signal is developed at output #2. Circuit function is best illustrated by the timing diagram, Fig. 8-7b.

8-10 COMPLEMENTARY-SYMMETRY PHASE SPLITTER

An interesting form of phase splitter is often used to drive totem-pole OTL push-pull amplifier circuits. It is a form of complementary-symmetry design; a typical circuit is illustrated in Fig. 8-8a. The principle of operation of all such circuits is that the NPN transistor (Q_1) responds to the positive-signal alternation, whereas the negative alternation tends to cut it off. Conversely, the PNP transistor (Q_2) responds to the negative alternation but tends toward cutoff on the positive.

Note that Q_1 is connected as an emitter follower. It produces a positive-going output signal when responding to the positive alternation. On the other hand, its complement, transistor Q_2, is connected as a common-emitter amplifier, and the output is taken from the collector. Phase inversion occurs in this stage. The same positive-going signal alternation is therefore inverted to become negative-going at output #2. The phase splitter produces two signals of opposite polarity at the same instant. These facts are illustrated in the timing diagram, Fig. 8-8c.

8-11 VOLTAGE ACROSS PN JUNCTION—VARIATION WITH TEMPERATURE

The bias network for the complementary-symmetry phase splitter (Fig. 8-8b) is somewhat elegant in conception. It is electrically simple, provides bias voltage of correct polarity to each transistor, adjusts the bias voltage when temperature changes, and couples signal voltage to each transistor base. How is all this done?

To fully understand its operation it is necessary to be aware of an important characteristic of the solid-state diode (PN junction). Once barrier potential is overcome in a PN junction (about 0.5 V for germanium and 0.7 V for silicon), there is relatively little voltage variation across a forward-biased junction. The current may change considerably, but voltage variation is not appreciable as long as the temperature is constant. A second important characteristic of PN junctions is that when temperature rises, leakage current increases.

When a forward-biased diode is conducting, a certain voltage drop in the neighborhood of approximately 0.5 to 0.7 V is measurable across it. If the temperature rises, so also does leakage current. Electricity sees this as a drop in resistance and as a result the voltage drop across the diode is reduced. It falls approximately 2 mV for each degree Celsius rise in temperature.

Figure 8-8b shows typical voltage expected across diodes D_1, D_2, and D_3. This voltage decreases slightly when the temperature rises. It may appear to be a trivial decrease, but even a slight change is significant when dealing with small voltages, both bias and signal.

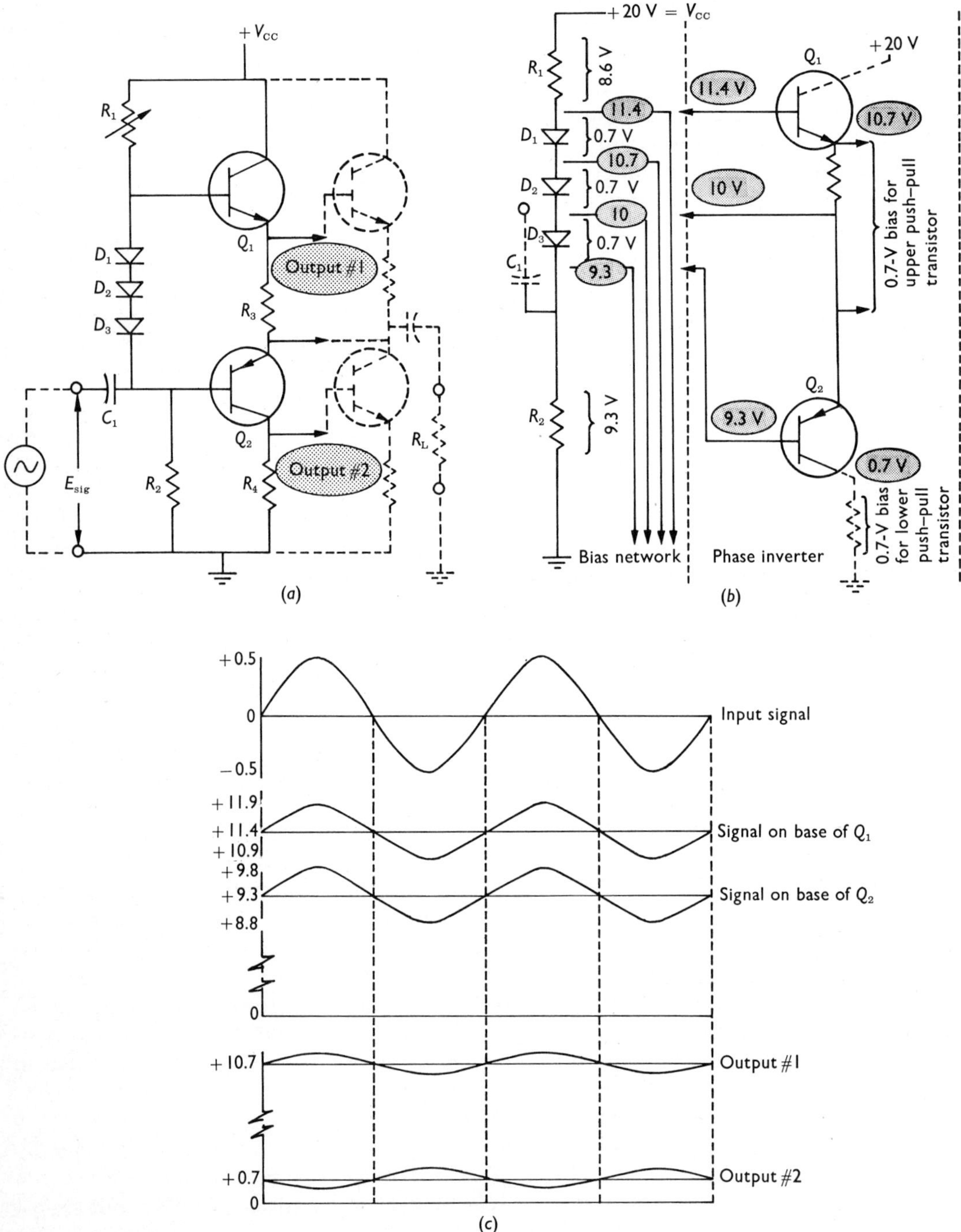

Fig. 8-8 Complementary-symmetry phase splitter: (*a*) Schematic diagram; (*b*) Simplified diagram, showing typical quiescent voltage distribution at various circuit points; (*c*) Timing diagram with sine-wave input signal.

8-12 DIODE-TYPE BIAS NETWORK

In a balanced OTL circuit the common or output connection has quiescent voltage midway between power-supply potential and ground. The common point (Fig. 8-8b) is therefore at +10 V with respect to ground in this example. The diodes will have approximately 0.7 V across them as shown because each of them is forward-biased and conducting. Resistor R_1 is adjusted until the voltage drops and values at the various circuit points shown in Fig. 8-8b appear. With these values, bias will be correct for each of the transistors Q_1 and Q_2. For transistor Q_2 the base is approximately 0.7 V more negative than its emitter. For transistor Q_1 the base is approximately 0.7 V more positive than its emitter.

Once the quiescent or operating point is established by the bias levels, try to visualize how a signal voltage superimposed on R_2 reacts on the circuit. Signal current arriving through capacitor C_1 will alternately add and subtract from the quiescent current flowing through R_2. The varying current in R_2 will cause its IR drop to fluctuate. At all times the sum of all voltage drops in the network R_1, D_1, D_2, D_3, and R_2 totals 20 V (Kirchhoff's law). The voltages across the diode remain fixed at approximately 0.7 V each regardless of other voltage fluctuations. The only possible way in which Kirchhoff's law can remain at all times satisfied is for the voltage drop across R_1 to fluctuate. It varies by an amount to compensate for the fluctuations on R_2. The base of transistor Q_1 therefore receives the same signal fluctuations as the base of Q_2. To further clarify these concepts, study the timing diagram in Fig. 8-8c.

8-13 TEMPERATURE COMPENSATION IN THE BIAS VOLTAGE

Before describing in detail how bias is temperature-adjusted automatically, examine first how it is done in the less-complicated circuit shown in Fig. 8-9.

Transistor Q is a power type, mounted on a heat sink. Diode D_1 is also mounted on or near the same heat sink so that it senses heat-sink temperature changes. Suppose that the ambient temperature rises. A temperature increase will cause a greater collector-leakage current which would appear as a shift in the transistor's quiescent or operating point. How-

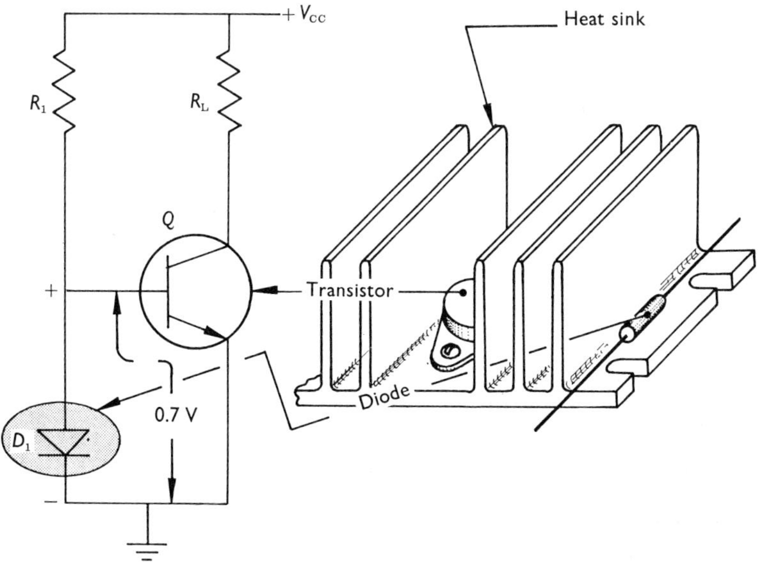

Fig. 8-9 Using a temperature-sensing diode to compensate for bias changes.

ever, diode D_1, whose voltage drop is supplying the base bias for the transistor, also senses the temperature increase. Its voltage drop decreases, the bias is reduced, and as a result quiescent collector current is reduced to near normal.

In the phase inverter in Fig. 8-8a, bias is temperature-stabilized in the same manner as in the circuit in Fig. 8-9. The diode string D_1, D_2, and D_3 can be mounted on the power-transistor heat sink. If the temperature were to rise, the diode-voltage drops would fall, developing less bias for the phase-splitter transistors Q_1 and Q_2. Outputs #1 and #2 would fall accordingly, and since these are connected directly to the bases of the power transistors, the quiescent current in these transistors would be reduced. The increase in collector-leakage current due to temperature rise is matched by a similar decrease in collector current due to the drop in bias voltage, and again quiescent current remains near normal.

8-14 PRACTICAL PHASE-SPLITTER CIRCUITS

In practical circuits the bias diode string is often replaced by a transistor as shown in Fig. 8-10. The circuit has the advantage that bias

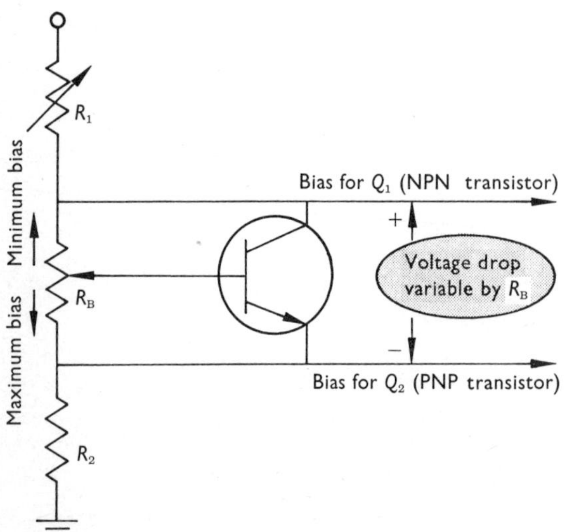

Fig. 8-10 Using a transistor to develop bias for a complementary-symmetry phase splitter.

voltage can be adjusted whereas with diodes it is fixed and depends on the characteristics of the diodes chosen for the task. In other respects, such as temperature compensation, circuit operation is similar to that of diodes.

The phase splitter shown in Fig. 8-8a has been simplified for purposes of explanation. In practice, the signal is not normally injected via a resistor R_2 as shown. The resistor is replaced by a transistor amplifier as shown in Fig. 8-11. Less signal power is required to drive the phase splitter because of the additional gain provided by transistor Q. With the higher signal level it is possible to add other compensating features, such as positive and negative feedback. A practical OTL circuit is discussed in Appendix F.

Summary

Moving-coil-type speakers require a signal source with low-voltage, high-current drive capability, i.e., a low-impedance source.

Emitter-follower circuits provide a low-impedance output and can be used to drive speakers (and other low-impedance energy-consuming loads) directly.

The last amplifier in a cascade chain is known as the output stage, usually a power-output stage.

In audio systems high-impedance output stages require an output transformer to match the stage to the speaker impedance.

Low-impedance output stages such as emitter followers require no output transformers to couple them to a speaker. Such stages are known as *"output transformerless"* or "OTL" circuits.

When amplifying devices are "stacked" in series to make push-pull OTL circuits the name "totem-pole" circuit is often used.

Totem-pole OTL circuits are often designed with an output-coupling capacitor. The capacitor prevents dc shorting of the output stage and does not require the stage to have a positive and negative power supply with respect to ground.

One disadvantage of OTL circuits using an output capacitor is that a "plop" is heard from the speaker when power is first turned on. This results from the initial capacitor-charging current which flows through the speaker voice coil.

Push-pull stages must be driven in antiphase. This can be done either with center-tapped secondary driver transformers or with a phase-splitter stage.

The constant voltage characteristics of a forward-biased PN junction are used to provide bias for push-pull OTL circuits.

In addition to providing the required bias, PN junctions are temperature-sensitive; they automatically adjust the bias to compensate for temperature change.

Questions and Exercises

1. Why is it necessary to keep to a minimum the mass of the moving coil assembly in a speaker?
2. How is it possible to obtain high magnetization from a speaker coil which has relatively few turns?
3. A speaker coil is an example of a low-impedance load. What is a feature of a low-impedance load?
4. Discuss two methods which may be used to connect a low-impedance load to a high-impedance source.
5. How much power is delivered to the speaker of a guitar amplifier system if it has an 8-Ω voice coil and 3 A of signal current flow through it?
6. Why is transistor power dissipation relatively high at low signal levels in a class A power amplifier?
7. Draw a schematic diagram of a complementary-symmetry emitter-follower-type output transformerless (OTL) output stage. Use direct coupling (no capacitor) to the speaker. With arrowheads indicate the signal-current flow in the speaker circuit for a positive input-signal excursion. (You must trace the complete current path and return to the point where your arrowheads begin.) Next, with arrowheads of a different color, show signal current for the negative input-signal excursion.
8. What is the purpose of capacitor C_1 in Fig. 8-2b?
9. What are some possible problems associated with a complementary-symmetry circuit?
10. Draw a schematic diagram of a transformer-driven totem-pole-type OTL circuit. With different color arrows indicate speaker current for both positive and negative input-signal excursions.
11. How is crossover distortion reduced in an OTL circuit?
12. Why must an output-coupling capacitor be used in the OTL circuit in Fig. 8-4a?
13. State two disadvantages of using an output-coupling capacitor.
14. What limits the charging current on the output-coupling capacitor when power is first turned on?
15. Why is a high initial charging current in the output-coupling capacitor undesirable?
16. ***a.*** What voltage value (what percentage of power-supply voltage) does the charge across the coupling capacitor assume?
    b. Does this voltage change appreciably on positive or negative signal excursions?
17. ***a.*** Suppose a 250-μF coupling capacitor were used in an OTL circuit where the combined charge or discharge resistance (including the speaker and transistors) is 10 Ω. How long would it take for the capacitor to charge or discharge to 63 percent of any sudden voltage change, such as a change in signal alternation? (*CR* time constant in seconds is $t = CR$.)
    b. Would the capacitor have sufficient time to change its charge during the positive or negative signal excursions of a signal with a frequency of 20 Hz?
18. Why must the driver stage supplying drive signal to an OTL stage also be a power amplifier?

19. What two basic functions must the driver stage perform?

20. What is meant by the term "balance" as applied to push-pull stages?

21. List some reasons why coupling transformers are avoided between the driver and the OTL stage.

22. What is the function of a phase inverter or phase splitter?

23. What component does a phase inverter replace?

24. What is the maximum theoretical voltage gain which the phase inverter in Fig. 8-6a can provide if both outputs are to be equal? (Hint: Refer to Sec. 6-12.)

25. Can you suggest one advantage the phase inverter in Fig. 8-7a has over the phase inverter in Fig. 8-6a?

26. The phase inverters as shown in Fig. 8-6a and Fig. 8-7a cannot be used in dc amplifiers, but the circuit shown in Fig. 8-8a can be. Why?

27. What functions do the diodes D_1, D_2, and D_3 in Fig. 8-8a perform?

28. If the temperature increases, will the voltage drop across diodes D_1, D_2, and D_3 (Fig. 8-8a) increase, decrease, or remain the same?

29. **a.** What are two important characteristics of a PN junction diode when it is forward-biased?

b. Draw a freehand sketch of a *current versus voltage* graph to illustrate these characteristics.

30. Why are biasing diodes often mounted on the transistor's heat sink?

31. What advantage does the bias circuit shown in Fig. 8-10 possess over the diode network in Fig. 8-8a?

32. Why do most practical phase-inverter circuits use an amplifier similar to that shown in Fig. 8-11?

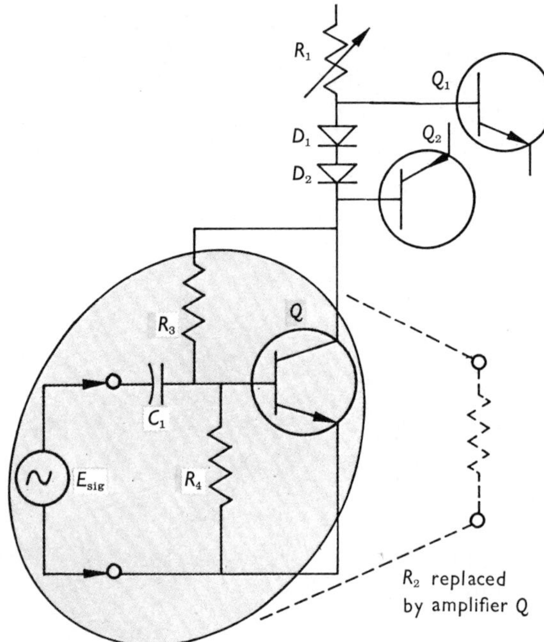

Fig. 8-11 Input to complementary-symmetry phase splitter.

9-1 REQUIREMENTS FOR OSCILLATION

An oscillator is essentially an amplifier circuit with positive feedback. It produces an output signal but requires no external input signal to do so.

Any amplifier can be converted to an oscillator if two conditions are met. First, the amplifier's output must be connected to its input so that positive feedback occurs. Second the amplitude of the feedback signal must be large enough to generate a self-sustaining signal. When these conditions are met, the amplifier feeds on its own output, regenerating the signal passing through it continuously, and oscillations are sustained.

9-2 TYPES AND APPLICATIONS

The output-signal waveform from an oscillator may be à single frequency (i.e., a pure sine wave) or it may be rich in harmonics and appear as a rectangular-, trapezoidal-, or other-shaped waveform. Some oscillators are designed to produce a burst of waves and then rest for a predetermined period. Others produce just a single cycle and then only when fired by a triggering signal or pulse.

The majority of oscillators are used to generate a low-power signal with certain characteristics. In many instances the power level of the signal may be raised by subsequent power amplifiers, as in radio transmitters. There is also a large class of applications in which oscillators operate at high power levels producing a high power-output signal. In this class are oscillators which are used for high-frequency heating in industry, medicine, and in power converters.

Traditionally, oscillators have been classified into (1) those which produce an uninterrupted string of sine waves, i.e., *continuous-wave* (cw) oscillators, and (2) those which produce nonsine waves or a fractional sine wave, i.e., *relaxation* oscillators.

9-3 TIME- AND FREQUENCY-DOMAIN REPRESENTATION

Electrical signals, regardless of origin, are rarely pure sine waves. Some approach a sine wave in purity, but most are complex and are therefore composed of many sine waves of differing amplitudes and frequencies. It is a well-established fact that all waveforms, regardless of complexity, are the resultant of many pure sine waves. It can be shown graphically and by a method known as Fourier analysis that the shape of a wave is determined by the number and the relative amplitude of the sine waves composing the complex wave.

Fourier analysis requires mathematical expertise beyond the level of this book. Fortunately, graphical summation, though not as powerful, serves well to show that a complex waveform is made of many parts. Figure 9-1a shows a timing diagram of three waveforms. If their amplitudes are added algebraically point by point in time, the composite waveform shown in Fig. 9-1b results. Figure 9-1b is a representation of a signal in the *time domain*. In time-domain representation, time is plotted along the horizontal or *x* axis and amplitude along the vertical or *y* axis. It is the time-domain representation that is displayed on most oscilloscopes. Also, all timing diagrams are time-domain displays. Although a time-domain display is interesting, it gives only the most general information as to the frequency and amplitude (spectral) content of a signal.

A *frequency-domain* display has frequency plotted along the horizontal axis and amplitude along the vertical. This display therefore shows a spectrum of frequencies and their relative amplitudes. Figure 9-1c shows a frequency-domain display of the signal in Fig. 9-1b. Each point on the horizontal axis represents a frequency found in the original composite waveform. Instruments which display a signal in the frequency domain are known as *spectrum analyzers*.

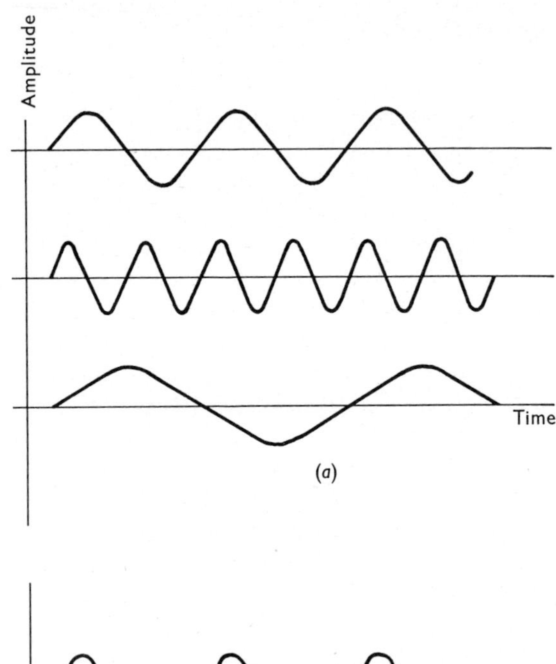

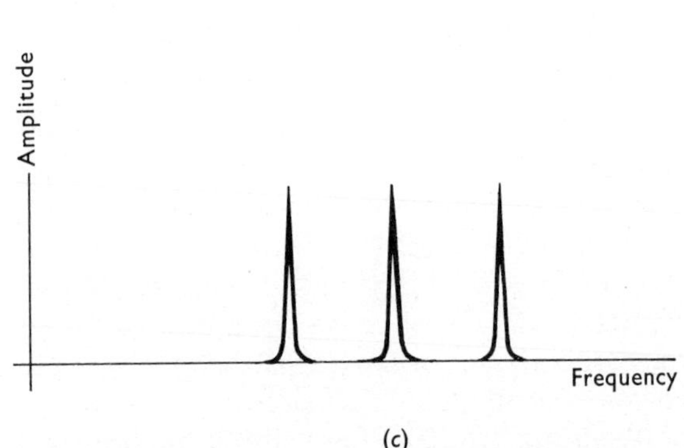

Fig. 9-1 Types of signal display: (*a*) Three composite signals summed together; (*b*) Time-domain display of composite waveform. Amplitude vs. time, as displayed on oscilloscopes equipped with a time base; (*c*) Frequency-domain display. Amplitude vs. frequency, as displayed on spectrum analyzers.

9-4 CLASS OF OPERATION AND SPECTRAL PURITY

As in amplifiers, the transistors, tubes, or FETs used in oscillators may operate at various bias levels. In general, oscillators operating class A can produce a signal of greatest spectral purity. It is difficult to operate an oscillator stage in class A. To do so requires that the feedback signal have sufficiently small amplitude to operate the amplifying device on the linear part of its characteristic curve. Against this is balanced the requirement that the signal have sufficiently large amplitude to sustain oscillations.

In a large variety of applications the class of operation is not critical because the distortion produced is removed by a filter. The oscillator is then operated class C, i.e., enough feedback signal is available to drive the amplifying device between cutoff and saturation.

An overabundance of feedback is preferable to too little feedback. With ample feedback signal, the oscillator will continue to operate even though external factors such as power-supply voltage and temperature are varied.

9-5 ACHIEVING POSITIVE FEEDBACK

A first requirement for oscillation is a positive feedback path. In a single-stage common-emitter amplifier the output or collector voltage is 180° out of phase with the input-signal or base voltage. If the output is to be fed to the input to form an oscillator, the signal must be phase-inverted. Then it will arrive at the base such that reinforcement or regeneration rather than cancellation or degeneration occurs. One means of obtaining the 180° phase shift is to connect a series of capacitors and resistors known as a *lag line* as shown in Fig. 9-2a. Each of the *CR* circuits phase-shifts the signal approximately 60°, as shown in Fig. 9-2b. A three-stage *CR* lag line produces the required 180° as shown in the accompanying phasor diagrams, Fig. 9-2c.

9-6 SINGLE-STAGE *CR* LAG-LINE OSCILLATOR

Note that only a single frequency will receive exactly 180° phase shift after passing through the lag line. Other frequencies arrive at the end of the lag line with other than 180° phase shift and do not produce as much regeneration. For this reason a lag-line oscillator produces essentially a single-frequency signal, i.e., a sine wave.

Figure 9-3 shows schematic diagrams of lag-line oscillators using a single transistor and FET amplifier. A comparable vacuum-tube circuit is shown in Fig. G-1 (Appendix G). When only a single stage is used, the amplifier must have a voltage gain of 8 or greater. The phasor diagrams in Fig. 9-2c illustrate that the signal frequency is attenuated by a minimum factor of 8 after passing through the lag line. In practice the attenuation would likely be greater because of losses in the components. Practical upper-frequency limits for lag-line oscillators are in the neighborhood of 1 to 2 MHz. At higher frequencies the losses in the lag line become too great to sustain reliable oscillation.

Although the lag line was explained on the basis of three identical *CR* sections, it is important to realize that the capacitors and resistors may all have differing values. Under such conditions the net phase shift is still 180°, although one *CR* section may shift more or less than 60°. If for example the first section shifts 55°, the other two must between them shift 125° to give a net shift of 180°. Regardless of the values chosen, a three-section lag line will invert one and only one frequency exactly 180°. In practice one or more of the resistors may be variable, thereby allowing the user to adjust the oscillator for a variety of output frequencies.

9-7 OUTPUT FREQUENCY

Frequency of oscillation depends primarily on the values of capacitors and resistors used in the lag line. In practice it is very difficult to

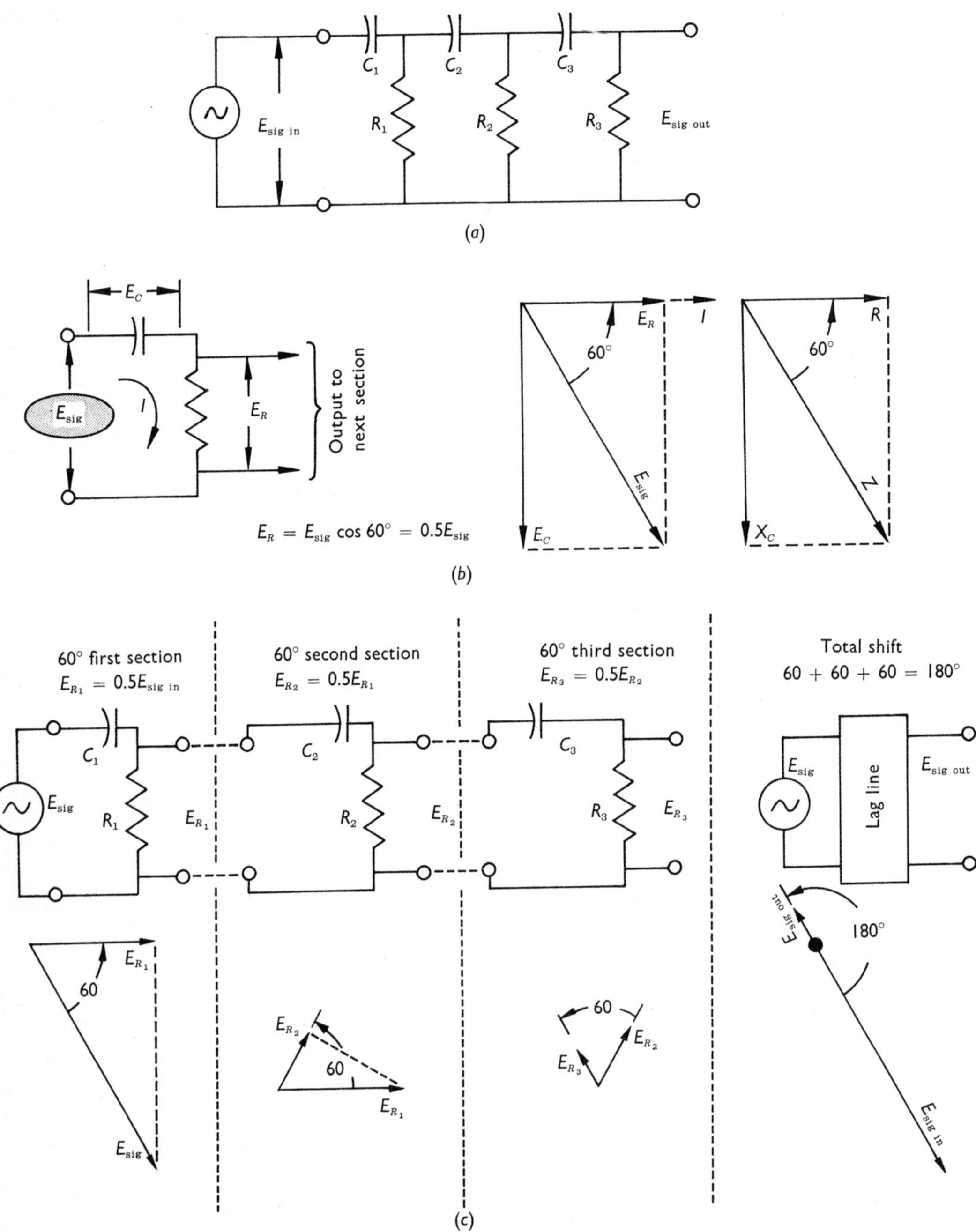

Fig. 9-2 Operation of a *CR* lag line: (*a*) Schematic diagram of a three-section *CR* line; (*b*) At signal frequency, values of *C* and *R* are chosen so that the phase angle between E_R and E_{sig} is 60°. Then $E_R = E_{sig} \cos 60^\circ = 0.5E_{sig}$ (signal voltage drops by half for each section); (c) input and output voltage phasors for each section and for all sections combined, giving a net phase shift of 180°.

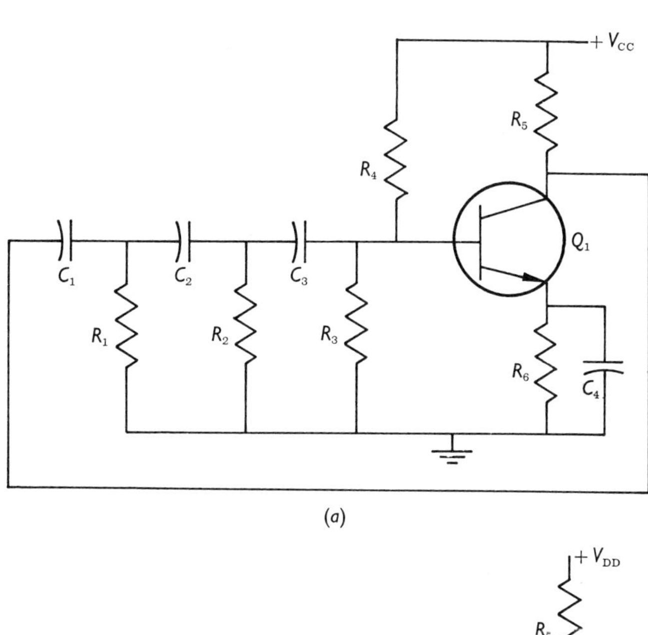

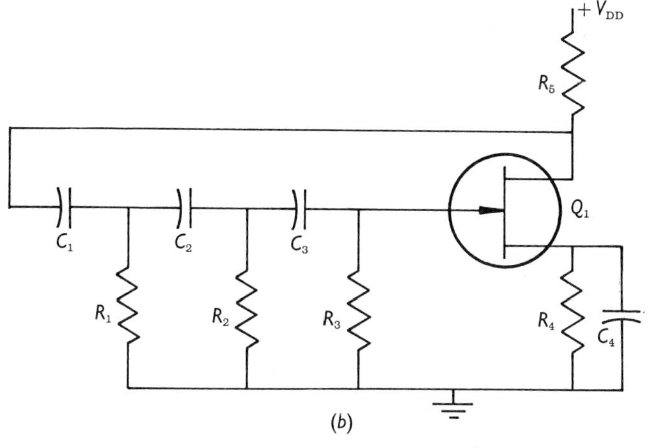

Fig. 9-3 *CR* lag-line oscillator circuits: (*a*) Transistor type; (*b*) FET type.

calculate because of variations in amplifier output resistance and input loading effects. These have the same effect as additional resistors and capacitors connected to the lag line. A very rough approximation of the output frequency can be estimated by assuming three identical sections and ignoring any loading effects. Assume that each network shifts exactly 60° and, under these conditions, reference to the phasor diagram in Fig. 9-2b shows that

$$X_C = 1.73R \left(\text{because } \frac{X}{R} = \tan 60° = 1.73\right)$$

The equation for capacitive reactance is

$$X_C = \frac{1}{2\pi fC}$$

Substituting for X_C gives

$$\frac{1}{2\pi fC} = 1.73R$$

from which

$$f = \frac{1}{2\pi(1.73)CR}$$

or finally

$$f = \frac{0.092}{CR}$$

The following example illustrates how the equation may be applied to estimate the approximate frequency of oscillation of a lag-line oscillator. (More accurate equations are available, but in practice it is usually preferable to construct a circuit and trim it with variable resistors.)

EXAMPLE 9-1 What is the approximate frequency of oscillation of a *CR* lag-line oscillator constructed from three 0.01-μF capacitors and three 10-kΩ resistors? Applying the equation and substituting numerical values $C = 0.01\ \mu$F and $R = 10$ kΩ gives

$$f = \frac{0.092}{(0.01 \times 10^{-6}) \times (10 \times 10^{3})}$$

$$= \frac{0.092}{1 \times 10^{-4}}$$

$$= 0.092 \times 10^{4}$$

ANSWER $= 920$ Hz

9-8 MINIMIZING LOADING EFFECTS

In the frequency equation in Example 9-1, note that when *C* and *R* are small numbers, the frequency is large. This means that to obtain high-frequency oscillations, low values of resistance are required. As a result the lag line loads the output of the amplifying device. Excessive loading drops the voltage gain to the point where oscillations will not be sustained. One means of minimizing the loading effect is to insert an emitter follower between the output of the oscillator and the input of the lag line. Figure 9-4 illustrates a typical circuit. Although it has no voltage gain, the emitter follower requires very little drive and is able to supply considerable driving power to the lag line.

9-9 PHASE INVERSION IN CASCADED AMPLIFIER

A basic requirement for oscillation is positive feedback. The output being fed back must be in phase with the input. As was expressed in the previous section, a single-stage amplifier inverts the signal, and a lag line may be used to further invert so that a total inversion of 360° occurs. In this way the input and output signals are in phase.

It is possible to use two cascaded amplifier stages as shown in Fig. 9-5 to obtain the required phase shift for all frequencies. Some form of frequency filter circuit must be included to ensure that a specific frequency is generated.

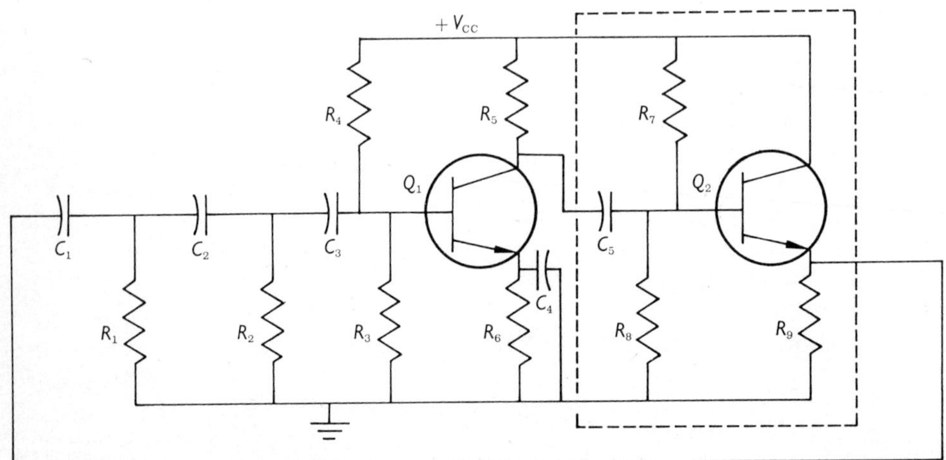

Fig. 9-4 *CR* lag-line oscillator with an emitter follower.

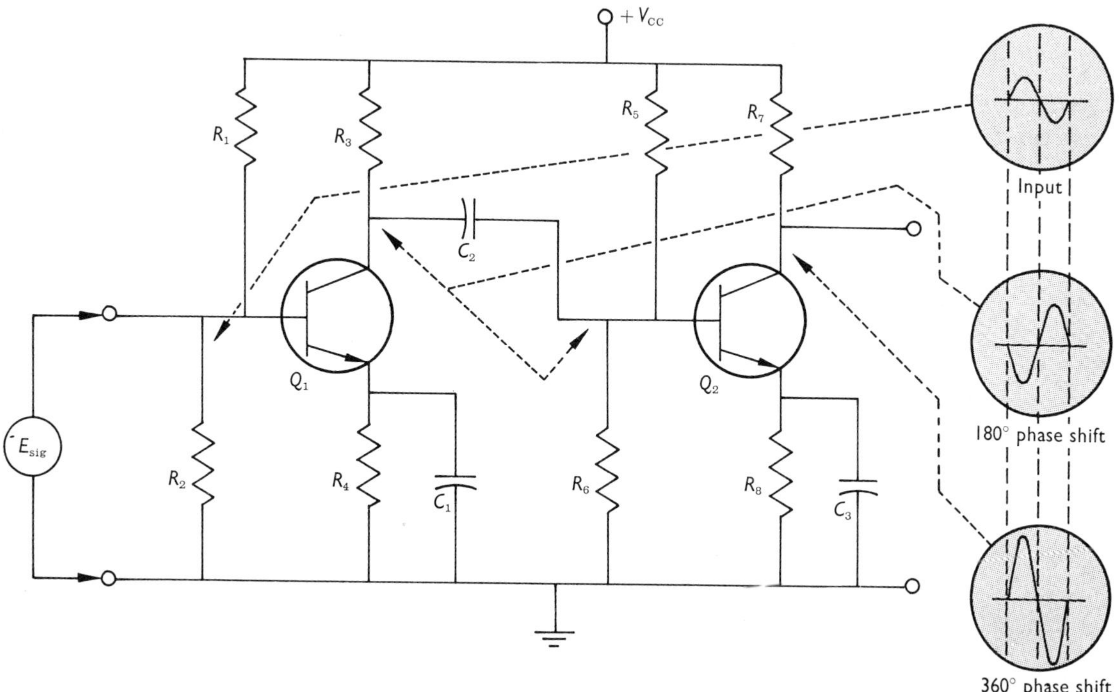

Fig. 9-5 Voltage phase relationships at various points in a two-stage voltage amplifier.

9-10 THE WIEN BRIDGE

The circuit shown in Fig. 9-6a is known as a *Wien bridge*. There is one frequency and only one which will experience no phase shift as the signal passes from input to output. If this zero-phase-shift frequency is arbitrarily called the *center frequency,* then all frequencies except the center frequency will experience some phase shift. The Wien bridge can be used as a frequency filter to exclude all but the center frequency from returning to the input of the oscillator.

How does the Wien bridge accomplish zero phase shift for the center frequency but not for others? A study of the phasor diagrams will help you understand the process. It is important to master this concept, because many electronic circuits use various forms of *CR* filters to achieve similar results. Using *CR* filters and amplifiers with feedback, it is possible, but not necessarily practical, to simulate and surpass in performance combinations of reactive elements, both capactive and inductive.

9-11 WIEN-BRIDGE PHASOR DIAGRAM

The phasor diagram for the Wien bridge is constructed by employing fundamental knowledge of ac theory. First construct the current phasor diagram for the parallel *CR* circuit consisting of C_2 and R_2. Since this is a parallel section and there is identical voltage (E_o) across C_2 and R_2, the currents I_{C_2} and I_{R_2} in the component parts will be 90° out of phase, as shown in Fig. 9-6b. The sum of these currents is I_{in}, the source current arriving through R_1 and C_1. Its phasor is shown in Fig. 9-6c.

The phase difference between I_{in} and E_o (the output voltage) is 45° for the center frequency (Fig. 9-6c, d). This relationship occurs because the voltage across R_2 (E_o) and the current through R_2 are in phase but I_{R_2} and I_{in} are 45° out of phase.

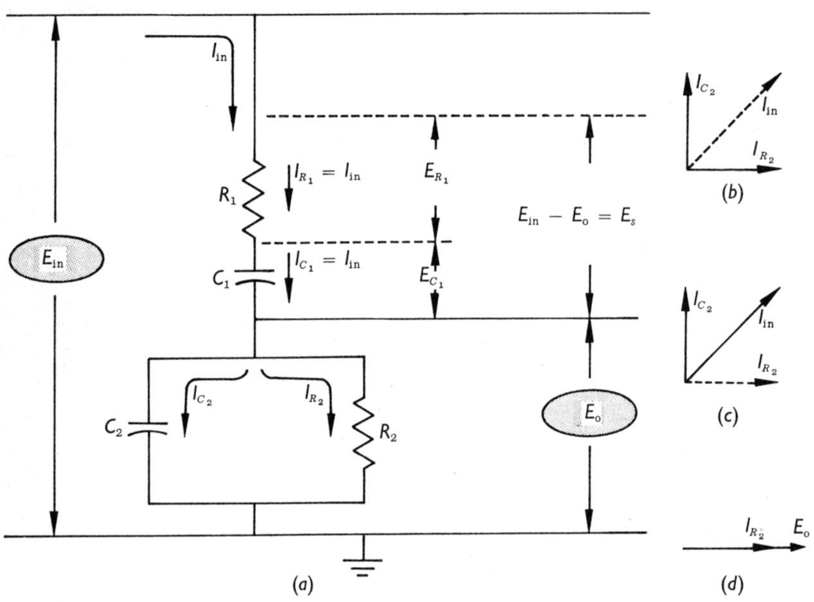

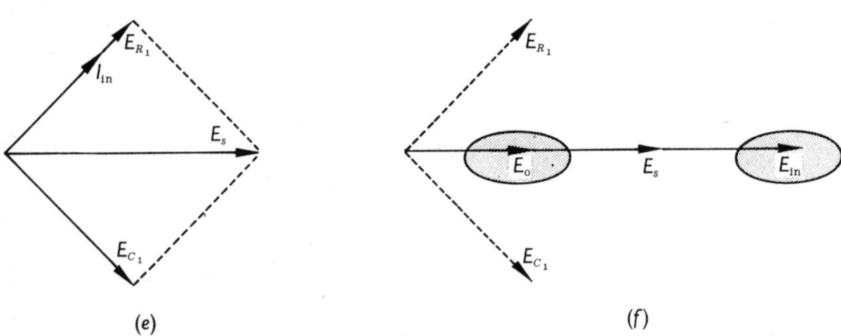

Fig. 9-6 Wien-bridge circuit and its phasors (not to scale): (*a*) Wien-bridge circuit; (*b*) Current phasors in the parallel section composed of R_2 and C_2; (*c*) Resultant current phasor through parallel section, or bridge current I_{in}; (*d*) Output voltage phasor—in phase with the current through R_2; (*e*) Voltage phasors for series section of the bridge composed of R_1 and C_1; (*f*) Wien-bridge voltage phasor diagram.

Next analyzing the series section, note that the voltage across R_1 (E_{R_1}) is in phase with I_{in}, but the voltage across C_1 (E_{C_1}) lags I_{in} by 90°. These facts are shown in the phasor diagram in Fig. 9-6e. The diagram also shows the resultant sum voltage ($E_{in} - E_o = E_s$) across the series section. By Kirchhoff's law the applied voltage E_{in} is the phasor sum of the voltage across the series section E_s and the parallel section of the bridge E_o. A voltage phasor diagram for the center frequency is shown in Fig. 9-6f. For any other frequency, voltage phasors E_{R_1} and E_{C_1} will be unequal, as will be current phasors I_{C_2} and I_{R_2}. Then E_o will no longer be precisely in phase with E_{in}.

An alternate way of stating off-frequency operation is to say that the center frequency is that frequency which receives the least attenuation by the bridge. Frequencies above the center frequency are bypassed to ground by the reduced reactance of C_2. Those below the center frequency are impeded from reaching the output connection by the increased reactance of C_1.

9-12 CALCULATING THE CENTER FREQUENCY OF A WIEN BRIDGE

At the center frequency, $X_{C_1} = R_1$ and $X_{C_2} = R_2$. For simplicity assume that the capacitors and resistors are of equal value. Then, ignoring loading effects, center-frequency capacitive reactance equals resistance, or

$$X_C = R$$

Substituting the expression

$$X_C = \frac{1}{2\pi fC}$$

gives

$$\frac{1}{2\pi fC} = R$$

Solving for the frequency term gives

$$f = \frac{1}{2\pi CR}$$

or

$$f = \frac{0.159}{CR}$$

The next example shows how this equation can be applied.

EXAMPLE 9-2 A Wien-bridge oscillator is constructed with two 20-kΩ resistors and two 0.05-μF capacitors in the bridge section. What will be its output frequency? Applying the equation $f = 0.159/CR$ and substituting numerical values gives

$$f = \frac{0.159}{CR}$$

$$= \frac{0.159}{0.05 \times 10^{-6} \times 20 \times 10^3}$$

$$= \frac{0.159}{5 \times 10^{-2} \times 10^{-6} \times 20 \times 10^3}$$

$$= \frac{0.159}{100 \times 10^{-2} \times 10^{-6} \times 10^3}$$

$$= \frac{0.159}{10^{-3}}$$

$$= 0.159 \times 10^3$$

ANSWER $= 159$ Hz

9-13 PRACTICAL WIEN-BRIDGE OSCILLATOR

In Fig. 9-7 the Wien bridge is shown connected to a two-stage amplifier to form a Wien-bridge oscillator. A manual gain control R_6 is provided to reduce the gain of the amplifier. Ideally the amplifier should provide just enough voltage gain to overcome the attenuation in the Wien bridge at the center frequency. Excessive gain will result in other than center-frequency signals receiving enough boost to regenerate, with resultant spectral impurity in the generated signal.

Practical circuits used in electronic equipment have some form of automatic gain control. One method is to include an incandescent lamp (I_1) in the emitter lead of transistor Q_1, as shown in Fig. 9-7. The lamp behaves as an unbypassed emitter resistor and therefore reduces the gain of Q_1. If, for any reason, the signal amplitude increases, the average emitter current of Q_1 also increases. The increased emitter current will raise the temperature of the filament of the lamp, whose resistance then increases. The increased resistance of the lamp causes more degeneration and the gain of Q_1 falls, thereby reducing signal amplitude.

9-14 AMPLITUDE CONTROL WITH NEGATIVE-FEEDBACK LOOP

The oscillator circuit shown in Fig. 9-8 employs a transistor (Q_3) to produce additional degeneration. There is at all times a certain level of degeneration produced by the unbypassed emitter resistor R_{10}. If for instance the base of Q_1 receives a momentarily positive-going signal, the increasing emitter current results in a degenerative voltage rise across R_{10}.

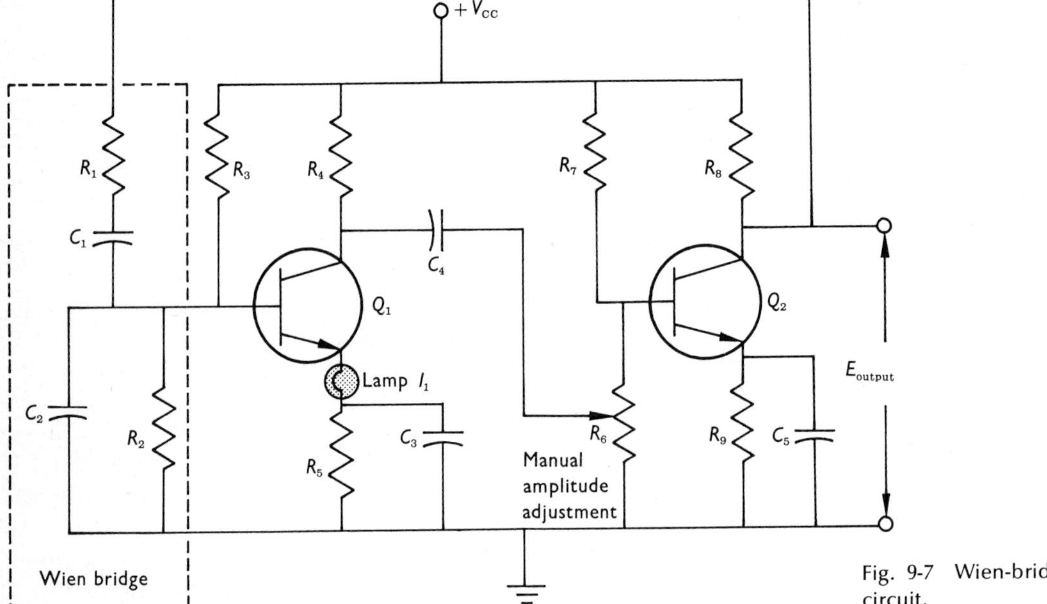

Fig. 9-7 Wien-bridge oscillator circuit.

After two inversions at the collectors of Q_1 and Q_2 the signal is passed to the output buffer emitter follower Q_4. The output signal across R_{17} is then in phase with the input signal entering the base of Q_1. Note that a fraction of the output signal is applied through R_{11} to the base of Q_3. A rising base voltage on Q_3 results in greater current flow in R_{10}, which produces even more degenerative voltage across R_{10}. Any tendency for the output to rise is opposed by the reduction in gain as a result of the increased degenerative voltage across R_{10}.

An advantage of this type of control over circuits using incandescent lamps is that response is nearly instantaneous. Incandescent lamps change resistance only after a temperature change has occurred and tend to react more slowly.

9-15 *CR*-TYPE NOTCH FILTER

The circuit shown in Fig. 9-9a is known as a *CR*-type *notch filter.* It has the characteristic that one and only one frequency receives maximum attenuation on its way through the filter. Operation of the filter can best be appreciated by considering how the various components behave at *low, high,* and some frequency which for purposes of explanation will be labeled the *notch* frequency.

First, assume that a signal of frequency lower than the notch frequency is applied to the input. It cannot pass through either of the capacitors, because at low frequencies both capacitors have high capacitive reactance. To a low-frequency signal the notch-filter circuit appears as in Fig. 9-9b, i.e., devoid of the capacitors.

Next assume that a signal of frequency greater than the notch frequency is applied. Capacitive reactance is now negligible, and the signal passes easily through the series capacitor C_1. Capacitor C_2 also has low reactance, but the resistors R_1 and R_2 prevent appreciable signal from taking this path. To a high-frequency signal the notch filter appears as in Fig. 9-9c.

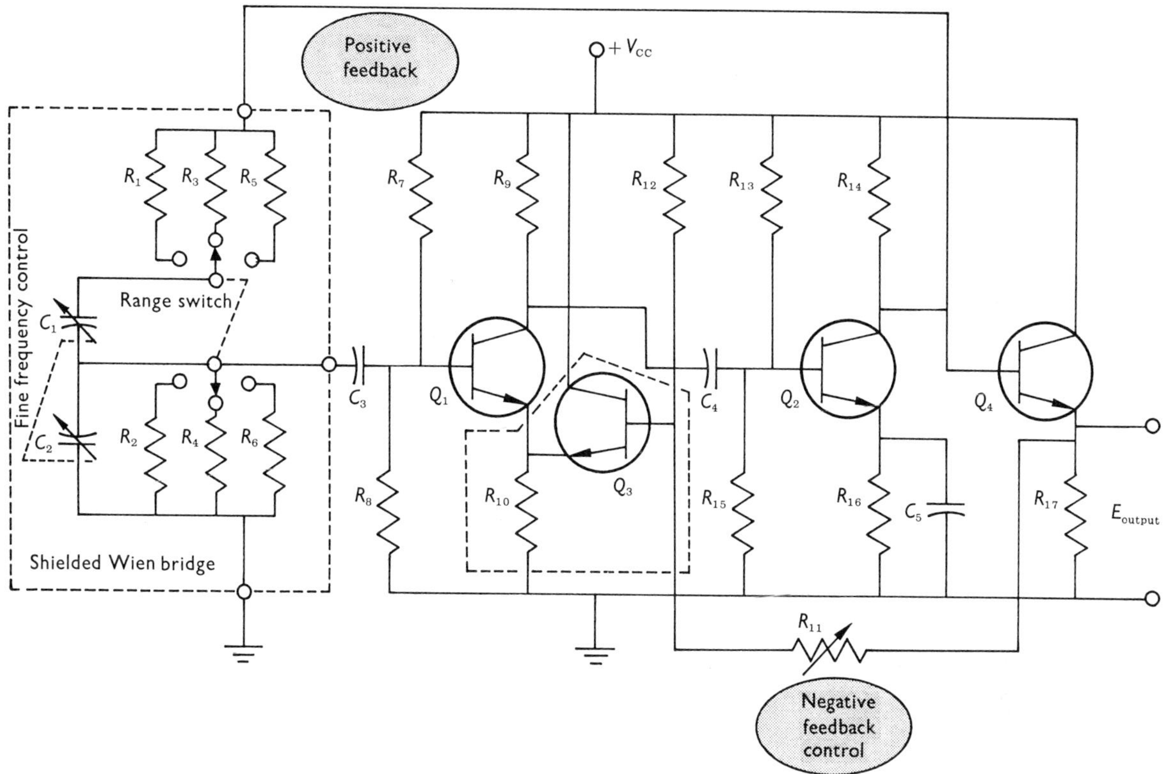

Fig. 9-8 Wien-bridge oscillator with three frequency ranges and automatic amplitude control.

Finally, at notch frequency, the filter will provide maximum attenuation. As frequency is increased from zero (disregarding the effects of capacitor C_1 for the moment) the bypass or shunt action of C_2 becomes increasingly more effective. It behaves like a jumper; a response curve (without C_1) is shown in Fig. 9-9d. Plotting a response curve with the shunt capacitor C_2 removed would give a curve similar to that in Fig. 9-9e. With both capacitors connected, the frequency-response curve shown in Fig. 9-9f is obtained. There is increasing attenuation as frequency is increased up to a certain value, after which output amplitude begins to rise again. A notch occurs in the response curve, hence the name notch filter. The notch frequency is given by the equation

$$f = \frac{1}{2\pi\sqrt{C_1 C_2 R_1 R_2}}$$

If $C_1 = C_2$ and $R_1 = R_2$, then the notch-frequency equation reduces to

$$f = \frac{1}{2\pi CR}$$

Upon a more critical examination you will note that if f and R are interchanged to give

$$R = \frac{1}{2\pi Cf}$$

the statement is equivalent to $R = X_C$. As in the Wien bridge we have again a CR network in

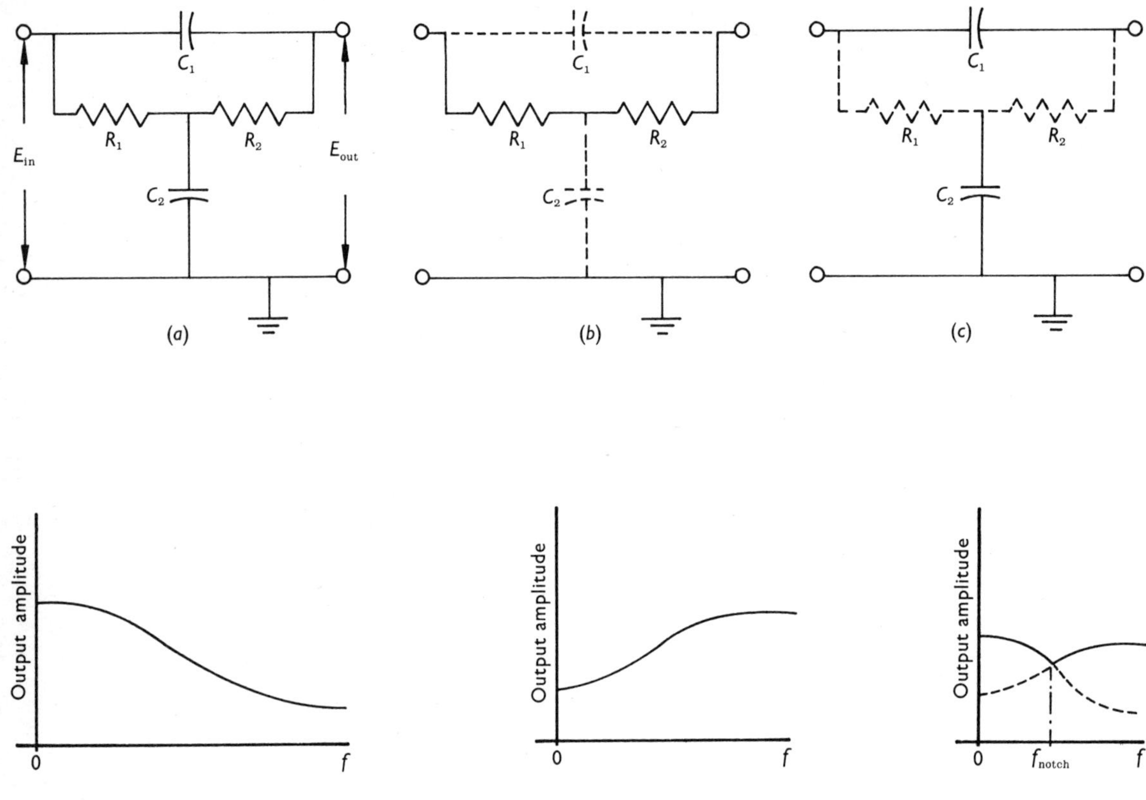

Fig. 9-9 Characteristics of a *CR*-type notch filter: (*a*) Schematic diagram; (*b*) Low-frequency equivalent; (*c*) High-frequency equivalent; (*d*) Response curve with C_1 removed; (*e*) Response curve with C_2 removed; (*f*) Composite response curve, showing notch frequency.

which capacitive reactance and resistance balance at center frequency.

9-16 *CR* OSCILLATOR WITH NOTCH FILTER

Figure 9-10a shows a two-stage amplifier connected as an oscillator. It will attempt to generate all frequencies, as it contains no frequency-discriminating network in its positive-feedback path. In practice such a circuit produces rectangular waves and is more commonly known as a free-running multivibrator. If the circuit is modified to include negative feedback via a notch filter as shown in Fig. 9-10b, only a single frequency signal is generated. Degenerative action because of negative feedback reduces amplifier gain for all frequencies except notch frequency. The notch filter rejects the notch frequency and as a result comparatively little degeneration takes place, allowing the circuit to oscillate at the notch frequency. A discussion of a notch-filter-type oscillator used for test purposes is found in Appendix G.

Summary

An oscillator generates a signal. It provides an output signal without the necessity of an input signal.

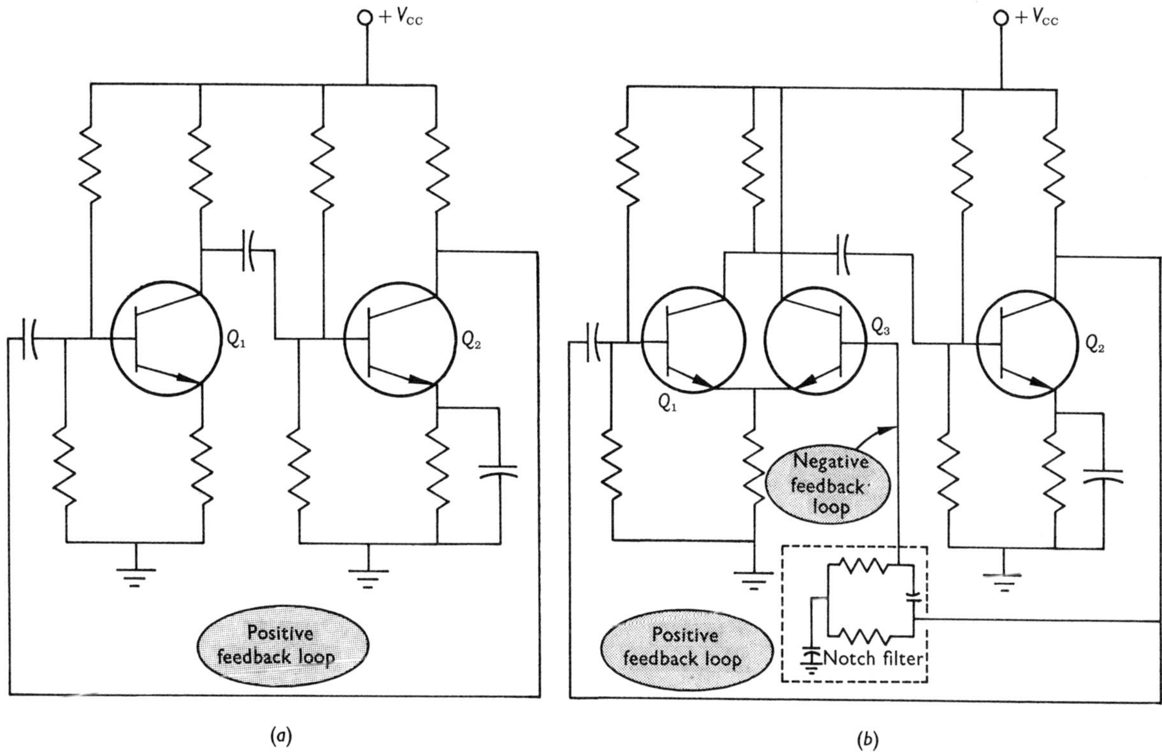

Fig. 9-10 Application of a notch filter in a *CR* oscillator: (*a*) Two-stage amplifier with positive feedback oscillates at "all" frequencies; (*b*) Two-stage amplifier with negative feedback introduced via a notch filter. Negative feedback kills all oscillations except that at the notch frequency.

An oscillator is essentially an amplifier circuit with positive feedback. A signal is generated because the amplifier uses its own output as its input signal. As long as the circuit has sufficient gain, oscillations are sustained.

Oscillators may be grouped into two types, (1) continuous wave and (2) relaxation.

Signals may be graphically displayed either in the time domain, which shows the waveshape, or in the frequency domain, which shows the spectral content.

A basic requirement for oscillation is that the feedback signal be in the correct phase to ensure regeneration.

A three or more section *CR* lag line is a phase-shifting network which may be used to couple the output to the input of an amplifier to produce an oscillator. Only one frequency receives exactly 180° phase shift. Therefore a *CR* lag-line oscillator is capable of producing nearly a pure sine wave.

A Wien bridge is a frequency-sensitive *CR* network, but at its center frequency it produces no phase shift. A Wien-bridge oscillator therefore requires two or more stages of amplification to obtain the required phase shift and gain. The *Wien bridge* is used as a frequency-selector circuit in the *positive-feedback loop.*

Oscillator amplitude control can be achieved automatically by using negative feedback.

A *CR*-type *notch filter* attenuates the center frequency. For this reason it can be used as a frequency-rejector circuit in the *negative-feedback loop,* to allow oscillation at the center frequency. All other frequencies are "killed" by negative feedback.

Questions and Exercises

1. What is an oscillator circuit?
2. How can an amplifier circuit be altered to make an oscillator?
3. What criteria are required for sustained oscillations?
4. Define the following and suggest one application of (*a*) a continuous-wave oscillator and (*b*) a relaxation oscillator.
5. Distinguish in your own words between time-domain and frequency-domain graphical representations of signals.
6. What electronic instruments are used for (*a*) time-domain display of signals and (*b*) frequency-domain display of signals?
7. How can spectral purity be obtained from oscillator circuits?
8. Explain why spectral purity may decrease when oscillation amplitude increases in a given circuit.
9. Why is it preferable to provide too much rather than too little feedback in an oscillator circuit?
10. Draw a schematic diagram of a single-stage *CR*-coupled transistor amplifier. Will the circuit oscillate if a capacitor is connected between collector and base (output to input)? What important consideration, if any, is being overlooked?
11. What is the purpose of a *CR* lag line?
12. Explain why any given lag line provides 180° of phase inversion for one and only one unique frequency.
13. Disregarding its application in oscillator circuits would you classify a *CR* lag line as (*a*) a low-pass filter, (*b*) a band-pass filter, or (*c*) a high-pass filter?
14. *a.* What is a typical upper-frequency limit for *CR* lag-line oscillators?
    b. What imposes the limit?
15. Why would you avoid using a *CR* lag-line oscillator to generate a high power signal?
16. Calculate the approximate frequency of oscillation of an oscillator using a *CR* lag line constructed of three 1000-pF capacitors and three 20-kΩ resistors.
17. A lag line may shunt the output of an amplifier to such an extent that the loading becomes unacceptable. What can be done to provide the necessary signal power to the lag line without upsetting the phase relationships?
18. A *CR* lag line selects one frequency by shifting it exactly 180°. How does a Wien bridge perform frequency selection?
19. Without reference to the phasor diagram give a simple explanation of how a Wien bridge "selects" a frequency.
20. Refer to Fig. 9-7 and state what would occur under each of the following conditions: (*a*) C_1 is made larger, (*b*) C_2 is made smaller, (*c*) R_1 is replaced by a resistor of lower value, and (*d*) R_2 is replaced by a resistor of higher value.
21. What is the operating frequency of a Wien-bridge oscillator if the Wien bridge is constructed of two 0.01-μF capacitors and two 30-kΩ resistors?
22. What is the purpose of the incandescent lamp used in the circuit of Fig. 9-7?
23. Describe the operation of two types of automatic gain-control circuits which can be used in *CR*-type oscillators.
24. What is the superior feature of the automatic gain-control circuit in Fig. 9-8 as compared with the one in Fig. 9-7?
25. How does a *CR*-type notch filter perform frequency selection?
26. Both the Wien bridge and notch filter are for practical purposes three terminal devices (input, output, and common or ground). Why is it not possible to replace a Wien bridge in an oscillator circuit with a notch filter, leaving other parts of the circuit unaltered?

10-1 THE *LC* OSCILLATORY CIRCUIT

Although *CR* circuits are popular frequency-determining networks, their use is impractical at frequencies higher than 1 or 2 MHz. The resistors dissipate so much signal energy that network efficiency drops excessively.

An alternate frequency-determining circuit is the *LC* oscillatory circuit shown in Fig. 10-1a. It is sometimes known as a *tank* circuit. In oscillator circuits it may be regarded either as (1) a circuit which circulates a stored charge or (2) an electrical filter. Some circuits are more easily described in terms of the circulating-charge concept; other circuits are described in terms of filter action.

The circulating-charge concept is illustrated in Fig. 10-1b. It is assumed that somehow a charged capacitor is connected across a coil as shown in Fig. 10-1b(i). Capacitor discharge current drains off into the coil, creating a magnetic field around it. Ultimately the magnetic field rises to a maximum value, Fig. 10-1b(iii). Because there is no steady current to maintain it at this value, it begins to collapse. The collapsing field represents change in magnetic flux, which has the effect of inducing a current in the coil opposite in direction to the current which created the flux field initially, Fig. 10-1b(iv). The current so produced recharges the capacitor to reverse polarity, Fig. 10-1b(v). Again the capacitor discharges to create a magnetic flux about the coil, Fig. 10-1b(vii). Again the flux collapses to produce charging current for the capacitor. It becomes charged to its original polarity. In this way the initial charge oscillates back and forth until all its energy is dissipated. Capacitor voltage appears as an ac waveform of decreasing amplitude as shown in Fig. 10-1c. The frequency of this alternating voltage is that of the natural resonant frequency of the *LC* circuit and is numerically equal to

$$f = \frac{1}{2\pi\sqrt{LC}}$$

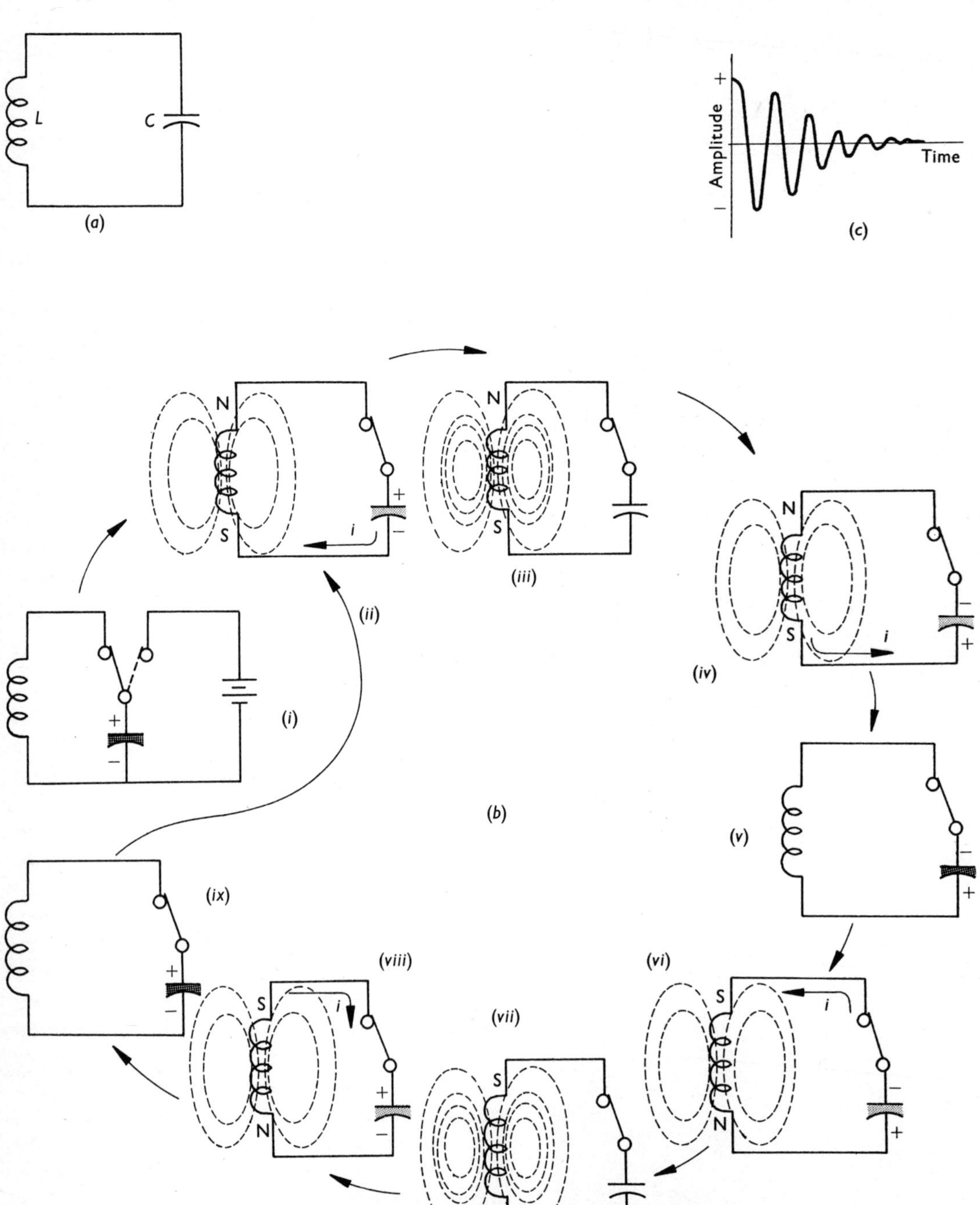

Fig. 10-1 Charge-circulating action of an *LC* circuit: (*a*) Schematic diagram; (*b*) Various phases in the oscillatory process; (*c*) Decreasing voltage amplitude across the capacitor.

10-2 SERIES- AND PARALLEL-CONNECTED *LC* CIRCUITS

An *LC* circuit is frequency-sensitive since both the coil and capacitor have reactance. At low frequencies the capacitor has high reactance while the coil has low. At high frequencies the reverse is true. At resonant frequency both components have equal reactance. These facts are shown in the diagram in Fig. 10-2.

Two types of circuit arrangements are possible with a coil and capacitor. They can be series-connected, in which case identical source current flows in each component, and voltage division takes place as shown in Fig. 10-3a. They may also be parallel-connected, in which case current division takes place as shown in Fig. 10-3b.

A series-connected *LC* circuit offers lowest reactance at resonant frequency. Since coil voltage leads current by 90° and capacitor voltage lags by the same amount, coil and capacitor voltage is 180° out of phase. The instantaneous net voltage drop adds up to zero. It is very important to realize that individually both the capacitor and the coil may have large voltage amplitude across them. These voltages may individually measure larger than the applied voltage, but because they are of opposite polarity a measurement across the coil-and-capacitor combination will show essentially zero.

A parallel-connected *LC* circuit offers maximum reactance at resonant frequency. Although there is a large circulating current passing between the coil and capacitor, very little line current passes from source to load (Fig. 10-3b). The line current is equal to the instantaneous sum of coil and capacitor current. These may be large but since both currents are 180° out of phase and nearly equal, the sum is essentially zero. Because of the low line current the load develops only a relatively small voltage drop. A measurement in the circuit in Fig. 10-3b would show nearly full applied voltage across a parallel *LC* circuit and nearly no voltage across the load.

10-3 FILTER ACTION OF *LC* CIRCUITS

An *LC* circuit may be connected to accept or reject a signal of resonant frequency. Various combinations with series- and parallel-connected *LC* circuits are shown in Fig. 10-4.

In Fig. 10-4a the series *LC* circuit allows maximum signal current to pass at resonant frequency. At lower frequencies current is blocked by high-capacitive reactance, whereas inductive reactance is dominant in blocking higher frequencies. By employing the series *LC* circuit as a shunt element as shown in Fig. 10-4b, a resonant-frequency signal is shunted. Although large source current may flow through the shunt, very little reaches the load.

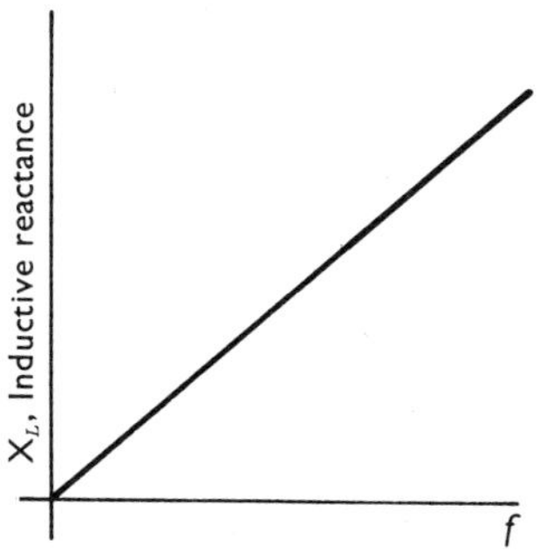

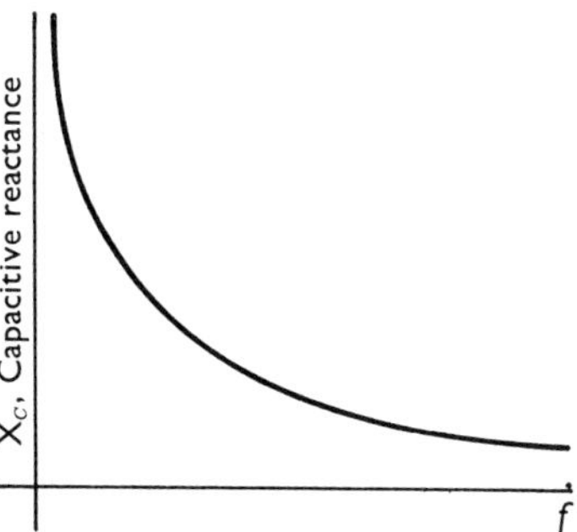

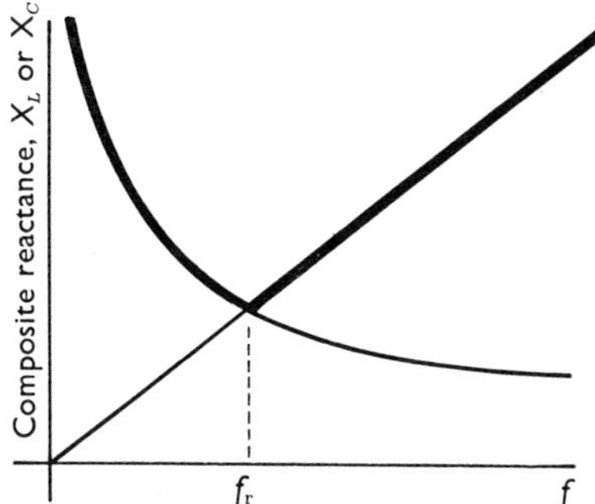

Fig. 10-2 Frequency response of an inductance, capacitance, and an *LC* circuit.

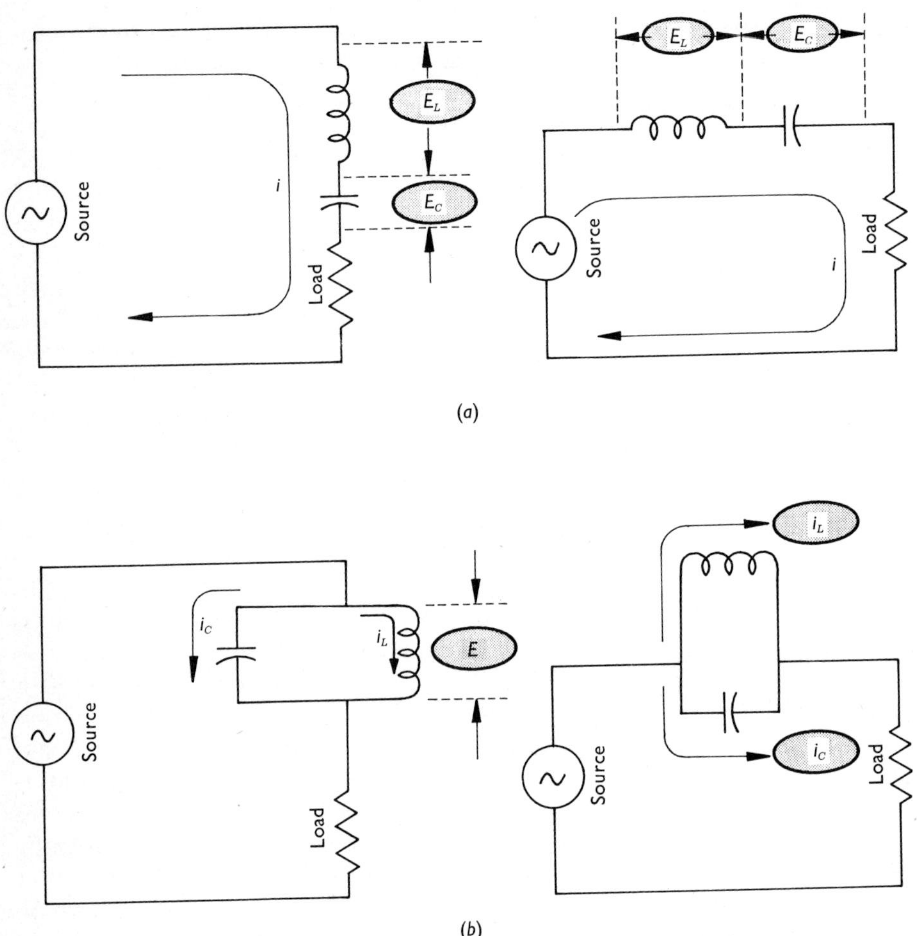

Fig. 10-3 Coil capacitor arrangement: (*a*) Series resonant circuit. Identifying feature is the identical current through *L* and *C*. Voltage division takes place; (*b*) Parallel resonant circuit. Identical voltage exists across *L* and *C*. Current division takes place.

A parallel *LC* circuit may also be used as a shunt element, as shown in Fig. 10-4c. It behaves as an acceptor circuit. The parallel circuit shunts all signals except those with resonant frequency. Low-frequency signals are shunted by the coil, high-frequency signals by the capacitor. A rejector can be constructed with a parallel *LC* circuit by including it as a series element in the line connecting source to load, as shown in Fig. 10-4d. In this instance low-frequency signals reach the load via the coil, high-frequency signals via the capacitor. Resonant-frequency signals receive equal opposition from both the coil and capacitor and are virtually prevented from reaching the load.

10-4 OBTAINING POSITIVE FEEDBACK WITH *LC* CIRCUITS

When an *LC* circuit is used in the construction of an oscillator, it is necessary to ensure positive feedback. In a common-emitter-

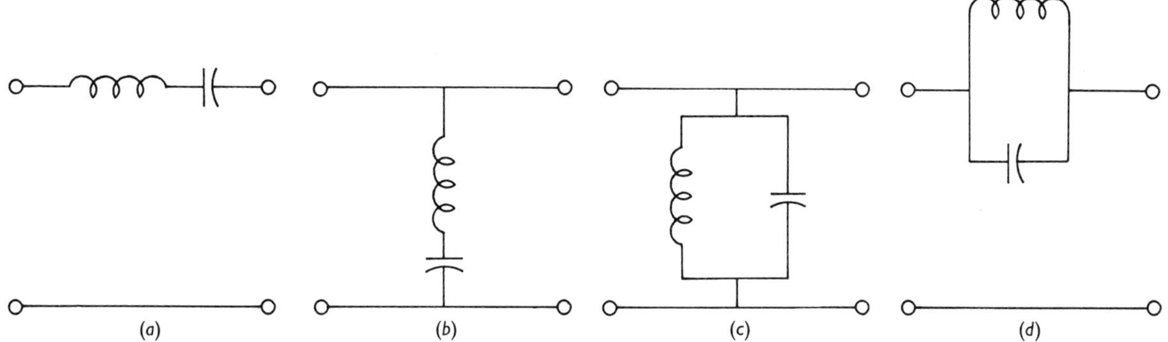

Fig. 10-4 Resonant-frequency filter circuits, each consisting of a single *LC* combination: (*a*) Series *LC* acceptor circuit; (*b*) Series *LC* rejector circuit; (*c*) Parallel *LC* acceptor circuit; (*d*) Parallel *LC* rejector circuit.

amplifier stage the collector and base signals are 180° out of phase. It is necessary to apply some of the collector signal to the base, but with 180° phase inversion. How is it possible to obtain the 180° phase inversion? Two means are available. One is the use of taps (electrical connections at other than the ends of a coil); the second is transformer action.

10-5 USE OF ELECTRICAL TAPS TO OBTAIN PHASE INVERSION

To comprehend how taps are employed to achieve phase inversion consider any coil segment as shown in Fig. 10-5a with three taps connected at the points *A, B,* and *C*. Assume that an ac signal flows through the coil and

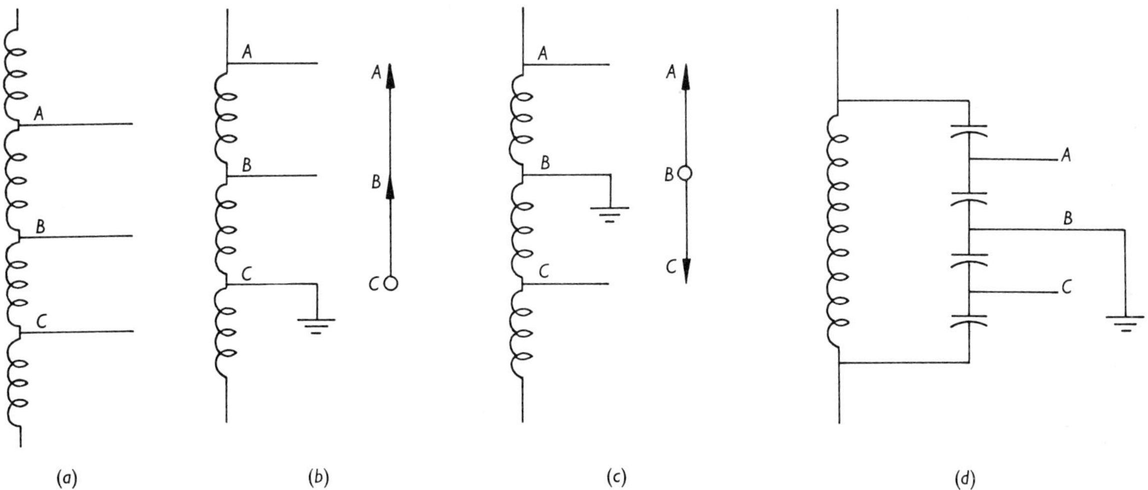

Fig. 10-5 Obtaining antiphase signal voltages by selecting taps on *L* or *C* voltage dividers: (*a*) Coil segment with taps connected to the windings; (*b*) Tap *C* selected as reference; (*c*) Central tap selected as reference, showing associated antiphase voltages; (*d*) Capacitor voltage dividers connected across the coil can perform the same function as taps.

produces a signal voltage across it. Some of this voltage can be measured between any two sets of taps, the relative amplitude being proportional to the number of turns between the two taps selected. This means that if one of the taps is selected as a common, or ground point, then all voltages can be expressed with reference to the grounded tap.

Suppose that tap *C* were grounded as shown in Fig. 10-5b. All taps on the same side of tap *C* would have voltages of differing amplitudes but of the same instantaneous polarity. This is shown in the accompanying phasor diagram. Next suppose that tap *B* is grounded (as in Fig. 10-5c). Now the voltages at taps *A* and *C* are of opposite polarity with respect to *B*. This fact can be stated as a principle. Whenever a voltage divider such as a coil is tapped, and one tap is grounded, all taps on one side of the grounded connection have voltages of opposite polarity (or phase) to those on the other side of the grounded tap.

Although the taps referred to in Fig. 10-5a are connections made to the wire on the coil, it is also possible to electrically tap a coil by using capacitors as shown in Fig. 10-5d. Signal voltage divides across the capacitors, and the voltage at point *A* is 180° out of phase with the voltage at point *C*.

10-6 TRANSFORMER ACTION TO OBTAIN PHASE INVERSION

A transformer can be connected such that primary and secondary coil voltage is in phase or 180° out of phase. To reverse the phase, reverse the wires connected either to the primary or the secondary (but not both). These facts are illustrated in Fig. 10-6 and will be used in the following section to describe one type of *LC* oscillator.

10-7 TICKLER OSCILLATOR

When transformer action is employed to obtain phase inversion between collector and base, the primary coil is known as a *tickler* coil. Figure 10-7 shows a schematic diagram of a tickler oscillator. Oscillation begins when collector voltage V_{CC} is applied. At this instant collector leakage current in the tickler coil induces a voltage in the secondary. The secondary-induced voltage forward-biases the base, causing increased collector current. A regenerative action is established whereby increasing collector current further causes increased base current, etc. Under these conditions the transistor is driven rapidly into saturation where no further increase in tickler-coil current takes place.

The action is not instantaneous; the time required depends partly on the amount of capacitance which must be charged in the transformer coils and partly on their inductance, i.e., upon the natural resonant frequency. When base voltage has reached its maximum positive value it begins to fall. The decreased base voltage (and current) causes a drop in collector current, which flows through the tickler coil. A reduction in tickler-coil current induces voltage of opposite polarity on the secondary causing it to fall even more rapidly. Secondary-coil voltage not only continues to fall but reverses polarity as secondary-coil oscillatory action completes its negative half-cycle. All the while, the reverse-biased base keeps collector current at zero. The instant the oscillatory action drives the base positive again, collector current begins to flow and the cycle repeats itself.

The output frequency of a tickler coil depends mainly on the effective resonant frequency of the secondary-coil circuit. In practice a capacitor of known value is added to the circuit as shown in Fig. 10-8. The base is connected to a tap to minimize shunting of the *LC* circuit. Output frequency is then given by the equation

$$f = \frac{1}{2\pi\sqrt{LC}}$$

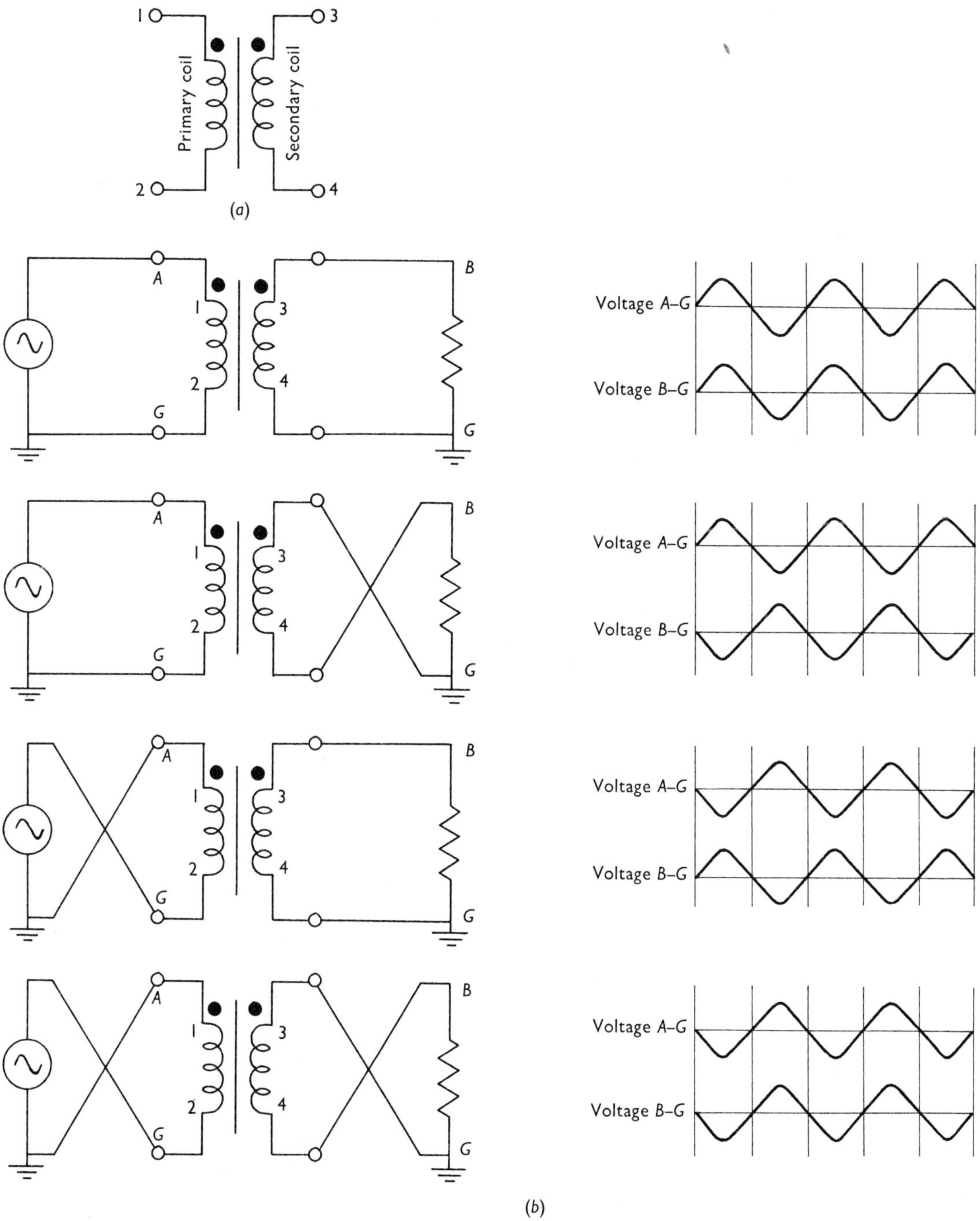

Fig. 10-6 Using a transformer to obtain phase reversal: (*a*) Transformer schematic diagram. Dots indicate that if connection 1 is positive with respect to connection 2, then connection 3 is positive with respect to connection 4; (*b*) Effect of reversing the connections to the primary and secondary coils.

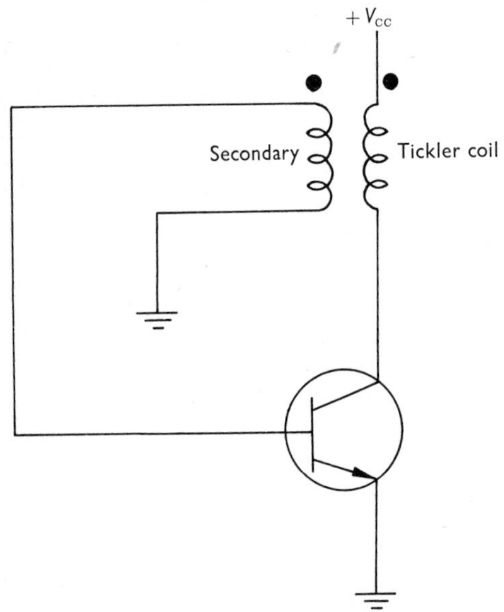

Fig. 10-7 Simplified tickler oscillator.

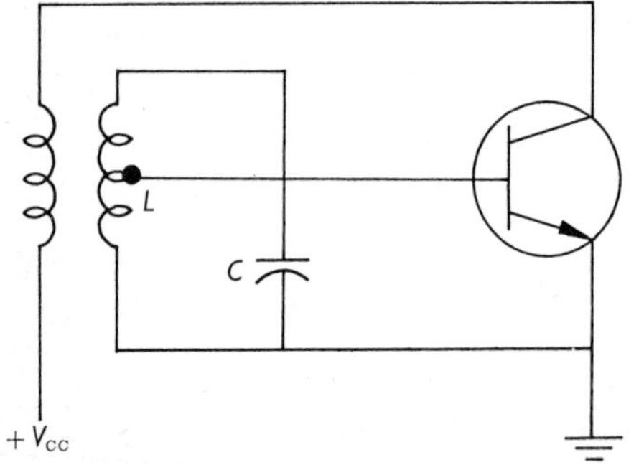

Fig. 10-8 Tickler oscillator in which the base is connected to a tap to minimize shunting of the *LC* circuit.

10-8 AMPLITUDE CONTROL

Tickler and other *LC*-type oscillators are normally operated class C. The bias voltage required for class C operation may be provided by a power source, but more commonly it is produced by the oscillator itself. Figure 10-9a shows a tickler oscillator with a biasing network consisting of C_b and R_b. To clarify its operation, the bias-network section of the oscillator circuit has been redrawn in Fig. 10-9b. Note that the circuit is very similar to that of a half-wave rectifier power supply (Sec. 2-5). The emitter-base junction serves as a rectifier in series with R_b, the "load" resistor, and C_b, the "ripple filter." After a few cycles of operation the capacitor C_b becomes charged to an average bias voltage (Fig. 10-9c). The average bias voltage depends on the values of C_b and R_b. A large *CR* time constant will result in relatively high bias voltage because very little discharge of C_b occurs between charge cycles. Sufficiently large C_b and R_b will accumulate and hold enough bias voltage to interrupt oscillator operation for several cycles (see Sec. 12-3).

A slightly different arrangement of C_b and R_b is shown in Fig. 10-9d. Here the resistor R_b is connected to the grounded side of the coil. There is no perceptible difference in circuit operation because both ends of coil L_s are essentially at ground for dc.

Automatic bias provides a measure of amplitude stability. If for any reason signal amplitude tends to change, the average bias voltage also changes. This adjusts the conduction time for the transistor during succeeding half-cycles until nominal amplitude is reestablished.

10-9 HARTLEY *LC* OSCILLATOR

A Hartley oscillator employs a tapped coil to obtain the required 180° phase inversion between collector and base. One arrangement is shown in Fig. 10-10a. A simplified diagram illustrating the feedback network appears in Fig. 10-10b. Note that tap *E* is located between taps *B* and *C*, therefore voltage *BE* is 180° out of phase with voltage *CE*. Self-bias is generated by the *CR* network consisting of C_b and R_b. Frequency of oscillation is given by the resonant-frequency equation

$$f = \frac{1}{2\pi\sqrt{LC}}$$

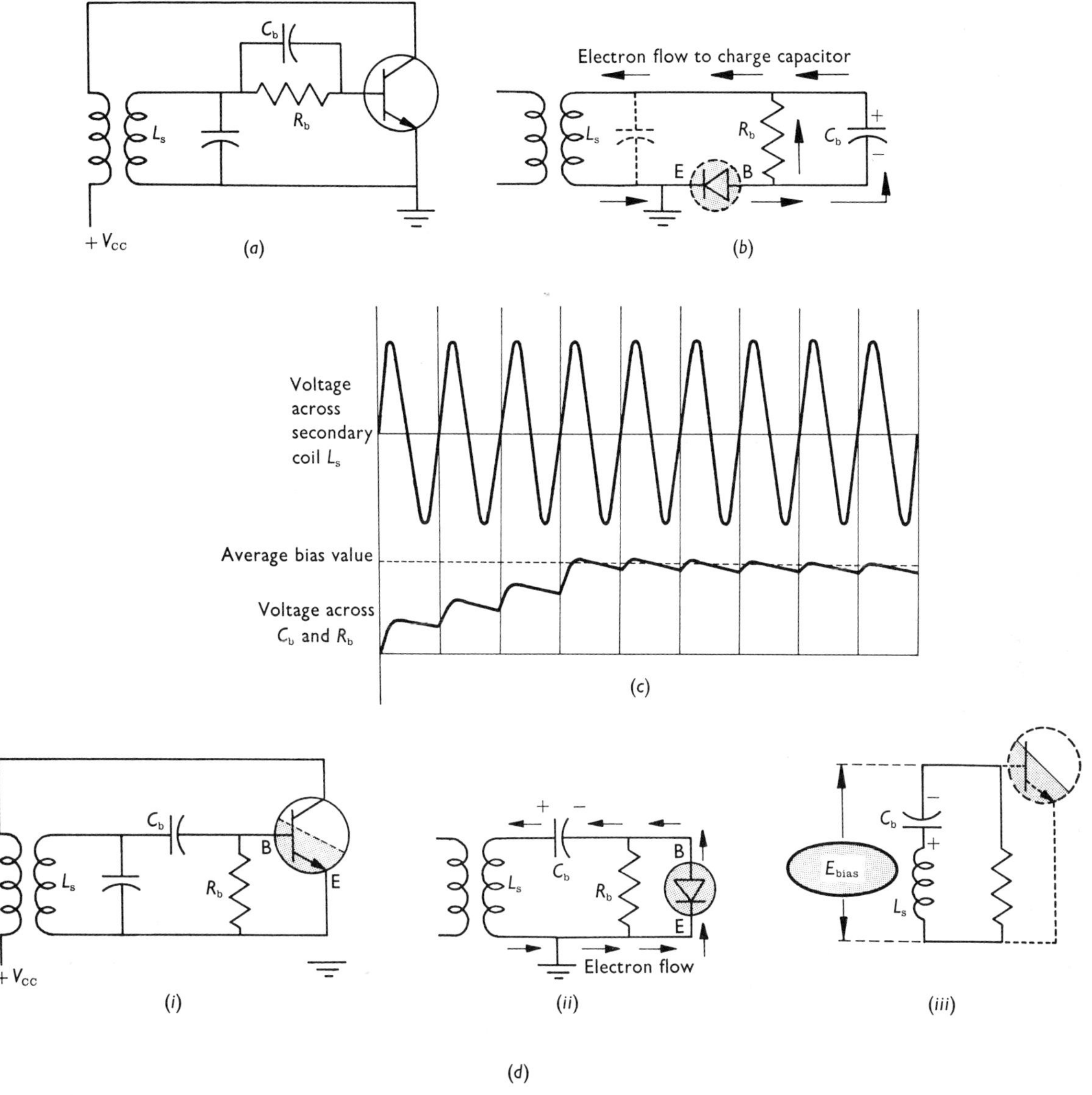

Fig. 10-9 Development of automatic bias in an oscillator: (*a*) Tickler-oscillator circuit with bias network R_b and C_b; (*b*) Equivalent circuit for the development of bias voltage; (*c*) Timing diagram of bias voltage buildup across C_b; (*d*) Slightly modified circuit, with R_b connected to the ground side of L_s.

Hartley oscillators are classified as to whether or not a dc component of power-supply current flows in the coil *L*. In the circuit of Fig. 10-10a all emitter current, both dc and signal, flows from ground, through tap *C*, and out of tap *E*. This then is a *series-fed* Hartley oscillator. The diagram in Fig. 10-11 depicts a *shunt-fed* Hartley oscillator. The dc component of the emitter current passes only through the choke coil (RFC). This route is blocked for signal current because of the high inductive reactance of the choke. Signal current finds relatively little opposition in the emitter-coupling capacitor C_e, therefore for signal frequencies the emitter is connected to a tap E on the coil L.

One advantage of a shunt-fed circuit is that dc components are not permitted to dissipate power in the coil. Unnecessary power dissipation in the coil causes its temperature to rise, which in turn causes a change in physical dimensions and inductance and therefore in output frequency.

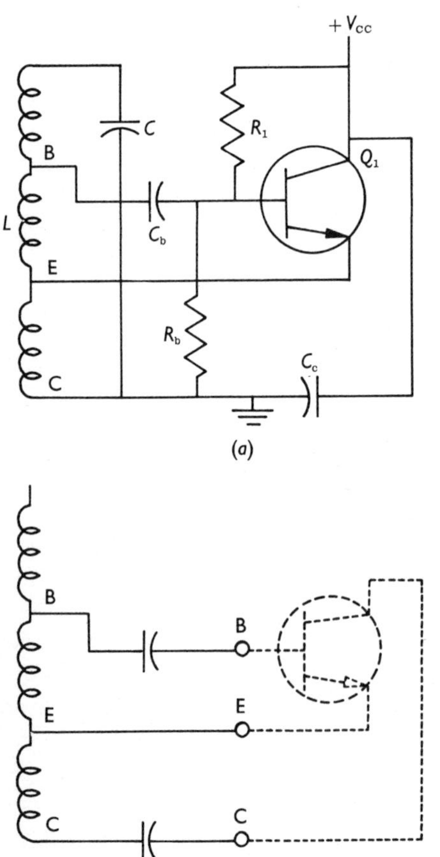

Fig. 10-10 Series-fed Hartley oscillator: (*a*) Schematic diagram; (*b*) Simplified diagram of feedback network.

10-10 COLPITTS OSCILLATOR

Some identifying features of a Colpitts oscillator circuit are:

1. The coil does not require physical taps.
2. Positive feedback is obtained by electrically tapping the coil with capacitors, as discussed in Sec. 10-5.
3. One end of the tank circuit is always connected to the collector, the other always to the base.
4. For signal currents the emitter must be connected to the coil by capacitive taps, even though the capacitors are not always obvious on schematic diagrams.

A fundamental Colpitts oscillator circuit is shown in Fig. 10-12a. Coil L is resonated with C_1 and C_2 in series, as well as capacitance associated with transistor Q_1 and the bias network. Note that no dc components are present in the coil of the tank circuit. In practice Colpitts oscillators have good stability and are favored in many applications. They can be constructed to cover a wide range of frequencies by including a provision for changing coils, a procedure which is simplified because only two connections are made to the coil.

Alternate forms of Colpitts schematic diagrams often appear in electronics literature. The diagrams in Fig. 10-12a, b, and c are all electrically identical yet at first glance appear different schematically. At very high frequencies (200 to 300 MHz) an arrangement similar to that in Fig. 10-12d is favored. The base of the transistor is grounded for signals and the upper frequency limit for a given transistor is thereby extended. A typical Colpitts oscillator of this type is depicted in Fig. 10-12e. The circuit, used in portable television receivers, operates in the frequency range 50 to 250 MHz. Note that different coil segments are switched in to vary the inductance. In this way a different frequency is obtained from each channel setting.

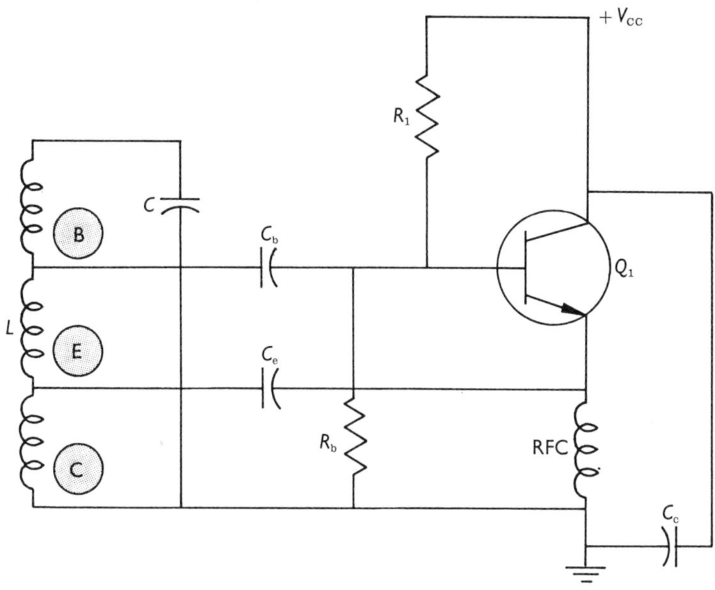

Fig. 10-11 Shunt-fed Hartley oscillator.

10-11 TPTG OSCILLATOR

The circuit shown in Fig. 10-13a is known as a *tuned-plate tuned-grid* (*TPTG*) oscillator. Its bipolar solid-state equivalent is shown in Fig. 10-13b. A FET circuit would be similar to the vacuum-tube circuit, and the inset (Fig. 10-13a and c) shows that a FET could be substituted for a triode vacuum tube.

The circuit is not notable for its application. Its importance lies in the fact that many amplifiers unwittingly take on this form of oscillation, unless precautions are taken to avoid it.

A TPTG oscillator obtains its feedback through the internal-tube grid-to-plate capacitance. A *tuned-drain tuned-gate (TDTG)* oscillator obtains its feedback through the gate-to-drain capacitance. In a *tuned-collector tuned-base (TCTB)* oscillator the collector-base junction capacitance provides the feedback. Feedback can become positive and of sufficient magnitude under the following conditions. (1) Both input and output are equipped with resonant circuits which can be tuned to the same frequency, and (2) the frequency is high enough that sufficient signal can be returned via the *interelectrode* or junction capacitance. In practical circuits additional capacitance can be added externally if desired to achieve oscillation. In bipolar transistor circuits junction capacitance depends also on collector-supply voltage (Fig. 10-13e). (3) Both input and output tank circuits must be inductive at the operating frequency. This means that the output frequency will be slightly lower than the resonant frequencies of the tank circuits. Inevitably one of the tank circuits is tuned more closely to the operating frequency; this tank has the most control over output frequency. The other controls the magnitude of the feedback signal.

Figure 10-13c and d shows the circuit as it appears electrically under normal operation. Capacitors C_g and C_b can be ignored at the operating frequency because of their low reactance. Although schematically these circuits appear like a Hartley, the important difference is that *no magnetic-flux coupling exists between coils* L_1 *and* L_2. For this reason the circuit resembles a Colpitts oscillator with inductive rather than capacitive taps.

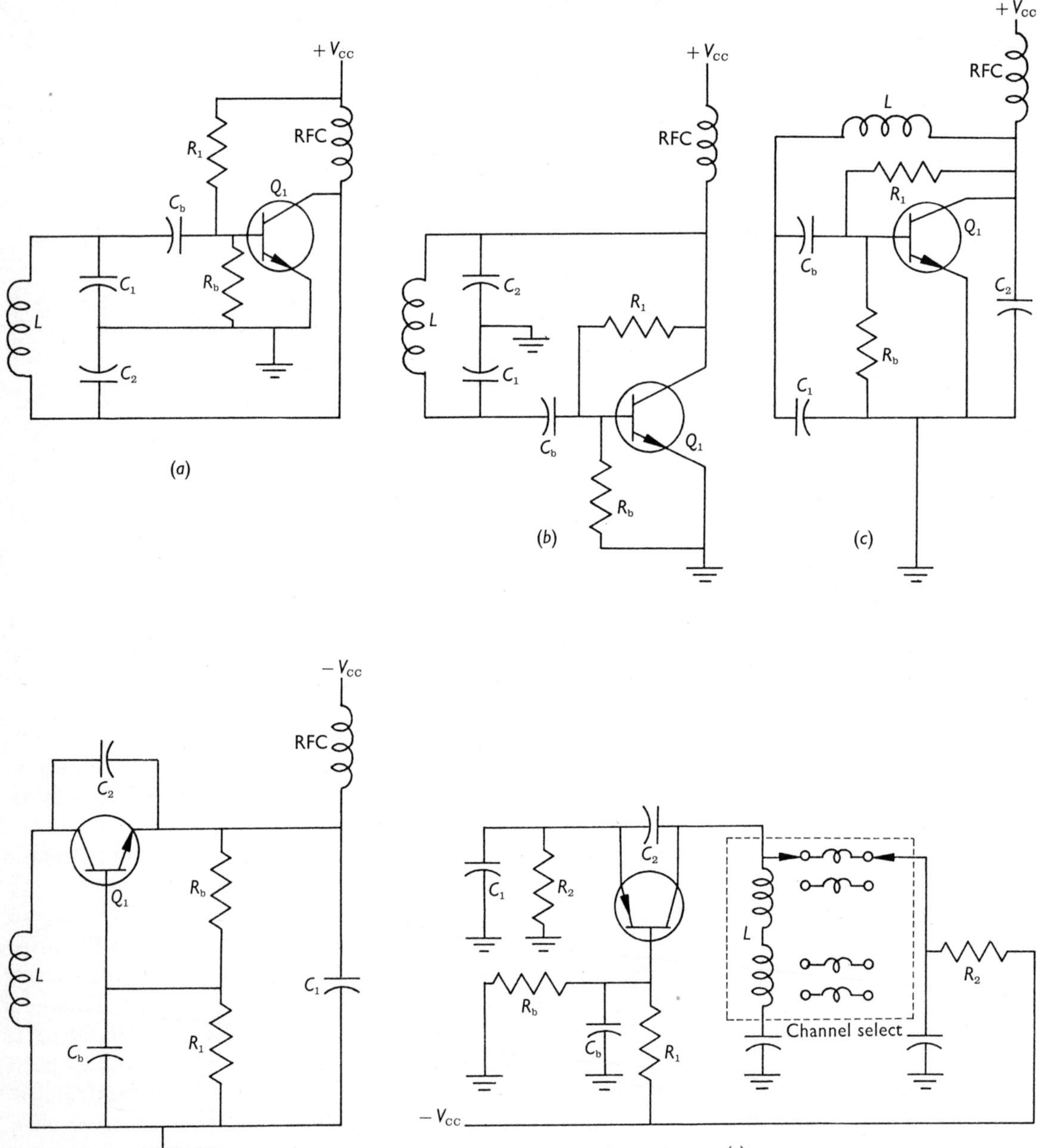

Fig. 10-12 Colpitts-oscillator circuits: (*a*), (*b*), (*c*) Typical conventional circuits. These are shown with bipolar transistors, but can be constructed with FETs and vacuum tubes; (*d*) Grounded base, high-frequency, Colpitts oscillator; (*e*) Typical high-frequency Colpitts oscillator found in television receivers.

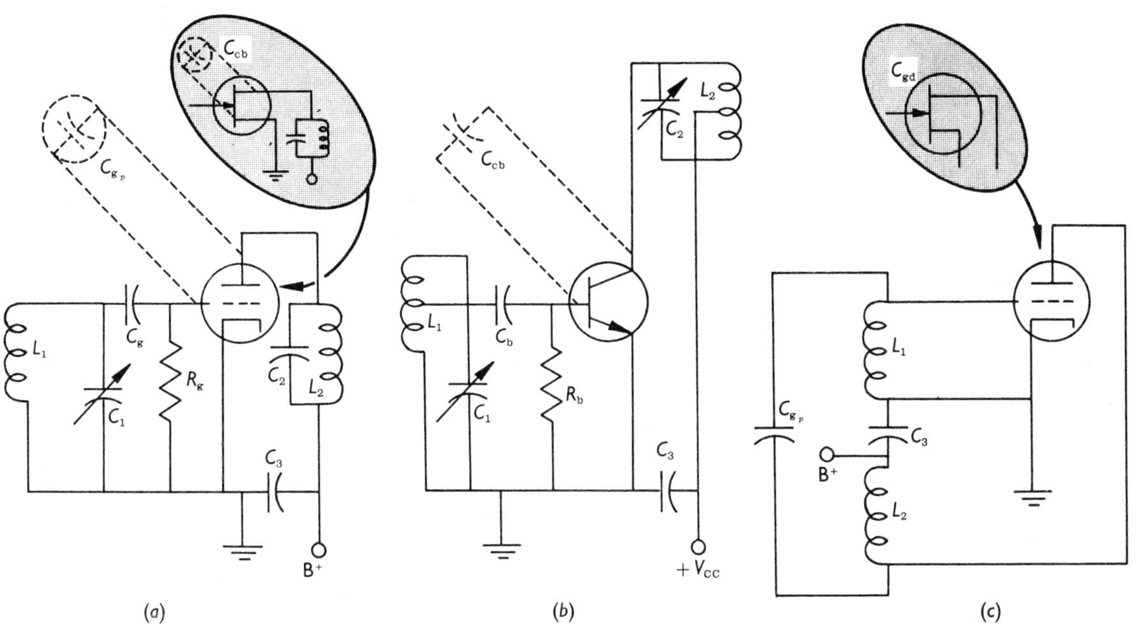

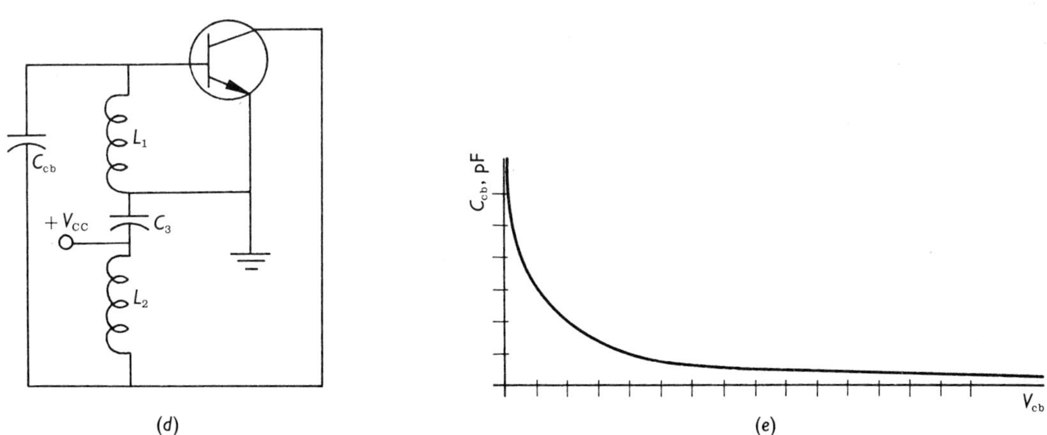

Fig. 10-13 Tuned-plate tuned-grid (TPTG), tuned-drain tuned-gate (TDTG), or tuned-collector tuned-base (TCTB) oscillator: (*a*) Vacuum-tube or FET circuit; (*b*) Bipolar solid-state circuit; (*c*), (*d*) Rearranged circuits, focusing on key components active during oscillation; (*e*) Collector-base junction capacitance plotted as a function of junction voltage.

10-12 CRYSTAL OSCILLATOR

Except for loading effects, frequency stability of an oscillator depends mainly on temperature and on the quality factor (Q) of the feedback network. The Q of an LC circuit depends primarily on the Q of the coil. In practice values of 200 or 300 are achieved.

A device which is similar in electrical behavior to an LC circuit is the *crystal resonator* shown in Fig. 10-14a. The structure consists of a thin

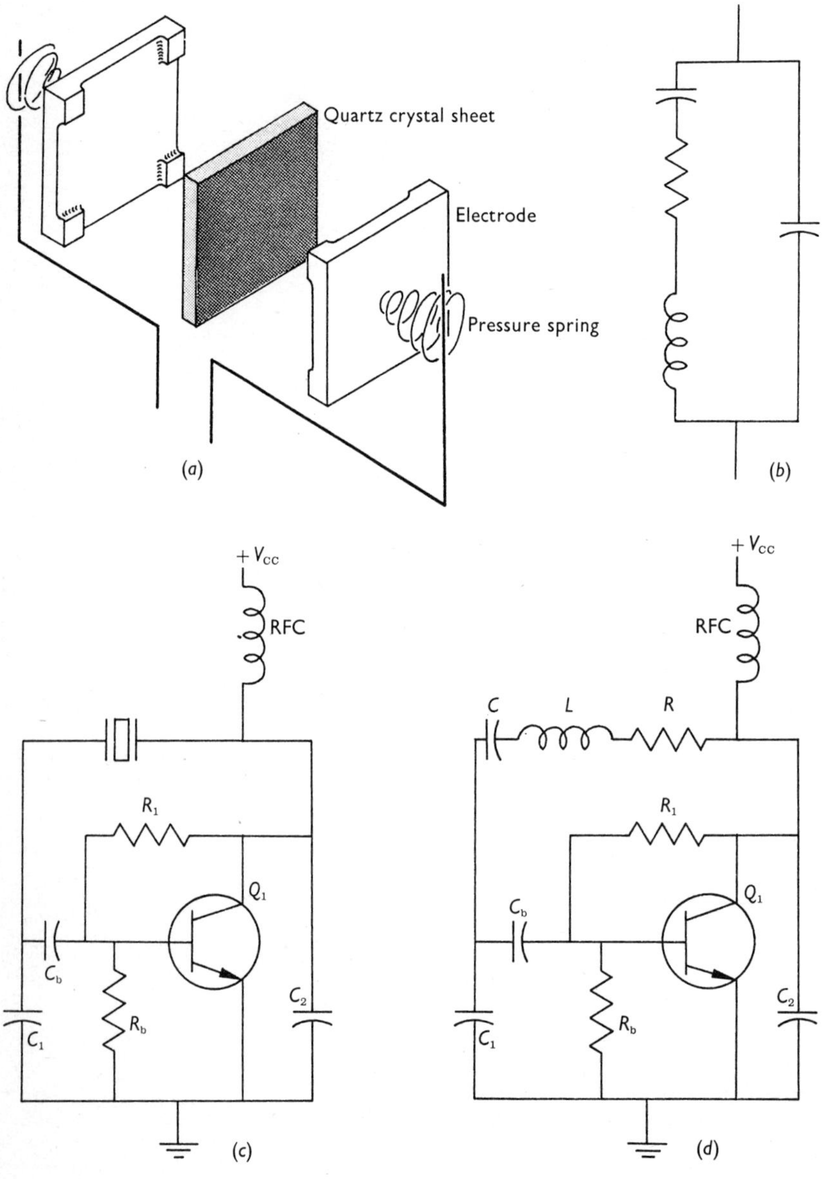

Fig. 10-14 Quartz crystal oscillator: (*a*) Structure of a quartz crystal resonator; (*b*) Electrical equivalent of a quartz crystal resonator; (c) Pierce crystal oscillator; (*d*) Clapp oscillator.

sheet of quartz suspended between two electrodes. When a voltage is applied to the electrodes, the physical dimensions of the crystal change slightly. The crystal may bend, stretch, or shear, depending on how it was cut from the block. An alternating voltage will cause the crystal to vibrate.

A vibrating crystal has a natural resonant frequency which depends on its mass and dimensions. At resonance the crystal exhibits electrical qualities like that of an *LC* circuit with a *Q* of several thousand. When used as a frequency-determining element in an oscillator, frequency stability may be improved several orders of magnitude. Figure 10-14b shows a schematic diagram of the electrical equivalent of a quartz crystal. An oscillator circuit is shown in Fig. 10-14c. Note the similarity to the Colpitts circuit in Fig. 10-12c, and the greater similarity to the Clapp oscillator in Fig. 10-14d.

Crystal oscillators are used in systems requiring signals with long-term high-frequency stability and spectral purity. Typical applications are signal generators for test purposes, master oscillators in transmitter circuits, frequency meters, and electronic wristwatches.

10-13 ULTRAHIGH-FREQUENCY TANK CIRCUITS

Oscillators which operate in the several-hundred-megahertz range require tank circuits which, although *LC* circuits electrically, are different in appearance from the familiar coil-capacitor combination. Ultrahigh-frequency tank circuits are shaped to take advantage of the distribution pattern of an electromagnetic wave. There is a simple relationship between the spatial length of an electromagnetic wave and the frequency of the wave.

Frequency is the number of oscillations or waves produced per second. High-frequency waves are radiated from a source at the speed of light c. (This constant can be expressed in whatever length units are convenient; 180 000 mi/s or 3×10^8 m/s are typical.) The first wave produced by the source is at a distance of c from the source 1 s later. During this interval the source has been producing waves at the rate of $1/f$, and at the end of 1 s there are f waves spaced out between the source and the first wave produced. Now f waves occupy a length c, therefore one wave occupies length c/f. The relationship between wavelength, frequency, and speed of propagation is therefore $\lambda = c/f$ (Fig. 10-15a).

When transmitted over a pair of parallel conductors, commonly referred to as a *transmission line*, electrical signals exhibit waveform behavior. High and low points of voltage and current are found at different positions on the line. If the line is open at the receiving end, no current can flow, but a high voltage can exist there. Electrically the open end appears similar to a parallel-resonant *LC* tank circuit. One-quarter wavelength from the end point (or 90° away, if a wave is expressed cyclically and measured in degrees) a maximum current point with minimum voltage exists. The line can be cut here and a shorting bar placed across the cut end. Such a section of line, known as a *Lecher line* (Fig. 10-15b), can be used as an *LC* resonant circuit.

Many Lecher lines may be connected in parallel, as shown in Fig. 10-15d(i). An infinite number of such sections form the basis of a *resonant cavity* shown in Fig. 10-15d(iii). Resonant cavities are used extensively in "microwave" applications where signals with wavelengths of a fraction of a centimeter are common.

10-14 LECHER-LINE OSCILLATOR

The open end of a Lecher line would normally be short-circuited if connected directly to the base of a transistor. It can, however, be

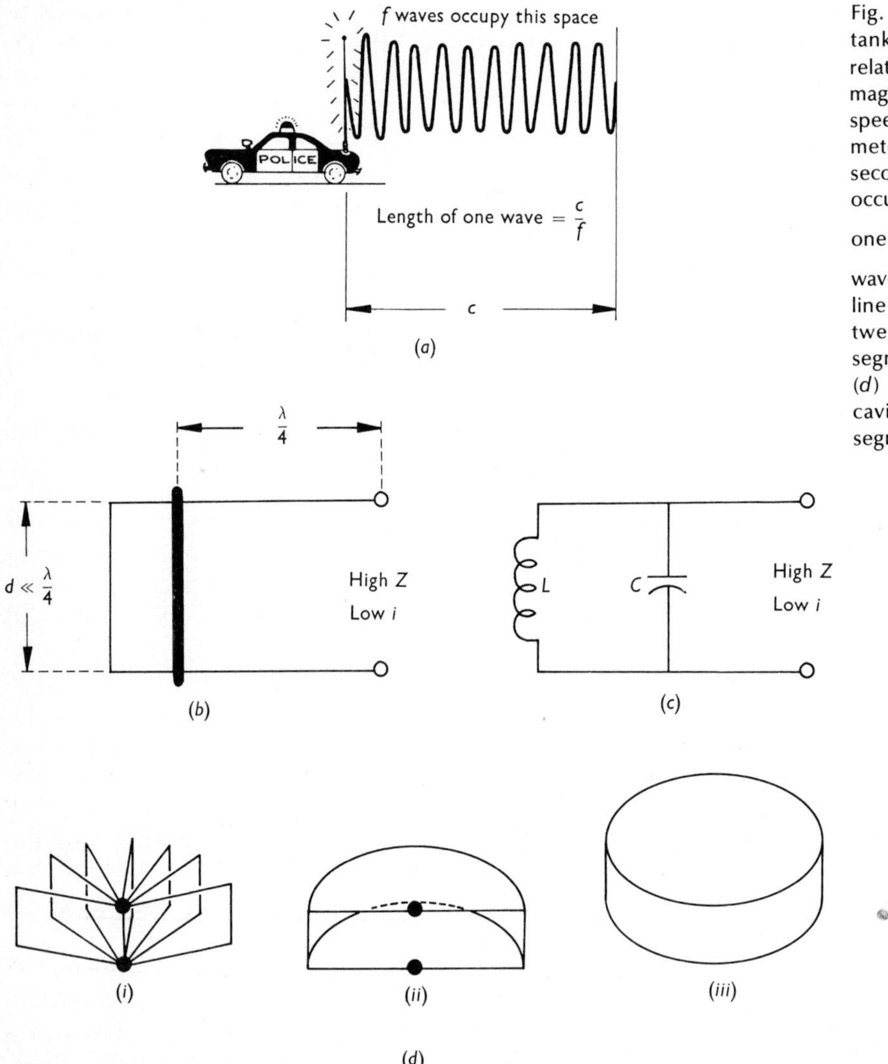

Fig. 10-15 Ultrahigh-frequency tank circuits and their spatial relationships to radiated electromagnetic-wave energy: (*a*) The speed of the wave is *c* miles or meters per second. After one second of transmission, *f* waves occupy distance *c*. The length of one wave is $\lambda = \frac{c}{f}$; (*b*) Quarter-wave segment of a transmission line; (*c*) Electrical similarity between a quarter-wave line segment and an *LC* tank circuit; (*d*) Development of a resonant cavity from quarter-wave line segments.

connected to the gate of a FET, which offers a high impedance to the line. The similarity between one type of Lecher-line oscillator and a Hartley circuit is shown in Fig. 10-16a to e.

10-15 CONCEPT OF NEGATIVE RESISTANCE

An *LC* oscillatory circuit of the type shown in Fig. 10-1a will not produce sustained oscillations unless energy is added to overcome circuit losses. If a resistor is placed across an *LC* circuit as shown in Fig. 10-17a, the damping action is increased and the oscillation amplitude drops even more quickly.

The resistor uses energy which comes from the *LC* tank circuit. A volt-ampere curve for a resistor is shown in Fig. 10-17b. Note that an increase in voltage across a resistor results in a proportional rise in current through it, as expected from Ohm's law.

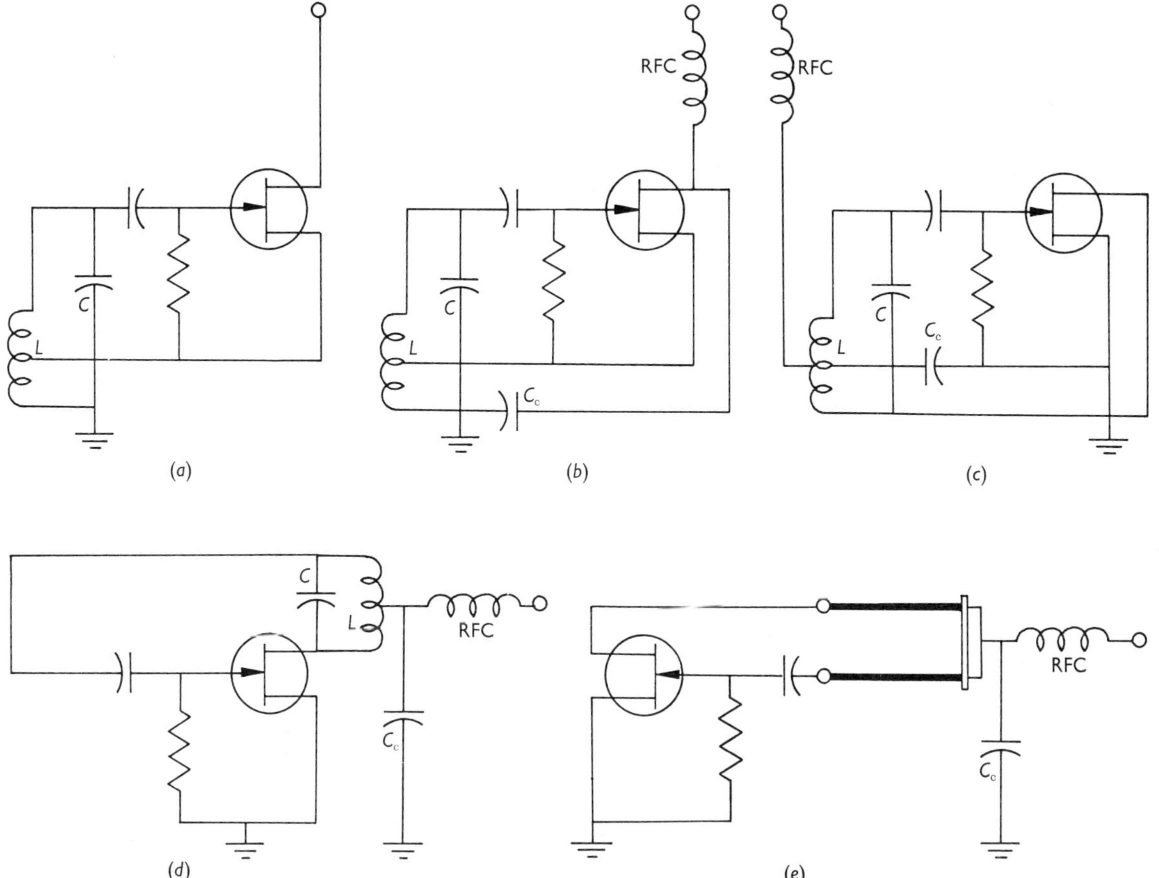

Fig. 10-16 Lecher-line oscillator: (*a*) Series-fed Hartley circuit with drain at signal-ground, via power supply; (*b*) Hartley circuit with drain-signal return path through C_c; (*c*) Modified Hartley circuit with dc drain current, rather than source current, in the tank circuit; (*d*) Circuit redrawn; (*e*) *LC* tank circuit replaced with a Lecher line.

Suppose, however, that it were possible to construct a device in which current would drop instead of rise when voltage was increased. Then if voltage decreased for any reason, the current would rise. The device would behave as an "antiload" or negative resistance, with a volt-ampere curve as shown in Fig. 10-17c. At a voltage V_s no current would flow through the device, but at zero voltage (i.e., the ends of the device connected together) a current of I_s A would flow. If such a device were connected across an oscillatory circuit, and if signal amplitude were falling, the device would inject current into the tank circuit, thereby helping to sustain oscillations. One device which displays negative-resistance properties is a *tunnel diode*.

10-16 TUNNEL DIODE

A tunnel diode is essentially a heavily doped PN junction. The heavy doping gives it the volt-ampere curve shown in Fig. 10-18. A curve for an ordinary diode is also plotted with a broken line for comparison.

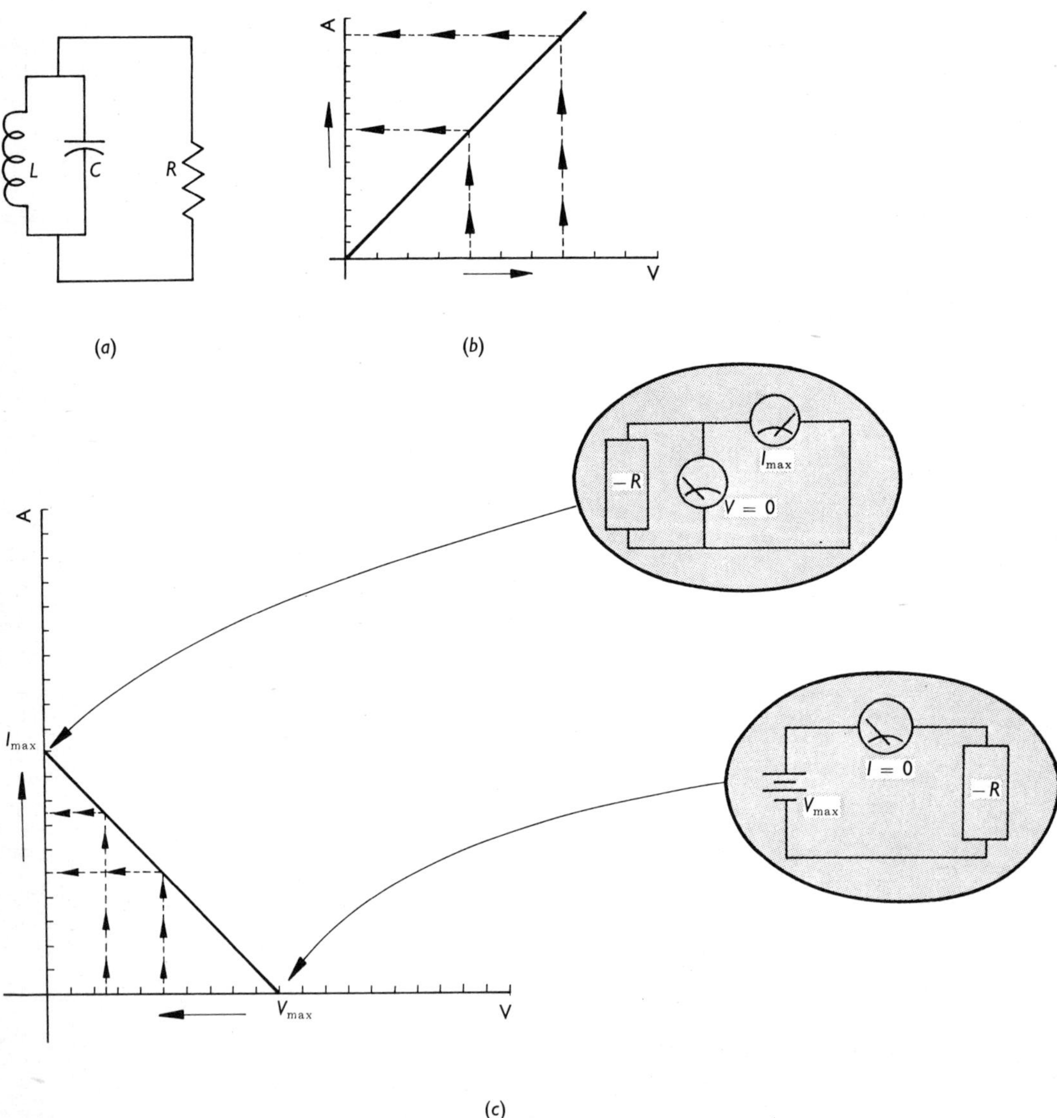

Fig. 10-17 Comparison of positive and negative resistance: (a) *LC* tank circuit shunted with resistance; (b) Volt-ampere curve for s resistor; (c) Volt-ampere curve for a fictitious *negative-resistance* device.

Note that as forward-biased voltage is increased from zero to a value V_1, tunnel-diode current also increases from zero to value I_1. At this point a further increase in bias voltage toward a value V_Q causes diode current to fall, i.e., negative resistance is being exhibited. By applying the correct bias (usually somewhat between V_1 and V_2 depending on the application) a tunnel diode can be operated in the negative-resistance region. Under this con-

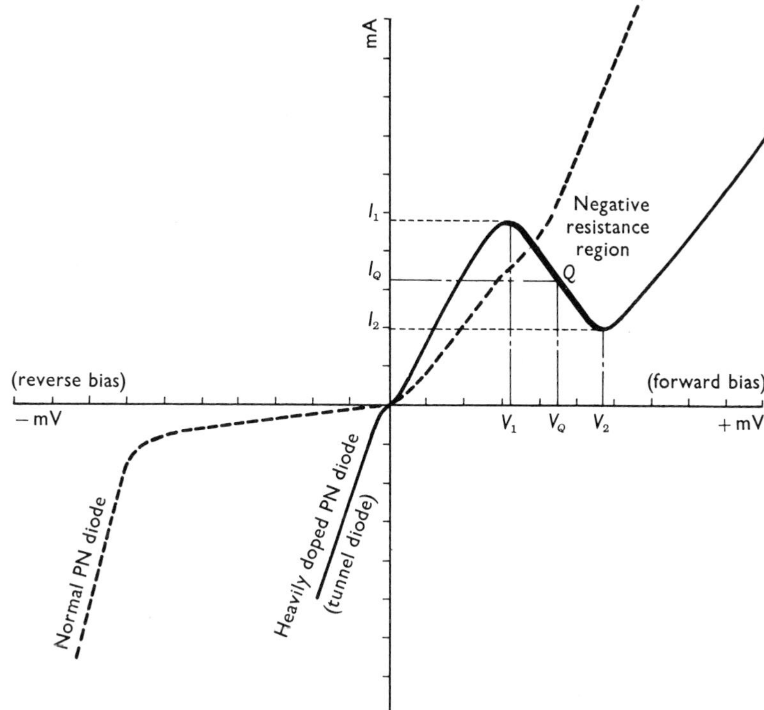

Fig. 10-18 Tunnel-diode, volt-ampere, characteristic curve.

dition it can be used in an oscillator circuit to sustain oscillations when connected across an *LC* tank.

A tunnel diode is an extremely fast device. This means that if forward-bias voltage is changed suddenly, diode-current change takes place within a few picoseconds. In practice this means that tunnel-diode oscillators can function at frequencies of several gigahertz.

10-17 TUNNEL-DIODE OSCILLATOR

Figure 10-19 shows a schematic diagram of a basic tunnel-diode *LC* oscillator circuit. Resistors R_1 and R_2 are biasing resistors which set the operating point. For maximum signal swing the operating point would be chosen in the middle of the linear portion of the negative-resistance curve. For a selected operating point the tunnel diode has a certain dc resistance (because current I_Q flows when voltage V_Q is applied). This resistance is obtained by using Ohm's law and is equal to V_Q/I_Q. The resistor R_1 is approximately $\frac{1}{3}$ this value, and R_2 is calculated to have an *IR* drop equal to $V - V_Q$.

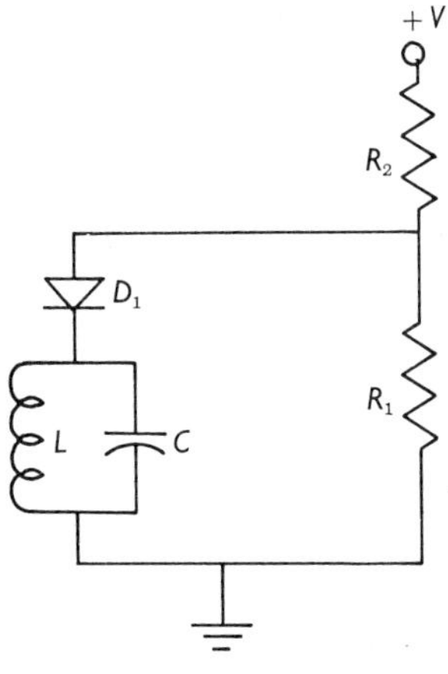

Fig. 10-19 Basic tunnel-diode *LC* oscillator circuit.

10-18 MAGNETRON, KLYSTRON, AND TRAVELING-WAVE OSCILLATORS

Tunnel-diode oscillators are capable of generating frequencies of several thousand megahertz, but at very low power levels. When high-power high-frequency oscillators are required, as in radar applications, certain types of vacuum-tube oscillators are used. Klystron oscillators operate on the principle of bunching electrons in flight. Magnetrons operate on the principle of interaction between a moving electron and a powerful magnetic field. Traveling-wave oscillators operate on the principle of imparting energy to an electromagnetic wave by means of an electron beam. These types of oscillators are well documented in literature and are too specialized to be dealt with here.

10-19 VACUUM-TUBE CIRCUITS

Equivalent vacuum-tube circuits for many of the circuits discussed in this chapter are shown in Appendix H.

Summary

An *LC* tank circuit contains only reactive elements and therefore does not waste signal power in resistance as do *CR*-type frequency-sensitive networks.

LC-type circuits and their ultrahigh-frequency equivalents, resonant cavities, function at frequencies to several gigahertz, whereas *CR*-type networks are not practical to use above 1 or 2 MHz.

LC circuits can be arranged to provide positive feedback. The usual methods are (1) by employing electrical taps (either direct connections or capacitive voltage dividers) and (2) by using transformer action.

LC oscillators can generate their own bias voltage which is adjusted automatically to maintain a constant output signal.

The fundamental *LC* oscillator circuits are (1) tickler, (2) Hartley, both series and shunt fed, (3) Colpitts, and (4) TPTG. Nearly all *LC* oscillator circuits can be identified as one of the above.

Crystal oscillators employ a quartz crystal which vibrates mechanically (like a high-frequency tuning fork) to produce a very stable output frequency.

A quarter-wave section of transmission line (a Lecher line) is used as the *LC* tank circuit in a Lecher-line oscillator.

Unlike a resistor which drains energy when connected across an *LC* circuit, a tunnel diode acts as a negative resistance and "feeds" energy to the circuit. Under proper bias conditions a tunnel diode is able to sustain oscillations.

Questions and Exercises

1. Why is it impractical to use *CR*-type frequency-selection networks in oscillators operating at frequencies above 1 or 2 MHz?
2. In oscillator circuits an *LC* tank circuit may be regarded in one of two ways. What are these?
3. What determines the rate at which an *LC* circuit "oscillates"?
4. Why will a current introduced into an *LC* circuit not circulate indefinitely?
5. Draw a schematic diagram showing how an *LC* circuit can be used as **(*a*)** an acceptor circuit for one frequency and **(*b*)** a rejector circuit for one frequency.
6. Draw a schematic diagram showing a signal source, a capacitor, a coil, and a load resistor all connected in parallel. Only the resonant frequency would produce maximum signal voltage across the load.
   a. What would happen to signals with frequencies above the resonant frequency?
   b. What would happen to signals with frequencies below the resonant frequency?
7. State two methods of obtaining phase inversion when using *LC* circuits in oscillator circuits.
8. What is meant by a coil tap?
9. When a coil tap is selected as the common

or ground tap, what can be said about the phase relationship of the voltages taken from other taps on either side of the common tap?

10. When using a transformer, if the secondary voltage is of the wrong phase, what can be done to reverse the phase?

11. Draw a schematic diagram of a simple tickler oscillator circuit and state briefly how it operates.

12. What output frequency will the circuit in Fig. 10-8 produce if coil *L* is 40 mH and capacitor *C* is 0.01 μF?

13. How can amplitude be controlled automatically in *LC*-type oscillators?

14. What would be the result in each case if in the circuit in Fig. 10-9a the following alterations were made: (*a*) C_b is greatly increased in value, (*b*) R_b is greatly decreased, (*c*) C_b is greatly reduced, and (*d*) R_b is greatly increased?

15. Explain how the amplitude of the oscillator signal is reduced by automatic bias if for any reason it attempts to rise above a nominal value.

16. How is the required phase inversion between input and output obtained in a Hartley oscillator circuit?

17. Distinguish between a *series-fed* Hartley oscillator circuit and a *shunt-fed* Hartley oscillator circuit.

18. Why is the shunt-fed arrangement usually preferable to the series-fed in Hartley oscillators?

19. What construction cost savings can be realized if a series-fed circuit is chosen instead of a shunt-fed Hartley circuit?

20. How can you identify a Colpitts oscillator circuit?

21. How is the required phase inversion obtained in a Colpitts oscillator circuit?

22. State one advantage of using a Colpitts circuit in a band-switching oscillator circuit to cover a wide range of frequencies.

23. Although the TPTG family of oscillators is rarely used as such, it is important to understand the operating characteristics of these circuits. Why?

24. How is the required feedback obtained in a TCTB oscillator circuit?

25. What is an important difference in a TCTB circuit and a Hartley, even though schematically the two forms appear to have similar features?

26. To what can we attribute the high degree of frequency stability in a crystal oscillator circuit?

27. Although it has superb frequency stability, can you suggest one limitation of a crystal oscillator in practical applications?

28. What length of Lecher line (in meters) would be required to form a resonant *LC* circuit at a frequency of 300 MHz? Assume the speed of light to be 3×10^8 m/s. (Remember that a Lecher line is one-quarter wavelength.)

29. With a series of sketches show that a resonant cavity can be considered to be made up of an infinite number of Lecher lines joined at the open end.

30. Although a bipolar transistor base would severely shunt the open end of a Lecher-line oscillator circuit of the type shown in Fig. 10-16e, can you suggest how a transistor could be used? Draw a schematic diagram of your suggested circuit. (Hint: Refer to Fig. 10-10 for some ideas.)

31. Discuss the concept of "negative resistance."

32. Because it displays negative resistance, a tunnel diode "feeds" energy to an *LC* circuit. Where does this energy come from?

33. If in the circuit in Fig. 10-19 *V* is supplied by a 1.34-V mercury cell, $R_2 = 750\ \Omega$, $R_1 = 18\ \Omega$, and the tunnel-diode bias voltage is 1.25 mV, calculate the following: (*a*) quiescent or bias current through the tunnel diode, and (*b*) total average current drain from the battery.

34. In what applications are magnetron, klystron, and traveling-wave tube-type oscillators used?

11 radio and home-entertainment circuits

11-1 PRINCIPLES OF RADIO TRANSMISSION

Every current has associated with it an electromagnetic field. An electric field is required to propel charge through an antenna, but the moving charge creates magnetic flux which surrounds the conductor as shown in Fig. 11-1. If the current reverses direction, so also does the electromagnetic flux.

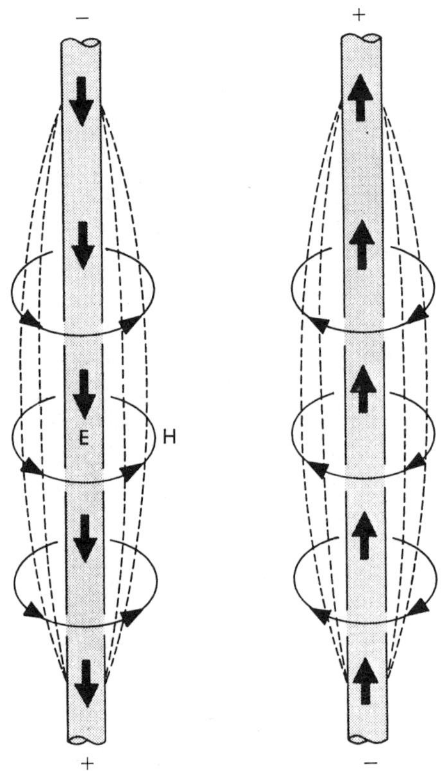

Fig. 11-1 Electromagnetic flux surrounding a segment of a current-carrying conductor. If the conductor is grasped in the left hand, so that the thumb points in the direction of electron flow, the fingers indicate the direction of magnetic flux.

Normally, very little energy is radiated into the surrounding space. If the current is removed, the electromagnetic flux collapses inward to the conductor. If the conductor carries alternating current, the flux alternately builds up and collapses. When the frequency

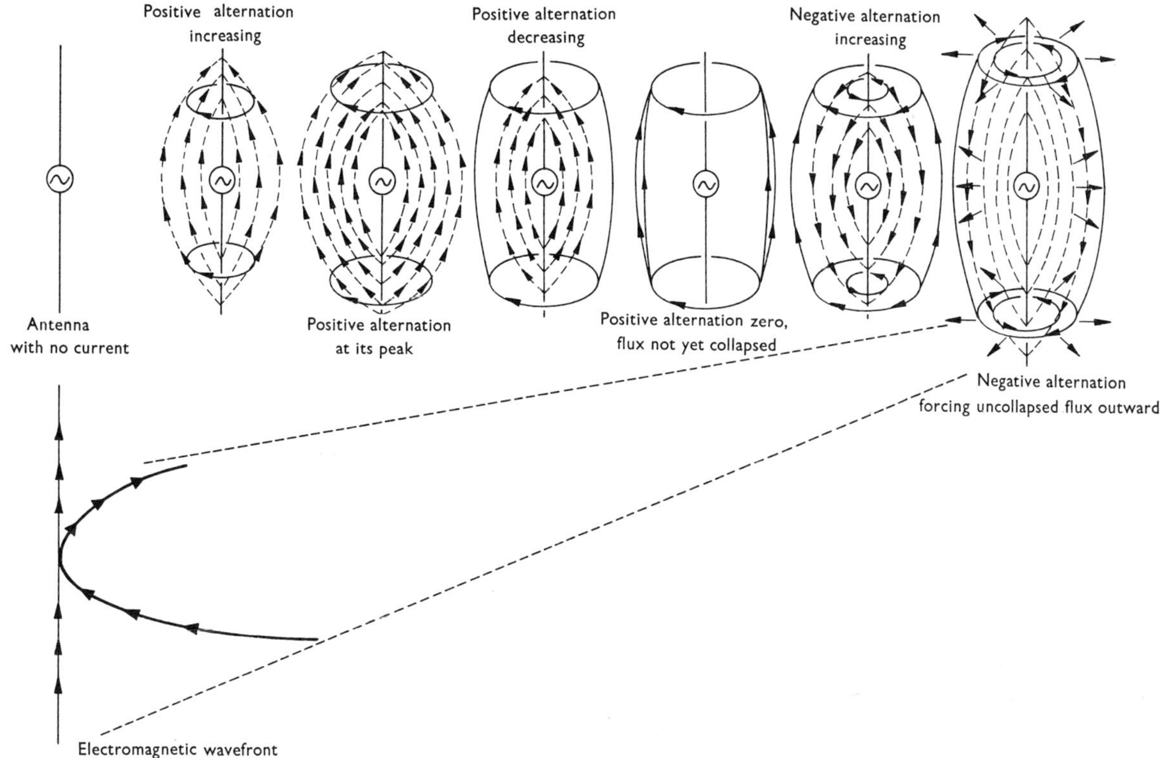

Fig. 11-2 The radiation process. At sufficiently high frequencies the collapsing electromagnetic flux field is not allowed to fall inward and is forced outward by increasing flux generated by the succeeding alternation.

of the alternating current is high enough, there is insufficient time for the field to fall inward. The collapsing flux is met by outward-bound flux from the next alternation, and a portion of the field is propelled into space as a radio wave. Figure 11-2 attempts to convey this process to the point where a portion of the electromagnetic field is about to leave the antenna. All frequencies which are high enough to be efficiently radiated are said to be in the radio-frequency or RF spectrum. The RF spectrum extends upward from approximately 15 kHz.

Radio reception is the inverse process of transmission. It consists of intercepting an electromagnetic wave with an antenna. A weak alternating-current signal is induced in it.

11-2 SIMPLE RADIO TRANSMITTER

Any device capable of generating a high-frequency signal can behave as a transmitter, often unintentionally. The earliest transmitters consisted of nothing more than an oscillatory circuit (see Sec. 10-1) energized by a sparking coil as shown in Fig. 11-3a. A spark is not a stable energizer and cannot be used to produce continuous waves. Later efforts included the use of high-frequency alternators to generate the signal (Fig. 11-3b), but the upper frequency limit of such devices was in the neighborhood of only 100 kHz.

The invention of the triode vacuum tube made it possible to construct continuous-wave

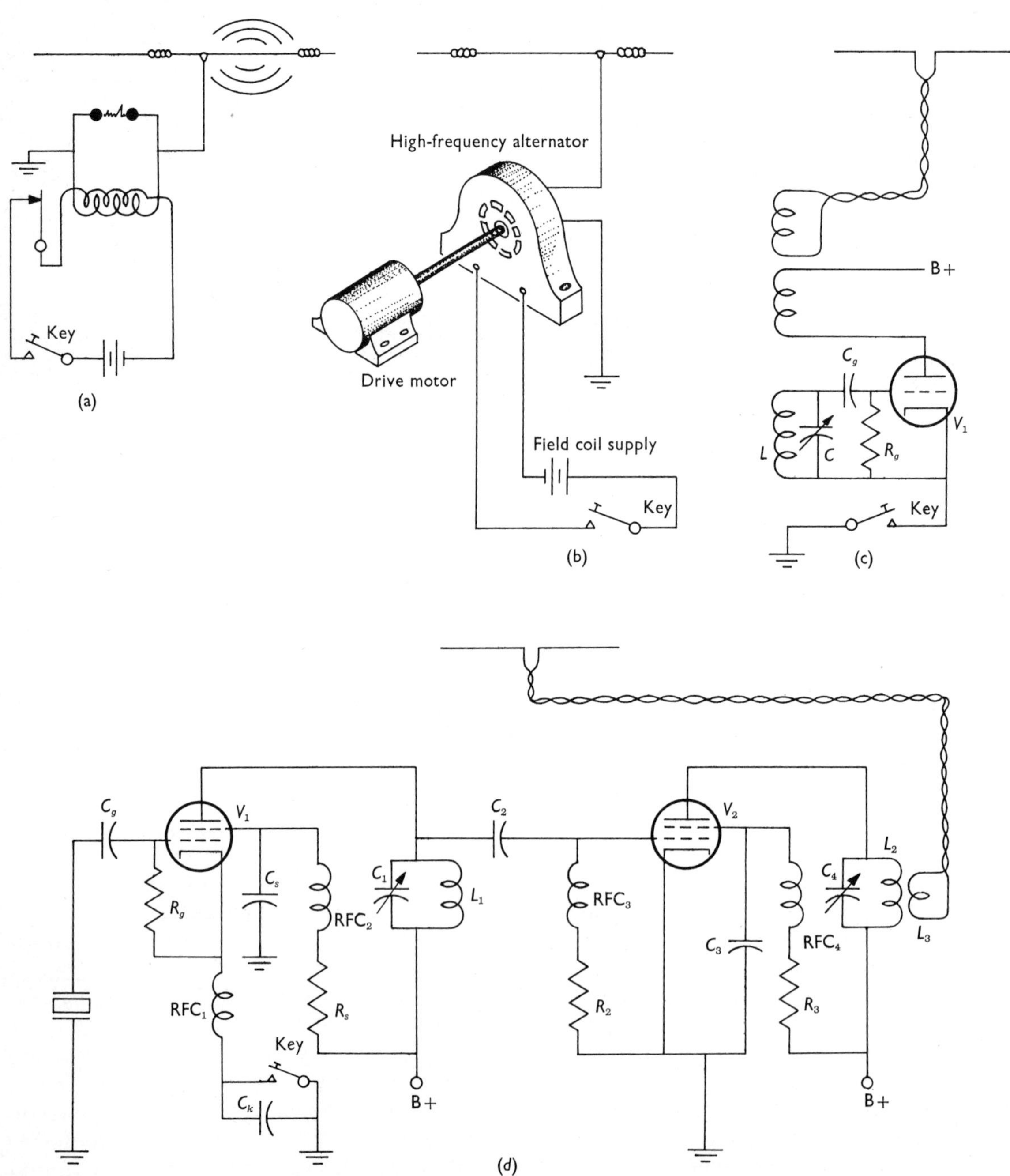

Fig. 11-3 Early types of transmitters: (*a*) Spark coil; (*b*) High-frequency alternator; (c) One-tube circuit consisting of oscillator only; (*d*) Master-oscillator power-amplifier (MOPA) type circuit.

oscillators and led to broadcasting. A simple one-tube oscillator used as a transmitter is shown schematically in Fig. 11-3c. The circuit is suitable for use as an emergency Morse code transmitter. Its frequency of transmission is determined by the equation

$$f = \frac{1}{2\pi\sqrt{LC}}$$

the resonant frequency of the LC tank circuit. In practice, such a circuit has poor frequency stability, especially if the antenna draws considerable energy.

A more suitable circuit has a separate oscillator followed by an amplifier, as in Fig. 11-3d. Such a circuit is known as a *master-oscillator power-amplifier transmitter (MOPA)*. Except for rare instances where small size, portability, and simplicity are prime requisites, all practical transmitters are of the MOPA type. In commercial and broadcast applications, the master oscillator is nearly always crystal-controlled (see Sec. 10-12).

11-3 MODULATION

The transmitters described in the previous section produce a train of continuous waves known as *carrier waves*. Signaling to a remote location is achieved by interrupting the carrier wave train with a Morse key, which in these examples disconnects the power supply from the oscillator. This is but one means of impressing information on the carrier. More generally, the process of impressing information on a carrier is known as *modulation*.

11-4 AMPLITUDE MODULATION—AM

For the alphanumeric characters the carrier pattern emitted by a continuous-wave (cw) transmitter using international Morse code is as shown in Fig. 11-4. Note that carrier amplitude is zero when the key is up and maximum when the key is down. It is the *amplitude* of the carrier which is being varied, i.e., *amplitude modulation* is taking place.

If an audio transformer T_2 is inserted in series with the power-supply lead as shown in Fig. 11-5a, voice and music may be used to modulate the carrier. The voice or music audio variations in the transformer secondary alternately add and subtract from the constant power supply feeding the power-amplifier stage. It is in effect being supplied power which varies at an audio rate. Carrier amplitude is directly proportional to the instantaneous power-supply voltage, as shown in the timing diagram (Fig. 11-5b).

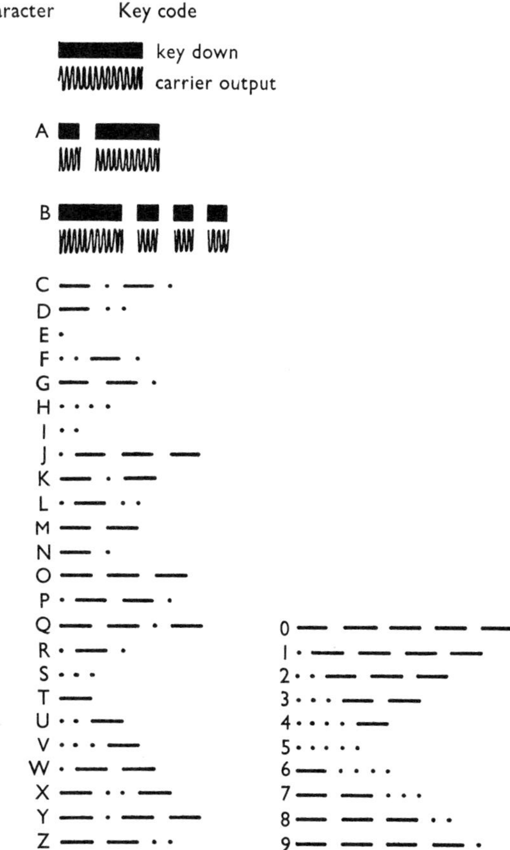

Fig. 11-4 CW (continuous wave) transmitter output. Carrier wave output shown for various alphanumeric characters of international Morse code.

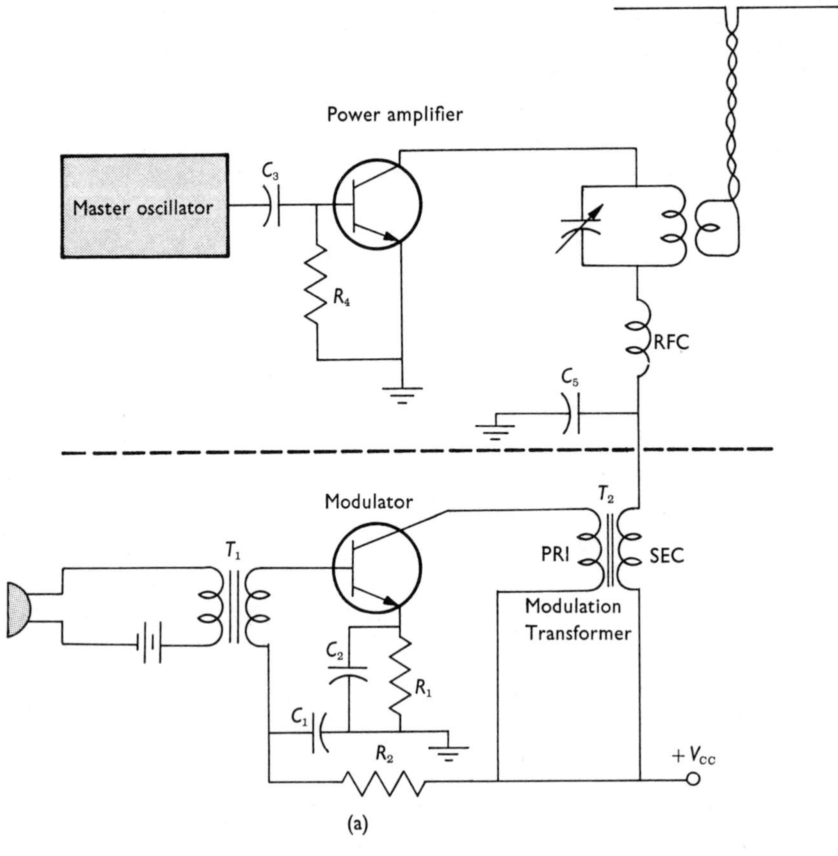

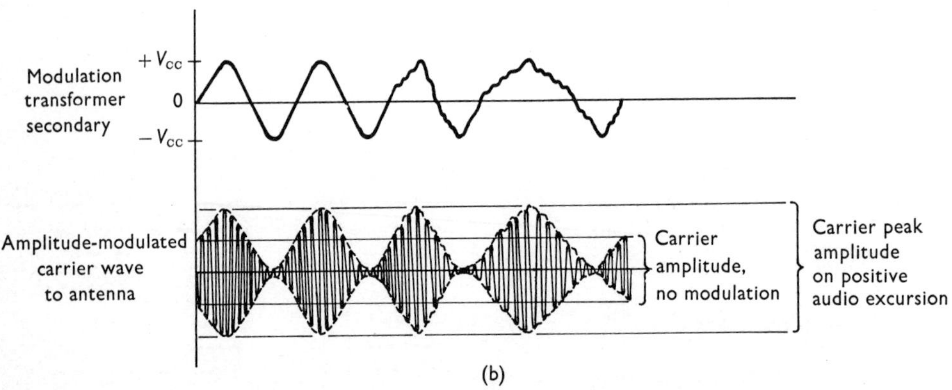

Fig. 11-5 Essentials of an amplitude-modulated voice or music transmitter: (*a*) Schematic diagram. Choke RFC and capacitor C_5 prevent carrier energy from entering the modulator, but do not impede the lower-frequency audio signal from reaching the power amplifier; (*b*) Timing diagram.

11-5 FREQUENCY MODULATION—FM

It is possible to transmit information by varying carrier frequency rather than carrier amplitude. This type of transmission is known as *frequency modulation* or *FM*. An FM receiver is purposely designed to ignore carrier-amplitude variations and to give an output signal only when it detects changes in carrier frequency.

An FM carrier can be produced with the circuit shown in Fig. 11-6a. Sound waves vary the output capacitance of a capacitor micro-

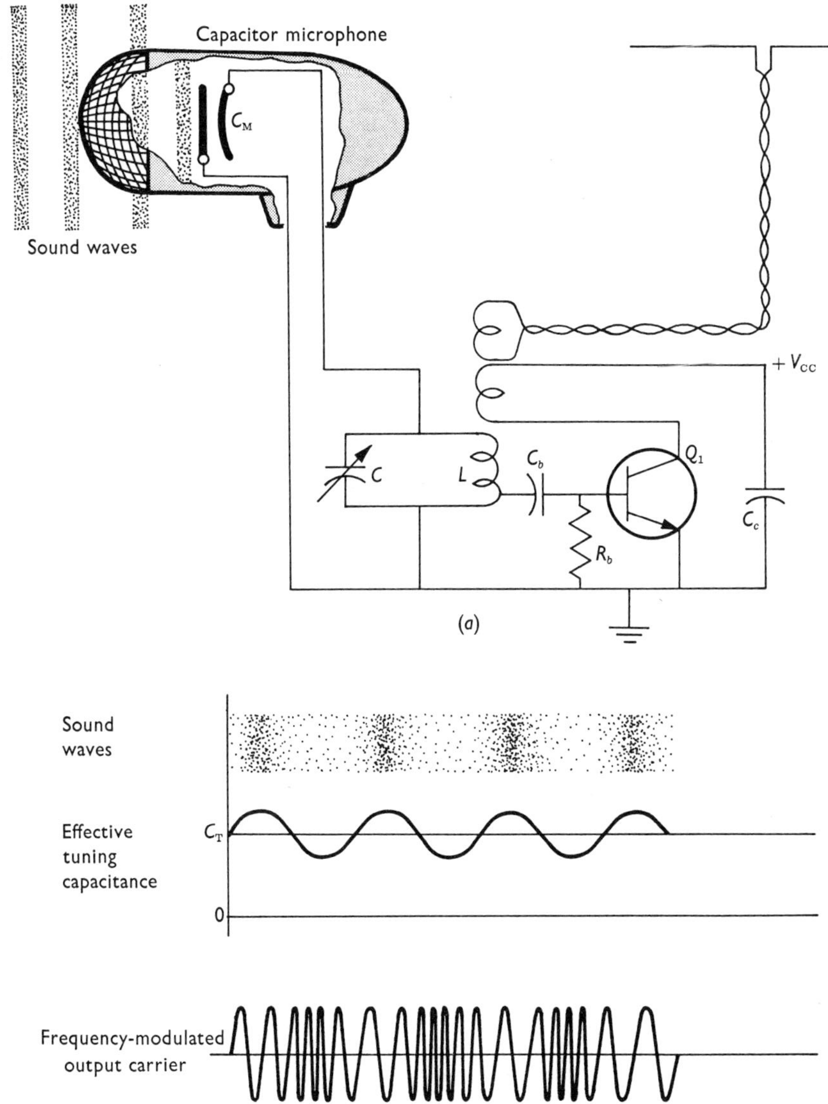

Fig. 11-6 Simple FM transmitter: (*a*) Schematic diagram; (*b*) Timing diagram.

phone which is connected across an oscillatory circuit. Frequency of oscillation therefore varies at an audio rate, as shown in the timing diagram (Fig. 11-6b).

A more practical circuit, shown in Fig. 11-7, uses a tuning diode D_1 (varactor diode) to vary the oscillator frequency. If the reverse-bias voltage across the diode is varied, its junction capacitance varies also, and the output frequency of the oscillator varies as a result. The microphone signal is first amplified by transistor Q_1 and the audio signal voltage plus diode-bias voltage is applied to diode D_1 via resistor R_1.

11-6 PRINCIPLES OF RADIO RECEPTION

Radio transmission is the process of modulating a carrier signal to produce new frequencies which contain the information to be transmitted. These new frequencies are known as sidebands.

Radio reception involves the interception of electromagnetic waves and the extraction of the information conveyed by the sidebands. The circuit which extracts this information is known as a *detector* or *demodulator.*

One of the earliest detectors used was a coherer. It consists of a glass vial of electrically conductive powder. When no transmitter signal is present at the antenna, the powder in the vial is loosely scattered, and conductivity is low. In the presence of a radio signal the particles cohere, and conductivity rises sufficiently to energize a relay which then completes the circuit to a buzzer.

If more than one transmitter operates within range of a radio receiver, there must be some way to select the desired signal and reject all others. An *LC* oscillatory circuit tuned to resonance at the desired signal frequency can function as a selector. The selected signal is then applied to a detector which extracts the modulating frequency. A block diagram of a receiver in its simplest form is shown in Fig. 11-8.

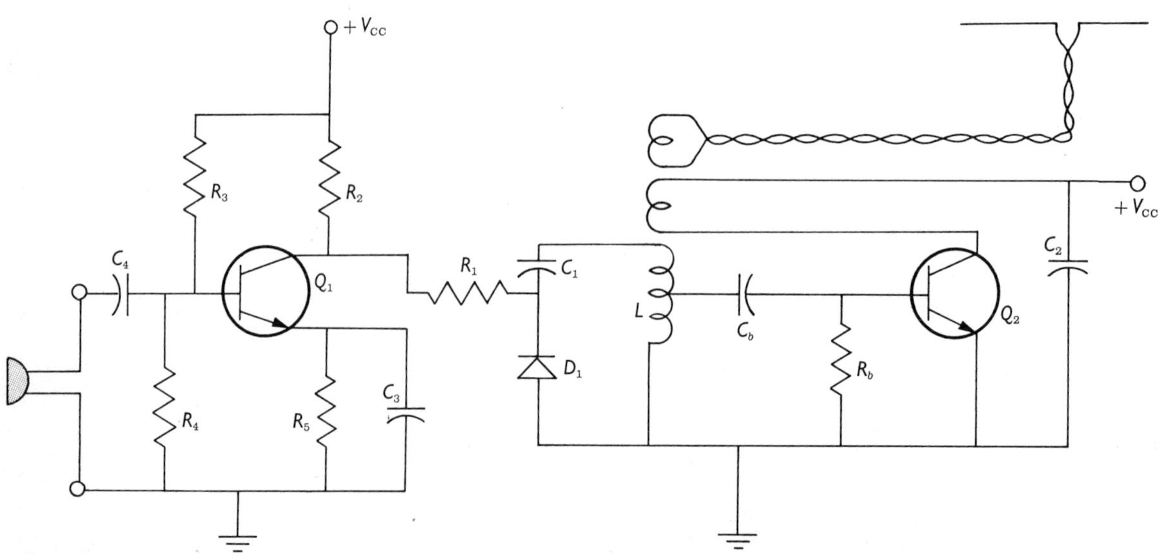

Fig. 11-7 Simple FM transmitter using a varactor diode (D_1) to obtain frequency modulation.

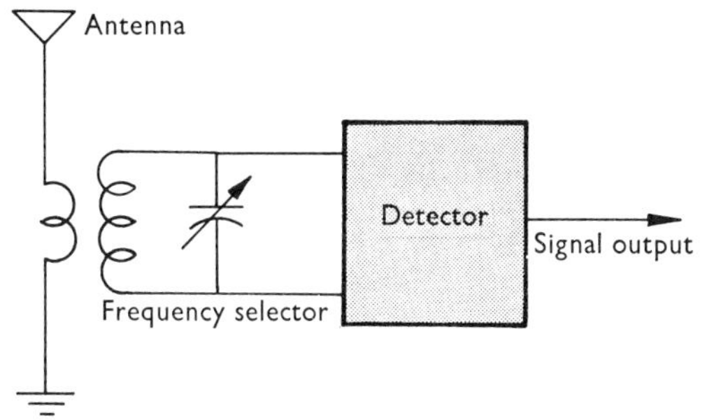

Fig. 11-8 Block diagram of the essential parts of a simple radio receiver.

11-7 RECEPTION OF AM SIGNALS

When mathematically analyzed, the demodulation of AM signals is similar to the amplitude-modulation process. The received carrier signal and the sidebands are injected into a detector. Signal products coming from the detector will be the sum and difference frequencies and will contain the modulating frequency. Unwanted frequencies are suppressed whereas the modulating frequency is amplified and fed to the output device.

11-8 SQUARE-LAW DETECTOR

In modern radio receivers the AM signal arriving at the antenna is first amplified to a high level and then demodulated. In the early days of radio good RF amplifiers were not available, so detector circuits of high sensitivity were required. One such circuit, the square-law detector, is shown in Fig. 11-9a.

The term "square-law detector" is derived from the shape of its transfer curve at the operating point. The mathematical equation for collector current is of the form $I_c = A + BI + CI^2$. It contains a "squared" term, hence the name. The main parts of the circuit are an *LC*-tuned circuit coupled to the antenna by transformer action. A signal developed in the *LC* tank is applied to the base of Q_1 which is biased to the desired operating point by resistor R_b. Capacitor C_c grounds the *LC* tank for RF but will not short-circuit the dc bias voltage to ground.

Sum and difference frequencies are produced during the detection process. These are listed in Fig. 11-9b. Capacitor C_f is used to filter out the residual RF frequencies ($2f_c + f_a$, $2f_c - f_a$, $2f_c$). At these frequencies it has low reactance and shunts them to ground. At the modulating frequency (f_a) it has a high reactance and thus almost no shunting effect. The modulation frequency, which is the transmitted intelligence, is therefore present at the output.

11-9 DIODE-DETECTOR CIRCUIT

The detector circuit used almost exclusively in modern radio receivers is the diode detector (Fig. 11-10a). Its operation can be explained in terms of heterodyning action of the carrier and its sidebands and the production of sum and difference frequencies. A more intuitive explanation can be produced by drawing attention to the diode's rectifying properties. Note the similarities between a diode-detector circuit and a half-wave-rectifier power-supply circuit, as shown in Fig. 11-10b. Recall that in the power-supply circuit (Sec. 2-4) only one of the input-voltage alternations is permitted to reach the load. The other is blocked by the diode. A diode-detector circuit operates in a similar fashion except that much higher frequencies are involved.

When the modulated carrier signals from many transmitters arrive at the antenna, signal currents flow in primary coil L_1 (Fig. 11-10a). These are induced by transformer action into the *LC* circuit which is tuned to resonance at the desired frequency. At resonance maximum signal voltage is developed across the tuning capacitor *C*, which is then applied to the diode-rectifier circuit. The diode permits only the positive-carrier alternations to flow in load resistor R_L. If capacitor C_f were not connected across R_L, these would develop pulsations as

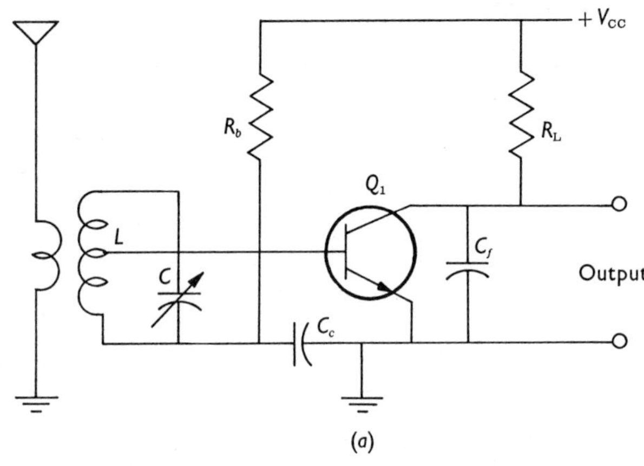

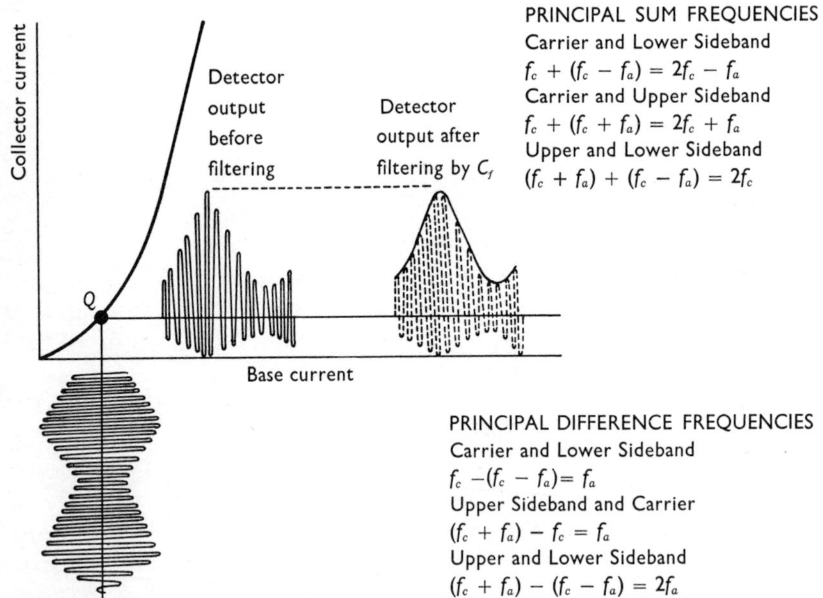

Fig. 11-9 Square-law detector circuit: (a) Schematic diagram; (b) Transfer curve showing output signal and frequencies generated in the heterodyne process.

shown in Fig. 11-10c(ii). Capacitor C_f behaves somewhat like a ripple filter. It charges to the peak value of the rectified carrier alternations. Between alternations some charge leaks off, but the amount is not appreciable [Fig. 11-10c(iii)]. The net result is that the voltage across C_f is a near replica of the modulation envelope of the RF carrier. This in turn is a replica of the original modulating frequency at the transmitter. In a broadcast receiver this would be program material, i.e., audio frequencies of speech and music. Figure 11-10d shows a schematic diagram of a diode detector with a potentiometer used as a load resistor. By ad-

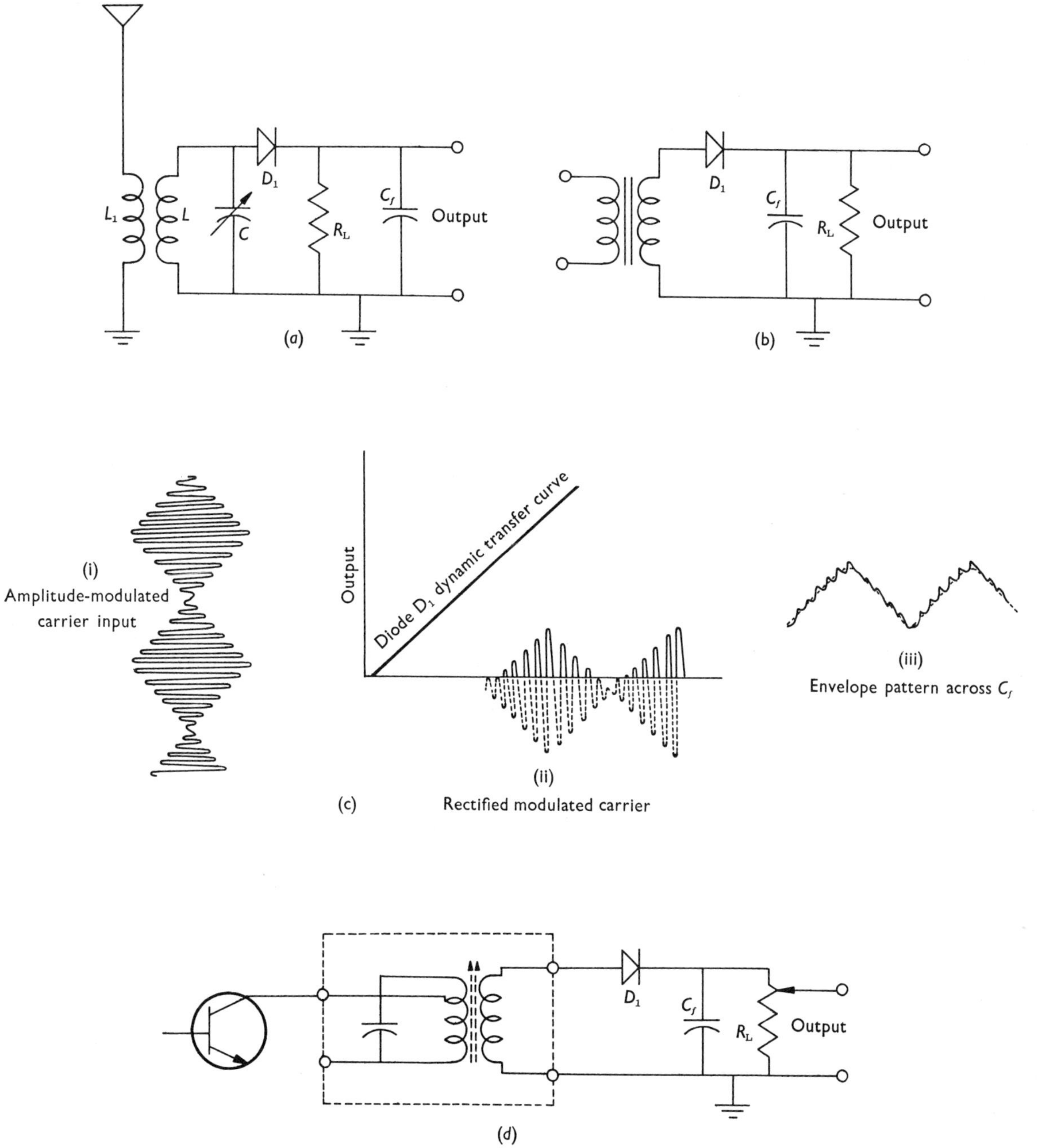

Fig. 11-10 Diode-detector circuit: (*a*) Schematic diagram; (*b*) Half-wave rectifier circuit to illustrate common features of both circuits; (*c*) Recovery of modulation frequency at the output; (*d*) Schematic diagram of detector circuit with adjustable output (volume) control.

justing the potentiometer (volume control), the output signal can be reduced to any desirable level before it is passed to the output-amplifier section of the receiver.

11-10 RF PREAMPLIFIER

A diode detector is not a very sensitive receiver. If a solid-state diode is used, the carrier signal must have a voltage amplitude greater than the barrier potential of the PN junction. In practice, a signal of several volts is desirable. At long distances from the transmitter, the signal voltage induced in the antenna must first be amplified before being applied to the detector. For this purpose, an RF amplifier stage such as is shown in Fig 11-11 may be placed before the detector.

The signal is picked up either by the ferrite-rod antenna or by an external antenna which may be coupled to the tuning coil L_1 by capacitor C_2. Coil L_1 with its ferrite-rod core has a high Q (of the order of 300), and ordinarily an external antenna is not required. The desired signal is coupled to L_2 by transformer action and then applied to the base of transistor Q_1 for amplification. Resistors R_1 and R_2 set the transistor operating-point bias on the linear part of its characteristic curve.

The amplified signal current at the collector is applied to a tap on L_3, which is tuned to the same frequency as L_1 by capacitor C_4. Coil L_4 applies a portion of the amplified signal to the diode detector.

A receiver which has one or more RF amplifier stages followed by a detector is known as a TRF (tuned radio frequency) receiver. Such circuits are rarely used because of the difficulty of designing RF amplifier stages capable of tuning over a wide frequency band. Tuning over the AM broadcast band, for example, represents a frequency ratio of approximately 3:1 (1600 kHz : 500 kHz).

11-11 SUPERHETERODYNE PRINCIPLE

The radio receiver most widely used is the superheterodyne* circuit shown in block diagram form in Fig. 11-12. A superheterodyne

*Its name (the heterodyne portion) suggests the mixing of frequencies to obtain new ones by some modulation process. The prefix "super" was added at the time of its discovery. It was intended to indicate its superiority as compared with TRF receivers of the day.

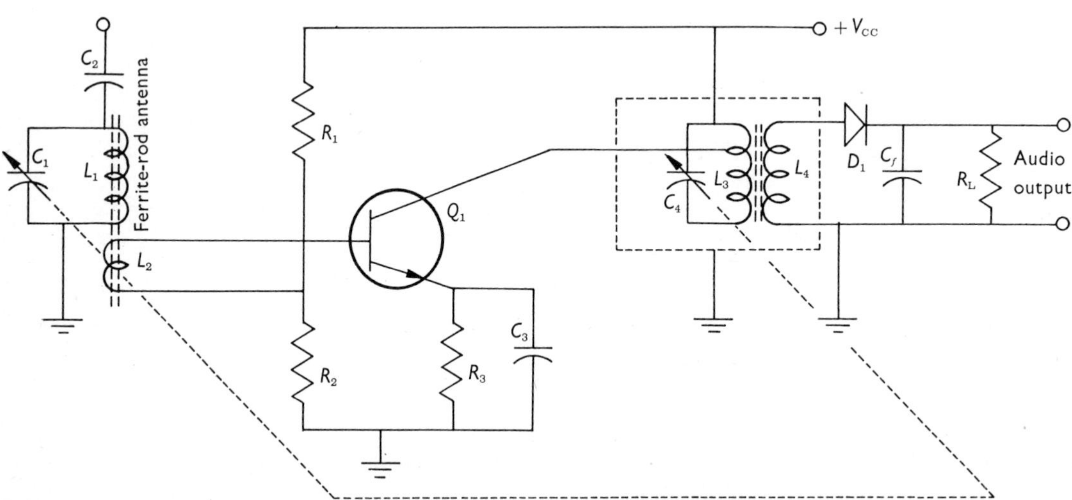

Fig. 11-11 Diode detector with RF amplifier stage.

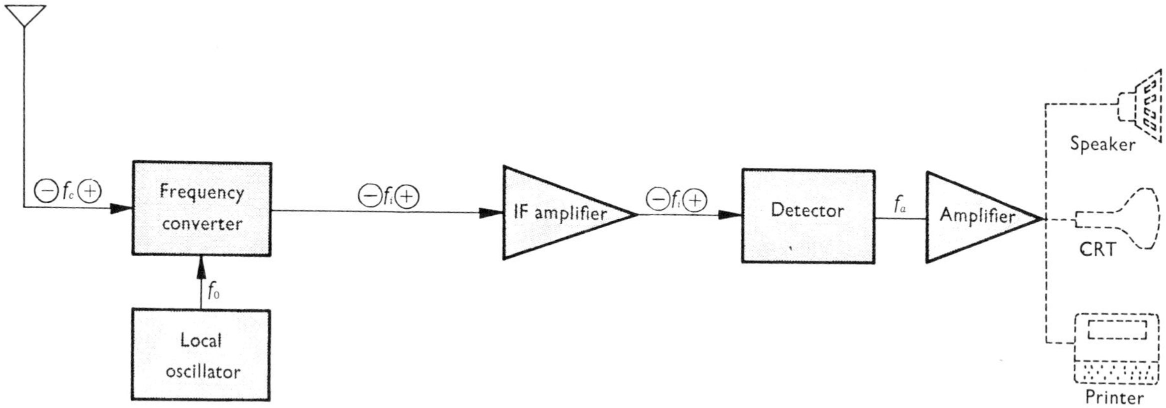

Fig. 11-12 Block diagram of the essential parts of a superheterodyne receiver.

circuit minimizes the compromises in RF stage design by changing all incoming signals to a single intermediate frequency (IF). The intermediate frequency and its sidebands are then amplified by an IF amplifier and applied to the detector. Here the modulating frequency (f_a) is recovered, amplified, and fed to an output device such as a speaker. These facts are depicted in Fig. 11-12. To condense the notation in the drawing, sidebands are indicated by ⊖ for the lower and ⊕ for the upper. Note that sidebands are always preserved in the heterodyning process in the frequency converter. In effect, the *converter* acts like a transmitter which *substitutes an intermediate frequency carrier signal* for the original incoming carrier.

11-12 SUPERHETERODYNE CONVERTER STAGE

The converter stage can be considered as a modulator, detector, or demodulator. Each point of view is at least partially correct, because all these devices have one function in common. They produce sum and difference frequencies from two given input signals. A converter stage is usually a form of square-law detector. One type of circuit is shown in Fig. 11-13. Here a separate oscillator produces the locally generated signal required for the heterodyning process. Converters can be designed to self-generate the local signal as well as performing the heterodyne function.

Signals arriving in the ferrite-core antenna (Fig. 11-13) are applied via L_2 to the base of Q_1. Also arriving at the base via capacitor C_3 is the locally generated signal from the oscillator. Oscillator frequency always differs from the received signal by the same amount because the local oscillator "tracks" the antenna-tuning circuit. Its tuning capacitor C_2 and the antenna-tuning capacitor C_1 are attached to the same shaft. If the radio dial is set to a high frequency, the local oscillator produces a high frequency. If a lower dial setting is selected, the local oscillator frequency is similarly reduced by the same amount. A numerical example using AM broadcast-band signals will clarify this concept.

Assume an IF frequency of 455 kHz, which is typical for broadcast-band receivers. Suppose first a signal of 560 kHz is tuned in. The local oscillator frequency is then $560 + 455 = 1015$ kHz. Table 11-1 shows incoming signal and oscillator frequencies for two dial settings. The graph in Fig. 11-14 shows local oscillator frequency and incoming frequency for any dial setting on the AM broadcast band. Note that with proper tracking, the intermediate frequency is always 455 kHz.

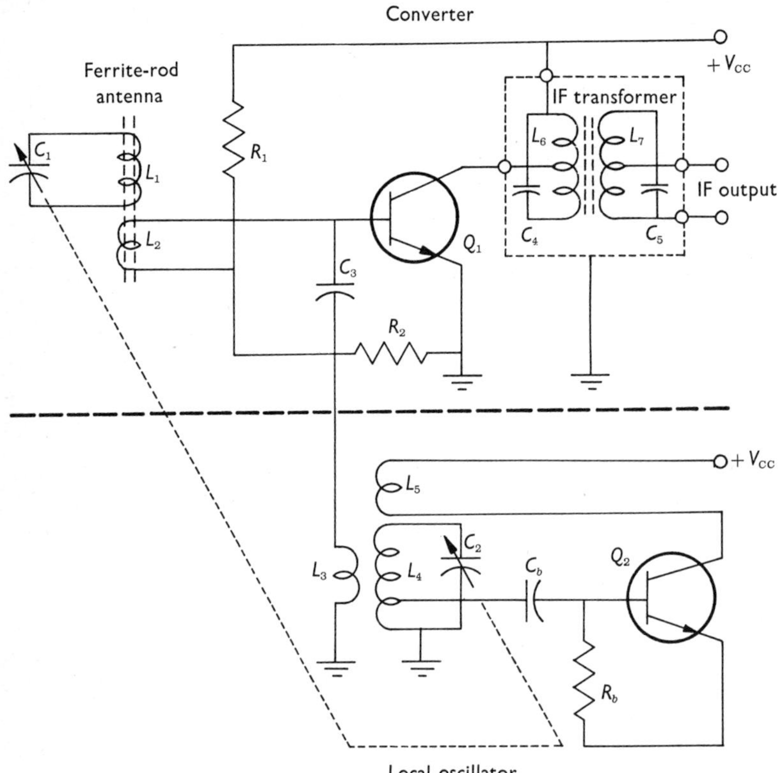

Fig. 11-13 Transistor converter circuit.

Table 11-1 Relationship between Local Oscillator and Incoming Signal Frequency

Received Signal (kHz)	*Oscillator Frequency* (kHz)	*Difference or IF Frequency* (kHz)
560	1015	455
1300	1755	455

In the converter's collector circuit are found the various sum and difference frequencies made up of the carrier, its sidebands, and the local oscillator signal. Only the intermediate frequency and its immediate sidebands are extracted by the tuned circuit consisting of L_6 and C_4. These two components make up the primary circuit of the IF transformer. Both primary and secondary are adjustable over a narrow range. Once tuned to the required IF, no further adjustment is required.

One weakness of a superheterodyne receiver is that more than one signal can combine with the local oscillator to produce the required intermediate frequency. Any undesired signals which produce an IF are known as image frequencies. If for example a receiver is tuned to a signal of 560 kHz, the local oscillator is producing a frequency of 1015 kHz to give a 455-kHz IF. A strong signal at 1470 kHz could enter the converter in spite of the tuned-antenna circuit's efforts to reject it. Once present in the converter it would beat with the local oscillator to produce an IF, and from this point on it would be treated like the desired signal. It is possible to have several image frequencies, especially if the local oscillator produces strong second or third harmonic frequencies. These, too, could beat with strong signals to produce image frequencies. An RF preselector stage helps to

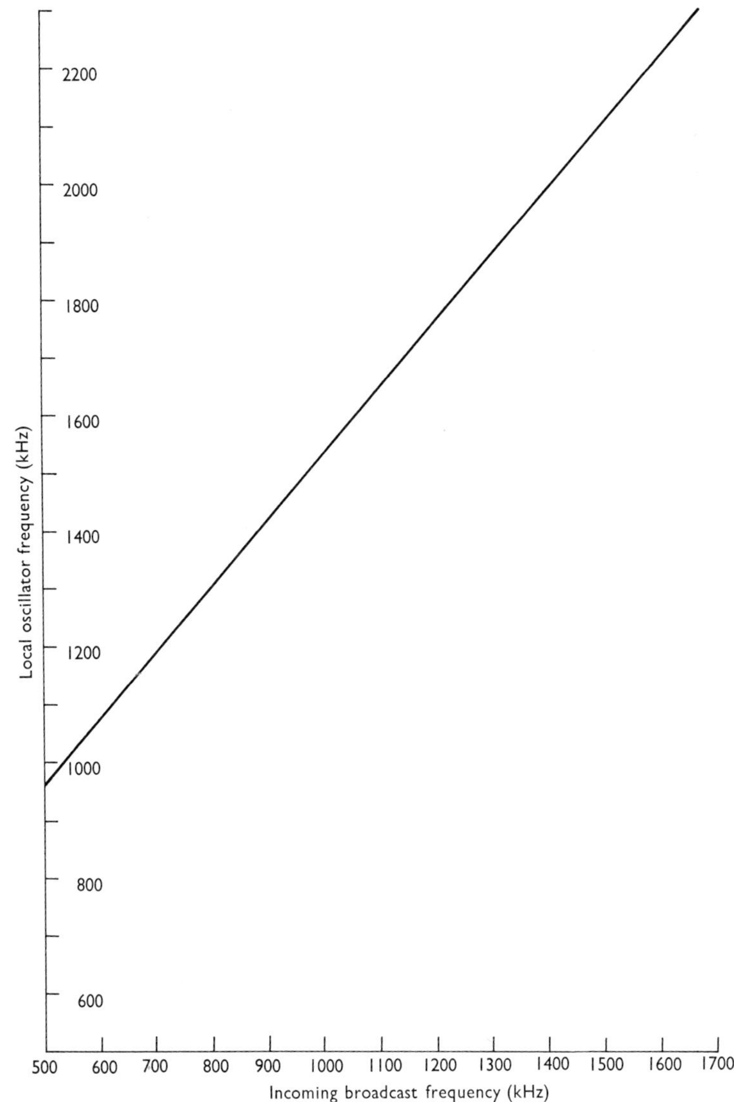

Fig. 11-14 Frequency relationship between local oscillator and incoming frequency (AM broadcast receiver, IF = 455 kHz).

reject image frequencies and is nearly always found on better-quality receivers.

11-13 AUTODYNE CONVERTER

Figure 11-15a shows a schematic diagram of an *autodyne* converter stage. This circuit not only heterodynes but functions as an oscillator as well. It features three tuned circuits as follows: (1) The antenna circuit consisting of L_1C_1, (2) the local oscillator circuit consisting of L_4C_2, and (3) the IF transformer primary circuit consisting of L_5C_5. Each is tuned to a different frequency, and each functions independently of the other. An isolated drawing of the oscillator section is shown in Fig. 11-15b. Electrically it is a tickler-type oscillator (Sec.

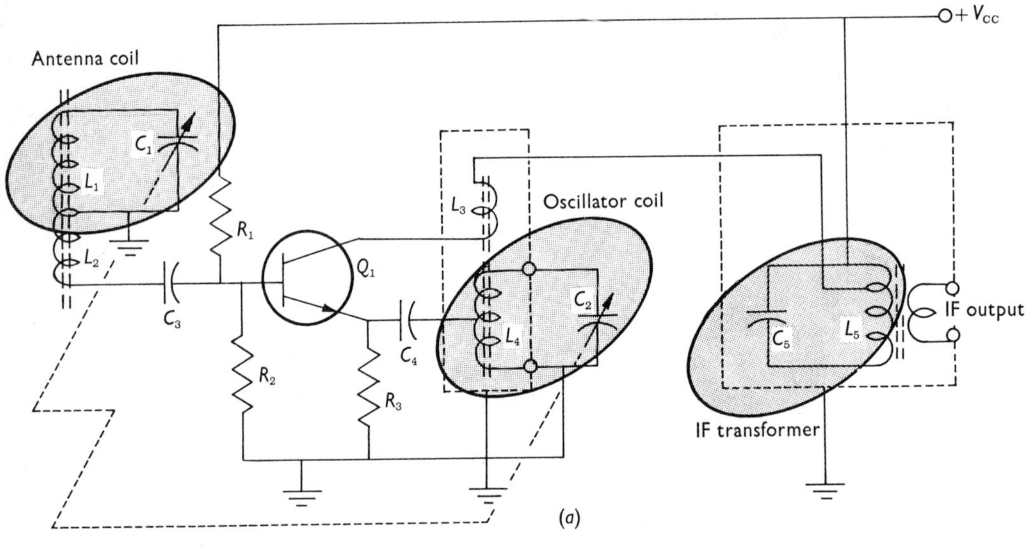

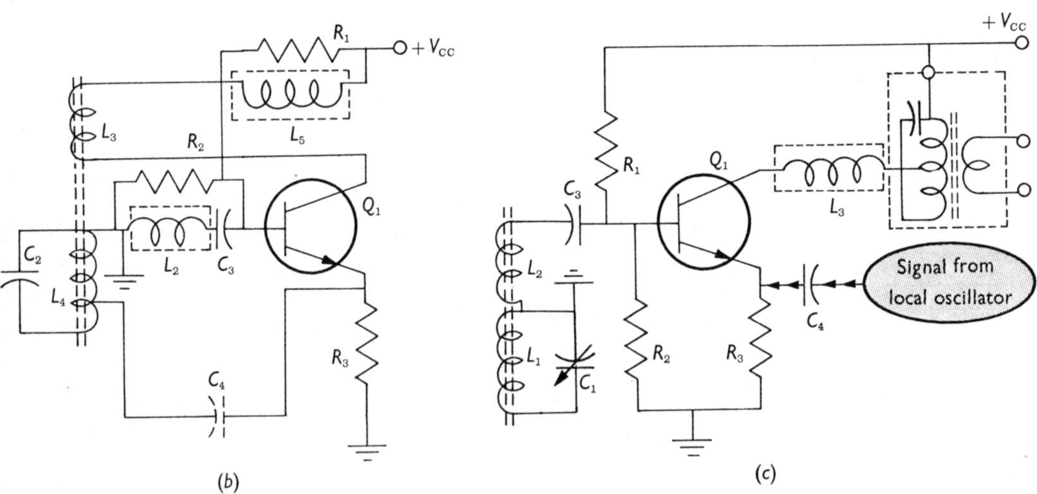

Fig. 11-15 Autodyne converter circuit: (a) Schematic diagram; (b) Isolated drawing of oscillator section only; (c) Isolated drawing of converter section only.

10-7). Although coils L_2 and L_5 are connected into the circuit, they do not interfere appreciably with oscillator operation. When viewed as a converter circuit (Fig. 11-15c) the circuit appears as a square-law detector biased into the nonlinear operating region by R_1 and R_2. Oscillator coils L_3 and L_4 although connected do not interfere with converter operation. The local oscillator frequency "enters" the converter section through capacitor C_4.

The autodyne circuit is very popular in battery-operated AM radios because only one transistor is required to perform conversion. One of its disadvantages is that it generates noise which can be heard as a hiss on weak signals.

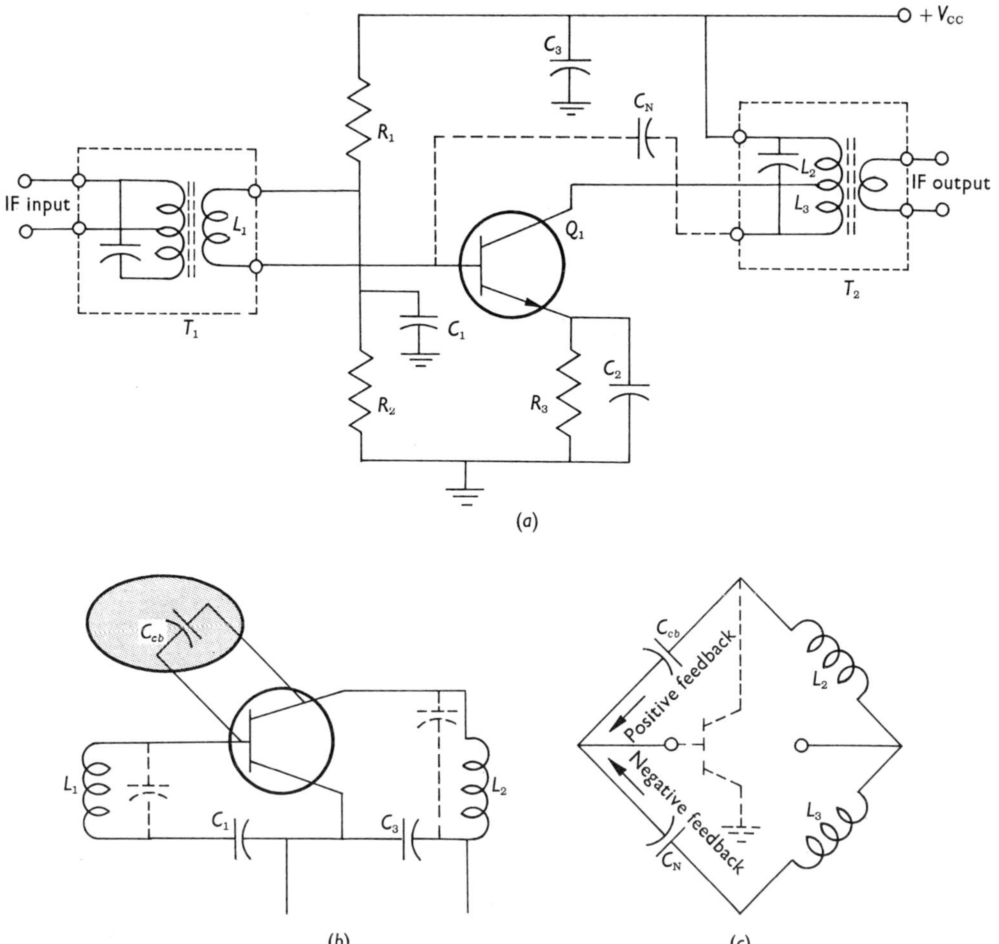

Fig. 11-16 IF amplifier stage: (a) Schematic diagram showing neutralizing capacitor C_N; (b) Circuit features which can make it possible for the circuit to become a TCTB oscillator; (c) Balanced feedback condition when neutralized.

11-14 IF AMPLIFIER

A typical IF amplifier circuit is shown in Fig. 11-16a. The input signal is applied via the secondary coil L_1 of IF transformer T_1. Resistors R_1 and R_2 fix the transistor operating point on a linear section of its transfer curve. The amplified signal is applied to L_2, the primary of the output IF transformer.

Since both T_1 and T_2 are tuned to the same frequency, there is a tendency for the IF stage to oscillate as a TCTB oscillator (Sec. 10-11). Figure 11-16b is a circuit simplified to show the features which make the IF amplifier resemble a TCTB oscillator. At reduced collector supply voltage, the collector-base junction capacitance (C_{cb}) increases, and in battery-operated receivers the risk of oscillation may increase as the batteries near exhaustion. To stabilize the amplifier against this type of oscillation, a neutralizing capacitor (C_N) is often included. (It is shown in Fig. 11-16a in broken lines to indicate that it is not always required.)

Neutralization is accomplished by feeding some antiphase voltage to the base to cancel the feedback signal voltage arriving via the collector-base junction capacitance. The antiphase voltage is obtained from a tap on the primary of transformer T_2, as shown in Fig. 11-16a. Electrically this has the effect of adding an arm to complete a bridge circuit, as shown in the simplified diagram in Fig. 11-16c.

11-15 IF STAGE FREQUENCY RESPONSE

Ideally an IF stage should amplify only the intermediate frequency and its information-bearing sidebands. An ideal frequency-response curve for an IF amplifier would appear as shown in Fig. 11-17a. It would have a flat top and sharp skirts or edges. The flat top would ensure that all sidebands receive equal gain. The steep skirts would ensure that all frequencies outside the IF channel receive severe attenuation. The IF stage gives a superheterodyne receiver its selectivity. In practice, the top of the curve is rounded and the skirts spread near the bottom, as shown in Fig. 11-17b.

Many techniques exist for tailoring the IF response curve to meet certain needs. Some receivers (for example, expensive communications receivers) are equipped with controls to adjust the IF selectivity at will. In this way maximum selectivity can be obtained when listening to Morse code in a crowded amateur band. Minimum selectivity can be selected when listening to music to obtain better fidelity because the higher-frequency sidebands which represent high notes can then pass through the IF amplifier.

Selectivity can be increased by the following means: (1) Decrease the coefficient of coupling between primary and secondary coils. This reduces gain, and an extra stage may be required. (2) Crystal or mechanical filters may be added to obtain a high Q in the IF stage. A "flat top" in the response curve to improve bandwidth can be obtained by *stagger tuning*. With this technique each of the IF transformers is peaked to resonance at slightly different frequencies. Whatever enhancement one tuned circuit provides to a specific frequency is offset by the other tuned circuits. Taken as a whole, they pass a wider band between them, as shown in Fig. 11-17c.

11-16 AUTOMATIC GAIN CONTROL (AGC)

Automatic gain control (AGC) or automatic volume control (AVC), as it is also denoted in broadcast radio receivers, adjusts the overall gain of the receiver to a nominal level. It is a form of negative feedback particularly useful for gain compensation with signals which fade occasionally. The primary purpose of AGC, however, is to prevent overdriving of the converter and IF amplifier stages when a powerful signal is tuned in. AGC is accomplished by developing a gain-control voltage proportional to the incoming signal strength. This control voltage can be used to vary the gain of one or more of the IF amplifiers, the converter, and the RF amplifier stage.

In bipolar transistors, the gain is dependent on quiescent emitter current as shown in a typical graph in Fig. 11-18a. To reduce gain the forward bias may be reduced. This results in lower emitter current and therefore reduced gain. Figure 11-18b and c show how the reduction in bias is achieved for strong signals.

If signal strength increases for any reason, the average output from the detector (Fig. 11-18b) tends to rise. The simplified diagram in Fig. 11-18c shows that the average output is a negative voltage obtained by rectifying the IF signal with detector diode D_1. It is applied via R_5 to a smoothing capacitor (C_1). The actual voltage across C_1 will depend on two applied voltages, the positive fixed bias, set by R_1, and the variable negative AGC voltage. If the AGC voltage increases, it cancels some of the positive fixed bias. Emitter current then falls, and IF amplifier gain is reduced.

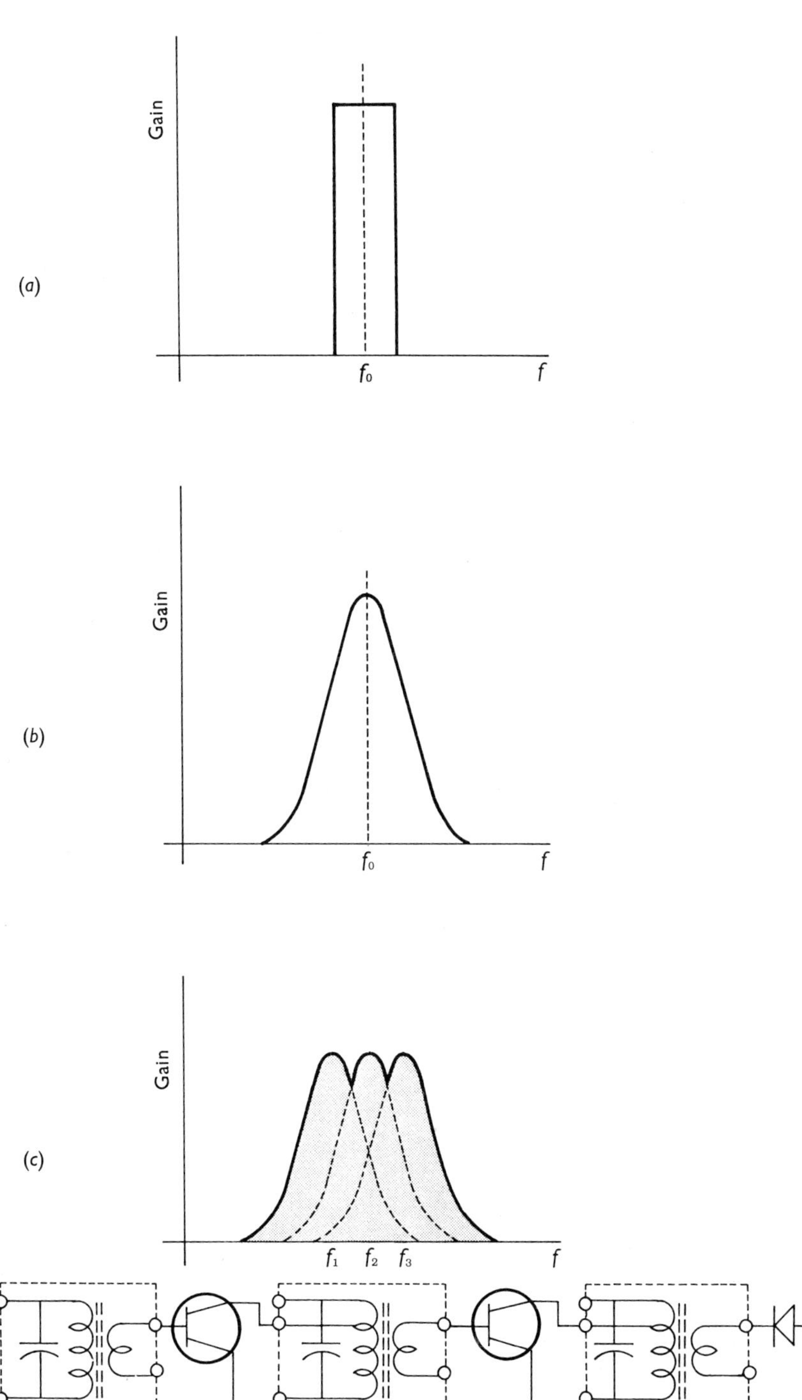

Fig. 11-17 IF amplifier frequency response: (*a*) Ideal; (*b*) Typical; (*c*) Obtaining broadband response by means of stagger tuning.

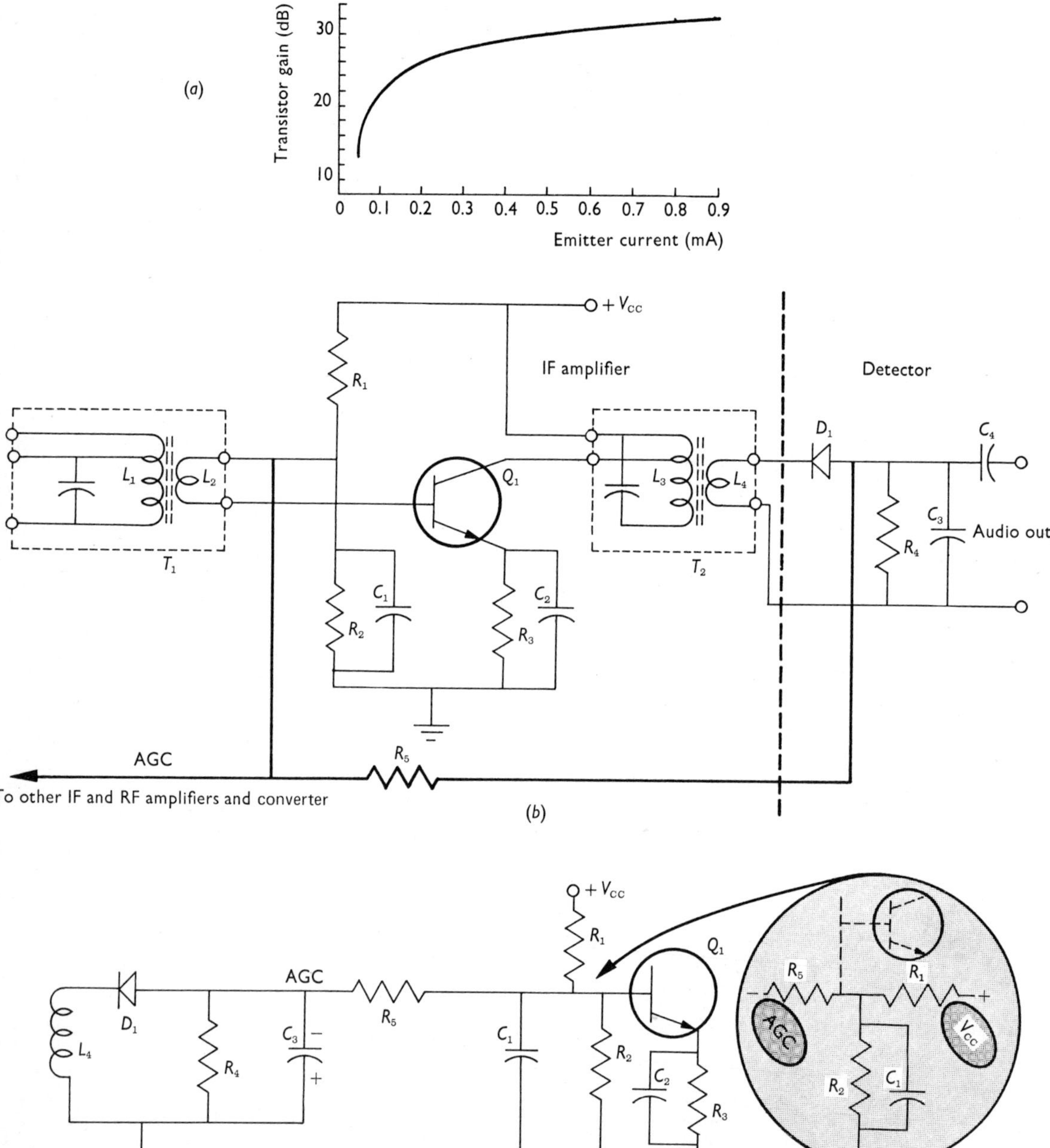

Fig. 11-18 Principles of automatic gain control: (*a*) Transistor gain vs. emitter current; (*b*) Typical AGC circuit; (*c*) Isolated diagram showing the source and load for AGC voltage. AGC and fixed bias voltages are summed at the base of Q_1.

When signal level falls for any reason the gain is increased because AGC voltage falls. This allows more positive bias to be developed for the transistor. Since gain is automatically adjusted up or down, the output-signal level remains relatively constant over wide signal strength variations.

11-17 DYNAMIC RANGE OF AGC

AGC is effective only within certain limits. If signal level becomes too great, overdriving of some stages can occur even with AGC. Generally, the more stages under AGC control, the more effective the AGC action.

Some circuits employ a dc amplifier to boost the AGC voltage. This also lowers the load on the second detector, thereby improving the effective Q of the IF output transformer. Typical circuits are shown in Fig. 11-19a and b.

The first circuit (Fig. 11-19a) employs a common-emitter dc amplifier. The detector diode D_1 is connected so that positive voltage from the detector is applied via R_4 and R_6 to the base of Q_2. If this voltage is low, Q_2 collector current is low, and collector voltage is high because of the small voltage drop across R_7. Bias for the AGC-controlled stages is taken from the collector of Q_2. An increase in signal strength causes a rise in detector output which increases Q_2 collector current. The collector voltage on Q_2 then drops with resulting loss of bias on Q_1, and the IF gain is reduced.

The second circuit (Fig. 11-19b) employs an emitter follower connected to the emitter resistor of the IF amplifier. Increased emitter current through Q_2 increases the voltage drop across R_3 and therefore reduces the bias on Q_1, reducing its emitter current and therefore its gain.

There is a limit to the amount of attenuation obtainable with AGC without signal distortion. The circuit shown in Fig. 11-19c can be used to "take over" to give further gain reduction after normal AGC action is at maximum. At maximum AGC the collector current of Q_1 will be at its lowest level with resultant minimum voltage drop across R_4. Diode D_1 then becomes forward-biased and begins to conduct. It then provides a low-resistance shunt path to the ground for the converter output signal. The shunt path is shown in the simplified diagram in the inset of Fig. 11-19c.

11-18 AGC RESPONSE TIME

All AGC circuits have delayed response. If they were to act instantaneously even output-signal amplitude variations would be nullified. For speech and music a response time of the order of $\frac{1}{10}$ s seems satisfactory. For video receivers a faster response is desirable. Response time is determined by the *CR* time constants of the resistors and capacitors in the AGC line.

11-19 RECEIVER OUTPUT STAGES

After preselection, conversion, and IF amplification, the received signal is fed to a detector where the signal intelligence is recovered and the AGC voltage generated. Detector output is usually a low-level signal and must be further amplified before being applied to the output device. In broadcast radio receivers an audio amplifier followed by an output power amplifier would be used to drive a speaker. In the video section of a television receiver a video amplifier would be used to drive the picture tube.

11-20 COMPLETE AM BROADCAST RECEIVER

Figure 11-20 shows a schematic diagram and pictorial layout of a typical battery-operated radio receiver. All circuits discussed in Secs. 11-9 to 11-18 are represented. A more complex receiver capable of AM and FM reception, typical of the types found in home hi-fi systems, is discussed in Appendix I.

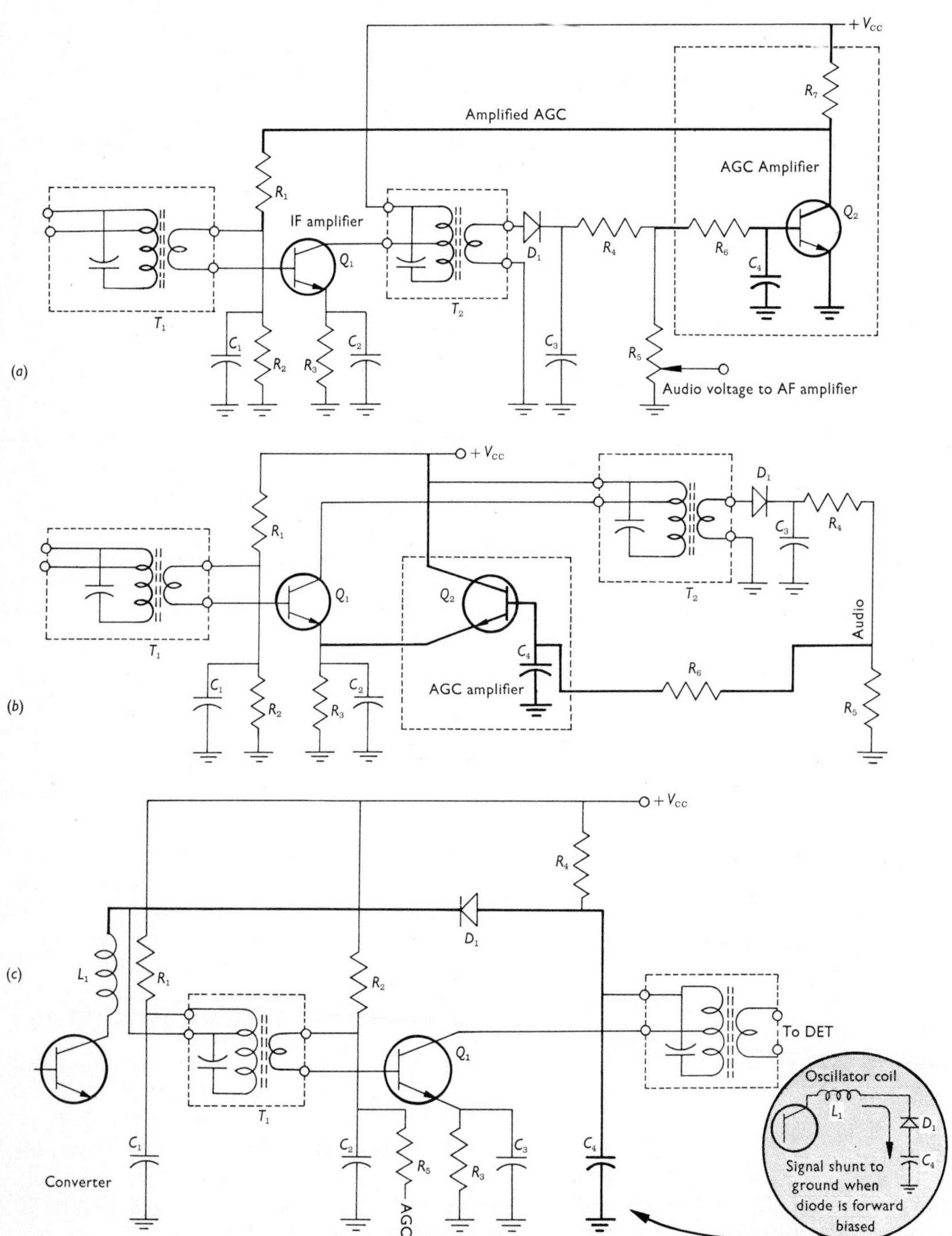

Fig. 11-19 Increasing the dynamic range of AGC circuits: (*a*) AGC amplifier, common-emitter connection; (*b*) Emitter-follower connection; (*c*) Shunt diode switched on when AGC voltage surpasses a threshold level.

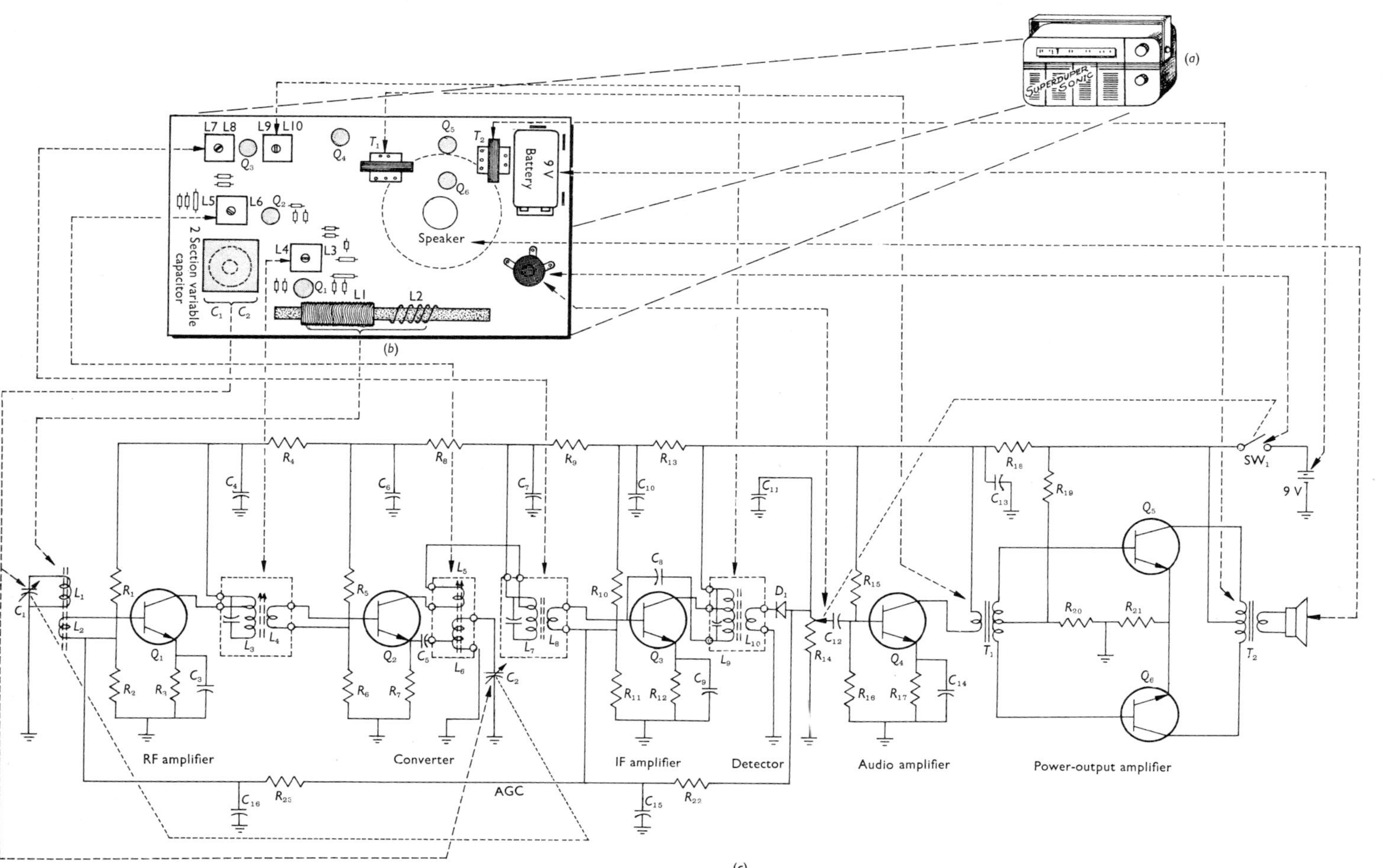

Fig. 11-20 Portable AM radio receiver: (a) Radio in carrying case; (b) Printed circuit board showing component layout; (c) Schematic diagram.

11-21 SUPERHETERODYNE ADJUSTMENTS

For maximum sensitivity, selectivity, and all-round performance the tuned circuits in a superheterodyne receiver must be set correctly. The procedure of tuning the circuit is known as *alignment*. It consists mainly of (1) tuning the IF transformers to resonance at the IF frequency specified, and (2) adjusting the local oscillator frequency to give the correct output at all dial settings (tracking adjustments).

In most circuits the IF transformers are adjusted by injecting the correct IF signal at the converter. The known IF is generated by a test instrument known as a signal generator. It is essentially a low-power transmitter simulator. It transmits a carrier which can be modulated with a tone. The IF transformers are adjusted for maximum output. Output can be judged by measuring the amount of AGC voltage being developed. Next the local oscillator frequency is checked to ensure that it is generating the proper IF when a known frequency is being received. Finally the RF amplifier is peaked to provide optimum gain at all dial settings.

11-22 FM RECEIVERS

FM receivers are of the superheterodyne type. The main difference between AM and FM receivers is in the detector circuit. Other differences are due to the operating frequencies, and a frequency comparison for receivers operating in the broadcast band is shown in Table 11-2.

Table 11-2 Frequency Range of AM and FM Broadcast Receivers

	Broadcast Band	*IF*	*Local Oscillator Tuning Range*
AM	535–1605 kHz	455 kHz	990–2060 kHz
FM	88–108 MHz	10.7 MHz	77.3–97.3 MHz

Because the IF in FM broadcast receivers is approximately twenty times greater than in AM receivers, the IF transformer coils contain relatively few turns of wire. At 455 kHz they offer no appreciable opposition to signals; the same amplifier circuit may be used for both AM and FM. A dual AM–FM IF amplifier is shown in Fig. 11-21. If a 455-kHz AM IF signal enters the circuit, it passes easily through the primary coil of T_1 and is impressed across the primary of T_3, its own IF transformer. The 455-kHz signal is then induced in the secondary of T_3 and passes easily to the base of transistor Q_1 where it is amplified. The 455-kHz collector current is transferred to the next stage by T_4.

If a 10.7-MHz signal enters the circuit, the 455-kHz transformer coils offer so much opposition that they appear as radio-frequency choke coils. They actually help to ensure that the 10.7-MHz signal is confined to the FM IF transformers.

Although AM–FM receivers often contain common IF amplifiers, they have separate converter and detector stages. Also the FM section usually requires more IF gain. The FM signal is greatly boosted and then clipped to remove amplitude variations. The clipping or *limiting* is done in the last one or two IF stages.

11-23 FM DETECTOR CIRCUITS

An FM detector is a circuit which changes frequency variations into voltage variations. Traditionally, some form of tuned *LC* circuit has been used as the frequency-discriminating device. One such discriminator circuit, the *balanced phase-shift discriminator,* is shown in Fig. 11-22a.

Essentially, the balanced phase-shift circuit compares the phase relationship between two voltages. Although both voltages are of the same frequency, they differ in phase. The phase difference is constant as long as there is no change in frequency. A change in frequency causes a change in phase and consequently in the output voltage. This is the way in which frequency variations are converted into voltage variations.

In the circuit in Fig. 11-22a the IF signal induced in the secondary coils (L_{2a} and L_{2b}) is

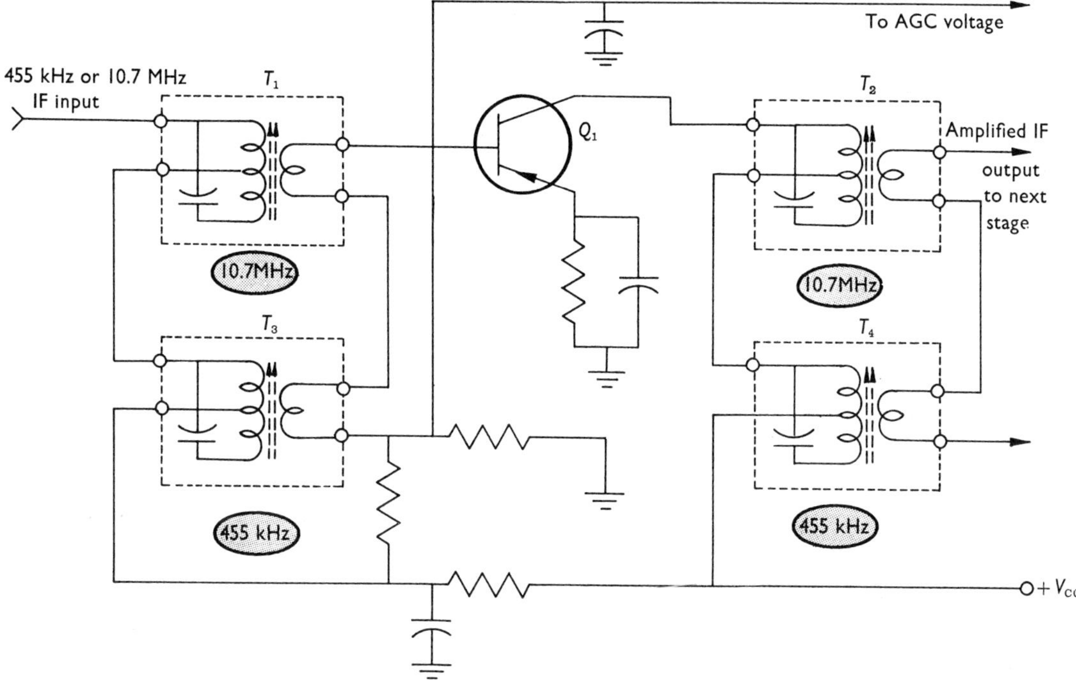

Fig. 11-21 Dual IF amplifier circuit. Operates at either 455 kHz or 10.7 MHz.

compared with the primary voltage which is coupled into the secondary circuit by means of capacitor C_3. This capacitor has a low reactance for IF signals and introduces virtually no phase shift. In the diagram the primary voltage is labeled E_{ref} and the secondary voltage E_{comp}. At the center frequency (i.e., at resonance) the primary and secondary signals are 90° out of phase.

If terminal *a* of the secondary coil L_{2a} is positive with respect to *b*, diode D_1 will be forward-biased into conduction. Under this condition E_{ref} and E_{comp} are added and the vector sum of their voltages charges capacitor C_1. On the next half-cycle C_2 is charged when diode D_2 conducts. At center frequency both capacitors should be receiving equal charging voltage, and net output should be zero because both are charged to opposite polarity. If the frequency deviates from center, there is no longer a 90° phase difference between E_{ref} and E_{comp}, which means that either C_1 or C_2 will receive the greater voltage, the imbalance being proportional to the deviation in frequency. A net output results as shown in the phasor diagrams and the simplified circuits in Fig. 11-22b, c, and d.

It is perhaps easier to visualize the operation of the discriminator circuit if it is redrawn as a bridge circuit, as in Fig. 11-23. In this way the interdependence of the primary and secondary voltages on the "balance" of the circuit is more evident.

The balanced phase-shift discriminator has the disadvantage that its output changes with signal amplitude as well as with frequency. An ideal FM discriminator should ignore amplitude variations.

A more popular FM detector, the ratio detector, is shown in Fig. 11-24a. The essential feature of this circuit is that the diodes are connected so that both conduct at the same time, but only for half a cycle. The other half-cycle is ignored by the circuit. Also note that E_{ref} is injected by a pickup coil (L_{ref}) rather than by a capacitor.

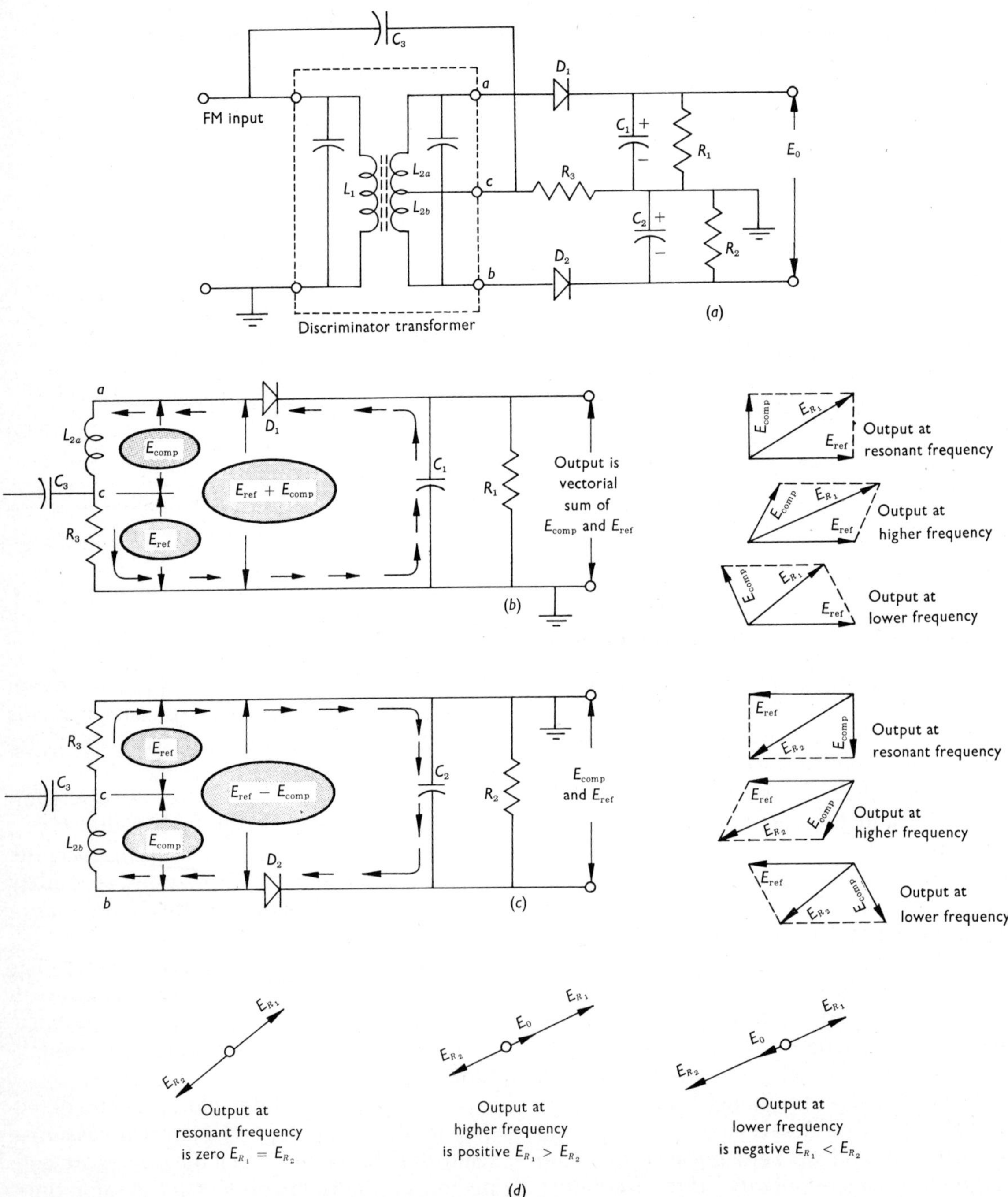

Fig. 11-22 Balanced phase-shift discriminator: (*a*) Essential parts of circuit; (*b*) Simplified circuit when D_1 is conducting; (*c*) Simplified circuit when D_2 is conducting; (*d*) Output voltage under frequency deviation.

At center frequency capacitors C_1 and C_2 divide the output voltage equally, and instantaneous output signal is zero. During a deviation in frequency E_{ref} and E_{comp} undergo a relative phase change with a resultant imbalance in voltage distribution across C_1 and C_2. A net output will then occur as shown in the phasor diagram in Fig. 11-24c.

An alternate form of discriminator-transformer arrangement is shown in Fig. 11-24d, and a complete FM receiver including the ratio detector is shown in Appendix I.

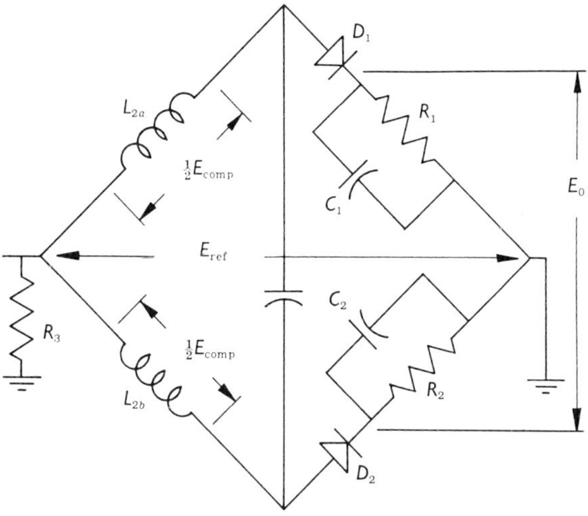

Fig. 11-23 Essential parts of a balanced phase-shift discriminator visualized as a frequency sensing bridge. At the center frequency (resonance) voltage vectors balance to give zero output.

11-24 REQUIREMENTS FOR HIGH FIDELITY

An ideal high-fidelity system must be capable of reproducing the entire audio-frequency spectrum in both pitch and volume. To add dimension the system should preserve the directional qualities of the original sound. This can be achieved with stereophonic reproduction. The ratio of power level emanating from a full orchestra to that from a single violin could be as great as $70\ \text{W}/4 \times 10^{-3}\ \text{W}$ or 17 500. The sound intensity range in decibels can be calculated from the equation

$$\text{dB} = 10 \log \frac{P_1}{P_2}$$

Substituting 17 500 for the power ratio P_1/P_2 gives an intensity range of

$$\begin{aligned}\text{dB} &= 10 \log 17\ 500\\ &= 10(4.2430)\\ &= 42.4\ \text{dB}\end{aligned}$$

Reproduction of such a wide range of loudness levels places great demands on a system. At low sound levels crossover distortion and background noise such as power-supply ripple must be minimized. At high sound levels the system must be capable of generating several hundred watts of signal, which when fed to low-efficiency speaker systems results in a few score watts of sound power.

Fortunately, domestic systems do not require the sound power needed to fill an auditorium. With a graduated series of compromises sound-reproduction systems are available in a wide price range.

11-25 SPECTRAL COMPROMISE

The audio-frequency spectrum ranges from 16 to 20 000 Hz. It is by no means necessary to reproduce the entire spectrum to offer pleasing performance to a majority of listeners. Adequate mid-frequency reproduction can be achieved relatively inexpensively. Distortionless low-frequency reproduction is more difficult to achieve than high-frequency. It has been found that a balance between low- and high-frequency reproduction is most pleasing. For example, a trimmed upper and lower response is more desirable than good response up to 20 000 Hz with poor low-frequency performance.

Always keep in mind that quite apart from technical capabilities a sound system is judged subjectively by listener tests. If in a given system the lowest and highest reproducible frequencies are multiplied together, and if the

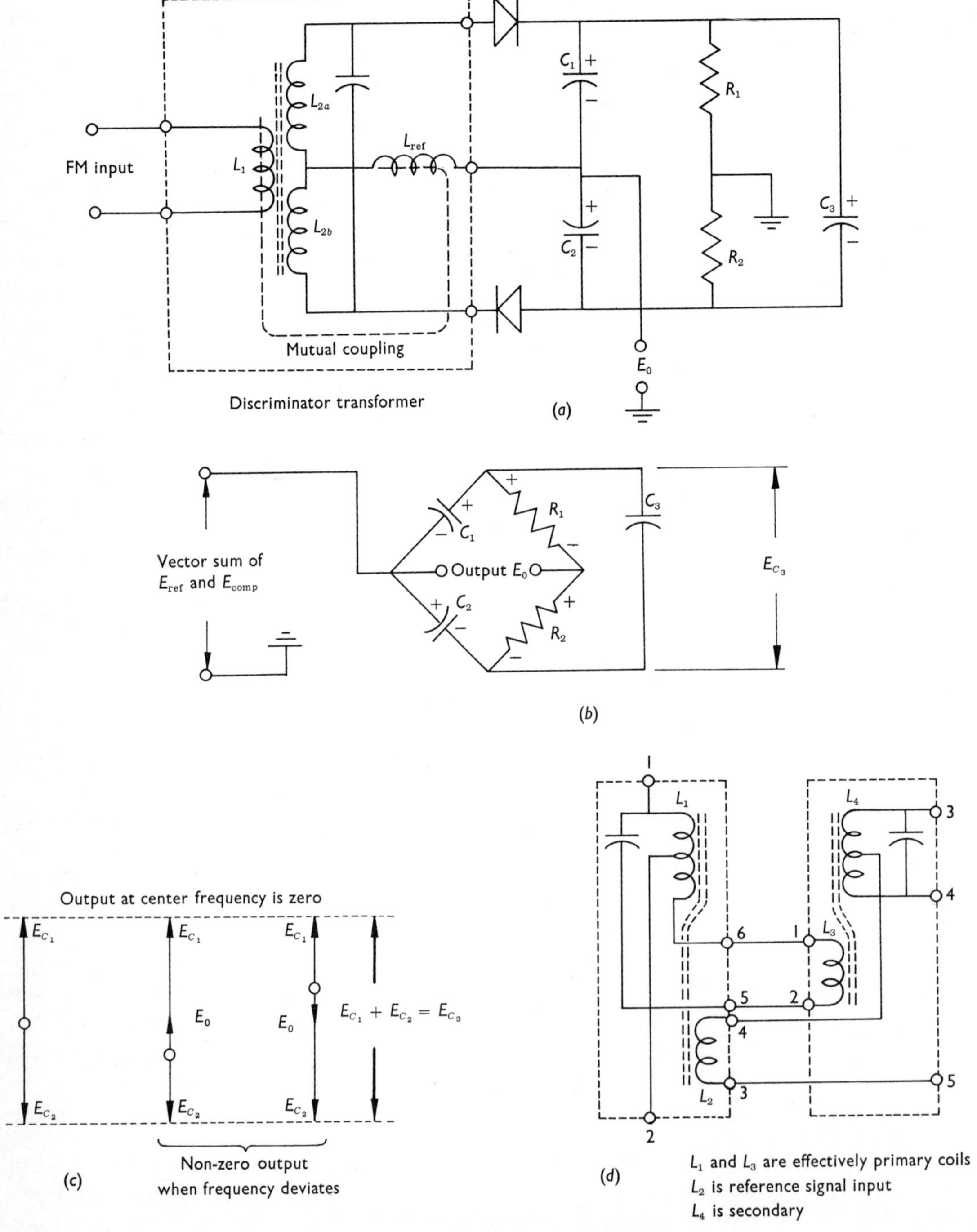

Fig. 11-24 FM ratio detector: (a) Typical schematic diagram; (b) Simplified circuit visualized as a bridge circuit; (c) Output-voltage vectors; (d) Alternate form of discriminator transformer.

product is in the range 400 000 to 500 000, the system will sound pleasing. Of course, the more octaves a given system can reproduce, the better its quality, even though the figure 400 000 can be obtained with a variety of upper- and lower-frequency combinations. A few of these are:

1. $400 \times 1000 = 400\ 000$ (very poor, not even adequate for speech)
2. $100 \times 4000 = 400\ 000$ (approximately AM standard broadcast quality)
3. $40 \times 10\ 000 = 400\ 000$ (very good, obtainable with many commercially available hi-fi sets)
4. $20 \times 20\ 000 = 400\ 000$ (excellent, obtainable with custom installations)

11-26 BASS-, TREBLE-, AND LOUDNESS-CONTROL CIRCUITS

Most home-entertainment systems and many portable and automobile receivers are equipped with some form of "tone" control circuit. Tone adjustments allow listeners to trim the frequency response of amplifiers to their preference.

One simple tone-control circuit is shown in Fig. 11-25a. It is essentially an adjustable low-pass *CR* filter. When R_5 is adjusted to maximum resistance, it inhibits signals from taking the shunt path to ground via capacitor C_5 (Fig. 11-25b). At minimum resistance, only the capacitor lies in the shunt path. Since it has low reactance at high frequencies, these are shunted to ground. The low-frequency signals continue on to the next stage.

Another form of tone control found in inexpensive transistor radio receivers is shown in Fig. 11-25d. It uses the principle of negative feedback to reduce gain at higher frequencies. With the switch in the open position the amplifier has a flat frequency response. When closed, capacitor C_3 allows higher-frequency signals to return from the collector to the base. They arrive out of phase and cancel some of the input signal. High-frequency gain is thereby reduced.

In better-quality hi-fi systems, two controls are provided for adjusting frequency response. One adjusts the low-frequency gain and is labeled the *bass* control. The other, the *treble* control, trims the high-frequency gain. With a well-designed circuit the mid-frequency gain should remain relatively constant at all settings of the bass and treble controls. A typical circuit is shown in Fig. 11-26a, and the circuit equivalent for maximum bass and maximum treble settings is shown in Fig. 11-26b and c.

At maximum bass setting the circuit is essentially a low-pass *pi section CR* filter as shown in Fig. 11-26b. High-frequency signals are shunted to ground by capacitors C_2 and C_4. In the maximum treble setting the circuit acts as a high-pass filter, because capacitor C_3 has a high reactance for low frequencies.

At low-volume levels the reduction in sound power is most noticeable at low frequencies. In some amplifiers a compensating circuit is used to attenuate high frequencies automatically if the gain-control setting is reduced. One such circuit is shown in Fig. 11-26d. It uses a gain control with a fixed tap (R_6) to which a low-pass *CR* filter (C_4R_5) is attached. At reduced gain settings, the gain-control slider arm resides in the vicinity of the fixed tap and the low-pass filter shunts high frequencies to ground. Such a circuit arrangement is often labeled a *"loudness" control* to distinguish it from an ordinary volume or gain control.

Figure 11-27 shows oscilloscope photographs of amplifier-frequency response when the bass-treble circuit in Fig. 11-26a is used. The amplifier is driven with a signal from a sweep-type audio oscillator with constant amplitude. The oscillator sweep voltage was also used to provide the horizontal deflection on the oscilloscope.

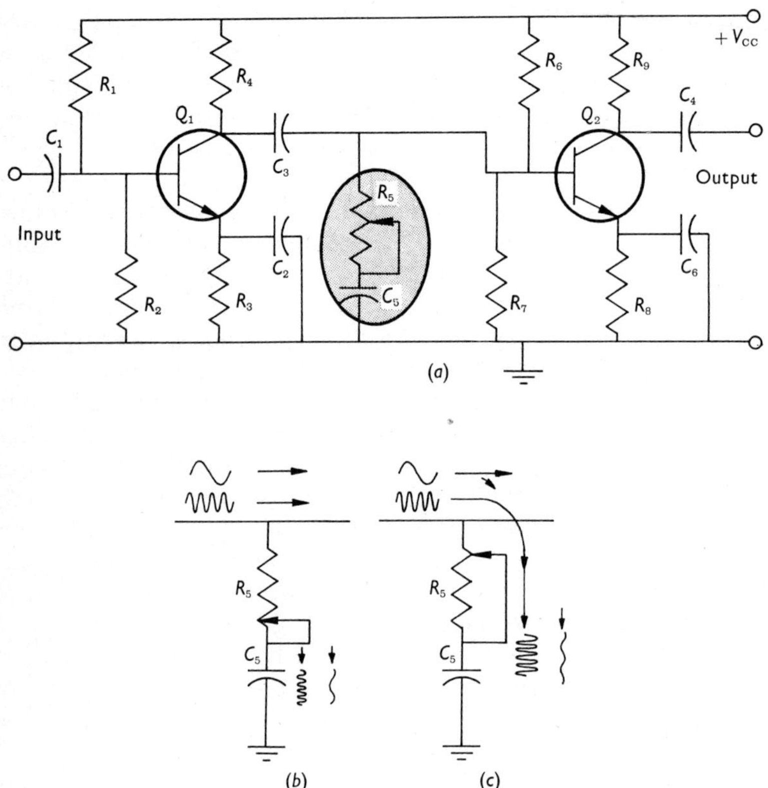

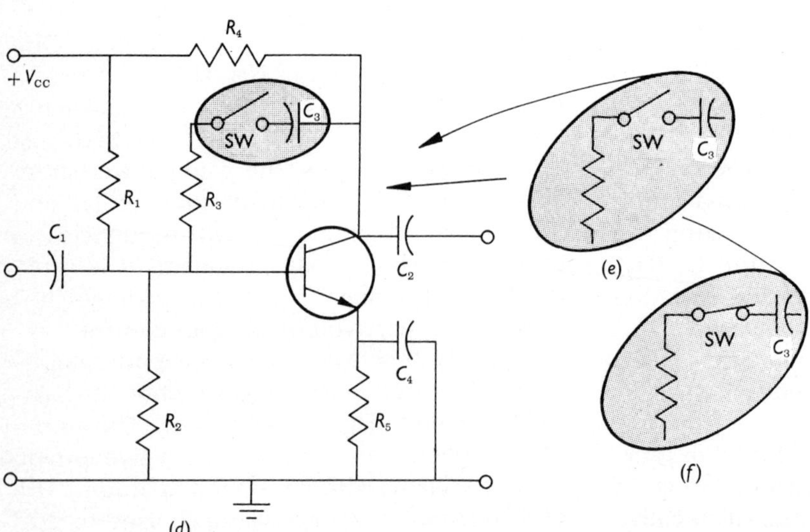

Fig. 11-25 Simple tone-control circuits: (*a*) Shunt-type low-pass filter; (*b*) Filter set for minimum high-frequency attenuation; (*c*) Filter set for maximum high-frequency attenuation; (*d*) Negative feedback tone control; (*e*) Switch open, flat frequency response; (*f*) Switch closed, high-frequency attenuation due to negative feedback.

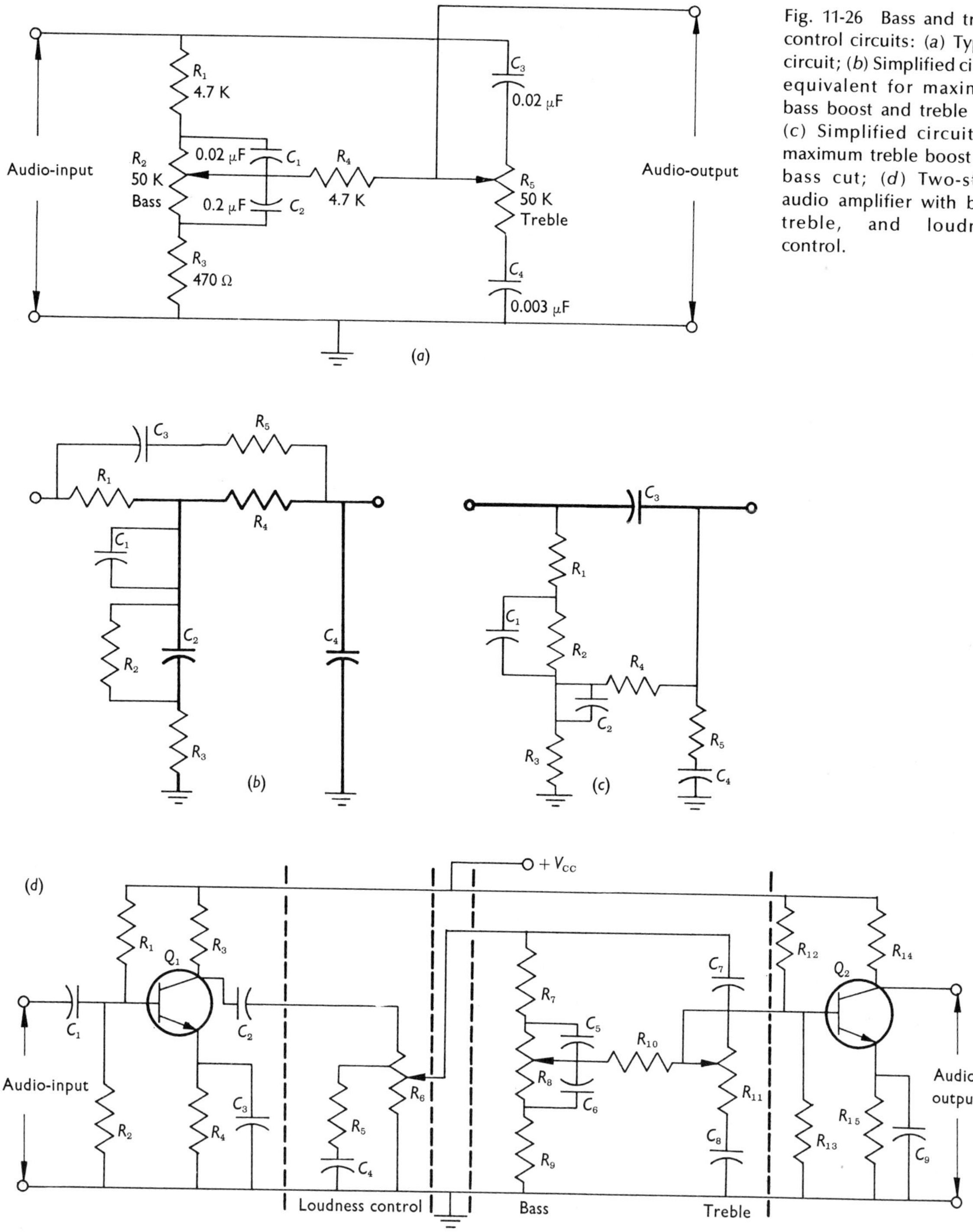

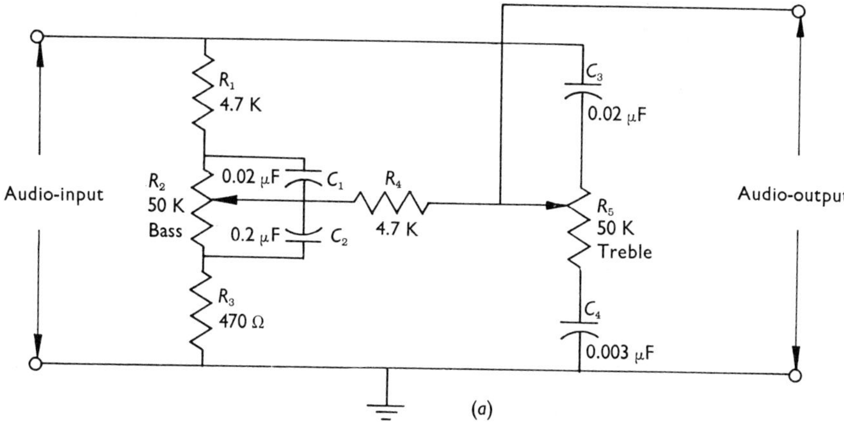

Fig. 11-26 Bass and treble control circuits: (*a*) Typical circuit; (*b*) Simplified circuit equivalent for maximum bass boost and treble cut; (*c*) Simplified circuit for maximum treble boost and bass cut; (*d*) Two-stage audio amplifier with bass, treble, and loudness control.

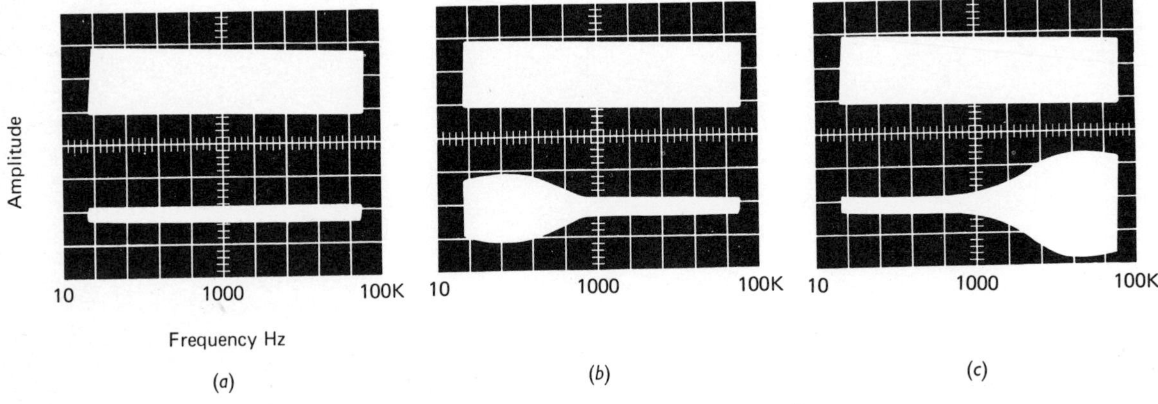

Fig. 11-27 Oscilloscope photographs showing amplifier frequency response for different bass-treble settings. Upper trace input signal, lower trace output: (*a*) Controls set for flat frequency response; (*b*) Maximum bass boost, minimum treble; (c) Maximum treble boost, minimum bass.

11-27 STEREOPHONIC SOUND REPRODUCTION

Stereophonic sound, or "stereo," gives direction to the sound source. The ears sense direction by comparing the phase difference between sound waves arriving at each ear. To preserve phase relationship in reproduced sound the system must be *binaural,* i.e., the original sound must be picked up by two microphones, from two locations. Two distinct recording tracks (one for each microphone) must be used when the sound is recorded. For playback a dual system is also required. Two amplifiers and speakers must be used, separated by enough space to give a binaural effect.

Stereophonic broadcasting has not been practical on the AM broadcast band because two distinct channels require more bandwidth than is allowed. Engineering studies suggest that AM stereophonic broadcasting could be feasible in the future. Most FM broadcasts are in stereo. The two channels are multiplexed (as explained in Appendix I) and transmitted on the same carrier.

Summary

Electromagnetic radiation occurs when the alternating current in a conductor has a sufficiently high frequency. Under these conditions the flux around the conductor does not have time to collapse entirely before the cycle reverses. A portion of the energy travels through space as an electromagnetic wave.

Radio reception occurs when a passing electromagnetic wave induces a current in an antenna.

Any device capable of generating a high-frequency signal can behave as a transmitter. (This also includes such undesirable transmitters as spark plugs in automobiles, rotor brushes in electric-appliance motors, relay contacts, etc.)

Although a simple oscillator can be used for transmitting, most practical circuits are of the *master-oscillator power-amplifier (MOPA)* type.

A fundamental signal produced in a transmitter is the *carrier wave.*

Modulation is the process of encoding information on a transmitter carrier.

Amplitude modulation is accomplished by varying the instantaneous carrier amplitude.

Frequency modulation is accomplished by varying the instantaneous carrier frequency. To some extent it can be accomplished by varying the phase of the carrier, but the process is then called *phase modulation*.

Every type of modulation produces *sideband frequencies* which contain the information. In many instances the carrier wave is suppressed at the transmitter and only sideband transmission takes place.

A *detector circuit* is used to demodulate a transmitted signal.

A *square-law detector* is sensitive but produces considerable distortion.

A *diode detector* is used almost exclusively in AM receivers.

An RF preamplifier stage serves two purposes: (1) It improves receiver selectivity and image frequency rejection, and (2) it raises the signal level before it enters the converter, which can result in an improved signal-to-noise ratio.

Nearly every receiver in use operates on the *superheterodyne* principle. In a superheterodyne receiver the incoming carrier and sidebands are converted to an *intermediate frequency (IF)*.

Tracking is important in a converter stage. At every dial setting the antenna-tuned circuit and local oscillator frequency must differ by the same amount, i.e., the intermediate frequency.

An *autodyne converter* does not require a separate local oscillator circuit.

The function of the *IF amplifier* stages is to select only the desired signal and to boost it to a level where efficient diode detection can take place.

Some IF amplifier circuits require *neutralization* to prevent oscillation at the intermediate frequency. IF-stage oscillation on AM broadcast receivers causes loud squealing sounds on all signals. This occurs because the signal IF and IF produced by oscillations beat in the detector. If they differ by an audio frequency, this audio frequency is reproduced at the output and is stronger than the desired signal.

IF stages are *stagger-tuned* to give a more nearly ideal frequency-response curve.

Automatic gain control (AGC), a form of negative feedback, keeps the receiver output signal constant. This is accomplished by reducing receiver gain automatically for strong signals.

The process of adjusting the tuned circuits in a receiver is known as *alignment*.

FM receivers operate at higher frequencies than broadcast AM receivers. Otherwise they have many common features.

An FM receiver has a *frequency-discriminator* circuit in place of a detector. The discriminator changes frequency variations to voltage variations. The *ratio detector* is the most popular form of discriminator circuit.

An ideal high-fidelity sound system preserves the frequency response, relative loudness levels, and the directional property of the sound source. In all home-entertainment systems there is a price–performance compromise.

A sound system will sound pleasing if the product of the low- and high-frequency cutoff lies in the neighborhood of 400 000 to 500 000.

Audio-amplifier frequency response can be trimmed by the listener with *bass* and *treble* tone controls.

Better-quality systems use frequency compensation on the volume control. Such controls are then usually labeled *loudness controls*.

Stereophonic systems are *binaural* throughout, from sound pickup to sound reproduction. It is not possible to produce "stereo" from a "mono" source.

Questions and Exercises

1. Explain briefly how electromagnetic radiation takes place.
2. If a powerful audio amplifier were connected to an antenna, very little transmission of signal would occur. Yet, under ideal conditions a signal from a "walkie-talkie" can be received miles away. Why?
3. If a pair of sensitive earphones were connected to a good antenna, no signals would be heard even with powerful transmitters operating in the community. Why?
4. What is meant by a MOPA transmitter and what advantages has it over a simple transmitter consisting of only an oscillator?
5. Why is it desirable to use a crystal oscillator in a transmitter?
6. What is a carrier wave?
7. Draw a sketch of a high-frequency carrier-wave amplitude modulated by an audio signal.
8. What is meant by the term modulation?
9. Draw a schematic diagram of an amplitude-modulated MOPA transmitter.
10. How is amplitude modulation accomplished?
11. With the aid of a waveform sketch, explain what is meant by frequency modulation.
12. Describe a practical circuit for producing a frequency-modulated signal.
13. How is the desired signal selected out of the numerous signals induced in a receiving antenna?
14. What are a good and a bad feature of a square-law detector?
15. Explain how a diode demodulates an AM signal.
16. What is the purpose of C_f in Fig. 11-10a?
17. What is one advantage of a diode-detector circuit?
18. In addition to raising the signal amplitude what other function does an RF preamplifier perform?
19. Explain the difference between a TRF and a superheterodyne receiver.
20. Why are TRF receivers seldom used?
21. What gives a superheterodyne receiver its superior performance?
22. What is meant by local oscillator tracking?
23. A portable broadcast receiver has an IF of 455 kHz. If it is tuned to a radio station operating on 980 kHz, what is the frequency of the local oscillator in the converter section?
24. List the three tuned circuits found in an autodyne converter.
25. Although an autodyne converter is economical in terms of components, it has an undesirable feature. What is this feature?
26. Why is neutralization required in many IF amplifier stages?
27. Describe briefly how neutralization functions in an IF stage.
28. Why may neutralization be more effective in a portable receiver when the battery is near exhaustion?
29. What is meant by stagger tuning?
30. Why is stagger tuning used?
31. What is meant by AGC?
32. Why is AGC used in radio receivers?
33. Explain how AGC operates.
34. What can be added to a receiver circuit to cope with powerful signals which the AGC action cannot sufficiently attenuate?
35. Outline the general procedure for aligning a superheterodyne receiver.
36. What is the main difference between AM and FM receiver circuits?
37. In the drawing in Fig. 11-21 explain why there is no interference from T_3 and T_4 when the circuit is amplifying an FM IF signal.
38. Why is the circuit arranged such that an applied signal passes through the primary of T_1 first, i.e., why cannot the positions of T_1 and T_3 be reversed?
39. Why is IF limiting possible in an FM receiver

but not in an AM receiver?

40. Why is a simple diode detector not adequate for demodulating an FM signal?

41. Explain briefly the operation of a balanced phase-shift discriminator.

42. If capacitor C_3 in Fig. 11-22a became open-circuited, how would this affect the operation of the discriminator?

43. Would the circuit in Fig. 11-22a cease to function entirely if diode D_1 became open-circuited? Consider your answer carefully.

44. The FM input signal to the circuit in Fig. 11-22a suddenly doubles in amplitude. What will happen to the output if (***a***) the incoming signal is at center frequency, and (***b***) the incoming signal is off center frequency?

45. How will the circuit in Fig. 11-24a react under the conditions stated in Question 44?

46. What are the requirements for high-fidelity sound reproduction?

47. In what ways will shortcomings be evident in sound systems operating at low sound levels?

48. For pleasing sound reproduction what should be the high cutoff frequency in an amplifier which has a low-frequency limit of 50 Hz?

49. Explain how negative feedback can be used for tone control.

50. How does a "loudness" control differ from a "volume" control?

51. Why must a stereophonic sound system be binaural from source to speakers?

12 relaxation oscillators, waveshaping, and special amplifier circuits

12-1 RELAXATION OSCILLATORS

Oscillator circuits whose output consists of interrupted oscillations are broadly termed *relaxation oscillators*. Whereas continuous-wave oscillators are widely used in communications systems, relaxation oscillators are more likely to be found in computing and control systems. The latter also require a variety of circuits to alter waveforms and to produce a particular response when receiving a given electrical stimulus.

12-2 PULSES AND LEVELS

The term signal, as used in the previous sections of this book, evokes an image of sinusoidal voltages. In control and computing circuits, signals consist also of *pulses* and *levels*.

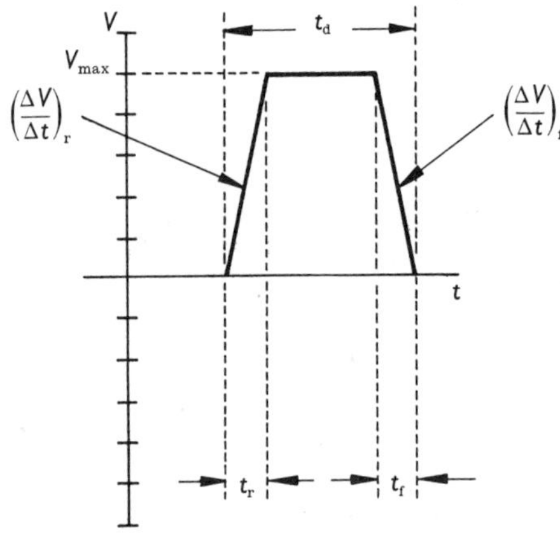

Fig. 12-1 Timing diagram illustrating certain signal characteristics. Shown are the amplitude (V_{max}), the duration (t_d), the rise time required to reach a specified percentage of the final amplitude (t_r), the fall time required to fall to within a specified percentage of the initial amplitude (t_f), the rate of amplitude rise $\left(\frac{\Delta V}{\Delta t}\right)_r$, and the rate of amplitude fall $\left(\frac{\Delta V}{\Delta t}\right)_f$.

A pulse is a rapid amplitude change followed by amplitude restoration. It has a specified *rise time* (t_r), *duration* (t_d), and *fall time* (t_f), as indicated in Fig. 12-1. In some instances the rate of amplitude change is of importance as well as the total rise and fall times. Pulses may be positive- or negative-going, although only a positive-going pulse is shown in Fig. 12-1.

A level is a defined amplitude of unspecified duration. Level duration can be a function of many variables and is usually dependent upon the completion of a given task in a portion of a system. Important characteristics of a level change are rise and fall times (t_r and t_f), settling time (t_s), and rate of change of amplitude ($\Delta V/\Delta t$). These are depicted in Fig. 12-2.

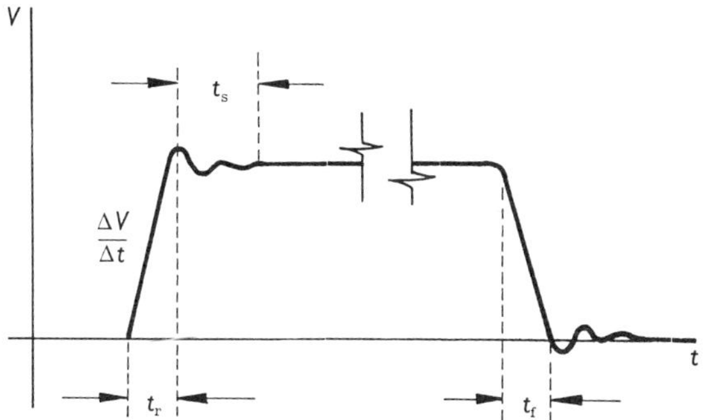

Fig. 12-2 Timing diagram of a level change. The duration is unspecified, but the rise and fall time, settling time (t_s), and rate of amplitude change are important characteristics.

12-3 BLOCKING OSCILLATOR

A blocking oscillator is a circuit which produces a high-energy pulse by means of regenerative action. The main application of blocking oscillators is in the production of pulses for the initiation of a given action in various circuits. Because of the high-energy output from blocking oscillators they are ideally suited for driving heavy loads.

Blocking oscillators are of two kinds, *astable,* or free-running, and *monostable,* which must be triggered into action by an external signal.

A free-running circuit is shown in Fig. 12-3a. Its circuit resembles a tickler *LC* oscillator (Sec. 10-7). Transformer T_1, known as a *pulse transformer,* has a tightly coupled primary and secondary to ensure a high degree of positive feedback. A ferrite core is commonly used. It must not hamper high-frequency response, because of the rich harmonic content present in a sharp pulse.

Oscillation begins as in any tickler oscillator, but the high degree of regeneration causes transistor Q_1 to saturate quickly. Maximum collector current flows and collector voltage drops to a minimum because of the large voltage drop across the load (primary of T_1). Points of minimum collector voltage and maximum collector current are shown as E_{C_2} and I_{C_2} on the timing diagram (Fig. 12-3b). At the instant when forward bias on the base is at maximum (E_{B_2}), there is large base current (I_{B_2}) which charges capacitor C_b.

When coil-flux change ceases, induced voltage falls and regenerative action quickly drives the transistor to cutoff. Capacitor C_b has accumulated sufficient bias voltage to hold the transistor at cutoff for time *t,* after which the cycle repeats itself. Time *t* depends on the value of capacitance C_b and the magnitude of the discharging resistance (mainly R_1 and R_b in parallel).

12-4 TRIGGERED BLOCKING OSCILLATOR

One type of monostable, collector-triggered blocking oscillator is shown in Fig. 12-4a. When a positive trigger pulse is applied to the base of Q_2 via C_1 it draws collector current through the primary of the pulse transformer (Q_2I_C in Fig. 12-4b). Primary current induces sufficient voltage on the secondary to send Q_1 into conduction. Once conduction begins, the blocking oscillator completes an oscillatory cycle, and a pulse is generated as shown in the timing diagrams (Fig. 12-4b).

12-5 MULTIVIBRATOR

A multivibrator is essentially an oscillator circuit without a frequency-determining network. It therefore produces a signal which

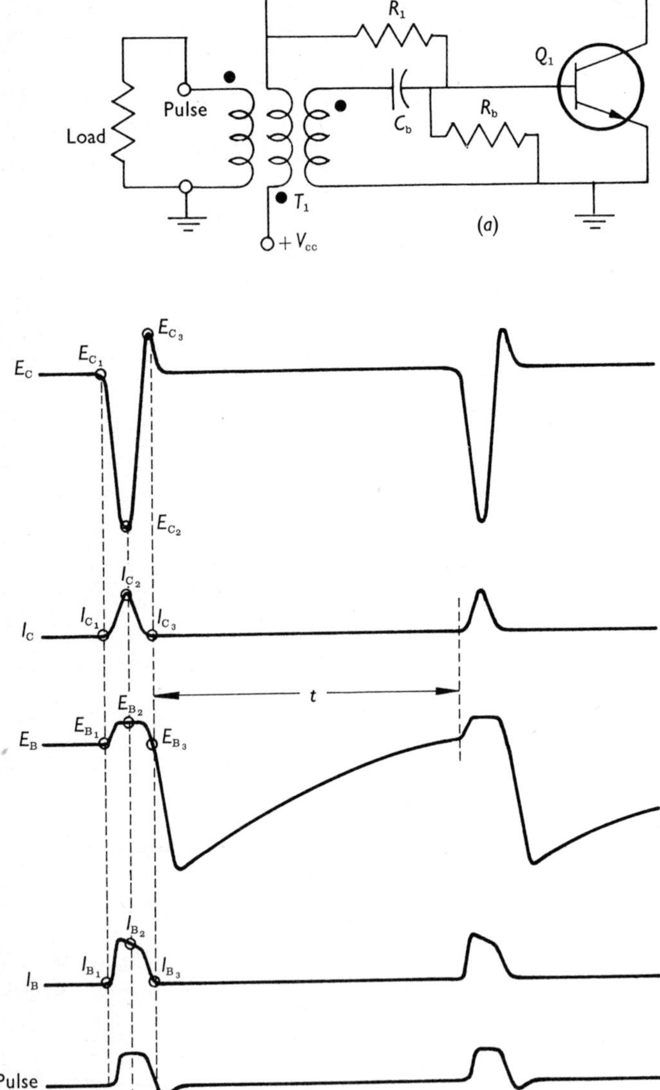

Fig. 12-3 Free-running blocking oscillator: (a) Schematic diagram; (b) Timing diagram.

contains "all" frequencies, i.e., a rectangular waveform.

Multivibrators are of three kinds: free-running or *astable, monostable,* and *bistable.* A free-running multivibrator produces a continuous stream of rectangular waves. A monostable produces a single waveform upon being triggered. A bistable undergoes an output level change upon being triggered.

In terms of circuits the three types of multivibrator consist of two cross-connected amplifiers, as shown symbolically in Fig. 12-5. The

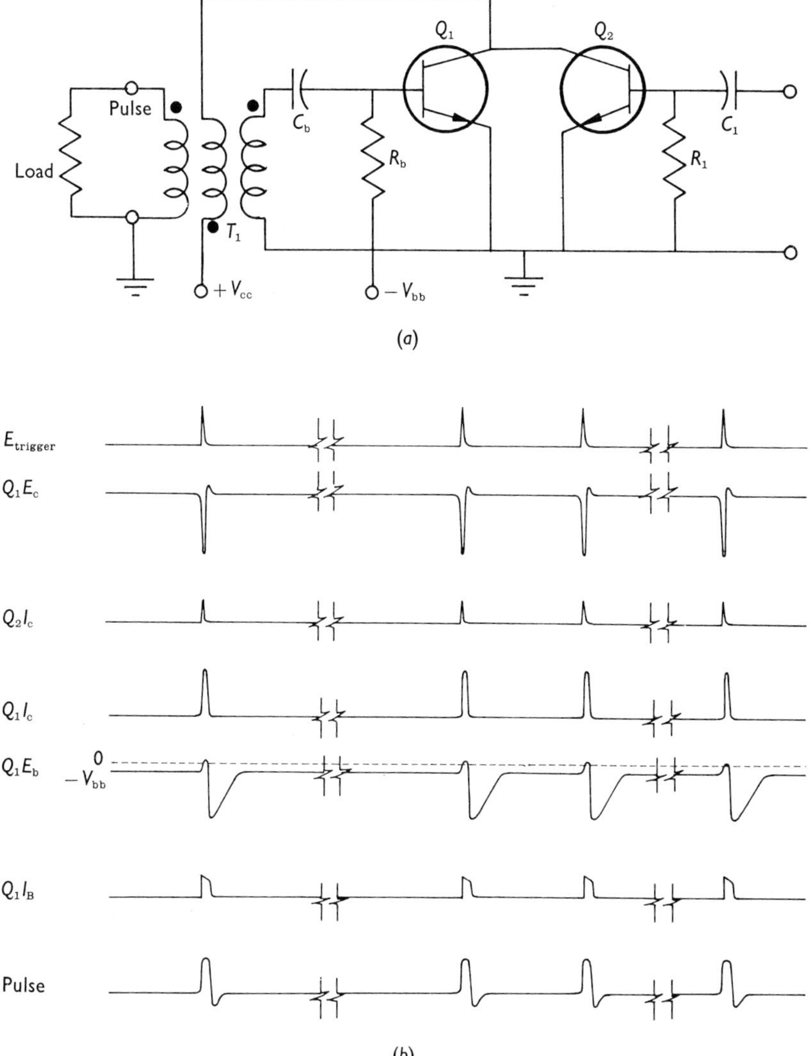

Fig. 12-4 Collector-triggered blocking oscillator: (a) Schematic diagram; (b) Timing diagram.

type of coupling network between the amplifiers determines whether the circuit will perform as a free-running, monostable, or bistable circuit.

12-6 FREE-RUNNING MULTIVIBRATOR

A free-running multivibrator consists of two amplifiers coupled with capacitors as shown in Fig. 12-6a.

When collector-supply voltage V_{CC} is applied, both transistors begin to draw current. Assume that the voltage on the collector of Q_2 rises slightly more quickly than the voltage on the collector of Q_1. This positive-going rise is coupled via capacitor C_1 to the base of Q_1, causing the base current to increase. The positive-going signal being applied to the base of Q_1 is amplified and inverted at the collector and appears as a large negative-going signal.

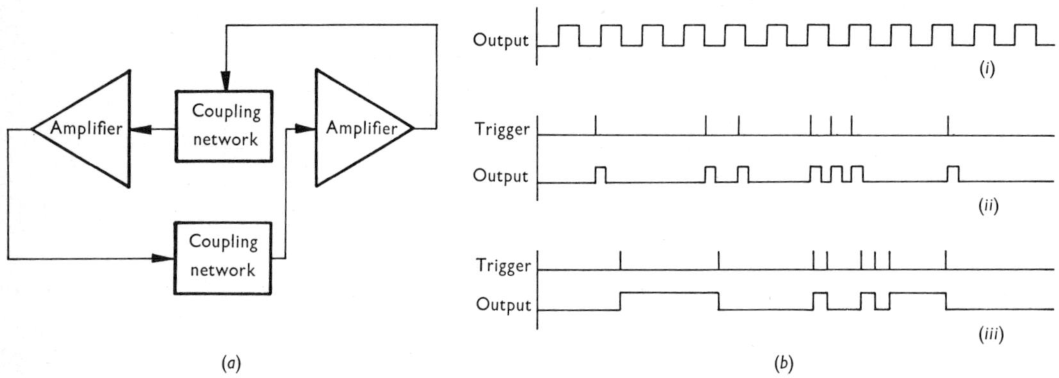

Fig. 12-5 Multivibrator: (*a*) Block diagram; (*b*) Outputs from (i) astable, (ii) monostable, and (iii) bistable multivibrators.

Capacitor C_2 transfers the negative-going signal to the base of Q_2 which cuts it off. With Q_2 cut off, output *A* is highly positive, and capacitor C_1 begins to charge. The charge path for C_1 is from ground through the emitter-base junction of Q_1, then through R_6 to $+V_{CC}$.

After a period of time any accumulated charge on C_2 leaks away through R_2 and R_4. Soon base current is again able to flow in Q_2. The collector of Q_2 then begins to fall. The negative-going falling edge is coupled via C_1 to the base of Q_1, cutting it off. Output *B* then goes high; capacitor C_2 charges via the emitter-base junction of Q_2 and R_5. Q_1 remains cut off until sufficient charge leaks off capacitor C_1 through R_1 and R_3. Then Q_1 again receives base current and the circuit flips once more to the state where Q_1 conducts and Q_2 is cut off. The circuit continues to flip back and forth with Q_1 and Q_2 alternately conducting. A timing diagram for the circuit is shown in Fig. 12-6b.

Cutoff duration is determined mainly by the *CR* time constant of the coupling components, i.e., capacitors C_1, C_2 together with their charge-discharge path resistances.

Outputs of a multivibrator are conventionally labeled as the "*one side*" and "*zero side*." The convention is carried over from computer applications where it is a convenient method of labeling the outputs. Either output may be labeled the "one" side. The other then becomes the "zero" side.

12-7 MONOSTABLE MULTIVIBRATOR

A *monostable* multivibrator is a form of "flip-flop" which remains "stuck" in one position unless triggered into action. When triggered, it "flips," remains "flipped" for a predetermined duration, and then returns to its "stuck" or monostable position.

Figure 12-7a shows a schematic diagram of a monostable multivibrator. In its steady state transistor Q_1 is conducting, because it receives a fixed forward base-bias current through R_4 from the power-supply source (V_{CC}). Its collector is then essentially at ground potential because nearly all the supply voltage is dropped across R_3. The transistor behaves like a closed switch as shown in the simplified diagram of Fig. 12-7b, and transistor Q_2 receives no base current. With no base current it behaves as an open switch and no collector current flows. Capacitor C_1 then charges to V_{CC} with polarity as shown in Fig. 12-7b. Transistor Q_3 is also open because it receives no base current.

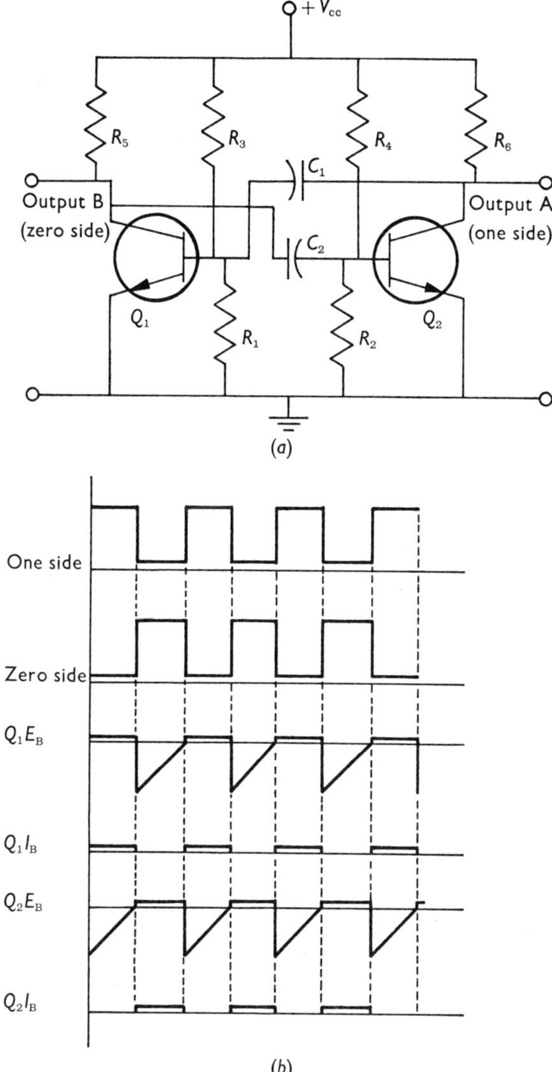

Fig. 12-6 Free-running multivibrator: (*a*) Schematic diagram; (*b*) Timing diagram.

When a trigger signal is applied to the base of Q_3, a pulse of base current flows, causing a sharp rise in the collector current. Whenever Q_3 conducts during a trigger signal it acts like a switch which closes and connects the positive side of C_1 to ground, as shown in the simplified diagram of Fig. 12-7c. Immediately, the voltage across C_1 is applied to the base of Q_1, driving this transistor into cutoff. Its collector then rises sharply, applying forward base bias to Q_2 via R_2, turning Q_2 on. With Q_2 on, capacitor C_1 has its positive side grounded whether a trigger is being applied or not, and the circuit remains with Q_2 conducting and Q_1 cut off until the charge on C_1 leaks away (mainly through R_1). The duration of the "flipped" state is determined principally by the *CR* time constant of C_1 and R_1.

In some publications this circuit is also referred to as a "one-shot" or "single-shot" flip-flop. Its applications include "pulse stretching," i.e., generating a wide pulse from a narrow trigger; pulse delay (generating a pulse at the end of its timing cycle); and many other timing applications. Although a trigger transistor is shown in Fig. 12-7a, it is not strictly required. The circuit may also be base-triggered by applying either a positive trigger signal to the base of Q_2 or a negative trigger signal to the base of Q_1.

12-8 BISTABLE MULTIVIBRATOR OR FLIP-FLOP

A *bistable* multivibrator, as its name implies, is stable in two states. It consists of two dc coupled amplifiers whose inputs and outputs are cross-connected.

In the circuit shown in Fig. 12-8a transistors Q_1 and Q_2 depend on each other for the supply of base current. Transistors can be made to simulate the behavior of a single-pole single-throw switch as shown in Fig. 12-8b. When sufficient base current is applied, a transistor conducts and behaves like a closed switch connected between collector and emitter (ground). In practice a transistor, even when saturated, is not a perfect conductor, and the collector remains a fraction of a volt above ground potential. When base current is removed, the transistor opens and the output is "pulled up" toward $+V_{CC}$.

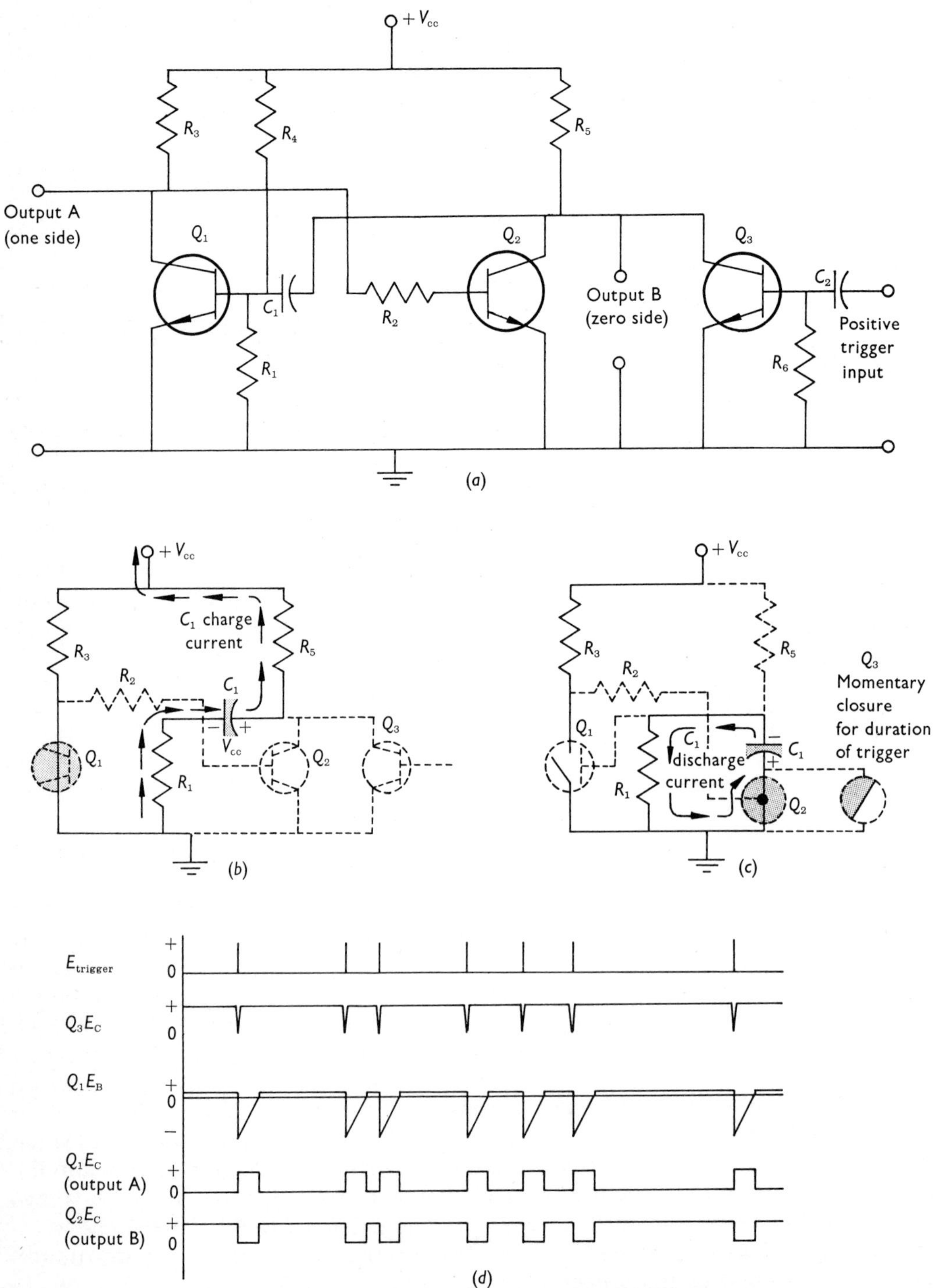

Fig. 12-7 Monostable multivibrator: (*a*) Schematic diagram; (*b*) Simplified circuit, in recovered state; (*c*) Simplified circuit, in triggered-timing state; (*d*) Timing diagram.

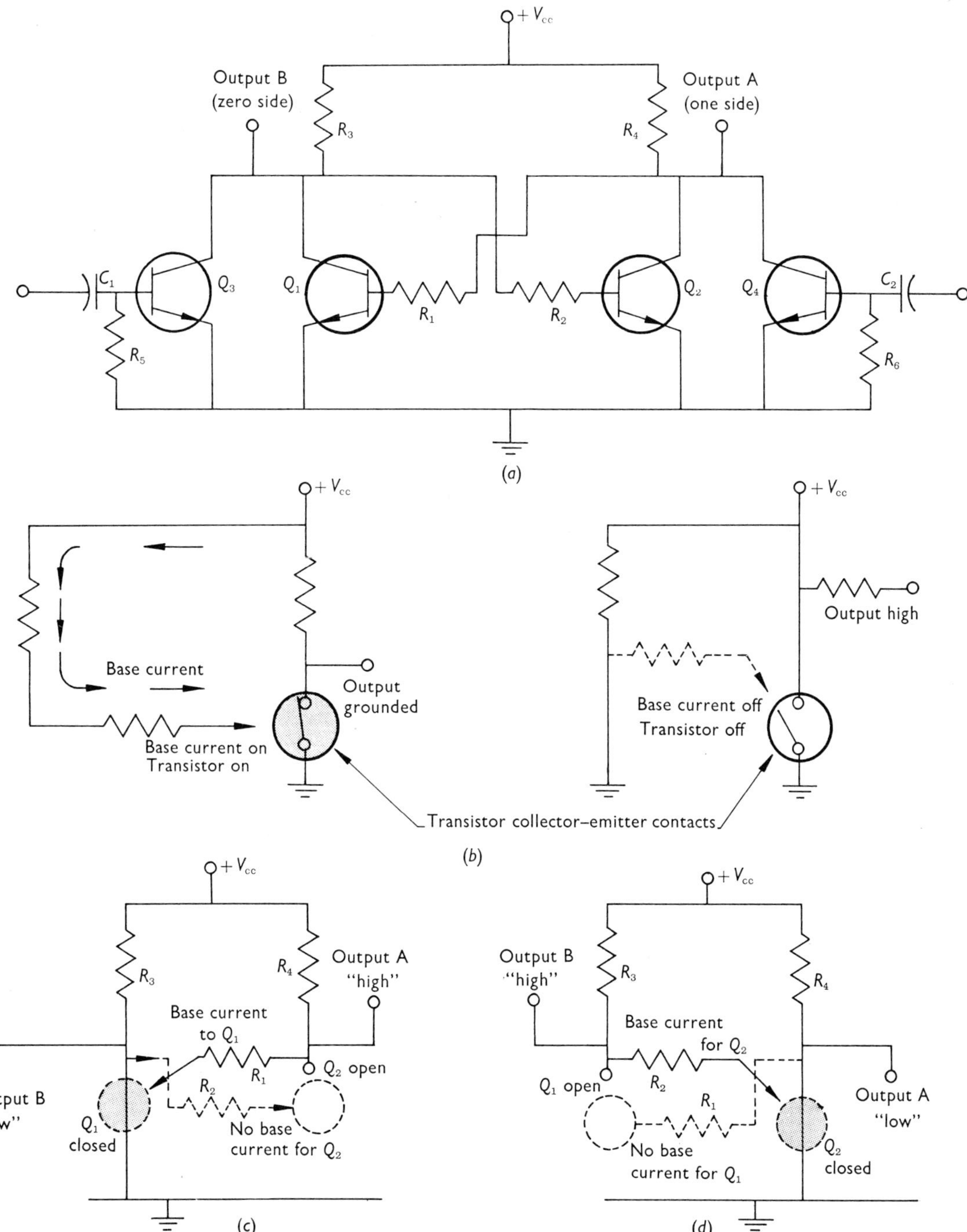

Fig. 12-8 Bistable multivibrator, or "flip-flop": (*a*) Schematic diagram; (*b*) Simplified diagram. The transistor is regarded as a single-pole, single-throw switch controlled by the base current; (*c*) Simplified diagram, showing Q_1 on and Q_2 off; (*d*) Simplified diagram, showing Q_2 on and Q_1 off.

In the circuit in Fig. 12-8a, assume transistor Q_1 is conducting. Output *B* will therefore be low, or "ground," and no base current will be applied to Q_2 through R_2. Transistor Q_2 is therefore cut off and output *A* is high, but this high output supplies base current to Q_1 via R_1 to hold it on. The circuit will remain in this state indefinitely as long as $+V_{CC}$ is applied.

If a positive trigger pulse is applied to the base of Q_4 via C_2, the transistor is momentarily switched on. Output *A* becomes "grounded" long enough to remove base current from Q_1. It opens and allows base current to be applied to Q_2. The circuit now remains with Q_2 conducting and Q_1 switched off. To flip it back requires a trigger pulse on the base of Q_3. Although triggering transistors have been shown in this circuit, they can be omitted, and trigger pulses can be applied directly to the bases or collectors of Q_1 and Q_2.

Bistable flip-flops are used to generate control levels in digital computers and various logic circuits. They may be regarded as switches which can be turned on and off electronically.

12-9 LIMITING

Limiting circuits may be used in a variety of applications, two of which are (1) amplitude limiting as in FM receiver circuits and (2) waveform distortion as in sine-to-square wave converters.

Basically a limiter is an overdriven amplifier circuit as shown in Fig. 12-9a. When a sinusoidal ac input waveform is applied, only the positive alternations (of sufficient amplitude) cause base current in Q_1. Whenever Q_1 is switched on, base current to Q_2 is interrupted, and the output is high. These relationships are shown in the timing diagrams (Fig. 12-9b).

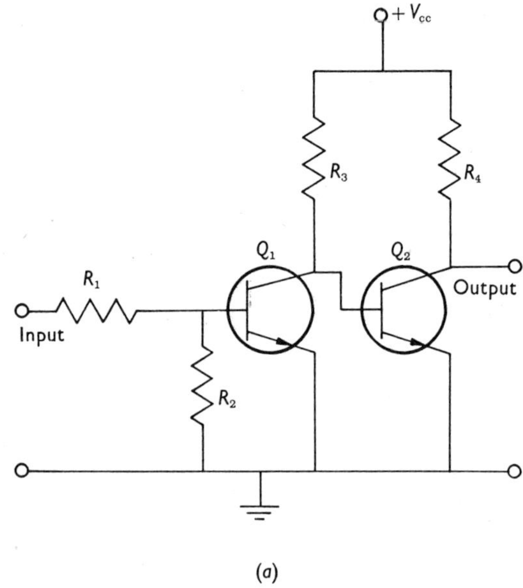

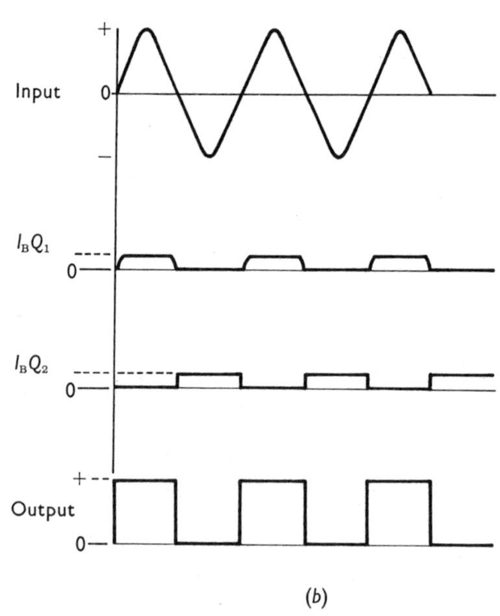

Fig. 12-9 Limiting circuit: (*a*) Schematic diagram; (*b*) Timing diagram.

12-10 CLAMPING

Clamping circuits are used to ensure that amplitude does not exceed a predetermined level.

Diodes are used extensively as clamps, with typical circuits shown in Fig. 12-10. Clamping refers to the output-signal level and means

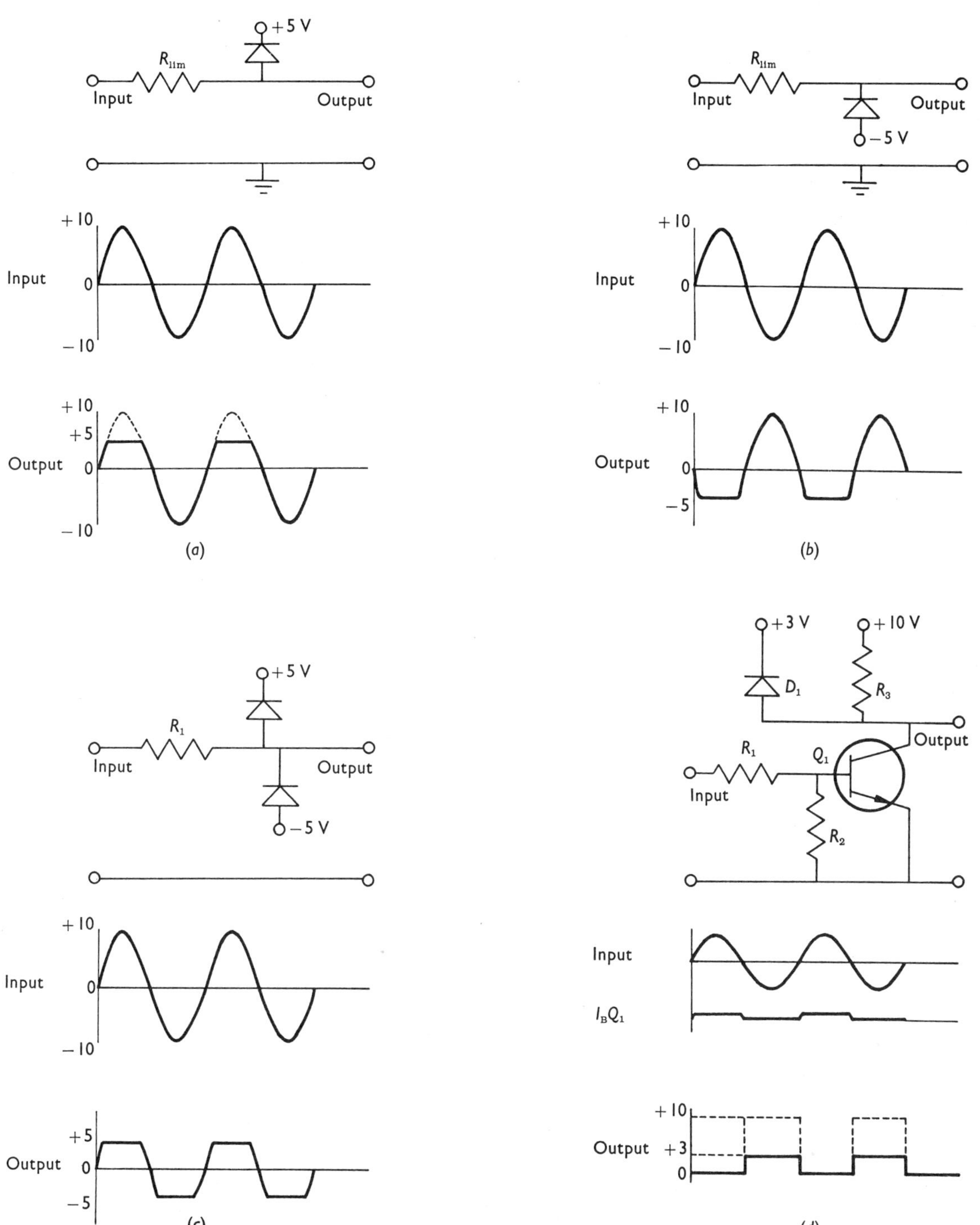

Fig. 12-10 Clamp circuits with timing diagrams: (a) Positive clamp; (b) Negative clamp; (c) Positive and negative clamping; (d) Transistor-output clamping.

that it cannot rise above a certain level, i.e., it is *"clamped."* Clamping is achieved by connecting a diode in parallel with the load. The diode is held off by reverse bias until the signal reaches sufficient amplitude to overcome the bias. Then the diode becomes "switched" on and connects the clamping level (the bias) to the load. After this point is reached, further signal amplitude increase has no effect because the load is connected to a constant voltage source, i.e., the bias or clamping level.

Choice of a value for R_{lim} is significant. It must be low in comparison with the load resistance but high in comparison with the forward resistance of the diode. Before the diode becomes switched on, there is relatively little signal loss across R_{lim} and virtually all signal is transferred to the load. Once the diode switches on, R_{lim} drops all the signal voltage in excess of the clamping level.

12-11 DC RESTORATION

A *dc restorer* circuit changes ac to dc *without distorting the ac waveform.* This statement is not strictly correct, because some waveform distortion occurs while the capacitor (Fig. 12-11a and e) draws charging current.

Negative restoration is illustrated in Fig. 12-11b and c and the accompanying timing diagram, Fig. 12-11d. During the first positive input alternation the capacitor C_1 charges to the peak value of input-voltage amplitude and retains its charge. The small amount of leakage is replenished on subsequent positive alternations. During negative alternations the instantaneous input voltage and the accumulated capacitor voltage are applied in series to the load. These facts are shown graphically in the timing diagram, Fig. 12-11d.

Positive restoration is accomplished in a similar manner to that of negative restoration. The essential difference is that the diode connections are reversed. It is illustrated in Fig. 12-11e, f, g, and h.

Restorer circuits are widely used at the cathode of a television picture tube. Before the video signal is applied to the cathode, it may be restored to become a ground referenced dc signal.

12-12 DIFFERENTIATING CIRCUIT

In many instances a circuit is required to respond to the rate of change in a signal (Fig. 12-12a). One application for such a circuit is to generate a pulse from the leading or trailing edge of a level (see Sec. 12-1 for a definition of pulses and levels).

In higher mathematics the name given to the term which expresses rate of change is the *derivative*. The process of finding the derivative is known as *differentiation*. The parallel, in electronics, is that *a differentiating circuit produces an output amplitude which is directly proportional to the rate of change of the signal waveform*. Rectangular waveforms give the greatest output as shown in Fig. 12-12b, because at the leading and trailing edges the rate of change is very large.

Circuit operation is briefly as follows. When input-signal amplitude undergoes a sudden change (from 0 to +10 V as shown in Fig. 12-12b), the capacitor C_1 begins to charge through R_1. The capacitor's initial charging current causes an *IR* drop of 10 V across R_1. As the capacitor accumulates charge, the charging current diminishes exponentially, and so does the *IR* drop. After five *CR* times, charging current (and *IR* drop) is virtually zero so that there is no longer any output. Output remains zero unless there is a further change in input amplitude. This next occurs when the input drops again from +10 to 0 V. Now the capacitor discharges through R_1. The discharge current flows in the opposite direction to the charge current, and R_1 experiences an *IR* drop of opposite polarity. After five *CR* times discharge is virtually complete and output is again zero.

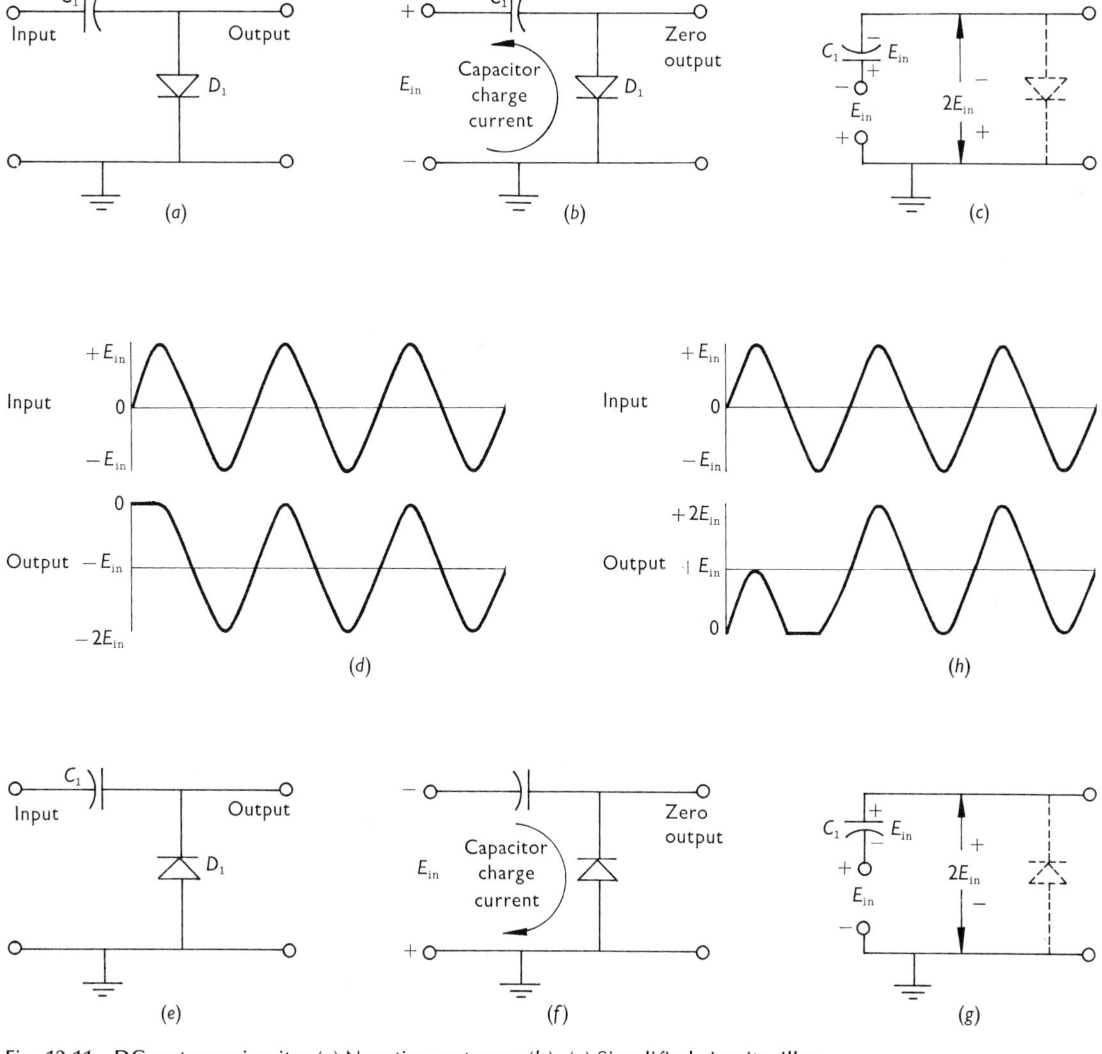

Fig. 12-11 DC restorer circuits: (*a*) Negative restorer; (*b*), (*c*) Simplified circuits, illustrating operation during positive and negative alternations; (*d*) Timing diagram; (*e*), (*f*), (*g*) Positive restorer and operation; (*h*) Timing diagram.

It is the *CR* time constant of C_1R_1 which determines how "thin" the output pulse will be. A short *CR* time (either a small capacitor, resistor, or both) means that the capacitor will charge and discharge very quickly so that output will be zero except at leading and trailing edges of the input waveform.

12-13 INTEGRATING CIRCUIT

An *integrating circuit* has output amplitude proportional to the input-signal *amplitude and duration* (Fig. 12-13a). Occasional pulses of short duration produce virtually no output. A long pulse or a rapid succession of short pulses will produce an output.

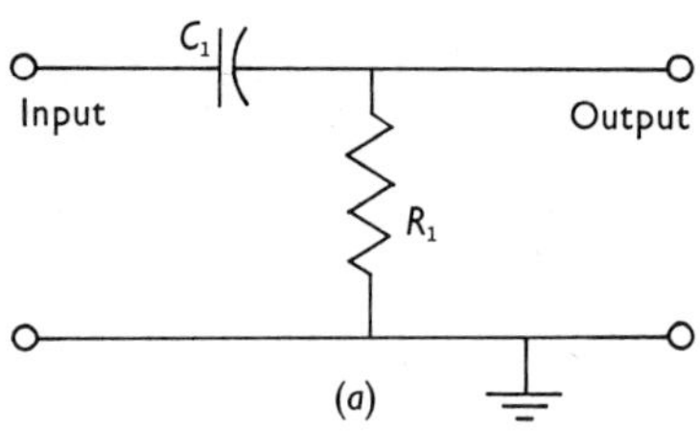

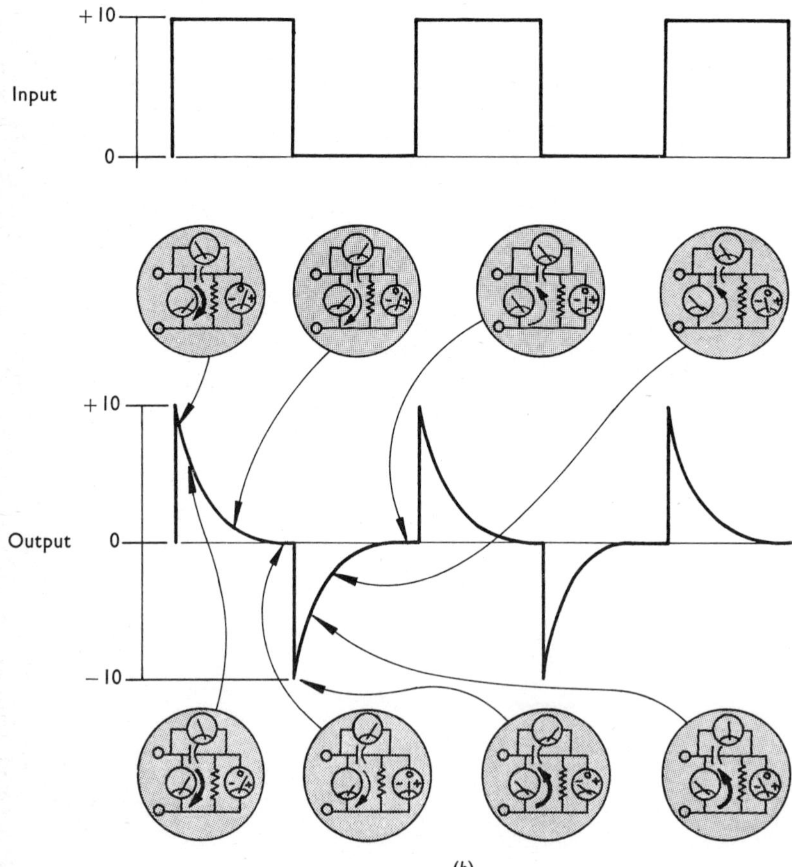

Fig. 12-12 Differentiating network: (a) Schematic diagram; (b) Timing diagram. Voltage is displayed graphically as a waveform and pictorially with imaginary voltmeters connected across the input, C_1, and R_1, the output.

In higher mathematics, the area enclosed under a section of curve is found by a process known as *integration*. When a pulse is displayed graphically, as in a timing diagram, the area under the pulse is proportional to the energy content of the pulse. Many pulses in rapid succession represent much more energy than a single pulse of the same amplitude and duration. An integrating circuit responds to the relative energy content of applied signal pulses, and its output is proportional to the cumulative effect of many pulses.

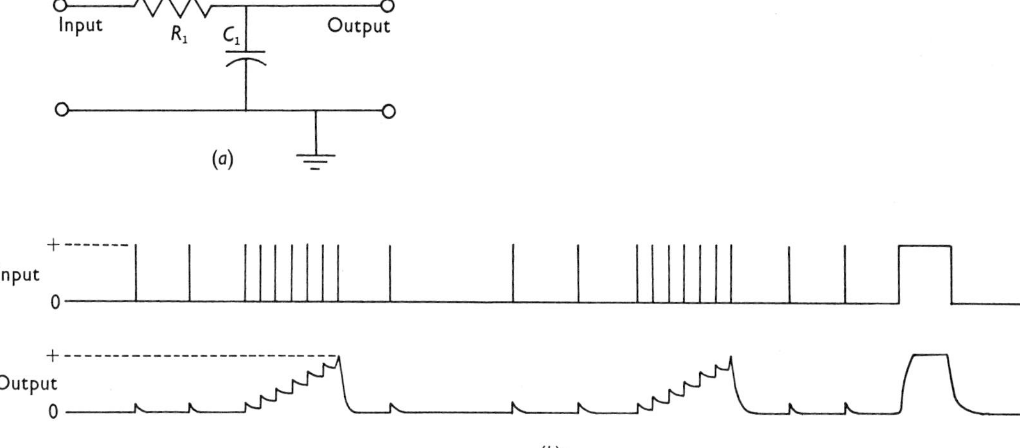

Fig. 12-13 Integrating *CR* network: (*a*) Schematic diagram; (*b*) Timing diagram.

Figure 12-13b shows that for a single pulse there is virtually no output. For substantial output to occur, capacitor C_1 must charge to a given voltage level. Unless subsequent pulses arrive, the accumulated charge leaks off quickly into the load. Several pulses in rapid succession "pump" up the capacitor before substantial charge is able to leak away and there is a resultant output. One application of integrating circuits is in the sync separator section of television receivers. They recover the vertical synchronizing signal from a series of pulses transmitted along with the picture signal.

12-14 SCHMITT TRIGGER

A Schmitt trigger is a *bistable flip-flop whose state is determined by the instantaneous amplitude of the input signal.* When signal amplitude attains a certain critical value the bistable "flips." Then when the input signal falls again to a certain critical value (usually lower than the turn-on value), the bistable "flips" again to its initial state.

Refer to Fig. 12-14a when reading the following explanation of circuit operation. When power is first applied Q_2 receives base current from $+V_{CC}$ via resistors R_3 and R_4. Transistor Q_2 turns on and the output is low. Its emitter current flows through R_5 causing an *IR* drop with polarity as shown in Fig. 12-14b. Note how this *IR* drop (E_{R_5}) reverse-biases the base of Q_1 and holds it at cutoff (Fig. 12-14c). The normal state of the bistable circuit is therefore with Q_1 cut off and Q_2 conducting.

A signal with sufficient amplitude to overcome the reverse bias on the base of Q_1 will cause collector current. Collector current through Q_1 reduces the base current being delivered to Q_2 via R_4 and, as a result, the *IR* drop across R_5 decreases. This decrease reduces the reverse bias on the base of Q_1. The action is regenerative and the circuit quickly changes state. Now transistor Q_1 is conducting and Q_2 is held at cutoff because of lack of base current, as shown in Fig. 12-14d.

When input-signal amplitude falls back to the turn-on level, one would suspect that the circuit returns to its normal state. This is not the case, however, because once Q_2 is cut off it no longer contributes to the generation of reverse bias for Q_1. This means that less input is required to *hold* Q_1 on than was required to *turn it on*. The input signal must therefore fall

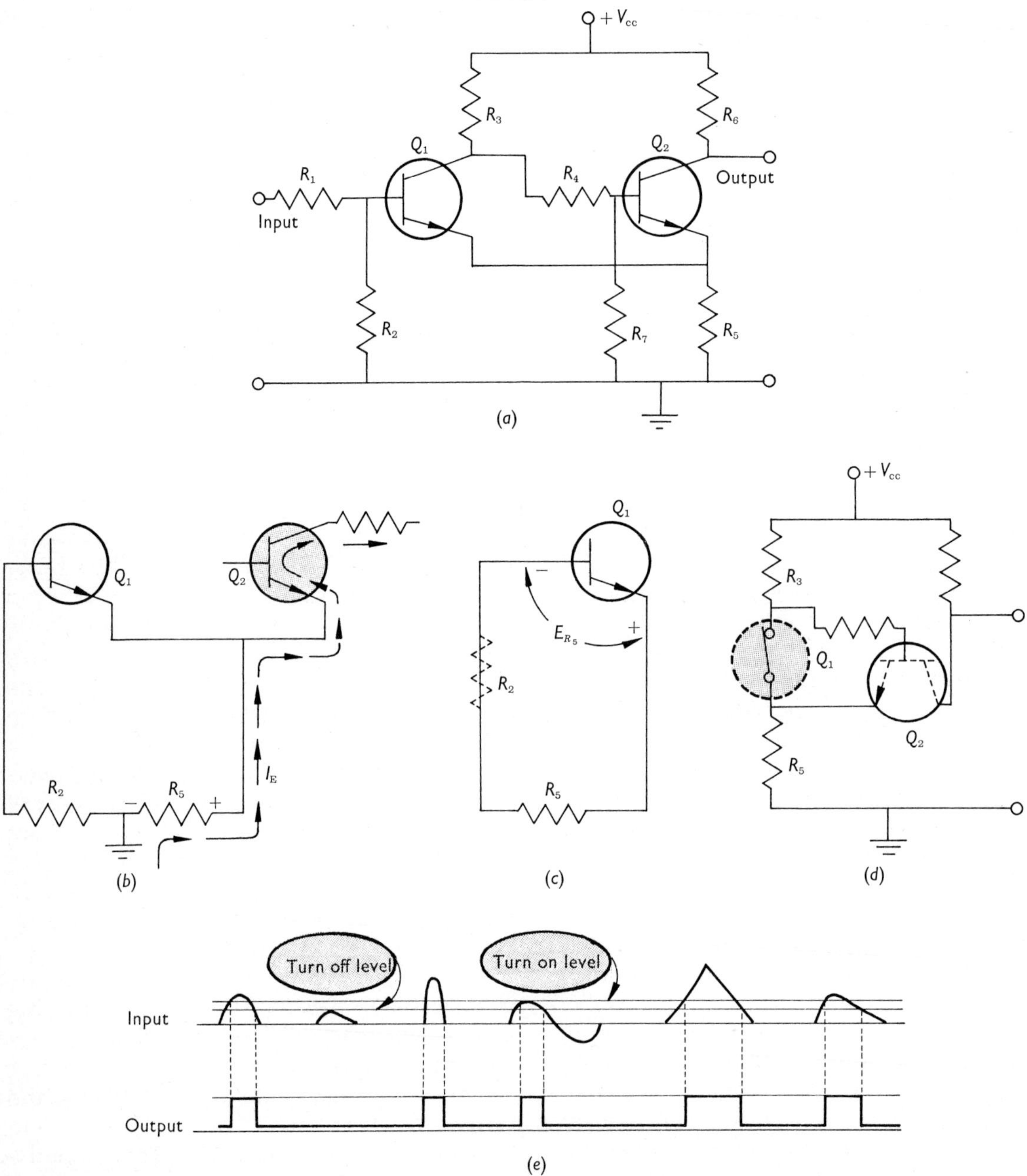

Fig. 12-14 Schmitt trigger: (*a*) Schematic diagram; (*b*) Simplified diagram, showing the Q_2 emitter current creating an *IR* drop in R_5; (*c*) The resultant *IR* drop cuts off Q_1—the normal state; (*d*) The input above the critical turn-on level amplitude turns on Q_1, which removes the base current from Q_2 and cuts it off; (*e*) Timing diagram, showing the circuit response to the various waveforms.

somewhat below the turn-on level before the circuit again flips to its normal state. These facts are shown in the timing diagram in Fig. 12-14e.

In summary, note that there are two critical input levels associated with a Schmitt trigger, i.e., the *"turn-on"* and *"turn-off"* levels. The turn-off level is less than the turn-on, i.e., the circuit exhibits a form of *hysteresis*. Schmitt triggers are used to provide signals of predetermined rectangular waveshapes from a variety of signals. Some typical example signals are shown in Fig. 12-14e.

12-15 DIFFERENTIAL AMPLIFIER

In certain applications there is a requirement to detect the magnitude of the difference between two signals in the presence of a third and possibly overriding signal. The *differential-amplifier* circuit shown in Fig. 12-15a is capable of performing this task.

As its name implies, the differential amplifier responds to the voltage difference which exists between its two inputs. The greater the difference, the greater the output. This is accomplished because of the similarity of the circuit to that of a Wheatstone bridge, Fig. 12-15b. When both inputs have identical amplitude, equal currents flow through collector resistors R_3 and R_5 and there is no potential difference across the load. Theoretically a signal of any magnitude applied to both inputs simultaneously will produce zero output. This is known as the *common mode* of operation; common-mode rejection capabilities are often quoted as one of the figures of merit of a differential amplifier. Figure 12-15f shows common-mode operation with the aid of a timing diagram.

The differential mode of operation is illustrated in the timing diagram in Fig. 12-15e. When input A is more positive than input B, transistor Q_1 conducts more heavily than Q_2 (as shown in the simplified diagram in Fig. 12-15c). Connection B of the load is therefore more positive than connection A. When signal polarity reverses, transistor Q_2 conducts more heavily than Q_1 and load connection A is more positive than connection B (Fig. 12-15d).

A differential amplifier may be operated in a single input connection by grounding one of the inputs. A fixed imbalance may occur under these conditions, but a varying signal applied to the ungrounded input will cause output variations in the load. The two outputs will have a 180° phase difference, and a circuit of this type can be used to drive push-pull amplifiers.

12-16 VACUUM-TUBE CIRCUITS

Equivalent vacuum-tube circuits for many of the circuits discussed in this chapter are shown in Appendix J, Fig. J-1.

Summary

Relaxation oscillators produce interrupted signals.

A *pulse* is an amplitude change followed by amplitude restoration.

A *level* is an amplitude change with no fixed duration.

A *blocking oscillator* produces at most one cycle of output signal. It may be *triggered* into oscillation or it may be *free-running*.

A *multivibrator* consists of two cross-coupled amplifiers. The coupling networks determine whether the circuit is *free-running, monostable, or bistable*.

A *limiting circuit* prevents signal amplitude excursions beyond a preset level.

Clamping circuits, like limiting circuits, confine amplitude excursions to a fixed value. Limiting is generally performed by an amplifier whereas clamping is performed at the amplifier output.

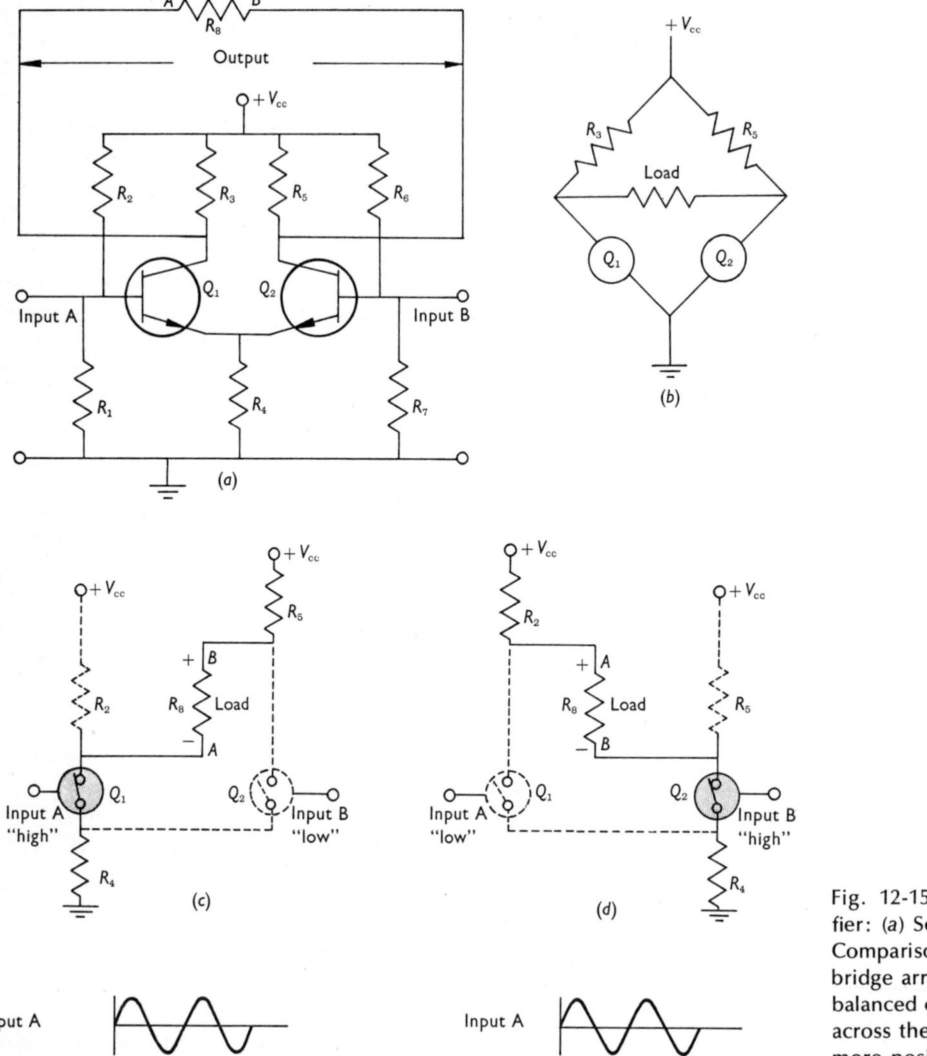

Fig. 12-15 Differential amplifier: (*a*) Schematic diagram; (*b*) Comparison to the Wheatstone bridge arrangement; (*c*) An unbalanced condition—the output across the load when input *A* is more positive than input *B;* (*d*) An unbalanced condition—the output across the load when input *B* is more positive than input *A;* (*e*) Output when operating in the differential mode; (*f*) Zero output when operating in the common mode.

Dc restoration changes an ac to a dc signal by shifting "the axis"; i.e., waveform distortion does not take place.

Differentiating circuits respond to rapid changes in signal amplitude but ignore relatively slow variations.

Integrating circuits respond to a persistent signal but ignore sudden changes.

A *Schmitt trigger* responds only to a preset amplitude. Turn-off amplitude is lower than the turn-on amplitude.

A *differential amplifier* responds to the difference in amplitude between two signals.

Questions and Exercises

1. How do relaxation and continuous-wave oscillators differ in terms of output signal?
2. What are some typical applications of (*a*) continuous-wave oscillators and (*b*) relaxation oscillators?
3. Distinguish between a pulse and a level.
4. Draw a graph of a 10-μs duration pulse of +2 V amplitude with a 1-μs rise time and 2-μs fall time.
5. Why is a blocking oscillator ideally suited for driving loads with relatively high energy requirements?
6. What is the difference between an astable and a monostable blocking oscillator?
7. How does a pulse transformer differ from other transformers?
8. How many cycles of oscillation occur in a blocking oscillator?
9. In a free-running blocking oscillator what determines the time period between oscillation?
10. What is the purpose of transistor Q_2 in Fig. 12-4a?
11. Which circuit components determine whether a multivibrator is astable, monostable, or bistable?
12. Prepare a sketch similar to Fig. 12-6b for the circuit shown in Fig. 12-6a, but with C_1 approximately ten times as great as C_2.
13. Which side of a multivibrator is the *one* side and which is the *zero* side?
14. What causes a monostable multivibrator to return to its initial state after it has been triggered?
15. What determines the duration of the on state of a monostable multivibrator?
16. Are transistors Q_3 and Q_4 essential for the operation of the bistable flip-flop in Fig. 12-8a? Explain your answer.
17. How long will the circuit in Fig. 12-8a remain in a given state?
18. Explain the statement "bistable flip-flops may be regarded as switches which can be turned on and off electronically."
19. Why does Q_2 in Fig. 12-9a receive no base current when the input is positive and of sufficiently high amplitude?
20. Draw a schematic diagram of a clamped limiter circuit. When a sine wave between 1 and 15 V amplitude is applied to the input it produces a rectangular wave of 2 V amplitude. (Hint: Combine circuits in Fig. 12-9a and Fig. 12-10a using proper bias voltages.)
21. Suggest a way to improve your circuit in Question 20 so that any sine wave from 1 to 100 V could be applied without danger to the input transistor base.
22. How can a positive restorer circuit be changed into a negative restorer?
23. What is required to ensure a sharp pulse of short duration from a differentiating circuit?
24. Give one application of an integrating circuit.
25. Which transistor in Fig. 12-14a is normally conducting?
26. If the instantaneous input on *A* in Fig. 12-15a is 560 mV and on *B* it is 640 mV, what is the effective input to the differential amplifier?
27. Will point *A* on the load be positive or negative with respect to *B* under the conditions in Question 26?

13-1 DIGITAL AND LINEAR CIRCUITS

Electronic circuits can be divided into two broad major categories, *linear* circuits and *digital* circuits. For practical purposes, the difference between these may be summed up as follows: Linear circuits are concerned with the preservation of the signal waveform, whereas digital circuits are designed to deal with fixed amplitude pulses and levels.

Digital circuits are not used solely in computers; they are found in a variety of control circuits. In many applications they have assumed functions which were traditionally performed by relays, stepping switches, and commutators. When used in certain combinations digital circuits are known as *logic* circuits.

13-2 PROPOSITION TABLES

Proposition tables or *truth tables* list in tabular form the responses of a circuit to all known input combinations. For example, suppose we wish to list all possible states of a lamp which is controlled by two single-pole single-throw switches as shown in Fig. 13-1a. The truth table in Fig. 13-1b shows the result. Note that the input conditions show all the different combinations possible with the given switches. The output column shows the response of the lamp under each of these combinations.

Note that there are four rows of output in this truth table. Are there always four? No, the number of outputs depends on two factors, (1) the number of inputs (i.e., the number of switches in this instance) and (2) the number of states each input can assume (i.e., two in this example, namely on and off). A general equation gives the number of outputs or rows in the truth table which can be expected for given inputs and conditions. It is:

$$(\text{No. of possible states})^{(\text{No. of Inputs})} = (\text{No. of output rows in truth table})$$

In the present example, substituting into the equation the values

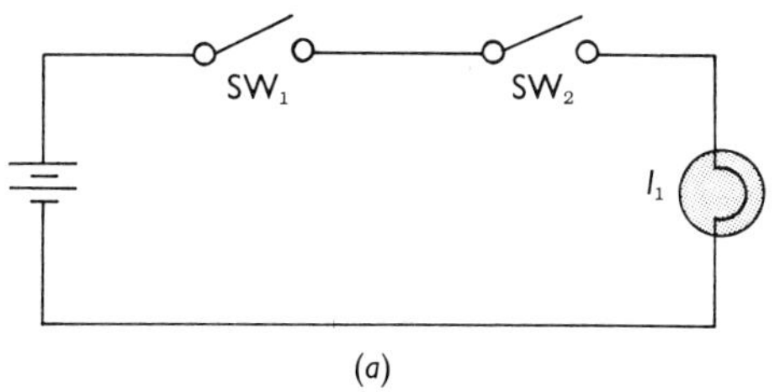

(a)

INPUT CONDITIONS (Switch Positions)		OUTPUT (Lamp)
SW_1	SW_2	
Off	Off	Off
Off	On	Off
On	Off	Off
On	On	On

(b)

INPUTS		OUTPUT
#1	#2	
L	L	L
L	H	L
H	L	L
H	H	H

(c)

Fig. 13-1 Simple control circuit: (a) Schematic diagram; (b) Truth table; (c) Truth table using H and L terminology.

No. of possible states = 2 (on and off)

No. of inputs = 2 (only 2 switches used)

gives

$$2^2 = 4 \text{ rows}$$

Another example will make this clearer.

EXAMPLE 13-1 How many output rows will be contained in a truth table representing a lamp controlled by four on-off switches? Applying the equation gives

No. of possible states = 2 (on-off)

No. of inputs = 4 (four switches)

ANSWER $2^4 = 16$ rows

13-3 ELECTRONIC GATES

The concepts expressed with the simple control circuit in Fig. 13-1a can be carried over to a class of electronic circuits known as *gates*. It is more descriptive to adopt a new terminology in place of on and off. One which is widely used is "high" and "low," where the terms refer not to switch closure but to voltage levels which in essence have the same effect as a switch turning on and off. In the most general sense the only required definition for high and low is that high is more positive than low. In practice, many integrated circuit gates operate with a "high" level of +2 to +5 V and a "low" level of 0 to +0.4 V (all with respect to ground). On and off conditions under this more general terminology then become "H" for high and "L" for low. A truth table using the HL terminology for the lamp circuit in Fig. 13-1a is shown in Fig. 13-1c.

13-4 THE AND CONDITION

Refer to the truth tables in Fig. 13-1 and note that one row is unique. It alone has an output H, whereas all others have output L. Note also that this is caused by switch #1 *and* switch #2 being on. Another way to state this more generally is to say that all inputs #1 AND #2, AND #3 . . . AND #*n* must be H to give an H output. This statement is known as a *logical* AND *condition*. In control-circuit diagrams, a logic symbol is used to designate a circuit which gives a high output only when all inputs are high. Such a circuit is known as a positive logic AND gate, and its symbol is shown in Fig. 13-2.

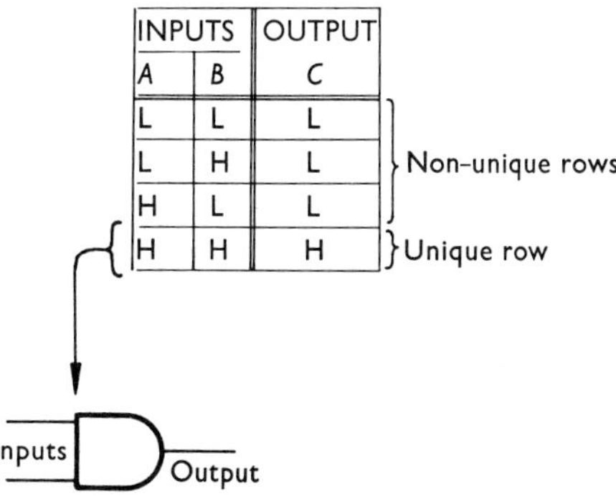

INPUTS		OUTPUT
A	B	C
L	L	L
L	H	L
H	L	L
H	H	H

Fig. 13-2 Logic symbol and truth table for a positive logic AND gate. The unique output row in the truth table defines the logical AND function.

13-5 POSITIVE AND NEGATIVE LOGIC

The gate shown in Fig. 13-2 is designated as a *positive logic* AND gate because a "high" is taken as an assertion. This means that H inputs are required to give an H output. It is also possible to construct AND gates which take L as an assertion and H as the lack of it. Then an L output will occur only when all inputs are L. Such a gate would be called a *negative logic* AND gate.

On logic symbols any function which is initiated by a low or L input is shown with a small circular loop on the input line. For example, a two-input negative AND gate requires two L inputs to give an L output (shown symbolically in Fig. 13-3).

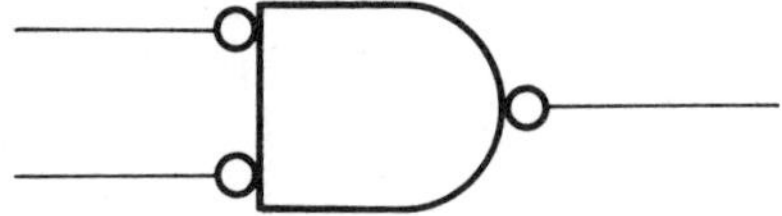

Fig. 13-3 Negative logic AND gate. Circular loops on the input and output lines indicate that two L inputs are required for an L output.

13-6 THE OR CONDITION

A simple control circuit using a lamp and switches may be constructed as in Fig. 13-4a to illustrate the OR function. Figure 13-4b shows the truth table which describes the OR situation in terms of on and off. A more generalized HL truth table is shown in Fig. 13-4c.

The OR situation is represented by the nonunique rows in a truth table. Put in words the OR situation is stated as: "The lamp is on if switch #1 OR switch #2 is on." In the more general case when gates are involved, output occurs if there is input #1 OR #2 OR #3 . . . OR #*n*. Figure 13-4d shows the logic symbol used to represent the *positive logic* OR gate. The symbol for the *negative logic* OR gate is shown in Fig. 13-4e.

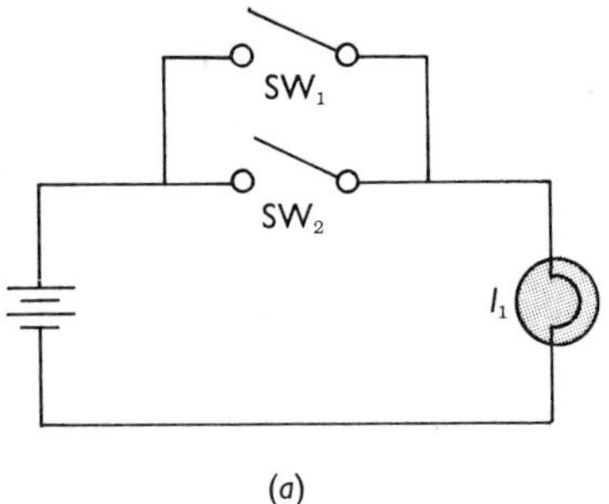

(*a*)

INPUT CONDITIONS		OUTPUT
SW_1	SW_2	
Off	Off	Off
Off	On	On
On	Off	On
On	On	On

(*b*)

INPUTS		OUTPUT
#1	#2	
L	L	L
L	H	H
H	L	H
H	H	H

(*c*)

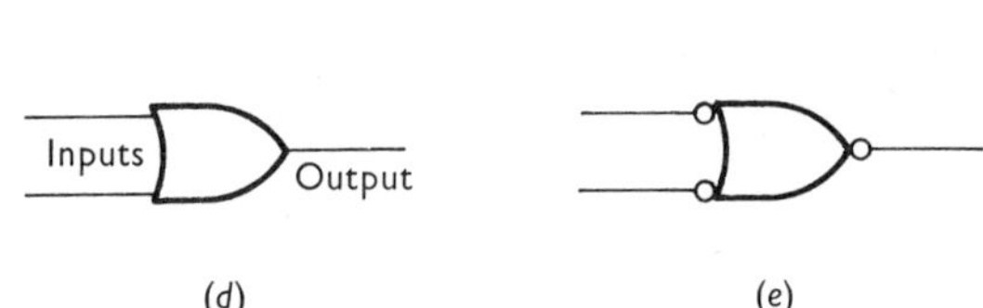

(*d*) (*e*)

Fig. 13-4 The OR condition: (*a*) Simple lamp and switch circuit which performs the OR function; (*b*) Truth table for circuit (*a*); (*c*) Truth table in terms of H and L; (*d*) Logic symbol for positive logic OR gate; (*e*) Logic symbol for negative logic OR gate.

13-7 RESISTOR-TRANSISTOR LOGIC (RTL) GATE

One of the simplest, most economical, and reliable electronic gate circuits is the *resistor transistor logic gate* or RTL gate. A schematic diagram for a two-input gate is shown in Fig. 13-5a. The inputs are labeled *A* and *B* and the output is labeled *C*.

To better understand the operation of the gate it is necessary to realize that in an electric circuit, output current delivered to a load can be interrupted in two ways. One is the well-known series-connected switch circuit shown in Fig. 13-5b(i). Another less obvious method is to insert a switch in parallel with the load as shown in Fig. 13-5b(ii). When the switch is on it shunts the load and prevents current flow to the load. Under this condition circuit current is confined to R_3 and the switch. The load has virtually zero current and voltage; i.e., output is zero.

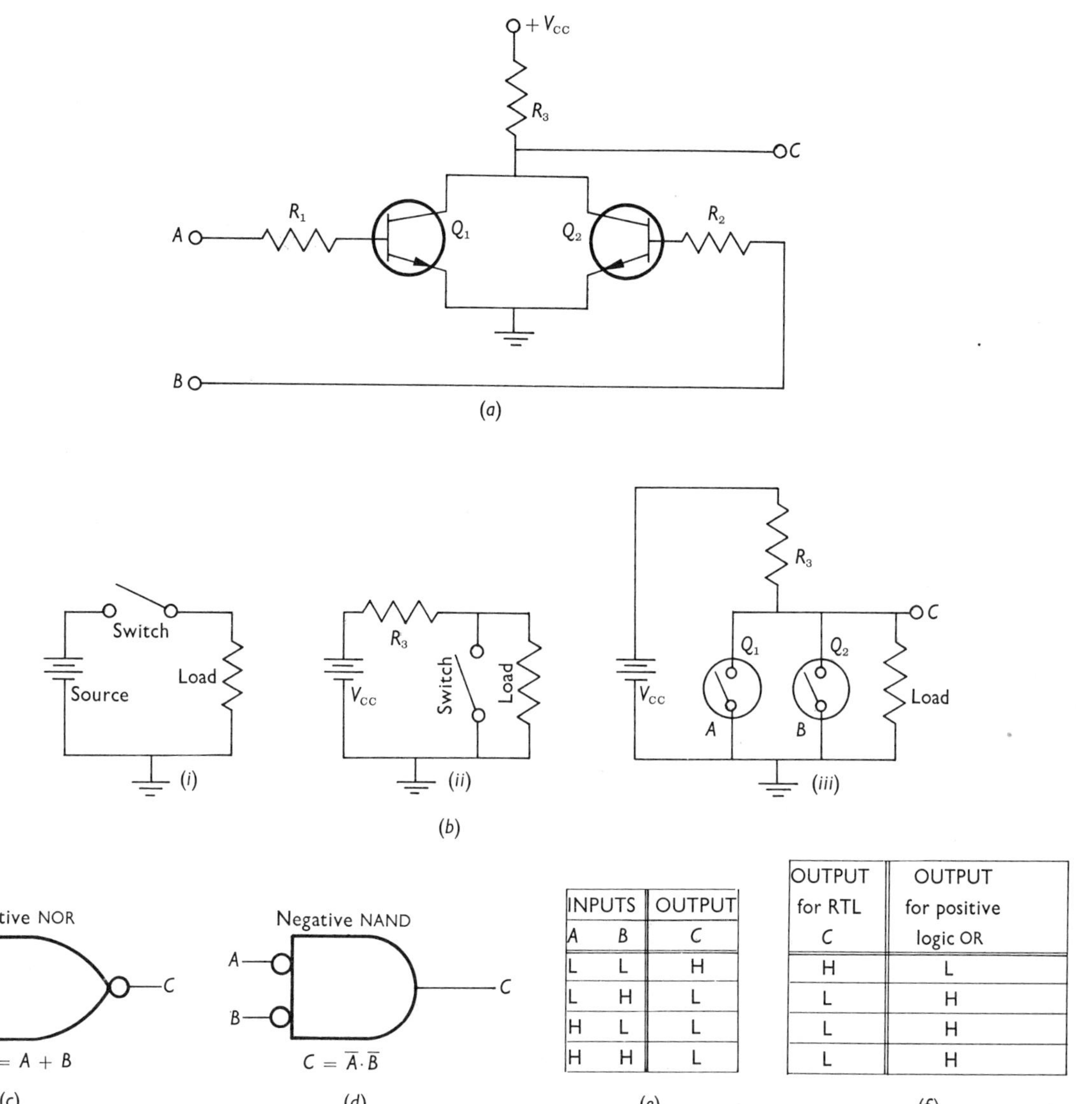

INPUTS A	INPUTS B	OUTPUT C
L	L	H
L	H	L
H	L	L
H	H	L

OUTPUT for RTL C	OUTPUT for positive logic OR
H	L
L	H
L	H
L	H

Fig. 13-5 RTL gate: (a) Schematic diagram; (b) Simplified circuits illustrating transistor switch action; (c) Positive logic symbol (positive NOR); (d) Negative logic symbol (negative NAND); (e) Circuit truth table in terms of H and L; (f) Comparison of RTL gate output with that of a positive OR gate.

Both transistors in the RTL gate circuit behave as switches shunting the output. A simplified diagram illustrating this concept is shown in Fig. 13-5b(iii). As long as an L level is applied to a given input, the base of the transistor receives no bias current and remains off. If both inputs A and B are L, both transistors are off and the output is H. An H applied to either A or B turns on either transistor Q_1 or Q_2, which drops the output to L. Gate function is illustrated by the truth table in Fig. 13-5e.

Logic symbols for a two-input RTL gate are shown in Fig. 13-5c and d. One is derived from the non-unique rows, the other from the unique. Consider first the one derived from the non-unique rows. A comparison of an RTL output and a positive OR gate output shows that both correspond except that they are inverse to each other (Fig. 13-5f). Whenever an L occurs on the OR, an H occurs on the RTL. Symbolically the RTL is shown as a positive OR with a circular loop at the output to indicate inversion. The name given to the circuit is positive NOR gate. The N indicates that the output is *not* like the inputs.

For negative logic the RTL gate behaves like an AND gate, i.e., has a unique output only when both inputs are L. Again, because of the inverted output the circuit is a negative NAND gate, shown symbolically in Fig. 13-5d.

13-8 PROCEDURE FOR DEDUCING LOGIC SYMBOL

Gate logic symbols can be deduced from truth tables by following a systematic procedure. Refer to Fig. 13-6 and note that two types of logic gates can be obtained from a given truth table. The two gates are in a sense complements of each other, and this duality will again appear in the Boolean algebra associated with the truth table (Sec. 13-11).

By isolating the row with a unique output * an AND-type input condition results. This is because all inputs must be the same, either all H or all L, to produce the unique output. The next step is to draw the input lines for the AND condition showing logic polarity. If the inputs are H for assertion, straight-line inputs are drawn (as in the example in Fig. 13-6). If L-type inputs are required for assertion, the input lines are distinguished by the addition of a circular loop to denote negation or inversion. To complete the symbol the output line is drawn with the proper polarity.

The name of a given gate is derived from two considerations. (1) Does the input perform an AND or an OR function? (2) Is there inversion at the output? If inversion takes place at the output (logic polarity of the input lines is different from that of the output), an N is added to indicate that input and output logic polarities are *not* the same. This gives us the names NAND and NOR.

13-9 BOOLEAN ALGEBRA EXPRESSIONS

An optimum logic circuit is one which performs the required function with the least number of gates. It is not only the least expensive in terms of components, but also in terms of power consumption, heat generation, and signal-propagation delay. *Boolean algebra* deals with logic propositions and is one of the tools which can be used to optimize logic circuits.

In ordinary high school algebra, the elements used are the natural numbers, and the operators are the familiar arithmetic operators multiplication, division, addition, and subtraction. In Boolean algebra the elements are logic assertions and the operators are the AND and the OR operations. (In mathematics these are called union and intersection.)

The AND operator is indicated with the same symbol as the multiply operator in ordinary

*If there is more than one unique row, more than one gate has been interconnected as a more complex logic circuit. The procedure outlined here applies to a single-gate truth table.

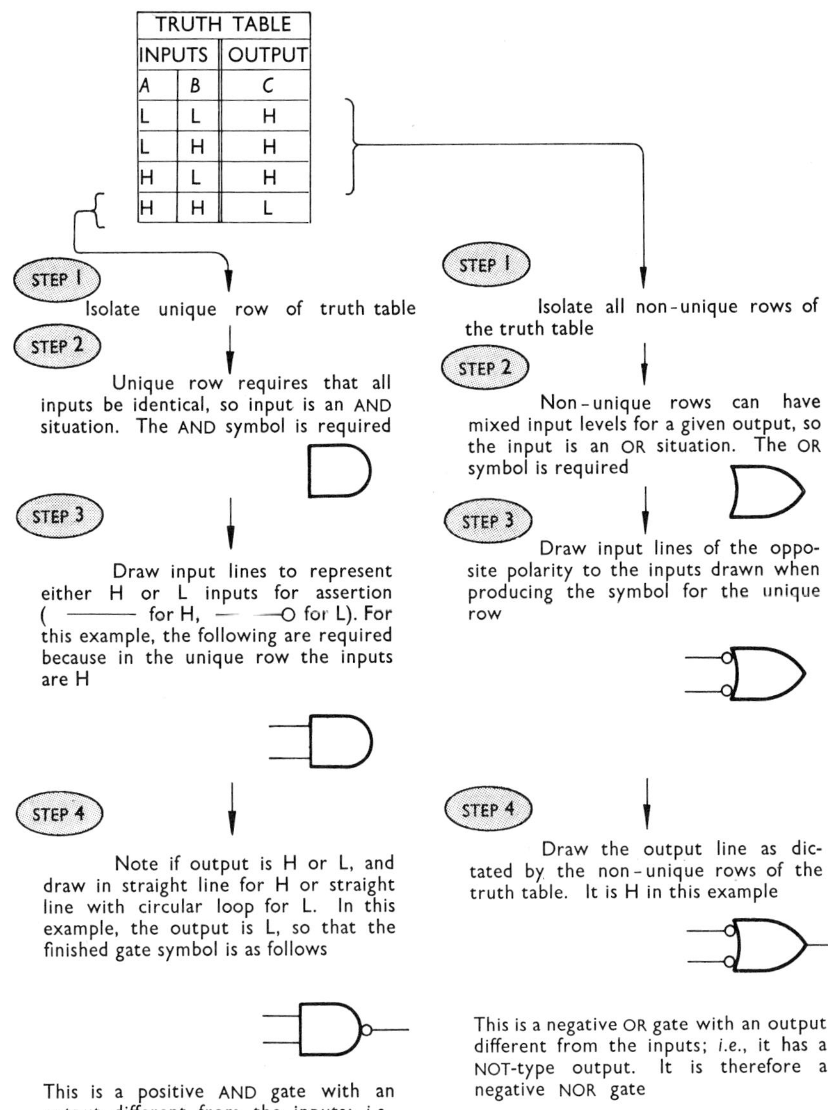

Fig. 13-6 Steps in determining the type of gate from a truth table.

algebra, i.e., a dot or an understood dot if it is omitted. A comparison of the meaning of the dot operator in ordinary and Boolean algebra is shown in Fig. 13-7.

The OR operator is indicated with the same symbol as the addition operator in ordinary algebra, a plus sign. A comparison of the meaning of the plus operator in Boolean and ordinary algebra is also shown in Fig. 13-7.

Although Boolean algebra expressions appear like ordinary algebraic equations the meanings of the elements and operators are different. In spite of this fundamental difference many of the familiar algebraic operations

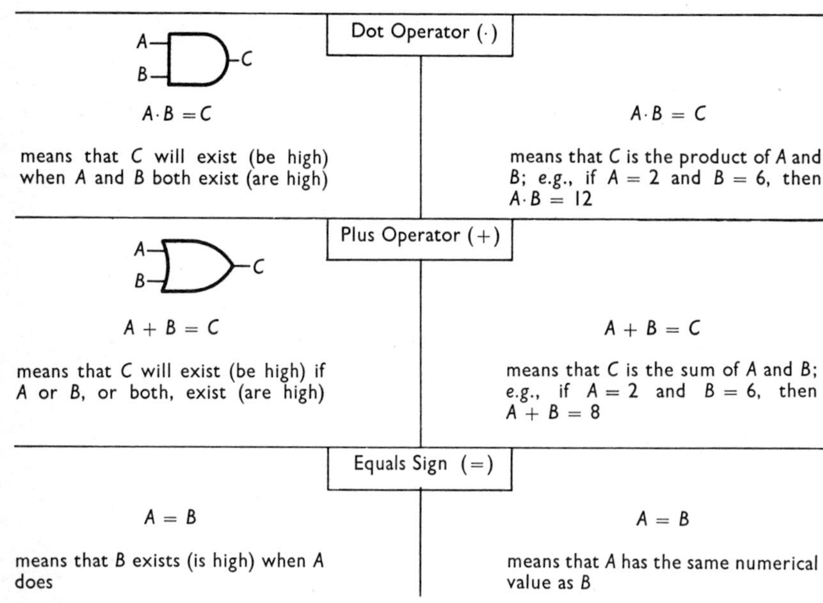

Fig. 13-7 Comparison of the dot, plus, and equals signs in Boolean and ordinary algebra.

such as expansion and factoring can be carried over to Boolean expressions. It is this feature which makes Boolean algebra so useful. One can start with a long and complex expression (which requires many gates to implement) and simplify it in much the same way as an ordinary algebraic expression. The result after simplification is the simplest logical expression required to perform the required function. Even so, it is not necessarily the least expensive because an alternate more complex expression could possibly be implemented with cheaper gates. The final decision rests with the designer.

13-10 THE ASSERTION AND ITS COMPLEMENT

Boolean expressions may contain as elements the letters or labels used to designate the input and output lines of logic gates. To indicate that a low is required for assertion, a bar or a prime is used over the letter. If A means that a high level is required on an input line for a given occurrence, then $\overline{A}$ or A' means that a low is required. $\overline{A}$ is known as the *complement* of A. This mathematical notation is analogous to the pictorial notation regarding the use of straight lines to indicate H-type assertions, and straight lines terminated by circular loops to indicate L-type assertions in logic symbols (Fig. 13-6, step 3).

In gate circuits the Boolean statement $A + B = C$ means that if the required voltage level (it can be H or L depending on whether positive or negative logic is used) is applied to input A *or* input B, then a similar level will be measurable on output C. Similarly, the expression $A \cdot B = C$ translates: If input A *and* input B receive the correct levels, then output C will have a similar level. A Boolean statement represents the information conveyed by one or more rows in a truth table.

These concepts can now be applied to the derivation of the Boolean expression for the switch circuit in Fig. 13-1. Referring first to Fig. 13-1 and designating the lamp-on condition as I_1 and lamp-off as $\overline{I_1}$, and a switch-on as SW while a switch-off is $\overline{SW}$, two Boolean expressions can be derived. The first results from the unique output row in the truth table, which gives: $I_1 = SW_1 \cdot SW_2$. The second arises from the non-unique rows, which give:

$$\overline{I}_1 = \overline{SW}_1 + \overline{SW}_2$$

The first statement, $I_1 = SW_1 \cdot SW_2$, translated into words says "The lamp is on when switch one *and* switch two are on." The second statement, $\overline{I}_1 = \overline{SW}_1 + \overline{SW}_2$, says "The lamp is off when switch one *or* switch two is off." Both statements are true or they could not appear in the truth table.

13-11 THE COMPLEMENT OF AN EXPRESSION —DEMORGAN'S THEOREM

Upon examination of the expressions

$$I_1 = SW_1 \cdot SW_2$$

and

$$\overline{I}_1 = \overline{SW}_1 + \overline{SW}_2$$

note that if $\overline{I}_1$ is the complement of I_1, then $\overline{SW}_1 + \overline{SW}_2$ must be the complement of $SW_1 \cdot SW_2$. This means that to find the complement of the expression $SW_1 \cdot SW_2$ you need only to change the connecting operators (a dot becomes a plus sign and vice versa) and then complement each of the elements. By this method the complement of $A \cdot B$ is $\overline{A} + \overline{B}$, and the complement of $C + D$ is $\overline{C} \cdot \overline{D}$. The formal statement of this procedure is known as *DeMorgan's theorem*. Briefly stated, it says *to complement a Boolean expression, exchange the operator and complement the terms.*

The theorem may be extended to prescribe a method for obtaining the complement of a gate (that gate which uses logic of opposite polarity for its assertions). To obtain the complement of a gate you need only to change the shape of the symbol, replacing an OR with an AND symbol and vice versa, and then invert all input and output lines. If this procedure is applied to the gate shown in Fig. 13-5c, the gate in Fig. 13-5d is the result.

13-12 GROUPING TERMS IN BOOLEAN EXPRESSIONS

Within certain limits, the terms in a Boolean expression can be grouped by bracketing to form a single term. Expressions can be "factored" and combined much like bracketed terms in ordinary algebra. For instance, the term $A = (B + C)D$ means logically the same as $A = (BD + CD)$, although electronically it represents a different combination of connected gates.

A bracketed term must be treated as a single term, even though it may contain numerous elements. Notation can be simplified somewhat by setting the expressions within brackets "equal" to a single term. For example, in the expression

$$D = \{(A + K) \cdot J\} + \{(A + K)(B + L)\}$$

notational simplification is achieved by putting $M = A + K$ and substituting M into the original expression to get

$$\begin{aligned} D &= \{(M \cdot J)\} + \{M(B + L)\} \\ &= (M \cdot J) + M(B + L) \end{aligned}$$

Insertion or removal of brackets can be done freely as long as the rules concerning brackets for ordinary algebra are observed. An obvious advantage is that there are no minus signs to take into account. Minus signs have no meaning in Boolean algebra.

When complementing a function by applying DeMorgan's theorem, each bracketed term is treated as a single element just as though it were a letter in the expression. For example, to complement the expression $A(B + C)$ gives

$$\overline{\{A(B + C)\}} = \bar{A} + \overline{(B + C)}$$

but $\overline{(B + C)} = \bar{B} \cdot \bar{C}$ so that $\overline{A(B + C)} = \bar{A} + \bar{B} \cdot \bar{C}$. Similarly, $A(B + C) = AB + AC$. Then

$$\overline{A(B + C)} = \overline{AB + AC} = \overline{(AB)} \cdot \overline{(AC)}$$
$$= (\bar{A} + \bar{B}) \cdot (\bar{A} + \bar{C})$$

13-13 BOOLEAN EXPRESSIONS FOR TWO-INPUT GATES

All gate circuits can be represented by Boolean expressions. For example, the Boolean expression for the RTL gate shown in Fig. 13-5a can be derived from its truth table, Fig. 13-5e. The elements in a Boolean expression derived from the truth table will be either A, B, C, $\bar{A}$, $\bar{B}$, or $\bar{C}$, since there are three terminals on the gate and each can have either a high or low logic level. The unique row in the table shows a low level applied to A, which means that $\bar{A}$ will appear as the assertion in the Boolean expression. Similarly, B has an L level applied and will appear therefore as $\bar{B}$, but C is shown at H level and will therefore have no bar. This gives the Boolean expression, $C = \bar{A} \cdot \bar{B}$, which describes the gate symbol in Fig. 13-5d.

The complement gate can be derived from the given Boolean expression by complementing both sides of the expression and applying DeMorgan's theorem to simplify the right side. This gives

$$\bar{C} = \overline{(\bar{A} \cdot \bar{B})}$$
$$= \bar{\bar{A}} + \bar{\bar{B}}$$
$$= A + B$$

[By DeMorgan's theorem the complement of the term $(\bar{A} \cdot \bar{B})$ is obtained by changing the operator and complementing each term.] The expression $\bar{C} = A + B$ describes the gate symbol in Fig. 13-5c.

In summary the following procedure may be followed to obtain the Boolean expression for a gate from its truth table:

1. Choose the unique row.
2. If an input or output has an L entry, the element in the Boolean expression has a bar or prime; otherwise it does not.
3. The unique row will always have the dot operator between elements, because it is the AND condition.
4. To obtain the expression for the complement gate, complement the given expression and use DeMorgan's theorem to simplify the right side.

LOGIC SYMBOL	BOOLEAN REPRESENTATION
 AND condition	• (AND operator)
 OR condition	+ (OR operator)
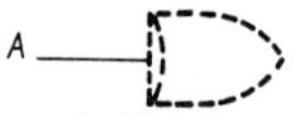 A — Input A must be high for assertion	A
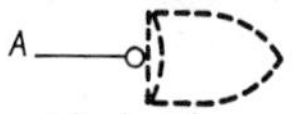 A — Input A must be low for assertion	$\bar{A}$ or A'
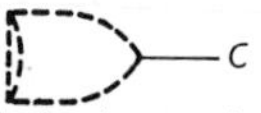 — C Output is high when conditions are met	$C =$ C exists if conditions on the right-hand side of the equals sign are present
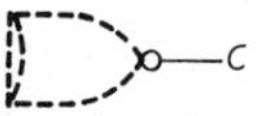 — C Output is low when conditions are met	$\bar{C} =$ Complement of C exists if conditions on the right-hand side of the equals sign are present

Fig. 13-8 Equivalence between the logic diagram and the Boolean representation as it applies to electronic logic gates.

13-14 BOOLEAN EXPRESSIONS OBTAINED FROM LOGIC SYMBOLS

A Boolean expression for a gate can be derived directly from its logic symbol without reference to the truth table. Figure 13-8 shows the relationship between the parts of a logic diagram and the elements and operators in a Boolean expression. Applying these criteria to the symbol in Fig. 13-9 gives the expression shown. It is the symbol and expression for a positive input NAND gate.

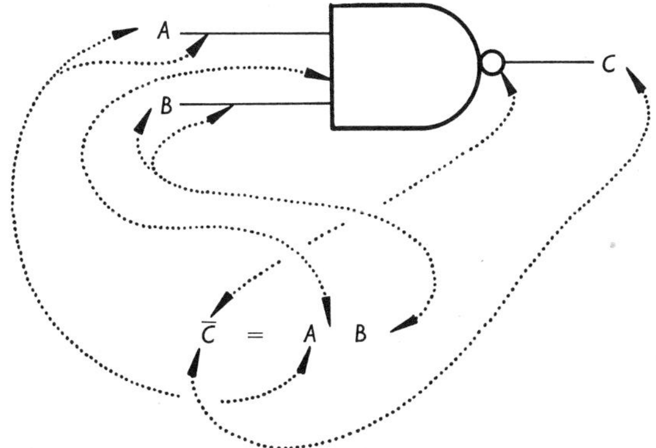

Fig. 13-9 Deriving the Boolean expression for a logic gate.

13-15 DRAWING THE LOGIC DIAGRAM FROM A GIVEN BOOLEAN EXPRESSION

To draw the logic diagram for a Boolean expression, the operator ($\cdot$ or $+$) indicates whether an AND or OR symbol is required. The remainder of the procedure is essentially that of drawing in the input and output lines, taking care to place a circular loop on all complemented elements (those with a bar or prime).

Figure 13-10 illustrates how the logic diagram for the expression

$$\bar{C} = (A \cdot B) + \overline{(D + E)}$$

is obtained. Note that whenever terms in a Boolean expression are grouped by bracketing, more than one gate will appear in the logic diagram.

13-16 DIODE-TRANSISTOR LOGIC (DTL) GATE

Although the RTL gate is economical and reliable, it has poor electrical noise immunity. A better, though more expensive gate is the diode-transistor logic gate shown in Fig. 13-11a.

When inputs A and B both receive a high-level transistor, Q_1 obtains base current through R_1 and it turns on. With Q_1 on, Q_2 obtains base current via diode D_3. It then turns on and shunts any output connected to C. Both inputs high give a low output.

When either input is low, the base current destined for Q_1 flows to ground through either D_1 or D_2, and Q_1 turns off. With Q_1 off, Q_2 also turns off for lack of base current. Output C then goes to a high level.

Circuit behavior is best portrayed by the truth table in Fig. 13-11b. The circuit is either a positive NAND or a negative NOR. Logic level requirements are not very critical. A low level can be any value from 0 V up to a point where the input diodes no longer conduct sufficiently to bleed off base current to Q_1. Typically, this point might be 0.5 V or more so that any value between 0 and 0.5 is taken as L. A high level can be any value higher than L which allows turn-on base current to flow to Q_1, typically $\frac{2}{3} V_{CC}$.

13-17 TRANSISTOR-TRANSISTOR LOGIC (TTL) GATE

Two unique features distinguish the TTL gate; one is a multiple-emitter input transistor, the other is the active "pullup" network in the output. This type of output is also known as a totem-pole circuit.

The multiple-emitter transistor (Fig. 13-12a) allows input signals to regulate the base current

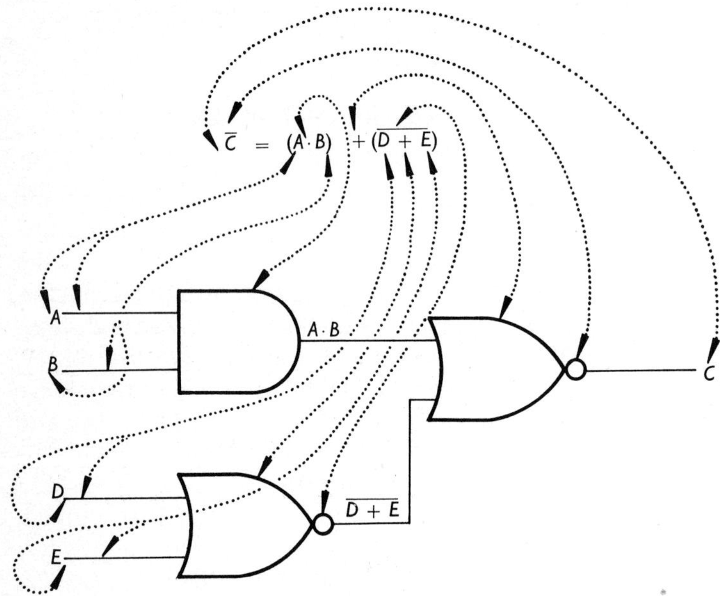

Fig. 13-10 Drawing the logic diagram represented by a Boolean expression.

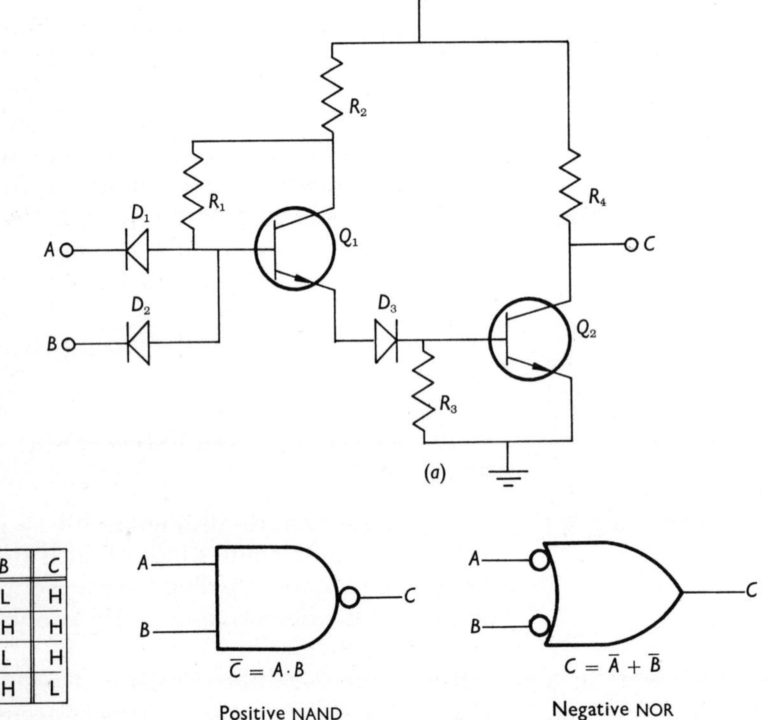

A	B	C
L	L	H
L	H	H
H	L	H
H	H	L

Fig. 13-11 Two-input DTL gate: (a) Schematic diagram; (b) Electronic truth table; (c) Logic diagrams and Boolean expressions.

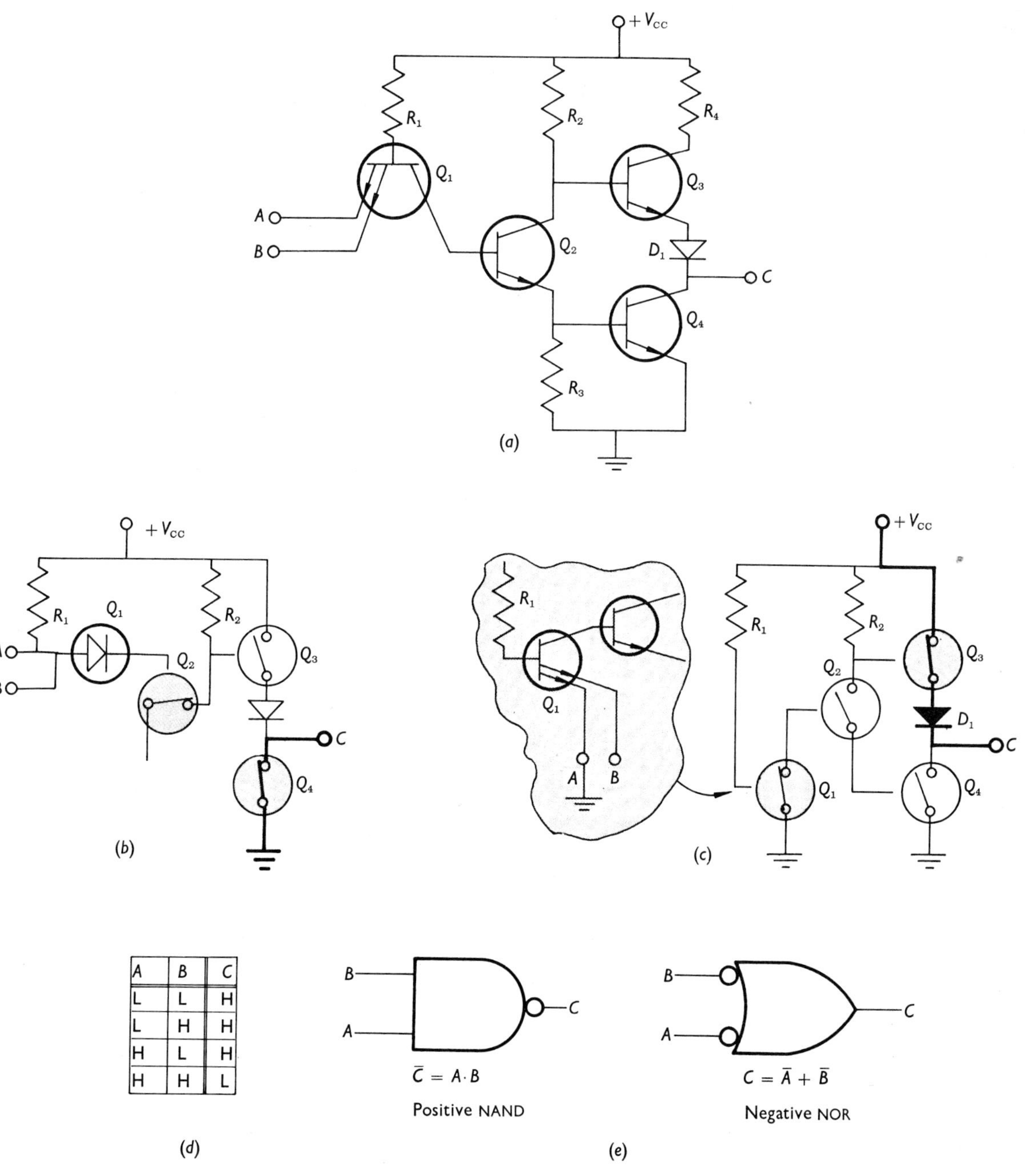

A	B	C
L	L	H
L	H	H
H	L	H
H	H	L

Fig. 13-12 Two-input TTL gate: (*a*) Schematic diagram; (*b*) Simplified circuit with both inputs high. Load is switched to ground; (*c*) Simplified circuit with one or both inputs low. Load switched to $+V_{CC}$; (*d*) Truth table; (*e*) Logic diagrams and Boolean expressions.

applied to transistor Q_2. When both inputs are high the base-collector junction of Q_1 behaves as a forward-biased diode which supplies base current to Q_2 via resistor R_1 (Fig. 13-12b). Transistor Q_2 therefore turns on and supplies base current to Q_4. It also shunts any base-to-emitter bias which might develop on Q_3, so that Q_3 is shut off for lack of base current. In this condition output C is low, as it is in effect connected to ground by turned-on Q_4 as shown in Fig. 13-12b.

If either of the multiple-input emitters in Q_1 is grounded (receive L input), the transistor becomes a forward-biased (turned-on) transistor receiving its base-bias current from R_1, as shown in the inset in Fig. 13-12c. Under these conditions the collector is low, and transistor Q_2 receives no base current. It opens, allowing base current to flow to Q_3 via R_2, but it blocks base current to Q_4. The load is therefore switched to $+V_{CC}$ as shown in the simplified diagram in Fig. 13-12c. A truth table for a TTL gate is shown in Fig. 13-12d. It is a positive NAND or negative NOR gate.

Two advantages consistent with moderate cost make the TTL gate a popular choice in a wide range of computer and control circuits. One is the high noise margin which is in the neighborhood of 1 V or more. This means that a low level can rise to the vicinity of +1 V accidentally without triggering the gate. Similarly, a high level could drop temporarily by the same amount before it is regarded as a low signal. Another advantage is the ability of the active-pullup output circuit to drive capacitive loads such as long transmission lines. Conductors of a transmission line behave somewhat like the plates of a capacitor, and require a high initial charging current before the signal can pass down the line. An active-pullup output circuit is able to supply this current surge more readily than a collector load resistor, which limits charging current. Because of this feature, however, a TTL gate's current drain from the power supply changes rapidly during its transition from one state to another. The power supply must have good voltage regulation and adequate filtering to minimize transients on the power bus.

13-18 EMITTER-COUPLED LOGIC

All the gates discussed in the previous sections operate with transistors in the saturated or near-saturated condition. A saturated transistor requires more time to switch off than one which is unsaturated because a large amount of electrical charge must be moved. *Current-mode* logic as represented by the *emitter-coupled gate* shown in simplified form in Fig. 13-13a is capable of higher switching speeds than gates with saturating transistors.

In essence the circuit is a differential amplifier, with one input (Q_1 base) held at a fixed level. The other input is allowed to follow the input logic signal. Several identical transistors may be connected in parallel to provide gating in the second input, although only two inputs (Q_2 and Q_3) are shown in Fig. 13-13a.

When both inputs are low, Q_2 and Q_3 are conducting lightly and output C_2 is high. (None of the transistors are allowed to turn off completely, or to saturate, so that speed will not be sacrificed.) With Q_2 and Q_3 conducting lightly, there is very little *IR* drop across R_e due to their emitter currents. Q_1 receives maximum permissible forward base bias and is in its maximum permissible conduction state. Output C_1 is therefore low.

If either or both inputs *A* or *B* go high, C_2 goes low. (This part of the circuit resembles an RTL gate, Sec. 13-7.) The additional *IR* drop across R_e has a polarity which reduces the forward bias on Q_1, placing it into a state of minimum conduction. Output C_1 therefore goes high. A truth table and logic symbols are shown in Fig. 13-13b and c.

The differential amplifier mode of operation offers two major advantages over other types of gates. (1) There is high common-mode

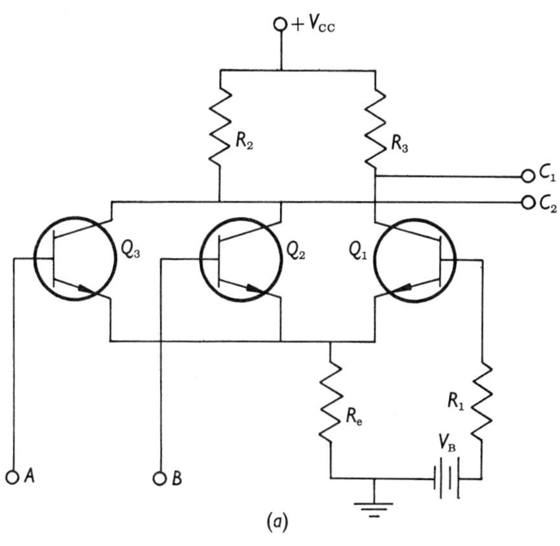

A	B	C_1
L	L	L
L	H	H
H	L	H
H	H	H

(b)

Negative AND/NAND

$\overline{C}_1 = \overline{A}\cdot\overline{B}$ using non-inverting output

$C_2 = \overline{A}\cdot\overline{B}$ using inverting output

Positive OR/NOR

$C_1 = A + B$ using non-inverting output

$\overline{C}_2 = A + B$ using inverting output

(c)

Fig. 13-13 Emitter-coupled logic gate (ECL): (a) Schematic diagram; (b) Truth table; (c) Logic symbols and Boolean expressions.

rejection so that the adverse effects of power-supply variations are greatly reduced. Furthermore, each gate draws a nearly constant current from the supply, even during transition from one level to another. (2) The inputs are high-impedance and therefore require negligible driving current from the signal source.

Practical ECL gates are of the monolithic integrated-circuit type, with emitter followers to prevent loading of the output. A fixed bias supply which adjusts for temperature variations is used to bias the fixed input side of the differential amplifier. Figure 13-14 shows a schematic of a typical ECL gate with self-contained bias supply. Note that in practice some manufacturers suggest that the positive power-supply input be grounded as shown.

13-19 CHOOSING THE TYPE OF GATE

The decision as to which type of gate is most suitable depends upon factors such as cost, speed requirements, power dissipation, and noise immunity. In parts of computer circuits, especially in some sections of large and expensive systems, speed is of utmost importance. Speed is also important in high-frequency counters used in test instruments. Table 13-1 lists the gates discussed and a comparison of their characteristics.

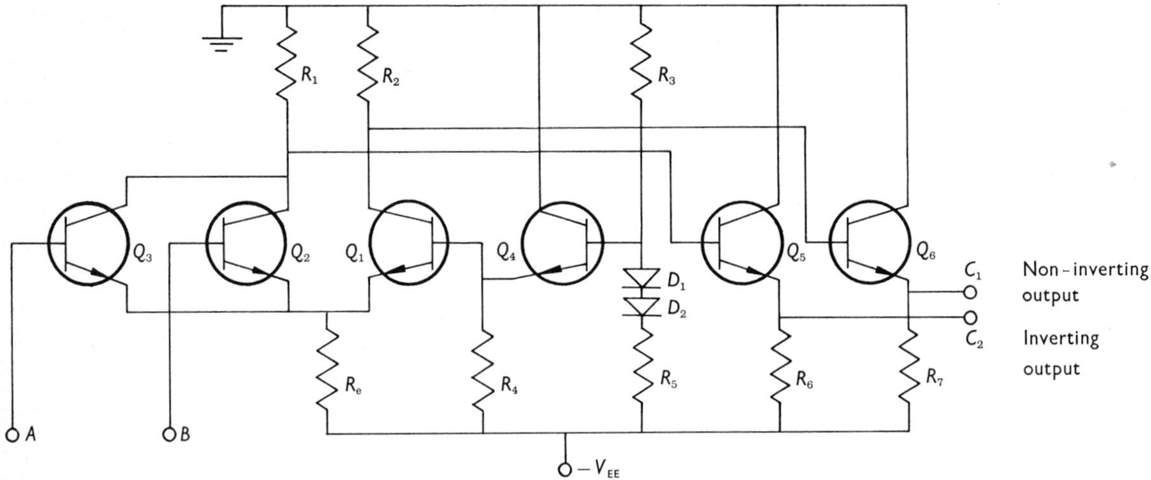

Fig. 13-14 Two-input ECL gate with self-contained bias supply and emitter-follower outputs.

Many large systems use a combination of gates in different sections. Environments which have a high electrical noise level, for example, industrial plants equipped with large electric motors and electric welding machines, usually require what is termed *high-threshold logic*. In this type there is a large voltage difference between the high and low level. Relatively large power-line disturbances will be ignored by gates which require large differences between H and L levels.

For applications in which minimum power dissipation is a prime requirement, field-effect transistors are preferable to bipolar types. One class of low-power gates is discussed in Appendix K.

13-20 PUNCHED-PAPER-TAPE DECODING

One application of electronic gates is to decode information arriving on a number of signal wires. In this example, logic signals are gener-

Table 13-1 Comparison of Electronic Logic Gates

Type	*Typical Signal-Propagation Delay Time*	*Typical Power Dissipation*	*Noise Margin*	*Typical Fan Out**	*Power-supply Voltage*	*Cost*
RTL	25 ns	10 to 15 MW	poor	5	3.6 V	low
DTL	30 ns	10 MW	good	8	5 V	medium
TTL	10 ns	15 MW	good	15	5 V	medium
ECL	6 ns	35 MW	fair	25	5.2 V	high

*Number of gate inputs which can be driven by one output.

ated by sensing holes punched in a paper tape. Tapes of this type are used to control a variety of machines, for example, numerically controlled machine tools. Each combination of punched holes represents a command to the machine, just as a combination of holes in a player-piano roll represents a group of notes to be played.

Figure 13-15a shows the essential sections of a paper-tape reader. A powered sprocket advances the tape into position so that a row of holes appears under the reading head. Then a group of pins (8 in this example) is released under spring tension to attempt passage through the tape. Only those pins which find a hole in the tape can move a sufficient distance to cause a set of contacts to close. Closure of the pin contact indicates that a hole has been sensed. After a predetermined time interval the pins are withdrawn and the sprocket steps to bring the next row of holes under the read head. Although a mechanical sensing arrangement using pins is described here, it is equally feasible to sense for holes optically. Mechanical readers are preferred in machine shops and wherever there is abnormal risk of dirt or dust accumulation during operation.

The switch contacts operated by each sensing pin distribute a high level to one of two wires. A gate input for each pin is connected to receive a high level for either a hole or no-hole condition. The wire and gate combination is known as a decoding matrix. Although a positive NAND gate is shown in Fig. 13-15b, other arrangements are possible. A low gate output occurs whenever the proper code is detected, i.e., when all gate inputs are high. If a high output is required whenever a code is detected, a second gate can be added which acts as an inverter (shown in dotted lines). Single-input gates available for this purpose are known as *inverters*. They are indicated with a triangular symbol as shown at the output of the REWIND code gate.

Decoding occurs whenever the proper combination of holes appears in the tape. At that instant all inputs to the decoding gate are high. A low occurs at the output and is used to actuate a mechanism.

Although only two gates are shown in Fig. 13-15, it is possible to connect as many gates as needed. The total number of different codes or combinations is given by the equation in Sec. 13-2. It is

$$(\text{No. of possible states})^{(\text{No. of Inputs})} = (\text{No. of combinations})$$

Substituting

$$(\text{No. of possible states}) = 2 \text{ (H and L)}$$
$$(\text{No. of inputs}) = 8$$

gives

$$2^8 = 256 \text{ possible codes with 8-hole tape}$$

13-21 DIGITAL INFORMATION STORAGE

If the push button shown in Fig. 13-16a is pushed momentarily and released, the lamp will light only while the button is depressed. The circuit has no "set" capability and will not retain, or remember, its on state. One way to give it set capability is to use a toggle switch in place of a push button. If a push button must be used, however, a relay circuit as shown in Fig. 13-16b can be constructed. When the push button is pushed, the relay closes, and the set of contacts shunting the push button allows coil current to continue even after the push button is released. The circuit has "set" capability.

To reset the relay circuit, power must be removed from the relay coil. A circuit with reset capability is shown in Fig. 13-16c. To achieve this, an additional push button which opens when pushed is used to break the relay-switch current and thereby turn off the circuit. We can think of the set-reset circuit as "storing" or "remembering" the last switch action per-

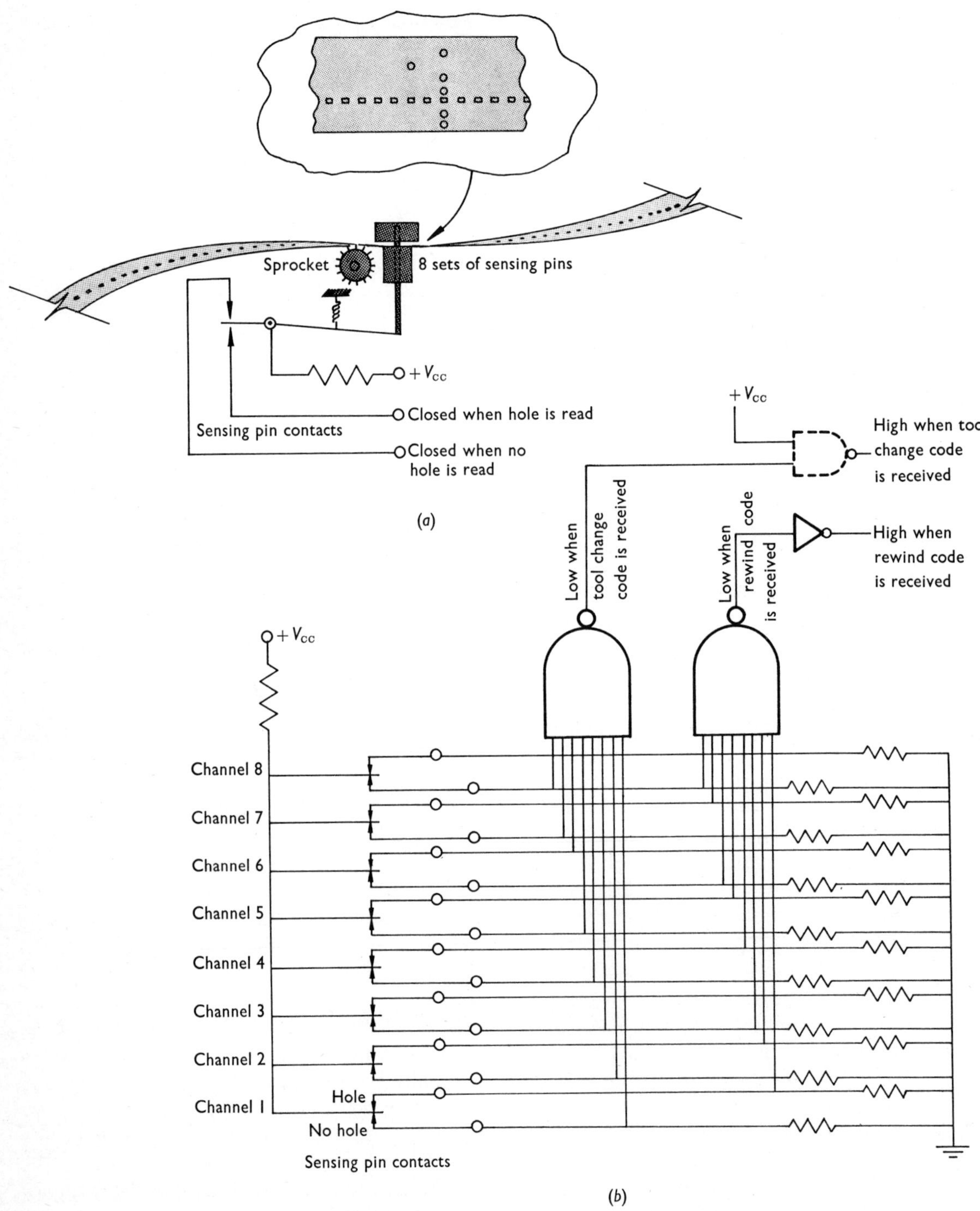

Fig. 13-15 Using gates to decode information punched on paper tape: (*a*) Electro-mechanical essentials of tape reader; (*b*) Decoding matrix.

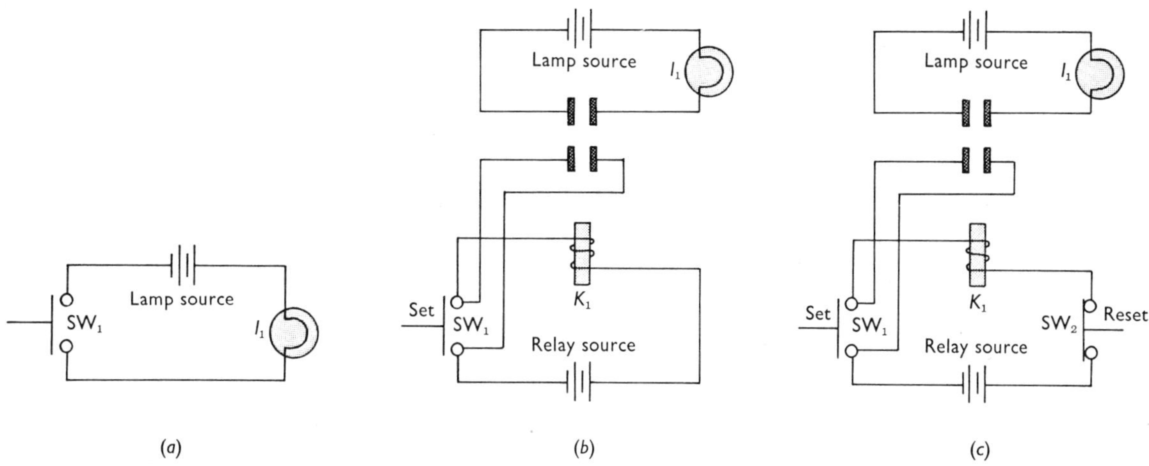

Fig. 13-16 Relay circuits with set and reset capability: (*a*) Momentary on—no set capability; (*b*) Set capability only; (*c*) Set and reset capability.

formed on it. To determine which switch action was last performed it is only necessary to note if the relay is energized or not, i.e., if the indicator lamp is on or off.

13-22 CROSS-COUPLED STORAGE ELEMENT—*RS* FLIP-FLOP

A circuit with set-reset capability can be constructed by cross-connecting outputs and inputs of two gates. The resulting circuit is a bistable flip-flop similar to that described in Sec. 12-8. A flip-flop with reset and set capability only is known as an *RS flip-flop*.

Figure 13-17a shows a diagram of two RTL-type gates connected to form an *RS* flip-flop. A transistor switch and lamp may be connected as shown to indicate when output Q is high. A high level on output Q provides base current to the lamp-switching transistor, turning it on. The letters Q and $\overline{Q}$ are labels and do *not* infer that output Q is high at all times and $\overline{Q}$ is always low. The labels do indicate, however, that the two outputs are always complements of each other.

For output Q to be in a high state, output $\overline{Q}$ must be low. If output Q is high, then a low output at $\overline{Q}$ is assured, because either input at high on this type of gate results in a low output. One way to make output Q go low is to push the reset push button. This places a high on one of the upper gate inputs which causes Q to fall. A low Q is fed to the input of the lower gate, giving a high output at $\overline{Q}$. This high is then fed back to the upper gate to hold Q low even after the reset button is released. The set button functions in the same way as the reset.

An *RS* flip-flop can also be constructed with TTL- or DTL-type gates as shown in Fig. 13-17b. These are NAND gates, and a low level is required for set and reset.

Assume $\overline{Q}$ to be low. This can happen only if Q is high because both inputs to the upper gate must be high to obtain a low output. With Q low, at least one input to the lower gate is low, ensuring that $\overline{Q}$ is high. Pushing the set button puts one low on the inputs of the upper gate, which makes output Q go high. Now the lower gate has two high inputs, and $\overline{Q}$ goes low, ensuring that Q is high. The reset button functions in the same way as the set.

The logic symbol for an *RS* flip-flop is shown in Fig. 13-17c. Pulsing the set input causes the Q output to go high. Pulsing the reset input makes output $\overline{Q}$ go high. The speed of transition from one state to the other depends on the propagation delay in the gates. Typically it is a few nanoseconds.

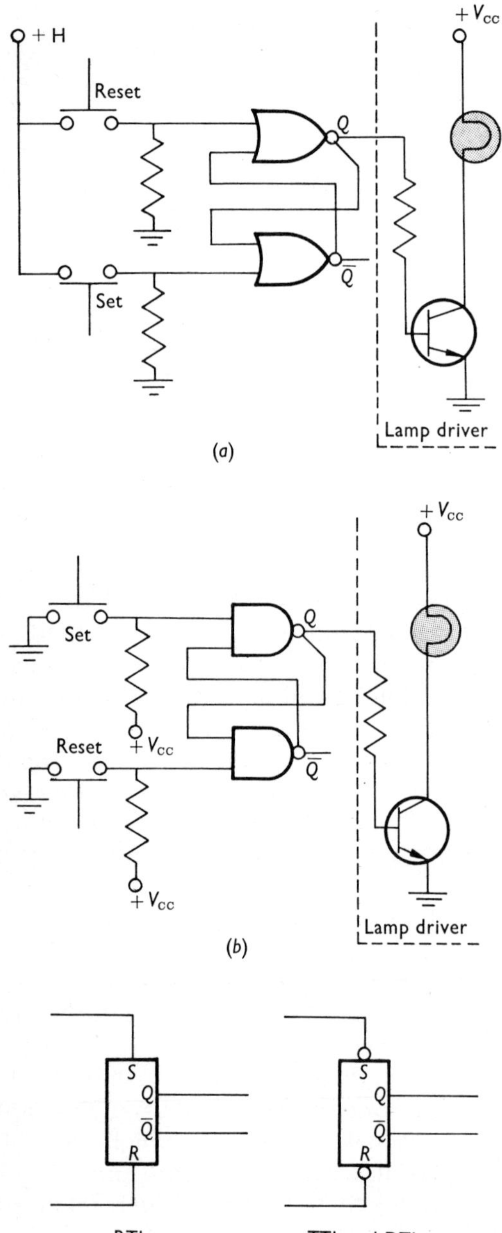

Fig. 13-17 Storage elements: (*a*) Cross-coupled positive NOR gate, *RS* flip-flop—requires a high for set or reset; (*b*) Cross-coupled positive NAND gate, *SR* flip-flop—requires a low for set or reset; (c) *RS* flip-flop symbols. Pulsing the set input makes Q go high. Pulsing the reset input makes $\overline{Q}$ go high.

13-23 SYNCHRONIZING DIGITAL CIRCUITS

In the majority of logic circuit applications set-reset capability alone is not enough. The set or reset must occur at a precise moment.

Most systems are internally synchronized by a master pulse termed a *clock pulse*. This pulse, and others synchronized by it, are distributed throughout a system to initiate a variety of actions. In some systems, the clock is a positive pulse with given rise time, amplitude, duration, and fall times. In others a negative pulse is used. With integrated-circuit logic of the type discussed in this chapter, a rectangular positive waveform provides an adequate clock signal (Fig. 13-18). This type of waveform allows for accurate triggering on both the leading and trailing edges, and it is relatively easy to generate and regenerate.

13-24 CLOCKED *RS* FLIP-FLOP

One type of clocked flip-flop in which setting and resetting takes place only at the positive-going edge of a clock pulse is shown in Fig. 13-18a. The flip-flop is composed of cross-coupled positive NAND gates as was the circuit in Fig. 13-17b. The difference here is that the set and reset signals are gated. Only if the set level *and* the clock go high together will the output of the set gate go low to change the state of the flip-flop. Similarly, the reset signal is dependent on the clock. These facts are better illustrated in the timing diagram, Fig. 13-18b.

One major drawback associated with this circuit is that if for some reason the set and reset levels both go high at clock time, both outputs Q and $\overline{Q}$ go high. Then the circuit ceases to function as a flip-flop because the output levels are no longer complementary. When the clock level falls (Fig. 13-18b), the flip-flop action is resumed, but which output will remain high, Q or $\overline{Q}$? It is not possible to state this with certainty. The condition where both set and reset levels go high simultaneously is an ambiguous condition and cannot be allowed.

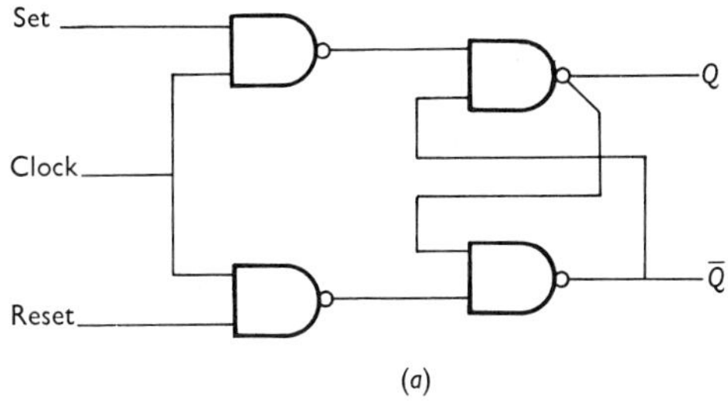

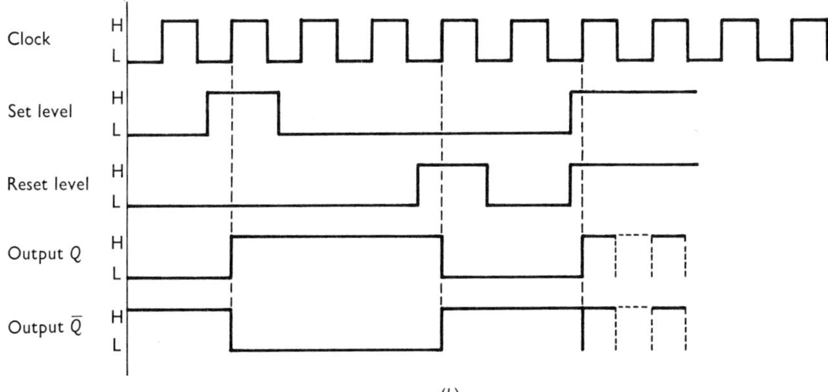

Fig. 13-18 NAND gate clocked *RS* flip-flop: (*a*) Logic diagram; (*b*) Timing diagram.

13-25 TYPE-*D* FLIP-FLOP

A clocked *RS* flip-flop is useful in many applications if the ambiguous condition is eliminated. One way to ensure that it cannot occur is to generate one of the levels, set or reset, as a complement of the other. This can be done as shown in Fig. 13-19a. Only a set level is required. The absence of a set causes reset to occur as shown in the timing diagram, Fig. 13-19c. Triple-input-type NAND gates are used to allow for unclocked *RS* capability when desired. The logic symbol for a type-*D* flip-flop is shown in Fig. 13-19b. Its truth table, Fig. 13-19d, shows the present output state (Q_n) and the output one clock pulse later (Q_{n+1}).

13-26 *J-K* FLIP-FLOP

The most versatile flip-flop is the *J-K* type whose general truth table is shown in Table 13-2. It has clocked set and reset capability, and the ambiguous condition is removed. The *J* and *K* inputs are used to determine what will take place when the clock signal arrives. If *J* and *K* are both low, the clock pulse has no effect. If *J* is high and *K* is low, the clock pulse will set output *Q* high. (If it was already high, it will remain so.) If *J* is low and *K* is high, the clock pulse will set output $\overline{Q}$ high. (If it is already high, it will remain so.) If both *J* and *K* are high, the clock pulse will flip or complement the flip-flop, regardless of the state it is in. You will note that there is a predictable occurrence for every combination of *J* and *K*. There are no ambiguous conditions.

Two kinds of *J-K* flip-flops are available. In one type the output responds after the clock pulse attains a certain amplitude on its positive edge; in the second it responds on the negative edge of the clock pulse. Most gates transfer signals when the clock pulse is in the high state. This means that in a negative edge-triggered

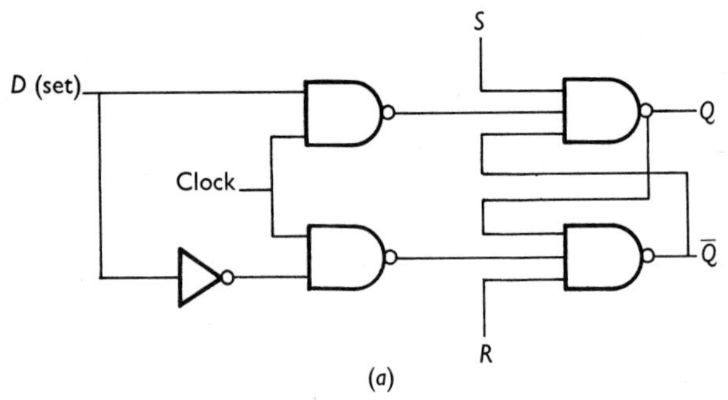

(a)

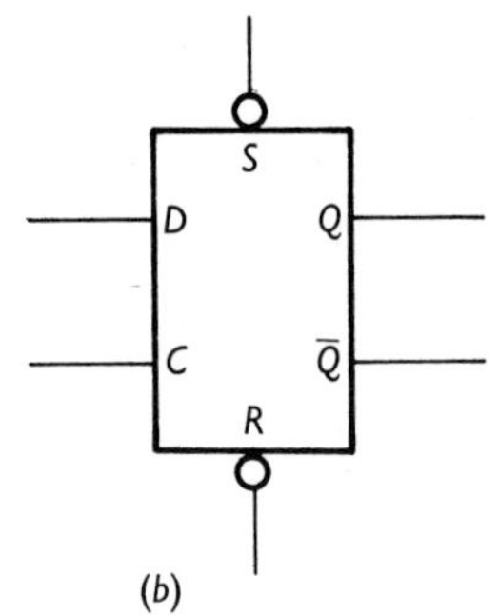

(b)

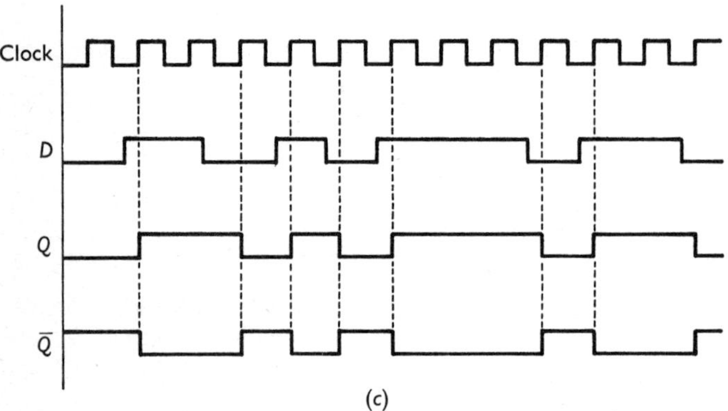

(c)

CHRONOLOGICAL TRUTH TABLE		
INPUT D	OUTPUT Q at instant n Q_n	OUTPUT Q at instant n plus one clock pulse Q_{n+1}
L	L	L
H	L	H
L	H	L
H	H	H

(d)

Fig. 13-19 Type-*D* flip-flop: (*a*) Logic diagram showing internal gates; (*b*) Logic symbol; (*c*) Timing diagram; (*d*) Chronological truth table.

J-K flip-flop, inputs *J* and *K* can be set up during clock pulse time, up to the moment when the clock signal begins to fall.

Table 13-2 Chronological Truth Table for a *J-K* Flip-Flop

Input J	*Input K*	Q_n *(Output Q at Time n)*	Q_{n+1} *(Output Q at Time n plus one Clock Pulse)*
L	L	H or L	No change
H	L	H or L	H (sets)
L	H	H or L	L (resets or clears)
H	H	L ⟶ H H ⟶ L	(complements)

13-27 *J-K* FLIP-FLOP CIRCUIT

A *J-K* flip-flop can be constructed from logic gates, but the circuit is complex. It is more practical to produce the unit in monolithic integrated-circuit form. In this way special gates with special properties can be fabricated as needed.

The logic gate equivalent for a typical negative edge-triggered *J-K* flip-flop is shown in Fig. 13-20a. It contains primarily TTL-type gates, and the internal circuit schematic is shown in Fig. 13-20b. Keep in mind that the logic circuit is an equivalent circuit, and therefore it is not always possible to physically identify the gates shown in Fig. 13-20a. For instance, gate G5 does not exist as such, but its functional equivalent is obtained by varying the logic level on the emitters of transistors Q_3 and Q_4 (Fig. 13-20b).

When both J and K inputs are low, the clock signal C has no effect. Both outputs of AND gates G_6 and G_7 are low, causing a continuous high at the outputs of NAND gates G_3 and G_4. The output flip-flop composed of NAND gates G_1 and G_2 remains in the state it was last placed (assume $\overline{Q}$ is high and Q is low, as shown initially in the timing diagram, Fig. 13-20c).

If input J goes high, indicating that the flip-flop is to be set, a high output occurs from gate G_6 when J and the clock signal *and* output $\overline{Q}$ are *simultaneously* high. This is shown in the timing diagram and the Boolean expression for the output of G_6 ($C \cdot J \cdot \overline{Q}$). The information is temporarily stored by the cross-coupled inverters until it is transferred to the output flip-flop. When C goes high the outputs of gates G_3 and G_4 are high, until clock signal C begins to fall. When clock signal C falls the output of G_4 remains high because of the low input applied to the base of Q_4 from the cross-coupled inverters and G_7. Output of G_3 goes low because of the falling clock voltage on the emitter of Q_3. The low output of G_3 sets the output flip-flop.

Reset action is similar to set, except that J is low and K is high. If both J and K are high, the outputs of G_6 and G_7 are determined solely by the state of the output flip-flop. Since J and K place no inhibiting restriction on which way the circuit will flip, it complements each time the clock signal goes low. These facts are summarized in the timing diagram, Fig. 13-20c.

13-28 NUMBER SYSTEMS

Probably the greatest contributing factor to our advances in mathematics is the invention of numerical notation. Its most useful feature is that the numerical value of a digit depends not only on the numeral but on its relative position in a number. For example, the numeral 6 can mean six, sixty, six hundred, etc., depending on the column in which it appears in a given number. Without this positional significance modern computers could not function.

One major advantage of using positional value is that even the largest numbers can be recorded with only a few symbols. The decimal system, which has been adopted universally, contains ten symbols: 0, 1, 2, 3, 4, 5, 6, 7, 8, 9. Any number whatever can be expressed by the proper combination of these symbols. Starting with the lowest integer, 1, we continue counting until 9 occurs. At this point we have run out of symbols. The next higher number is obtained by starting over again in the units column and adding 1 to the next column to the left, the tens column. A "carry" is always propagated to the next column whenever the result in the previous columns exceeds the available symbols.

Because of positional value and the carry feature, it is not really important how many symbols are used in a number system. Each time the count exceeds the number of available symbols in a given column, you start over and carry one to the next column. The *base* of a number system is defined as the factor by which a number increases its value as it moves to the next column. The decimal system has base ten. It also requires ten different numerical symbols. Although the decimal system is convenient for people, it is not the most convenient for use in computing circuits.

13-29 BINARY NUMBER SYSTEMS

Suppose you were limited to two numerical symbols instead of ten. This situation arises when it is necessary to represent numbers with electronic devices such as flip-flops, switches, lamps, etc. A flip-flop has two states, each of which can be represented by a numerical symbol. Rather than invent new symbols, a subset of the decimal numerals is used, namely, 0 and 1. Such a number system has base two. The value of a numeral doubles

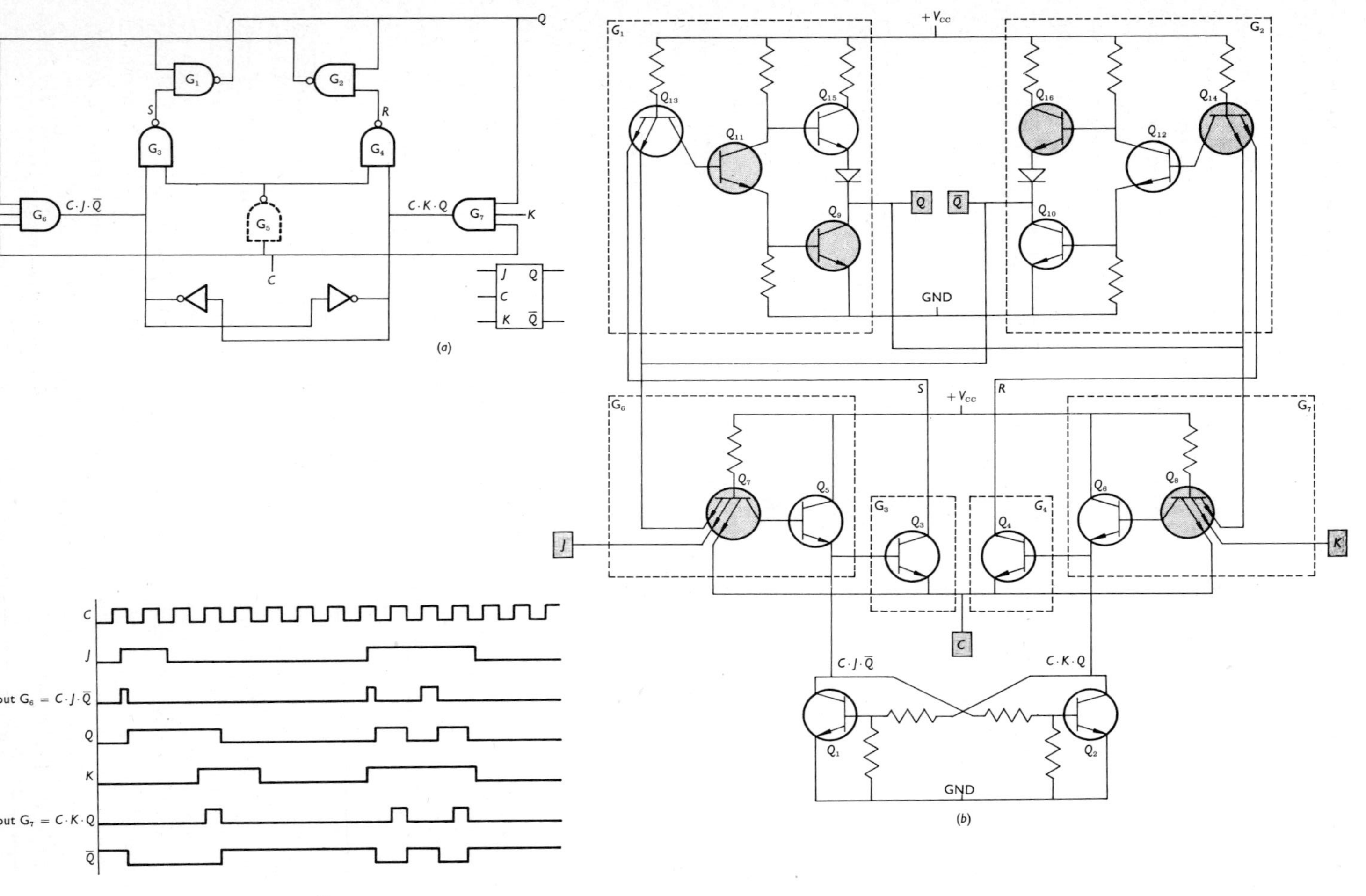

Fig. 13-20 Negative edge-triggered integrated circuit *J-K* flip-flop: (*a*) Logic gate equivalent circuit; (*b*) Schematic diagram; (c) Timing diagram.

when shifted left to the next column. Table 13-3 shows a listing of all the numbers from 0 to 15 in binary and decimal for comparison.

Table 13-3 Comparison between Decimal and Binary Numbers

Decimal	*Binary*
0	0
1	1
2	10
3	11
4	100
5	101
6	110
7	111
8	1000
9	1001
10	1010
11	1011
12	1100
13	1101
14	1110
15	1111

13-30 BINARY COUNTER

A number of *J-K* flip-flops can be connected as shown in Fig. 13-21a to form a *binary counter.* As each clock pulse arrives the status of one or more of the flip-flops changes. Note that both *J* and *K* inputs are maintained high so that a given flip-flop complements whenever its *C* input goes low.

The first flip-flop complements whenever a clock pulse goes low. The second flip-flop receives its *C* input from the first and complements only whenever the first flip-flop goes low. It will therefore complement at half the speed of the first. Each succeeding flip-flop in a binary counter complements at half the speed of the one beside it, as is shown in the timing diagram, Fig. 13-21b.

If indicator lamps are connected to the outputs as shown, the lamp I_1 will blink on and off at clock pulse rate. At high clock rates it would appear to glow continuously. Each succeeding indicator will blink at half the rate of the one beside it.

A high level from any flip-flop output signifies a binary one. The position of each flip-flop determines its true value. A high from the first flip-flop is worth one, a high from the second two, and a high from the third four. Another way to emphasize the positional value of each flip-flop is to state that the first flip-flop counts ones, the second twos, the third fours, etc. To determine the number of pulses which have been counted at any instant add the value of each of the turned-on flip-flops.

When a binary counter reaches its highest number (all flip-flops set to one), it resets to zero as shown in the timing diagram, Fig. 13-21b. Binary counters are used in a variety of control applications, test equipment, and in computers.

13-31 FLIP-FLOP REGISTER

A group of flip-flops connected to hold a binary number is known as a *register.* The binary counter is but one example of a register. There are many others, for example, a down counter, which begins with a high number and counts toward zero, and a shift register, which moves a binary number left or right or either way. With the proper arrangement of gates it is possible to design a universal register, shown in block diagram form in Fig. 13-22a. The accompanying function table (Fig. 13-22b) shows what operations can be performed on the data in the register.

13-32 DIGITAL COMPUTER

By interconnecting a number of registers with the appropriate gating, arithmetic can be performed on the binary numbers stored in the registers. Certain gates are set by the status of register flip-flops to issue control pulses at the right instant. In this way binary numbers may be added, subtracted, multiplied, divided, shifted, negated, etc. A large digital computer

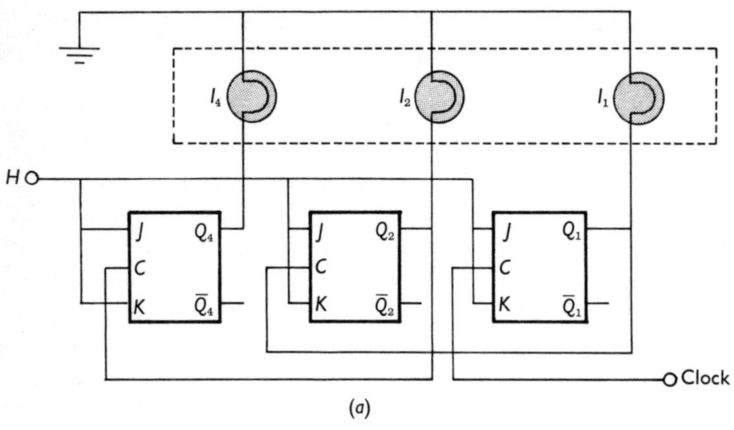

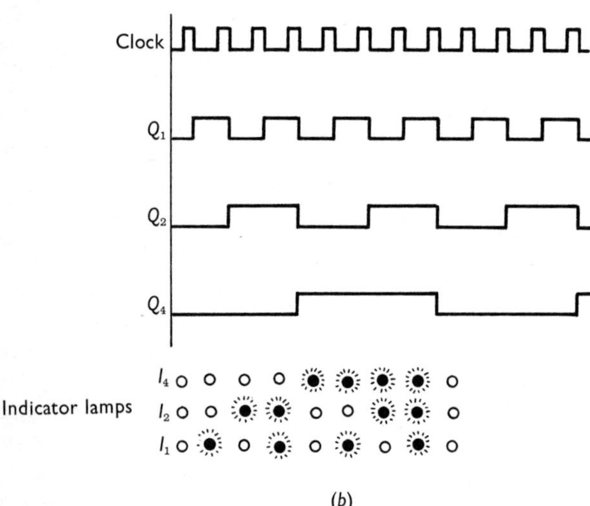

Fig. 13-21 Binary counter constructed of *J-K* flip-flops: (*a*) Logic diagram; (*b*) Timing diagram and status of indicator lamps.

contains gates and flip-flops connected to form an arithmetic element or data processor.

A digital computer requires much more data than can be stored in flip-flop registers. Data used by the processor are recorded as ones and zeros on magnetic tapes, magnetic discs, or in magnetic memory cores.

Summary

Most electronic circuits are either of the *linear* or *digital* type. Linear circuits preserve waveform, whereas digital circuits are generally two-state on-off devices.

Proposition or *truth tables* list the output responses of a circuit to various input combinations.

Circuits which respond with a given output when the correct input combination is present are known as *gate* circuits.

The AND condition results when a circuit does what the unique output row in a truth table indicates. If output and input have opposite polarity, the circuit is a NAND gate.

The OR condition results when a circuit does what the non-unique output rows in a truth

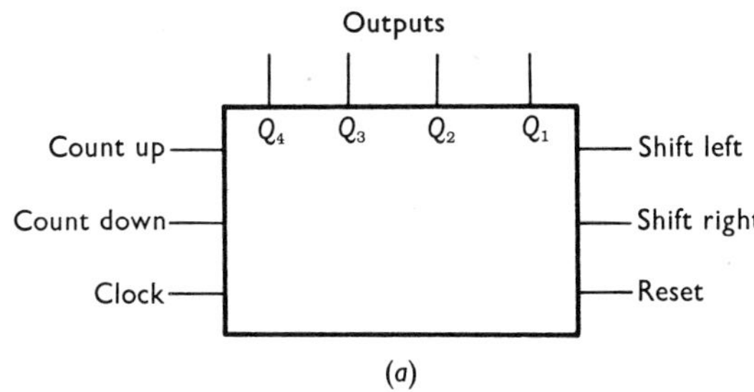

(a)

FUNCTION TABLE				
INPUTS				RESULT
Count Up	Count Down	Shift Left	Shift Right	
L	L	L	L	Binary number in register remains unchanged
L	L	L	H	$Q_{x+1} \rightarrow Q_x$, where $x = 1, 2, 3, 4$
L	L	H	L	$Q_x \rightarrow Q_{x+1}$, where $x = 1, 2, 3, 4$
L	H	L	L	Binary number reduces by 1 each time a clock pulse occurs
H	L	L	L	Binary number increases by 1 each time a clock pulse occurs
MORE THAN ONE INPUT HIGH				Ambiguous condition—not allowed

(b)

Fig. 13-22 Universal register: (a) Block representation; (b) Function table.

table indicate. If output and input have opposite polarity, the circuit is a NOR gate.

The *resistor-transistor logic* (RTL) gate is the most economical but has poor noise immunity. It is a positive NOR gate.

The *diode-transistor logic* (DTL) gate is more expensive than the RTL but has better noise immunity. It is a positive NAND gate.

The *transistor-transistor logic* (TTL) gate is more expensive than either the RTL or DTL but has excellent noise immunity. It is a positive NAND gate and is compatible with the DTL gate.

Emitter-coupled logic (ECL) is the fastest of the four types described. It is not compatible with the other types without the use of logic level changers.

Any inverting gate can be used as an *inverter* by placing an enabling level on all inputs but one. The gate output then depends only on the status of the remaining input.

Boolean algebra is useful for minimizing the number of gates required to construct a logic circuit. Although the manipulations in Boolean algebra are similar to those of numerical algebra, the elements and operators are not.

DeMorgan's theorem states that the inverse or complement of a logic expression can be obtained by complementing the terms and changing the operator.

Digital information can be stored in bistable flip-flops because they have set-reset capability, i.e., memory.

The commonly used flip-flops are the *RS, clocked RS, type D,* and the *J-K*.

The binary number system uses only two symbols, *0* and *1*. It can therefore take advantage of two-state devices such as switches and flip-flops to represent a number.

A group of flip-flops connected so as to represent a binary number is known as a *register*.

A digital computer consists of many registers with appropriate control flip-flops so that data processing can take place in the registers.

Questions and Exercises

1. Broadly speaking, what is the difference between linear and digital circuits?
2. What are logic circuits?
3. What information does a truth table convey?
4. How many rows would be in a truth table for a circuit as in Fig. 13-1a, but with four rather than two switches connected in series?
5. With switches it is customary to use the terminology "on" and "off." What two terms are used with electronic gate circuits?
6. What is meant by the AND condition?
7. Explain what is meant by positive and negative logic.
8. What is indicated by a small circular loop on a logic symbol?
9. What is meant by the OR condition?
10. Draw a schematic diagram of a lamp controlled by four switches such that any one of them can turn on the lamp.
11. Draw a logic symbol of a gate which is similar in behavior to the circuit in Question 10.
12. Describe two ways in which current can be prevented from flowing through a load.
13. If both inputs to a two-input RTL gate are high (H), what will the output be?
14. Draw a schematic diagram of (*a*) a two-input RTL gate and (*b*) a four-input RTL gate. Can a pair of two-input RTL gates be connected in any way to produce a four-input RTL gate? If so, draw the schematic diagram.
15. Deduce the logic symbols for each of the following truth tables. Two symbols, one for the AND and one for the OR condition, can be derived for each truth table.

Inputs			*Output*
A	*B*	*C*	*D*
L	L	L	H
L	L	H	H
L	H	L	H
L	H	H	H
H	L	L	H
H	L	H	H
H	H	L	H
H	H	H	L

(*a*)

Inputs			*Output*
A	*B*	*C*	*D*
H	H	H	H
H	H	L	H
H	L	H	H
H	L	L	H
L	H	H	H
L	H	L	H
L	L	H	H
L	L	L	L

(*b*)

16. Draw the logic symbol for the following truth table.

Input	*Output*
A	*B*
H	L
L	H

17. Draw the logic diagrams for the following Boolean expressions:
    a. $F = E + B + K$
    b. $R = ST$
    c. $Q = J\overline{K}$
    d. $E = RD + QC$
    e. $\overline{M} = (J + S)(R + T)$
18. Give the Boolean expressions for each of the symbols you derived in Questions 15 and 16.
19. Draw the logic diagram for the expression $Q = R(T + K) + M$ as it stands. Next attempt to simplify it and draw the logic diagram for the resulting expression. If all gates cost 20¢ per input, which circuit would be more economical to construct?
20. Write the expressions for $\overline{F}$, $\overline{R}$, $\overline{Q}$, $\overline{E}$, and $\overline{M}$ derived from the given expressions in Question 17. (Hint: Use DeMorgan's theorem.)
21. What would happen if the outputs of two TTL gates as shown in Fig. 13-12a are connected together, and one gate attempts to produce a high output while the other attempts to produce a low?

22. Which type of gate would you select for the construction of high-frequency flip-flops to be used in a frequency counter? (You want the highest possible counting rate.)
23. How many different codes is it possible to obtain from a 12-column standard Hollerith type card? (Assume that all 12 holes or any combination thereof can be punched at a time.)
24. Which type of flip-flop discussed in this chapter has set-reset capability only?
25. In order to set the flip-flops in Fig. 13-17c, which type requires a high?
26. What is meant by a "clocked" flip-flop?
27. How is ambiguity removed in a *D*-type flip-flop?
28. Make a truth table for a *J-K* flip-flop which responds on the falling edge of a positive clock pulse.
29. Draw a logic diagram of a binary counter which can count up to 15 before it resets to zero and begins counting again. (Use *J-K* flip-flops.)
30. Using TTL-type gates and *J-K* flip-flops draw a detailed logic diagram for the universal register in Fig. 13-22.

14-1 POWER-CONTROL CIRCUITS

Electronic power-control circuits are required when applied voltage must be accurately maintained under varying load and supply conditions. Certain applications require power to be switched on and off at various rates, such as motors, speed controls, and other industrial loads. Some circuits are controlled by the level of illumination and require light-sensitive devices. Others require accurate time intervals for their operation so that electronic timers must be provided. A few of the more common electronic control circuits are discussed in this chapter.

14-2 SERIES-TYPE VOLTAGE REGULATOR

The power-supply circuits discussed in Chap. 2 produce an output voltage steady enough for most radio, television, or home-appliance circuits. For more critical applications such as test equipment better voltage stability is required than is obtainable with a simple zener-diode regulator. An electronic regulator circuit is required.

A fundamental type of electronic regulator circuit is shown in Fig. 14-1a. Most medium-power regulator circuits are variations of this type. It is known as a series-type regulator, because the regulating device, transistor Q_1, is in series with the load.

A closer look at this circuit shows that it is essentially an emitter-follower circuit and that regulation occurs because of the negative feedback developed across the load. The load acts as the emitter resistor.

The object of the circuit is to maintain a constant voltage drop across R_L. Suppose that for some reason the load current increases, resulting in a greater IR drop across the load. The polarity of the IR drop is such that it decreases the net forward bias on the base of transistor Q_1. Its collector current therefore drops and decreased load current results. If load current falls, transistor bias is increased, and load current rises.

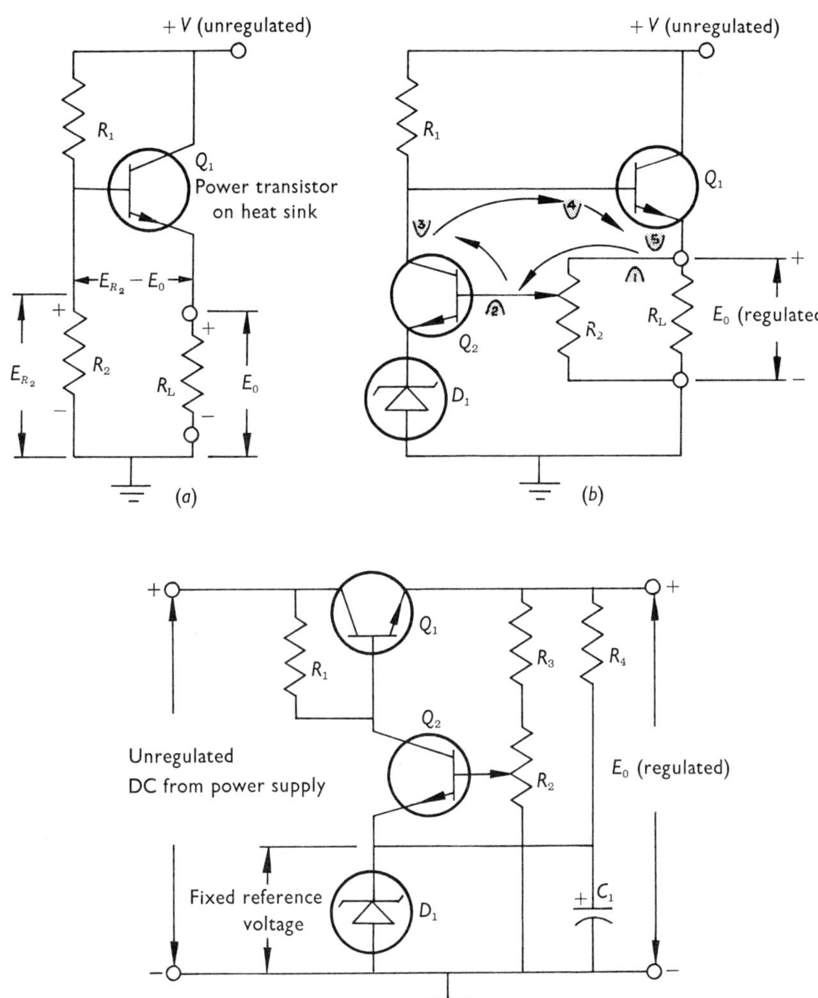

Fig. 14-1 Series-type electronic voltage regulator: (*a*) Simplified circuit to show emitter-follower properties; (*b*) Amplifier and reference voltage added to increase sensitivity; (*c*) Complete circuit.

Although the circuit in Fig. 14-1a works in principle it is barely more than a resistive voltage divider. Much greater sensitivity can be obtained by amplifying load-voltage variations. The amplified load variations are used to control the current through the series transistor. A circuit with an amplifier is shown in Fig. 14-1b. A fraction of the load voltage is applied to the base of Q_2 from the potentiometer R_2. The emitter of Q_2 is held at a fixed voltage by zener diode D_1 and is virtually free from variations.

If load voltage rises for any reason, the base-to-emitter voltage of Q_2 rises; i.e., it receives more bias. Its collector current increases resulting in a greater voltage drop across R_1 and reduced bias for the series transistor Q_1. Load current and voltage are reduced and regulation is achieved. The sine-wave segments at various points in the drawing show the phase relationship involved.

Figure 14-1a and b draw attention to the emitter-follower property of the circuit. Conventional schematic diagrams show the circuit as in Fig. 14-1c. Only the load terminals are shown, because the load is rarely just a resistor. In the reference power-supply section, resistor R_4 is added to increase zener diode current and make it less dependent on Q_2. Further stability is provided by capacitor C_1 which absorbs sudden variations. A stable reference voltage makes other voltage variations all the more pronounced.

14-3 SILICON-CONTROLLED RECTIFIER

A silicon-controlled rectifier (SCR) is a four-layer, three-junction device constructed as shown in Fig. 14-2. It has properties similar to a rectifier, but it has the added feature that it requires an activating signal to make it perform. Figure 14-3 shows how the activating signal turns on an SCR. The device is connected as a rectifier diode in Fig. 14-3a, but no current flow takes place across the PN junctions. In this turned-off condition an SCR can have a resistance of thousands of ohms regardless of the polarity. It behaves in effect like an open switch.

If a small amount of gate current is injected as shown in Fig. 14-3b, the gate-cathode "diode" begins to conduct. Electrons are injected into the control layer from the cathode to balance the holes injected at the gate. Now, even if gate current is removed, more electrons enter via the cathode to fill the spaces evacuated by the initial electron surge. The effect culminates in a hole-electron pair movement across all the junctions, and within a few microseconds the SCR goes into full conduction. Once turned on it remains so, even if gate current is disconnected. Only an interruption in the cathode-to-anode current can turn an SCR off.

In the on state, an SCR has a low resistance of only a few hundredths of an ohm, but only

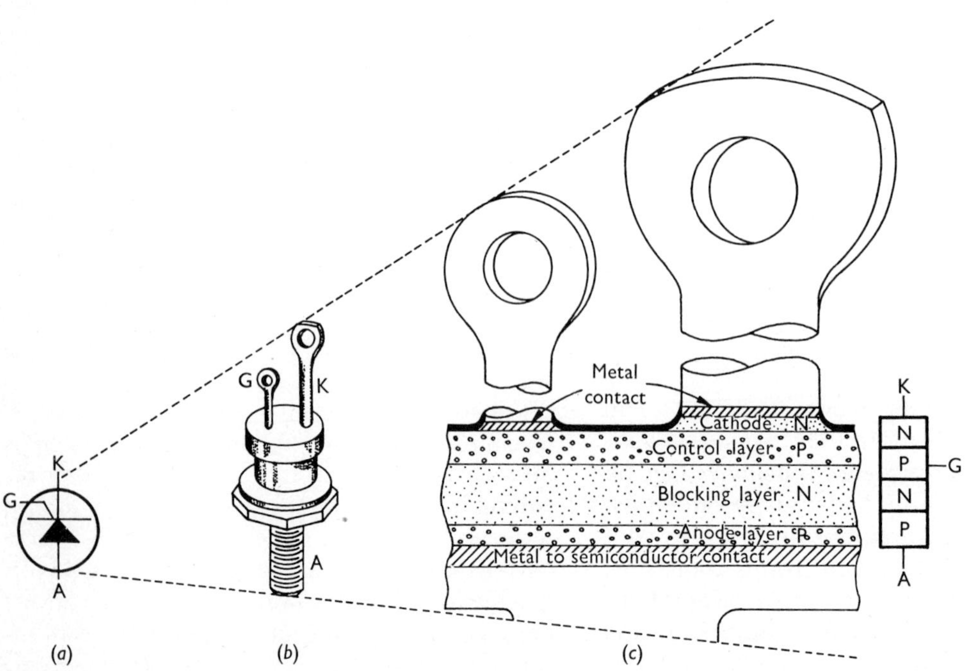

Fig. 14-2 Structure of a silicon-controlled rectifier (SCR): (*a*) Schematic symbol; (*b*) Pictorial diagram of typical medium power SCR; (c) Internal structure.

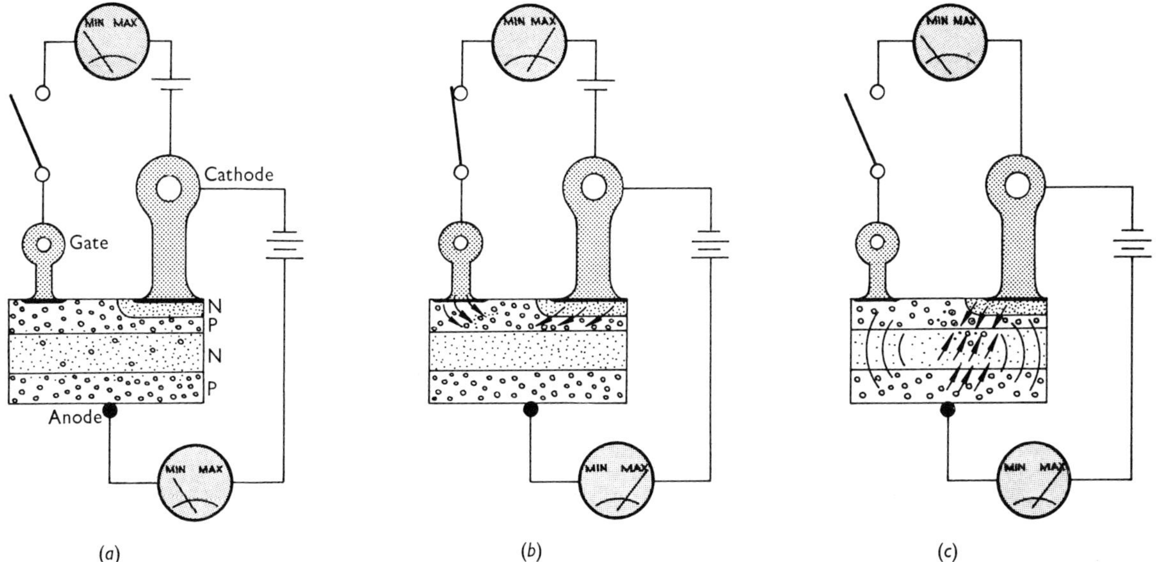

Fig. 14-3 Turn-on and conduction properties of SCR: (*a*) No gate current, no cathode-to-anode current; (*b*) Burst of gate current forces electrons out of cathode layer and initiates turn-on; (*c*) Gate current is not required to sustain SCR current after turn-on.

in the forward-biased direction. If reverse-biased (+ on cathode, − on anode) an SCR maintains its high resistance whether gate current is applied or not. It is this property which makes the SCR a "controlled" rectifier; i.e., it will rectify only if turned on by means of a gate signal.

14-4 SCR PHASE-CONTROL CIRCUITS

One of the most widely used applications of an SCR is in *phase-control* circuits. With phase control, power is applied to a load only for a fraction of each cycle of an alternating current. Of course, power can be controlled by other means such as with series resistance or inductance as shown in Fig. 14-4, but each has disadvantages.

An SCR phase-control circuit is shown in Fig. 14-5a. The SCR behaves as a half-wave rectifier and applies power to the load only on positive half-cycles. Diode D_1 is included to block the high reverse-gate voltage on the negative half-cycle. The control feature stems from the fact that the SCR turn-on may be regulated to occur at various points on the ac waveform. If R_1 is adjusted to a minimum value, there will be insufficient gate-firing voltage developed until the input waveform reaches its positive peak. Then the SCR conducts only for the remaining half of the positive alternation or a total of one-quarter cycle. Increasing the setting of R_1 results in earlier turn-on, and the SCR can be made to conduct for nearly the entire half-cycle. These facts are illustrated in the timing diagram in Fig. 14-5b and c. Since average power is proportional to the duration of applied voltage during each cycle, the SCR achieves power control.

14-5 UNIJUNCTION-GATE PULSE GENERATOR

A gate signal can be developed in a variety of ways. In the majority of applications the most satisfactory gate signal is a short pulse.

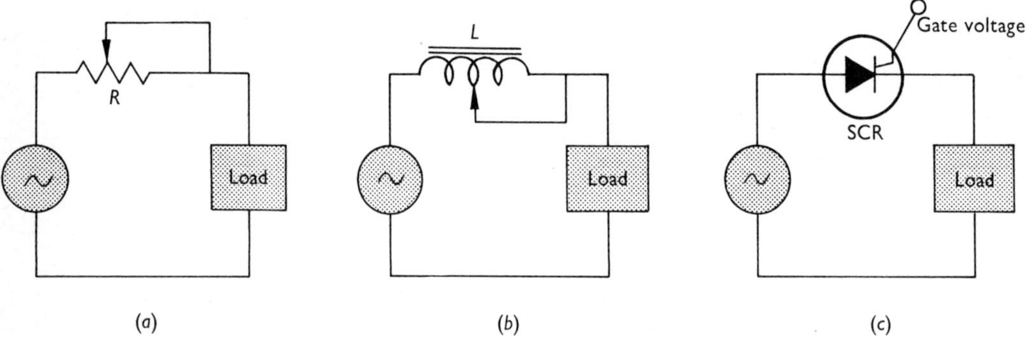

Fig. 14-4 Some methods of controlling ac power applied to a load: (*a*) Series resistor wastes power in I^2R losses; (*b*) Series inductance—bulky and expensive; (c)SCR—small, inexpensive, very efficient.

A simple and inexpensive way to obtain such a pulse is to use a unijunction oscillator circuit.

A unijunction transistor resembles a PN junction diode except that two connections are made to the N-type semiconductor block. Figure 14-6a shows how a PNP transistor must be structurally modified to give a unijunction transistor.

If connections are made to the emitter and only one of the base electrodes of a unijunction transistor, it behaves like a diode. If, however, a voltage is applied across the two base electrodes, voltage division takes place in the base block just as though it were a resistor. The emitter-diode junction appears to be "connected somewhere" along this resistance as shown in the equivalent circuit in Fig. 14-6b. Under these conditions, the N or cathode side of the emitter diode is at a positive potential with respect to ground.

To obtain emitter current, the emitter voltage must be raised to a point where the diode is forward-biased. The exact value, known as the *peak-point emitter voltage* (E_p), will depend on the voltage applied between the two base electrodes. When the emitter voltage is raised to

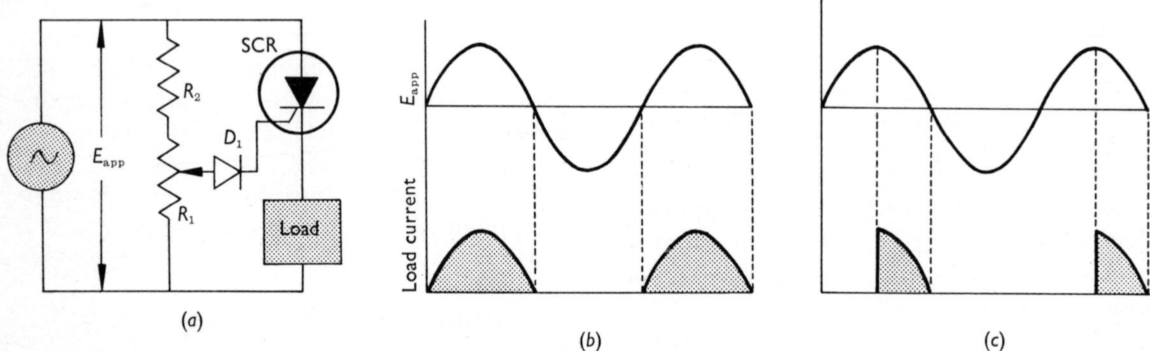

Fig. 14-5 Simple SCR phase-control circuit: (*a*) Schematic diagram; (*b*) SCR (and load) current for maximum setting of R_1. Gate-firing voltage occurs just after the ac waveform begins to go positive; (*c*) SCR and load current for minimum setting of R_1. Gate-firing voltage is developed only at the positive ac waveform peak.

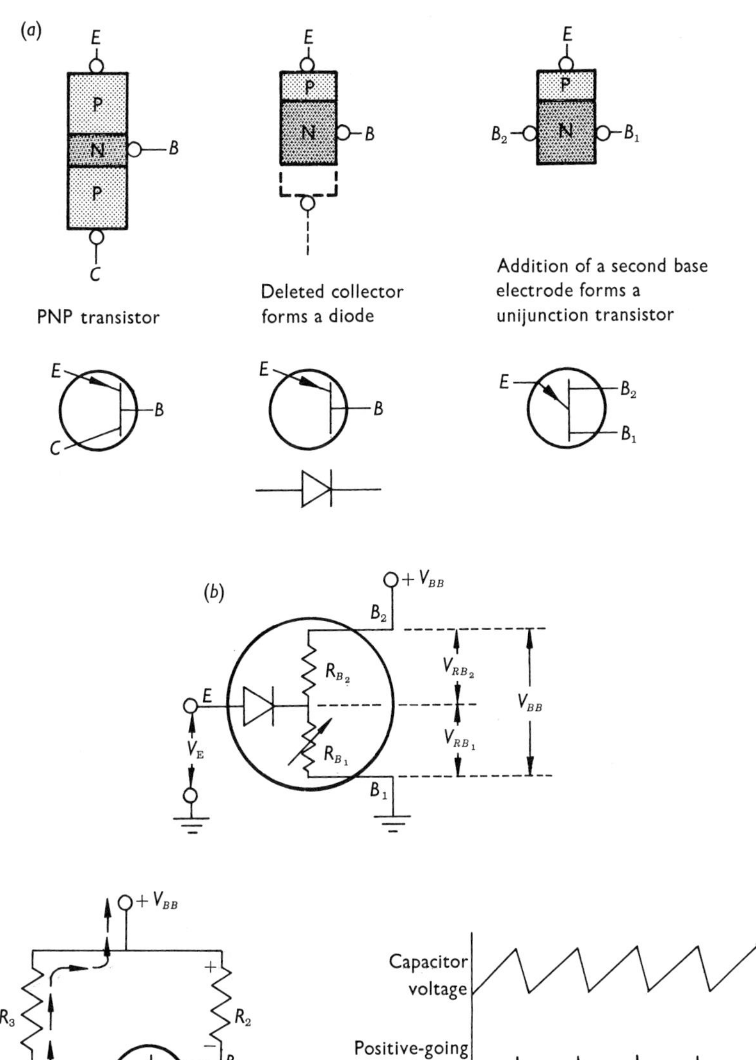

Fig. 14-6 Unijunction transistor: (*a*) Comparison between transistor, diode, and unijunction transistor; (*b*) Diode and resistive voltage divider equivalent circuit; (*c*) Unijunction pulse generator and timing diagram.

the peak point, the emitter diode not only conducts, it exhibits a negative-resistance region and emitter current actually increases if emitter voltage is decreased. There is then heavy current flow between the emitter and base, i.e., the unijunction transistor is turned on.

A unijunction oscillator circuit is shown in Fig. 14-6c. When power is first switched on, V_{BB} is applied to the base contacts via R_2 and R_1. It is also applied to the *CR* network consisting of resistor R_3 and capacitor *C*. Initially the capacitor has no charge and the emitter is at 0 V, so that the unijunction cannot turn on. Capacitor *C* begins to charge through R_3 as shown by the thin arrows in Fig. 14-6c. The rate of charge will depend on the *CR* time constant of *C* and R_1. After a period of time, the capacitor voltage will climb to E_p and the unijunction will turn on. Heavy capacitor-discharge current then flows between the emitter and base B_1. The capacitor is quickly discharged and the unijunction turns off, allowing the capacitor to begin another charge cycle. The intense capacitor-discharge current causes a voltage drop of short duration (a pulse) to appear across R_1. Another way of stating this is to say that when the unijunction turns on, the charged capacitor is connected across R_1 by the conducting emitter-base junction. These facts are portrayed in the accompanying timing diagram in Fig. 14-6c.

14-6 UNIJUNCTION-TRIGGERED SCR PHASE-CONTROL CIRCUIT

The positive-going pulses generated by a unijunction oscillator can be used to trigger the gate of an SCR. Figure 14-7 shows a typical SCR phase-control circuit of this type. The circuit can be used to vary lamp brightness, motor

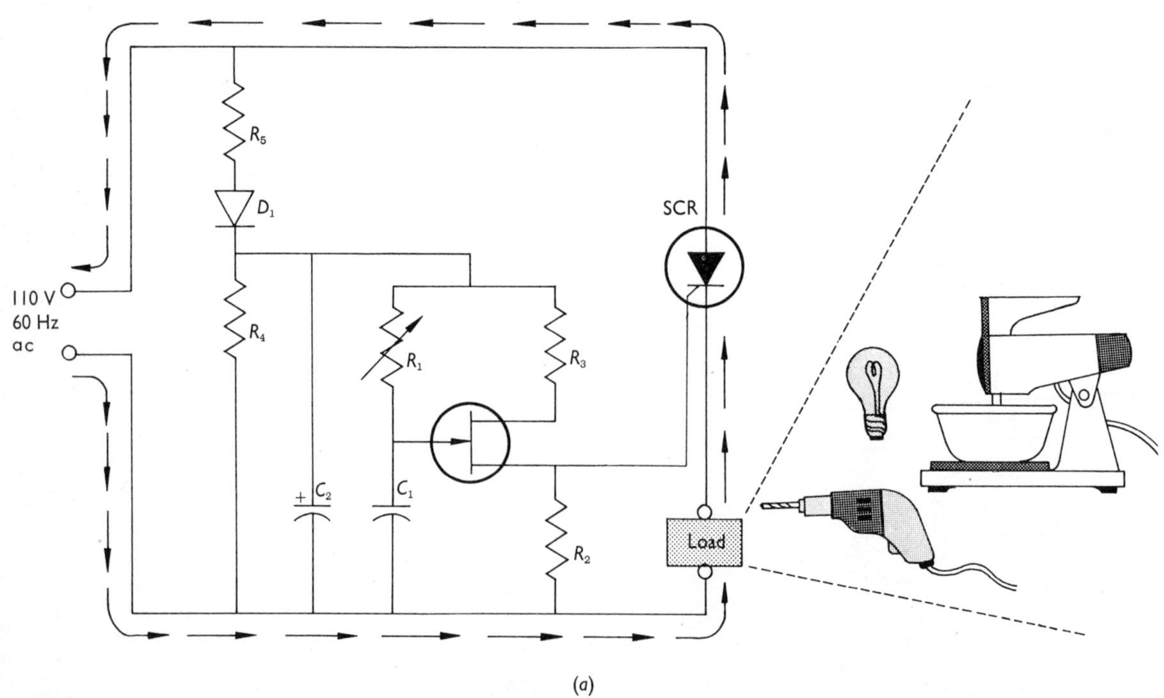

Fig. 14-7 Unijunction-triggered SCR phase-control circuit: (*a*) Schematic diagram; (*b*) Timing diagram for low and high trigger rates; (c) Oscilloscope photographs. Upper trace—gate triggering signal. Lower trace—load voltage across resistive load.

speed in appliances (e.g., power hand drills), and many other applications.

Resistors R_5 and R_4 form a voltage divider to reduce the applied voltage to a safe operating value for the unijunction transistor. Diode D_1 rectifies this voltage and capacitor C_2 filters the ripple so that essentially pure dc is applied to the unijunction oscillator circuit. Resistor R_1 is variable and is used to adjust the time constant of C_1 and R_1. This varies the charge time of C_1 and consequently the interval between unijunction-output pulses. If the interval is short (many pulses per second), the SCR is turned on immediately after the applied ac waveform starts its positive alternation.

The SCR therefore conducts for most of the half-cycle, applying maximum power to the load. If the interval is long, only a few pulses per second are produced. The SCR then may turn on near the completion of the positive ac alternation; consequently, very little power is applied to the load. These facts are illustrated in the timing diagrams (Fig. 14-7b) and also in the oscilloscope photographs in Fig. 14-7c.

Gate pulses occur only on positive power alternations if C_2 is omitted, resulting in "synchronized" operation of the unijunction oscillator.

14-7 ELECTRONIC TIMING CIRCUITS

In certain instances it is necessary to turn a circuit on or off for a short but accurate time interval. This can be done electronically to any degree of precision and accuracy desired. Two simple time-delay circuits are shown in Fig.

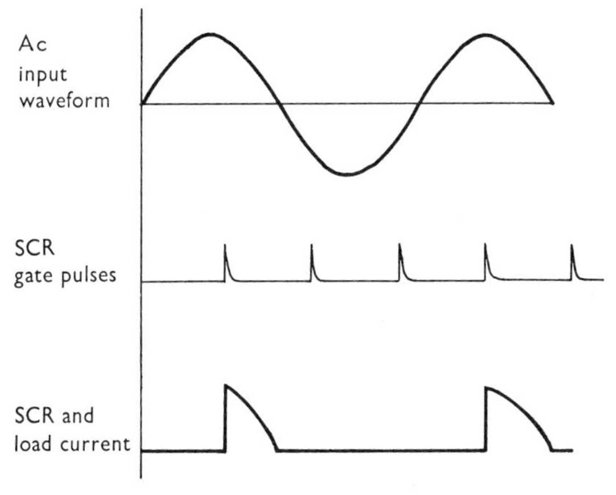

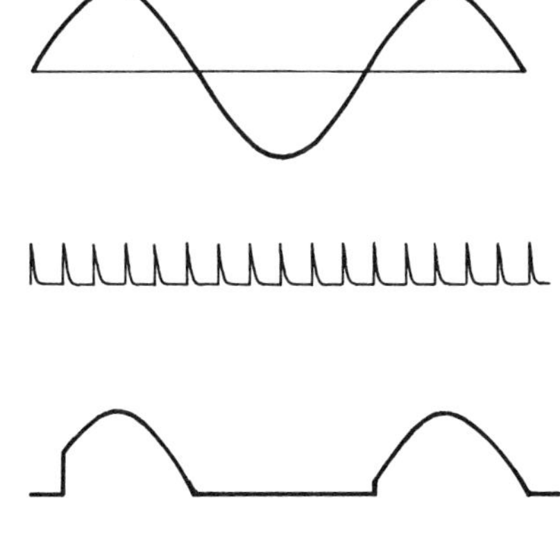

(b)

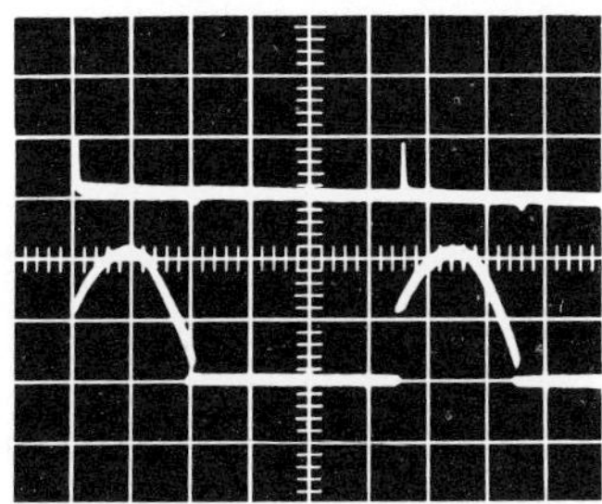

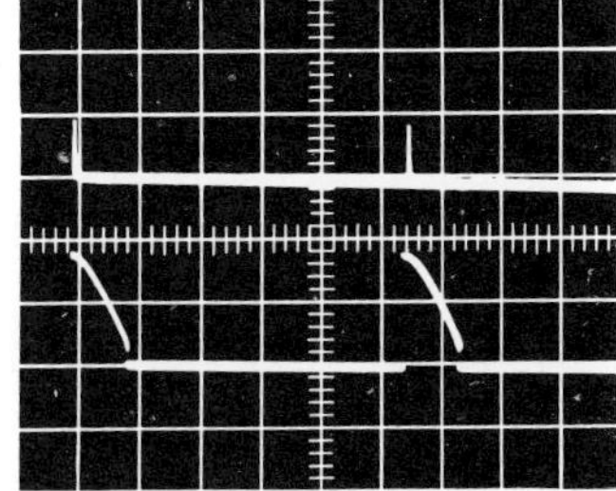

(c)

14-8. A typical application of such circuits would be to turn on a photographic enlarger for a preset time interval.

Nearly all electronic time-delay circuits make use of the *CR* time constant of a capacitor and resistor. In the circuits in Fig. 14-8a capacitor C_1 is charged by the battery. When the switch is placed into position 2 the capacitor discharges into the base circuit of transistor Q_1. While it is discharging it provides a source of base current for Q_1, and the transistor remains turned on for the duration of the discharge cycle. The relay coil is then energized by collector current and power is available at the timed outlet. As soon as the capacitor has discharged to a certain minimum level, the base and collector current are reduced. There is no longer enough current to energize the coil and its contacts open, switching off the timed power.

To initiate another timed cycle the switch must be again set to position 1 to charge the capacitor. Variable resistor R_1 is used to set the *CR* time constant to obtain the desired timing interval. An even greater range of time intervals can be obtained by switching in additional capacitors to change the net value of C_1.

A more economical timing circuit is shown in Fig. 14-8b. No relay is required, but the circuit has the disadvantage that the SCR blocks the

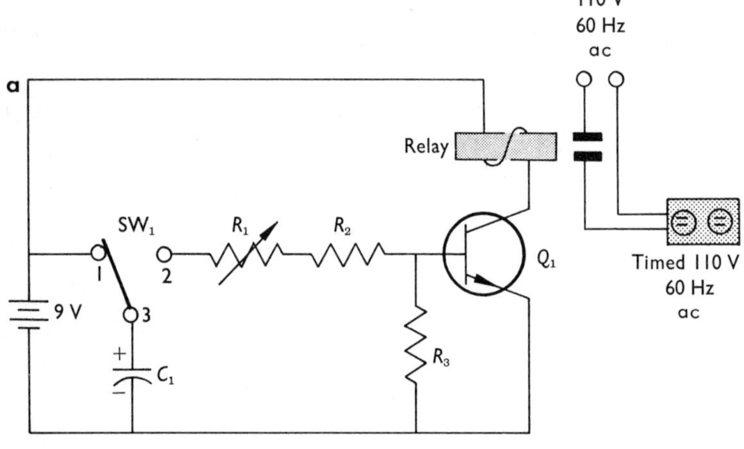

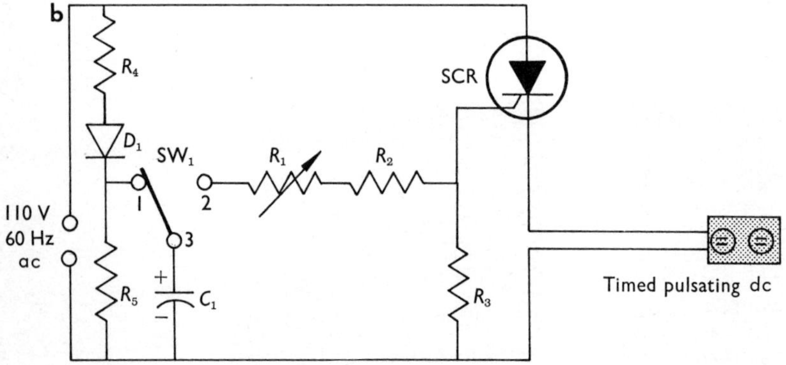

Fig. 14-8 Simple electronic timing circuits.

negative alternations of the ac waveform. If a lamp were used as a load, it would provide less illumination than obtainable with relay contacts.

The principle of operation is similar to that of the transistor circuit in Fig. 14-8a. In place of the battery a voltage divider (R_4 and R_5) and a rectifier diode D_1 provide the dc to charge C_1. With the switch in position 2, the discharging capacitor provides gate current for the SCR. The SCR will rectify only while it receives gate current. Otherwise it blocks current flow in both directions, and the power is in effect disconnected from the load.

Accurate time intervals of long duration cannot be obtained from these circuits. Greater precision and accuracy over wider intervals can be obtained by using field-effect transistors. Since they require no gate current, the timed interval depends purely on the values of the resistors and capacitors used in the *CR* circuits. Even greater precision can be obtained with an electronic counter triggered by a crystal oscillator. Most countries broadcast signals which can be used to calibrate or synchronize the oscillator so that it produces output pulses at highly accurate intervals.

14-8 LIGHT-SENSITIVE CONTROL DEVICES

There are many instances when electronic circuits are required to be light-sensitive. Some examples are circuits which turn on driveway lights at nightfall and automatic exposure devices in cameras. Examples of devices which sense changes in illumination are:

1. Photo resistor—changes resistance value when illuminated.
2. Photo cell—generates a current proportional to the illumination.
3. Photo transistor—has a light-sensitive base, collector current varies with change in illumination.
4. Light-activated SCR—has a light-sensitive gate, can be turned on by illumination.

A simple circuit which could be used to turn on a driveway light is shown in Fig. 14-9. It is small enough to be mounted inside a lamp-post column.

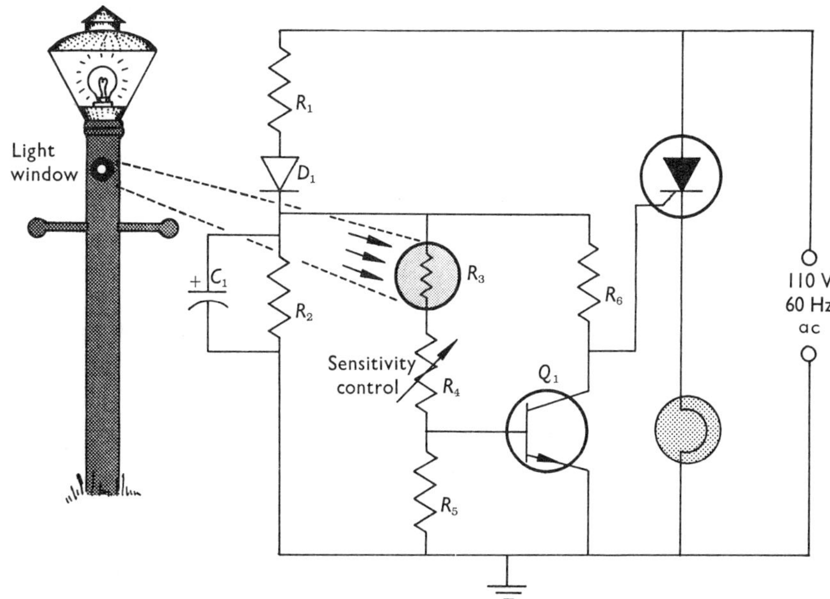

Fig. 14-9 Simple light-sensitive circuit for controlling a driveway light.

A photosensitive resistor is used (R_3) to regulate the base current to a switching transistor. It is mounted in a small window in the lamp post and typically has a dark resistance of 50 kΩ or more. In daylight it may have a resistance of only 1000 to 2000 Ω. During the daylight hours it passes enough base current to turn on Q_1. The collector (and gate of the SCR) will then be nearly at zero or ground potential, because most of the supply voltage will be dropped across R_6. After dark the resistance of the photo resistor increases and base current to Q_1 drops. The collector of Q_1 then rises and a positive-gate voltage is applied to the SCR. It turns on and supplies power to the lamp. In practice a time delay is included in such a circuit to prevent it from oscillating on and off during marginal light conditions.

Summary

Power switching and control circuits are required to regulate a variety of loads such as meters, relays, etc.

Electronic voltage regulators sense variations in output voltage and attempt to compensate for them.

A *silicon-controlled rectifier* can take the place of a relay in many applications. It is turned on by a short pulse applied to its gate electrode.

A *unijunction transistor* is an ideal source of SCR gate pulses.

Most electronic timing circuits utilize *CR* networks to develop the timing interval.

Many electronic controls require a light-sensitive device to initiate an action. Some common light sensors are *photo resistors, photo cells, photo transistors,* and *light-activated silicon-controlled rectifiers.*

Questions and Exercises

1. Explain the operation of the simple voltage-regulator circuit shown in Fig. 14-1a.
2. The circuit in Fig. 14-1a is only slightly effective. One reason is that no fixed reference voltage is used against which the transistor can compare variations. What is another reason for inadequate performance?
3. How does the inclusion of diode D_1 in Fig. 14-1b and c improve voltage regulation?
4. Suggest a reason why an SCR is more likely to be found in ac than dc circuits.
5. How is an SCR turned off?
6. Explain how the gate electrode needs only to start the turn-on action in an SCR.
7. Define phase control.
8. List some ways in which gate signals can be generated for an SCR.
9. What is the *peak-point emitter voltage* in a unijunction transistor?
10. How can phase control be accomplished using a unijunction device to generate SCR gate pulses?
11. On what principal feature do nearly all electronic timing circuits depend?
12. Assume that in Fig. 14-8a, the following components are used: $C_1 = 10\ \mu F$, R_1 and R_2 together are 2 MΩ, and R_3 and the base of Q_1 have negligible resistance. Also assume that the base current drops enough to cause collector current to release the relay if the capacitor voltage drops to 37 percent of its full charge. (This is not necessarily the case in a practical circuit.) How long will the relay be closed after SW_1 is moved from position 1 to 2 under these conditions?
13. List some light-sensitive devices which can be used in control applications.
14. Why would it not be practical to use a photo resistor directly to control the gate circuit of a sensitive SCR as shown in Fig. 14-9?
15. Draw a circuit similar to the one in Fig. 14-9 but omit the transistor and R_6. Connect the gate directly to the junction of R_4 and R_5 (where the transistor base is connected)

and then explain the operation of the circuit you have drawn.

16. Obtain an electronic components catalog and list applicable features of a zener diode and an SCR in a table to establish the range in which devices are available. List minimum and maximum anode current, voltage, and price.

17. From your components catalog list how many types of photo resistors are available, their ratio to light and dark resistance, and their wattage ratings.

15-1 TYPES OF TEST EQUIPMENT

Test equipment falls into two broad classes. One of these is *signal-generating equipment,* the other is *measurement equipment.* Usually both types are required for troubleshooting because an accurate measurement may be of little value unless a known test signal is available.

Steady-state phenomena are easiest to measure, because a reading can be taken as often as desired. An example of such measurement would be power-supply voltage or the value of a resistor. Time-varying signal measurements, especially those of high frequency, are more critical. Most difficult of all are measurements of a single event of short duration (for example, a 1-μs pulse which occurs only once in a long time period).

By consulting the test-equipment section of an electronic supply catalog you can readily see that a wide range of instruments is available. The more prominently used types are grouped and listed in Table 15-1, but this is by no means exhaustive.

Only two types of instruments are discussed in this chapter. These are the *cathode-ray oscilloscope* and the *volt-ohm milliammeter (VOM).* Both are fundamental measuring instruments. *Signal generators* are omitted because they are essentially oscillator circuits with special controls and calibrated output signals. Oscillator circuits are discussed in Chaps. 9 and 10.

15-2 CATHODE-RAY OSCILLOSCOPE

Probably the most useful test instrument is a good-quality oscilloscope. It can directly measure voltage, frequency, and time intervals. It can measure indirectly current, resistance, power, decibel level, and with special adapters can display a frequency spectrum. A simplified functional block diagram of an oscilloscope is shown in Fig. 15-1.

The signal to be measured is connected to the input of the vertical amplifier through attenuating resistors to prevent overdriving on large signals. A signal entering at the vertical input

Table 15-1 Common Test Instruments and Their Functions

Name	*Function*	*Typical Applications*
VOM (Volt-ohm milliammeter)	Measurement of voltage current and resistance	Service calls or lab bench. Units are small, portable, and handy to use
Electronic VOM Vacuum tube voltmeter (VTVM); transistor voltmeter (TVM); digital voltmeter (DVM)	Same as VOM. They contain internal amplifiers, are more sensitive, and cause less loading to the circuit under test	Lab bench, service bench. Some portable units are as easily used as VOM instruments
Bridge	Precise and accurate measurements not possible with VOM or electronic VOM	Lab bench for impedance or resistance measurements
Cathode-ray oscilloscope	Measuring and viewing signals in graphical form	Service calls and lab benches. Probably the most useful of all measuring instruments
Audio-signal generator	Generating audio signals of known output amplitude and typical frequency range to 100 kHz	Service bench when testing audio amplifiers, etc.
RF (Radio-frequency) signal generator	Generating both pure and modulated RF signals of known amplitude and typical frequency range to 200 MHz. Units are available for UHF	Service bench for radio and television receiver alignment
TV pattern generators	Generating a television screen pattern for black and white or color receivers	Service bench or field service calls. Troubles are diagnosed from the response to a known pattern

deflects the electron beam vertically only, and if no horizontal deflection is taking place it shifts the spot up or down from its rest position, as shown in Fig. 15-2a(ii) and (iii). The amount of this deflection is proportional to the peak input-signal voltage. Voltages can therefore be measured on an oscilloscope by noting the amount of deflection of the beam.

By deflecting the beam sideways in the correct pattern, signal voltage can be measured at various time intervals. If a ramp-shaped voltage waveform is applied to the horizontal deflection plates, the beam is swept from one side of the screen to the other at a uniform rate. At the end of its trace it flies back rapidly to begin another horizontal sweep. A blanking signal is applied to the cathode-ray tube (CRT) during flyback time so that the retrace back to the starting position is not seen. Figure 15-2b illustrates the type of display obtained by changing the sweep speed (the slope of the ramp voltage) when a sinusoidal voltage is applied to the vertical input.

15-3 CATHODE-RAY TUBE

Figure 15-3 is a simplified drawing which shows the structure of a cathode-ray tube of the type used in oscilloscopes. At one end of the

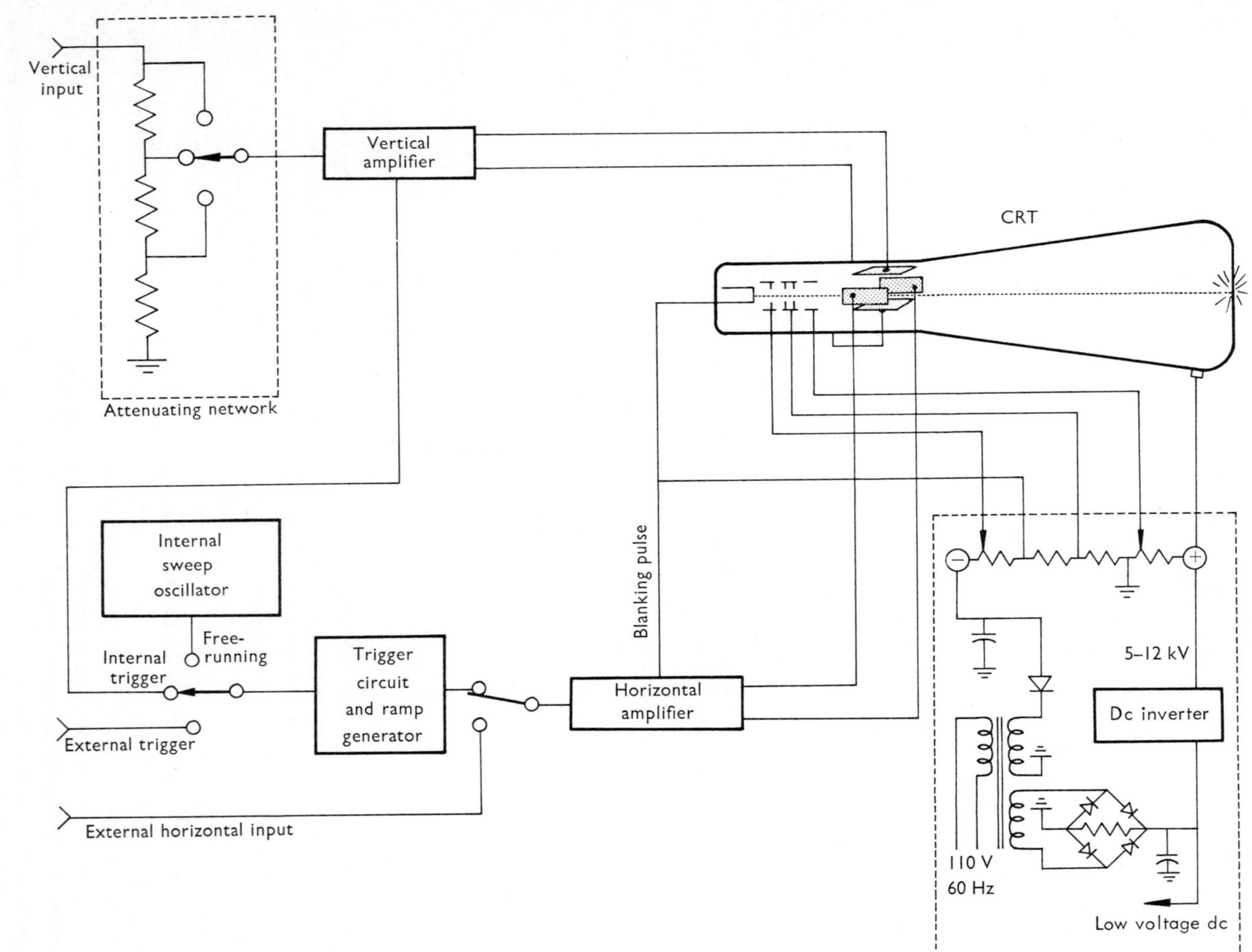

Fig. 15-1 Simplified functional blocks of an oscilloscope.

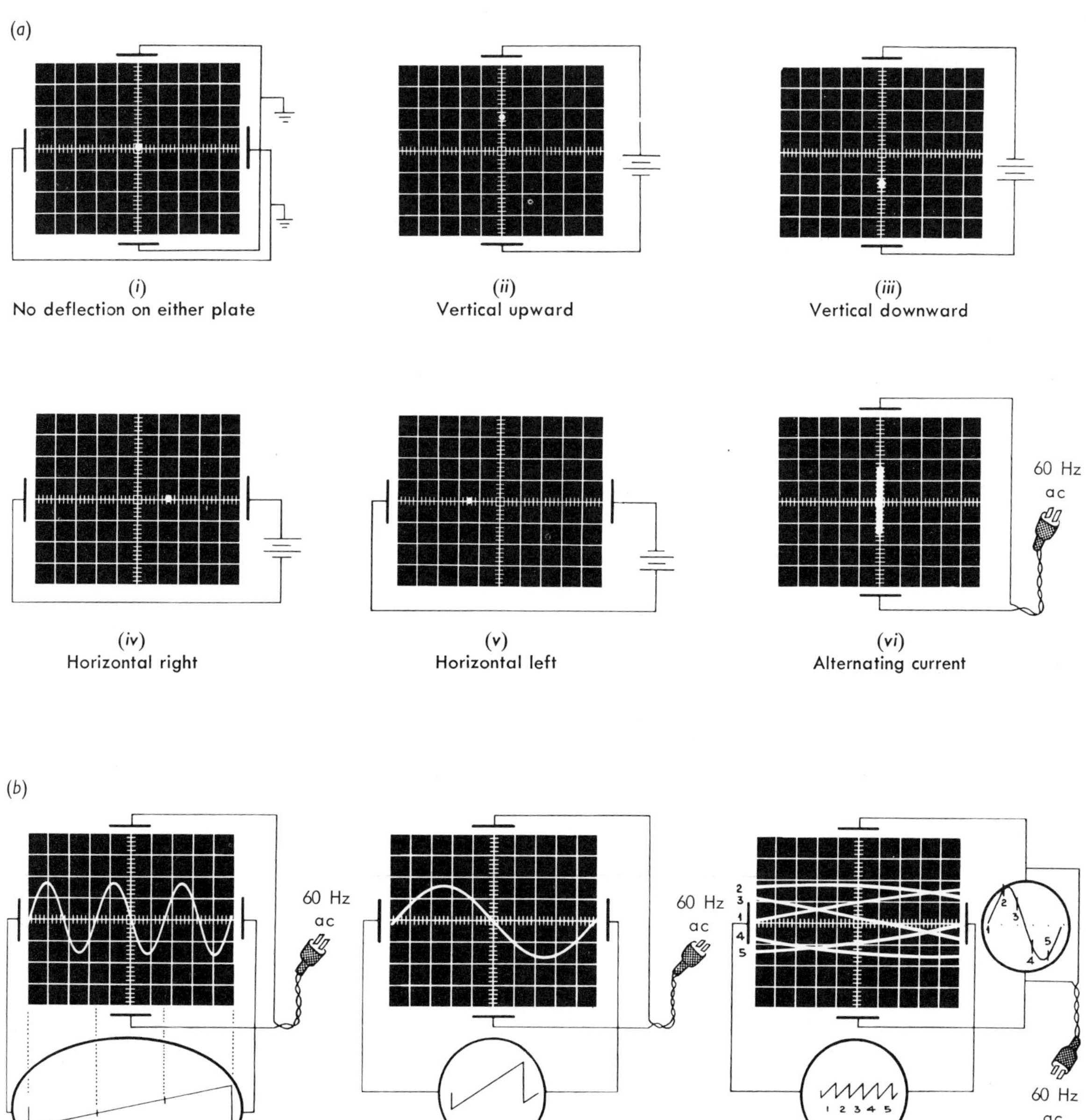

Fig. 15-2 Cathode-ray tube photographs showing beam deflections under various vertical and horizontal input voltages: (*a*) (i) to (v) Applying dc to one set of deflection plates only as shown; (vi) Alternating current on vertical plates only; (*b*) Sinusoidal voltage on vertical plates, ramp voltage on horizontal plates.

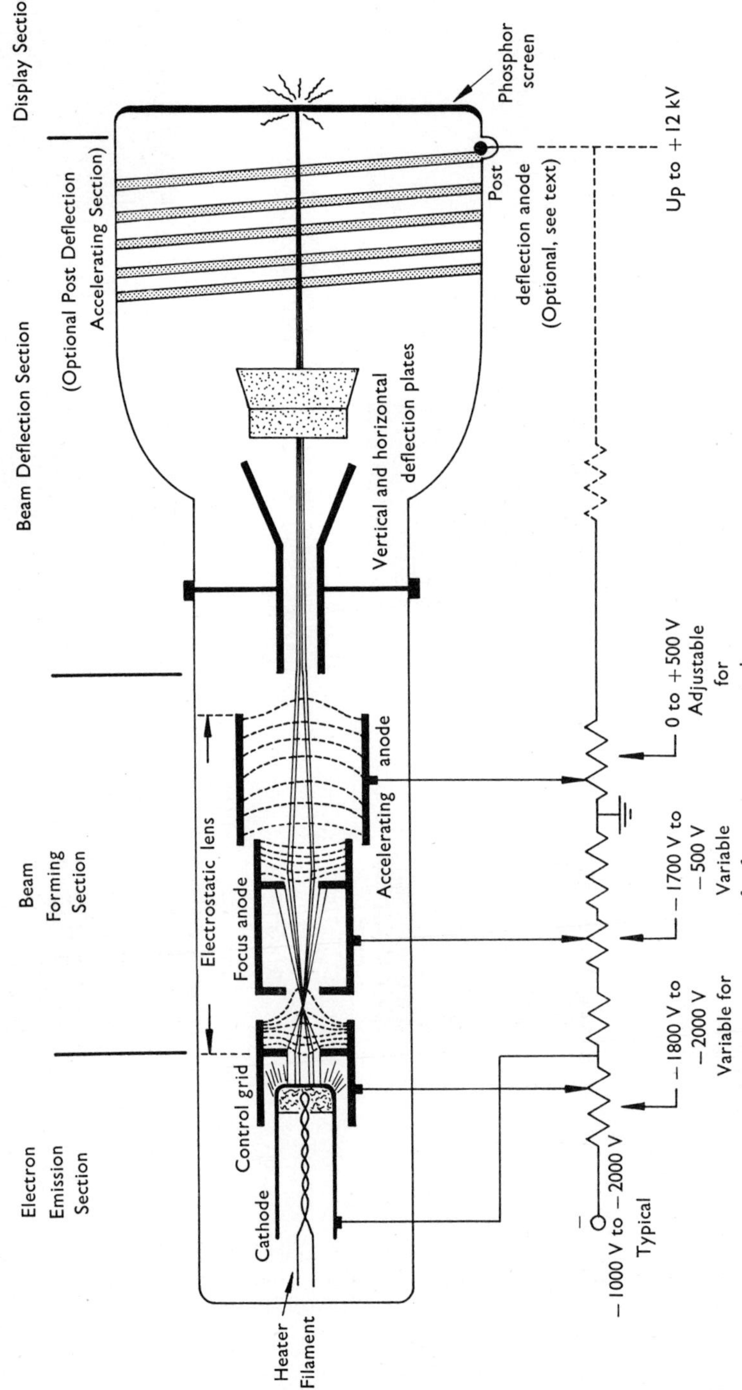

Fig. 15-3 Simplified diagram of a cathode-ray tube. Typical electrode voltages are shown with respect to ground.

tube is a cathode which emits electrons when heated by the filament. The electrons are emitted at various angles, but once they leave the cathode surface many are pulled toward the screen by the positively charged anodes. As they attempt to move toward the anodes, they must first pass through an aperture in the *control grid*. The "grid," which is actually a cylinder, has a negative charge with respect to the cathode. Many electrons are either repelled back to the cathode or change their paths sufficiently to pass through the aperture. If sufficiently negative, the grid can cut off beam current entirely.

Those electrons which pass through the aperture begin to accelerate rapidly under the influence of the electric field. The field is designed to focus the cloud of electrons into a fine beam. The broken lines in Fig. 15-3 represent *lines of equipotential*. A charged particle could move from one such line to another but not along the line because the potential difference between any two points on a line is zero. The electrons in the CRT tend to move in paths which are perpendicular to the lines of equipotential. By changing the relative voltage levels applied to the *focus* and *accelerating anodes*, it is possible to change the shapes of the lines of equipotential. In this way it is possible to direct most of the electrons to the same point. When they strike the phosphor screen a spot of light is seen.

Beam deflection can be accomplished either with a magnetic field as in television picture tubes or electrostatically by means of deflection plates. Electrostatic deflection has the advantage of speed because only the voltage between the deflection plates need be varied to change the beam angle. This voltage can be changed very rapidly. The only time limitation in the process is that the charge on the capacitance between the plates must change. With electromagnetic deflection there is counter emf to contend with, consequently, lower deflection speeds must be accepted.

In oscilloscope tubes operating at high deflection speeds, the connections to the deflection plates are very short and are brought out at the neck to reduce lead inductance. In slower tubes these connections are brought out at the base along with those of the other electrodes.

When very high deflection speeds (high writing rates) are used, the beam dwells only briefly at each spot on the screen. This means that a more intense electron beam is required to cause a visible trace. To achieve the necessary beam current a *post-deflection anode* is used. It consists of a helical coating of resistive material deposited inside the bell of the tube near the screen. A very high positive voltage is connected to this anode so that the electron beam is greatly accelerated after it leaves the deflection plates.

15-4 VERTICAL-AMPLIFIER CIRCUIT

The function of the vertical amplifier is to raise the input signal level to the point where it causes sufficient vertical deflection. Several volts are required to deflect the spot on the screen 1 cm, yet it may be desirable to view a signal of only a few millivolts amplitude. Since the oscilloscope is a general-purpose test instrument, it must be capable of accepting a wide range of signal amplitudes. At the input to the vertical amplifier it is therefore necessary to place a variable attenuating network. A simple voltage-divider-type attenuator (R_1, R_2, and R_3) is shown in Fig. 15-4. Its function is very similar to that of a volume control in a radio receiver. A fraction of the applied signal voltage is tapped off and passed to the vertical amplifier.

The input to the amplifier must be protected against inadvertently applied high-amplitude signals. Many oscilloscopes use vacuum tubes in the input because they can withstand temporary overdriving. The circuit in Fig. 15-4 uses two zener diodes connected as biased clippers. Assume that both are 5-V zener diodes for pur-

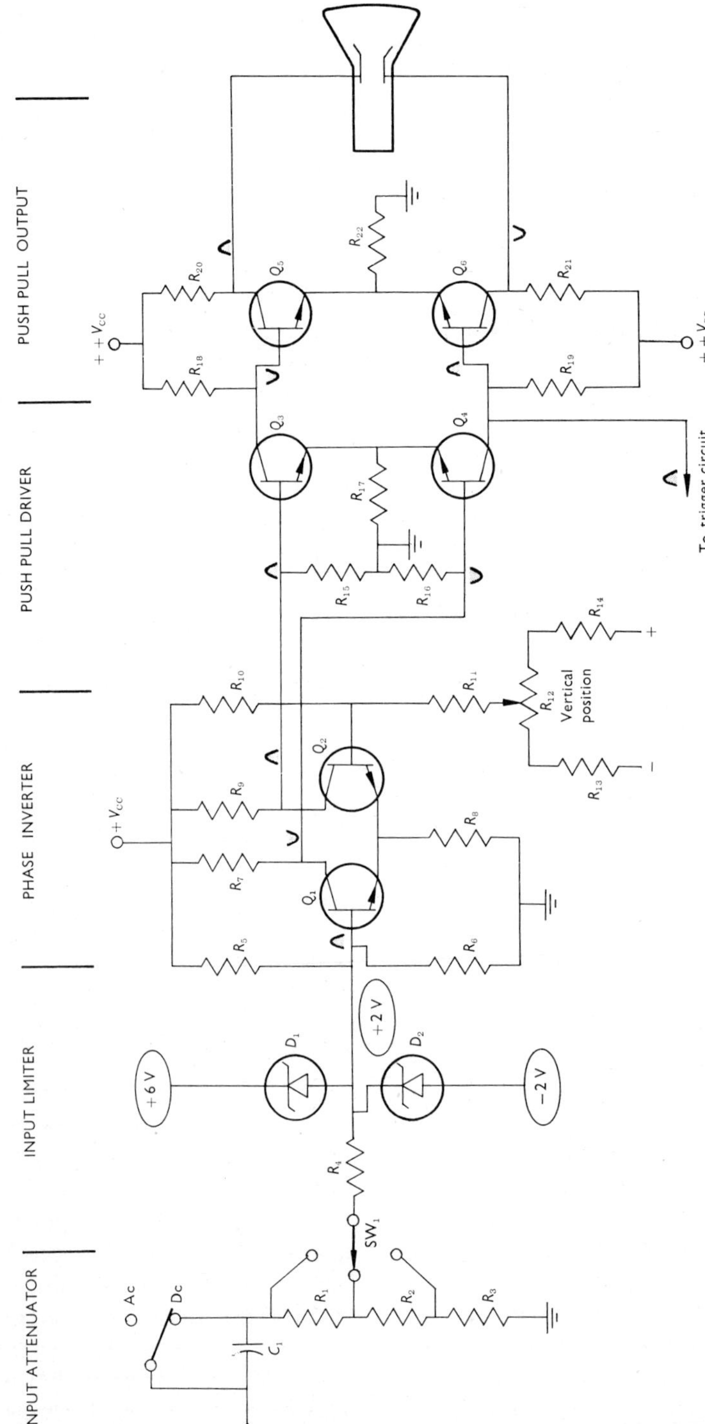

Fig. 15-4 Simplified vertical-amplifier circuit. Zener diodes D_1 and D_2 limit the input to transistor Q_1 to minimize overdriving on high-amplitude signals.

poses of explanation. Furthermore, assume that with no signal input, the base of Q_1 (and the junction between D_1 and D_2) is at a quiescent voltage of $+2$ V with respect to ground. Then neither zener diode will conduct, because each is only reverse-biased at 4 V [$6 - 2 = 4$ for D_1 and $2 - (-2) = 4$ for D_2]. If the positive signal excursion is greater than 1 V, D_2 will go into avalanche conduction, because the quiescent voltage plus the signal voltage across it will attempt to rise above 5 V. Similarly, D_1 shunts negative signal excursions exceeding 1 V. On the scope trace a signal clipped in this manner would either go off the screen at top and bottom or would appear clipped. The operator would then take corrective action by switching SW_1 to a position which causes more signal attenuation.

Transistors Q_1 and Q_2 are connected as a differential amplifier. When no signal is applied to the base of one of the inputs of a differential amplifier, it behaves simply as an emitter-coupled phase inverter. The "balance" or quiescent current to Q_2 is made adjustable to position the spot vertically. In this way, the display can be moved up or down on the screen of the CRT.

Transistors Q_3 and Q_4 are drivers for the push-pull output stage comprised of transistors Q_5 and Q_6. The circuit is deliberately simplified to show essentials. Actual circuits would appear more complex because of the added high-frequency compensation and negative-feedback arrangements.

15-5 SWEEP CIRCUITS AND HORIZONTAL AMPLIFIER

A simplified sweep-generator and horizontal-amplifier circuit is shown in Fig. 15-5. The horizontal-amplifier circuit consists of transistors Q_2 to Q_7 and is similar to the vertical amplifier described in the previous section.

Transistor Q_1, resistors R_1 to R_3, and capacitors C_1 to C_3 make up the ramp-voltage generator. The ramp signal is actually the voltage across a charging capacitor. Different ramp slopes are obtained by using varying combinations of C and R so that charge time can be changed. Charging cannot take place unless transistor Q_1 is turned off. Once a capacitor is charging and Q_1 turns on, it quickly discharges it. To the capacitor the transistor appears like a shunt switch which is suddenly closed. Whenever Q_1 turns on, flyback occurs.

The base of Q_1 is under the control of holdoff flip-flop. Its purpose is to prevent further sweep triggering once a sweep has started. An *RS*-type flip-flop (Sec. 13-22) shown here normally rests in the zero state. The high or H level from the zero side keeps Q_1 turned on, so that the ramp generator capacitors are normally shunted. When a sweep trigger appears at gate G_1 it sets the holdoff flip-flop to the 1 state. Transistor Q_1 base goes low and Q_1 opens, allowing a capacitor to charge.

In most sweep circuits the *CR* network is provided with a constant current source to ensure a linearly rising ramp. It is omitted here for simplicity. When the ramp reaches a certain amplitude, Schmitt trigger #2 (Sec. 12-14) turns on and produces a reset level for the holdoff flip-flop. It resets, turns on Q_1 to produce flyback, and also enables gate G_1 so that another sweep can take place. Note that once a sweep is in progress, it cannot be interrupted until it is complete (as signified by a preset ramp amplitude detected by Schmitt trigger #2).

Schmitt trigger #1 is used to select the exact amplitude point on the input signal at which sweep should begin. For example, it might be desired to display a sine wave at any starting position from when it first crosses the horizontal axis up to maximum amplitude. This is accomplished by introducing a variable bias into the Schmitt trigger so that various amplitudes may turn it on.

Normally the trigger for initiating a sweep is the input signal itself, and is obtained from a connection on the vertical amplifier. Occasion-

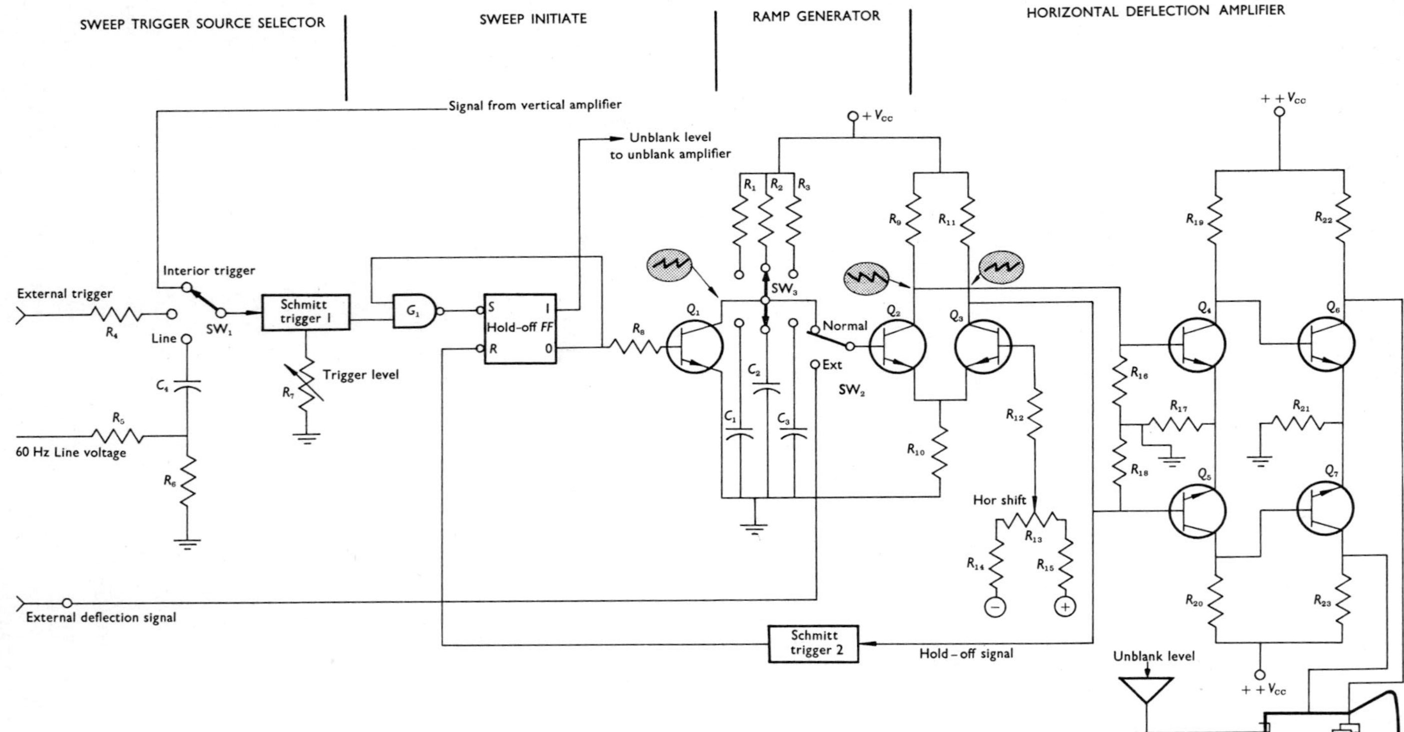

Fig. 15-5 Simplified circuit of horizontal sweep (ramp) generator and horizontal deflection amplifier.

ally, it is desirable to use an external trigger source, or to synchronize the sweep to the power-line frequency. These selections can be made with SW_1. Sometimes the internal sweep generator is bypassed entirely and horizontal deflection is accomplished by an external signal. This can be done by connecting the input of the horizontal amplifier directly to an external connector via SW_2.

15-6 OSCILLOSCOPE HIGH-VOLTAGE POWER SUPPLY

The high voltage required to operate the cathode-ray tube can be obtained from a high-voltage winding on the power transformer. For voltages in excess of approximately 2 kV the transformer becomes too bulky and expensive. Higher voltages are usually generated by some form of dc converter. The converter operates at a relatively high switching frequency so that the size of the transformer core can be reduced.

15-7 VOLT-OHM-MILLIAMMETER (VOM) CIRCUIT

A hypothetical though representative VOM is shown in the simplified pictorial drawing in Fig. 15-6. The meter has volt, ohm, and ampere scales on its face, and the measurement being performed is indicated by the function-selector switch. This is a four-deck rotary switch with each of the decks shown in the schematic diagram in Fig. 15-7. In the drawing the switch is shown resting in the maximum dc volts position.

It is difficult to trace the circuit in Fig. 15-7 for each switch setting. To simplify the task,

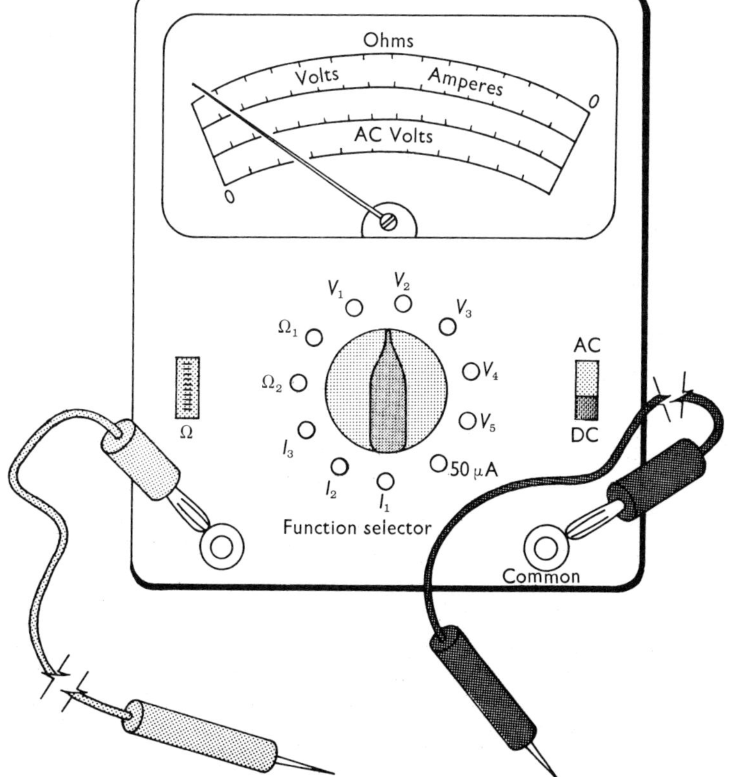

Fig. 15-6 Pictorial diagram of a typical volt-ohm-milliammeter (VOM).

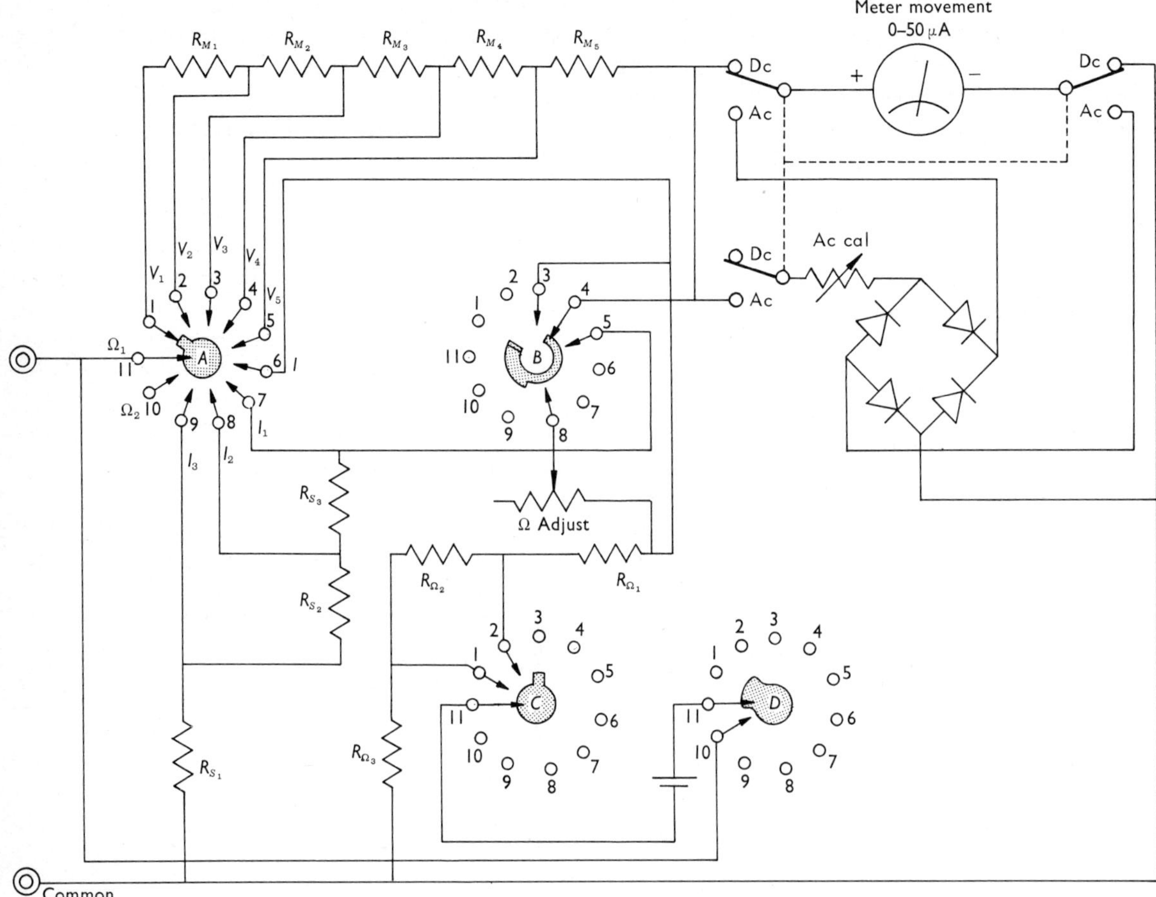

Fig. 15-7 Schematic diagram of a representative VOM circuit.

refer to the equivalent circuits in Fig. 15-8a, b, c, and d first, and then try to trace through the main circuit for each of the functions.

15-8 DC VOLTS FUNCTION

Five voltage ranges are available. On the highest range all the multiplier resistors are connected in series (Fig. 15-8a). Even with a high voltage applied to the input terminals only a small meter current flows. At no setting must the meter coil current exceed 50 μA, the full-scale deflection current for the meter.

The equation for voltmeter sensitivity is

$$\text{Ohms per volt} = \frac{1}{\text{full-scale deflection current}}$$

Substituting values gives:

$$\text{Sensitivity} = \frac{1}{50 \times 10^{-6}} = \frac{1\,000\,000}{50} = 20\,000\ \Omega/\text{V}$$

To obtain the loading resistance when a voltage measurement is made, multiply the voltage range setting by the sensitivity.

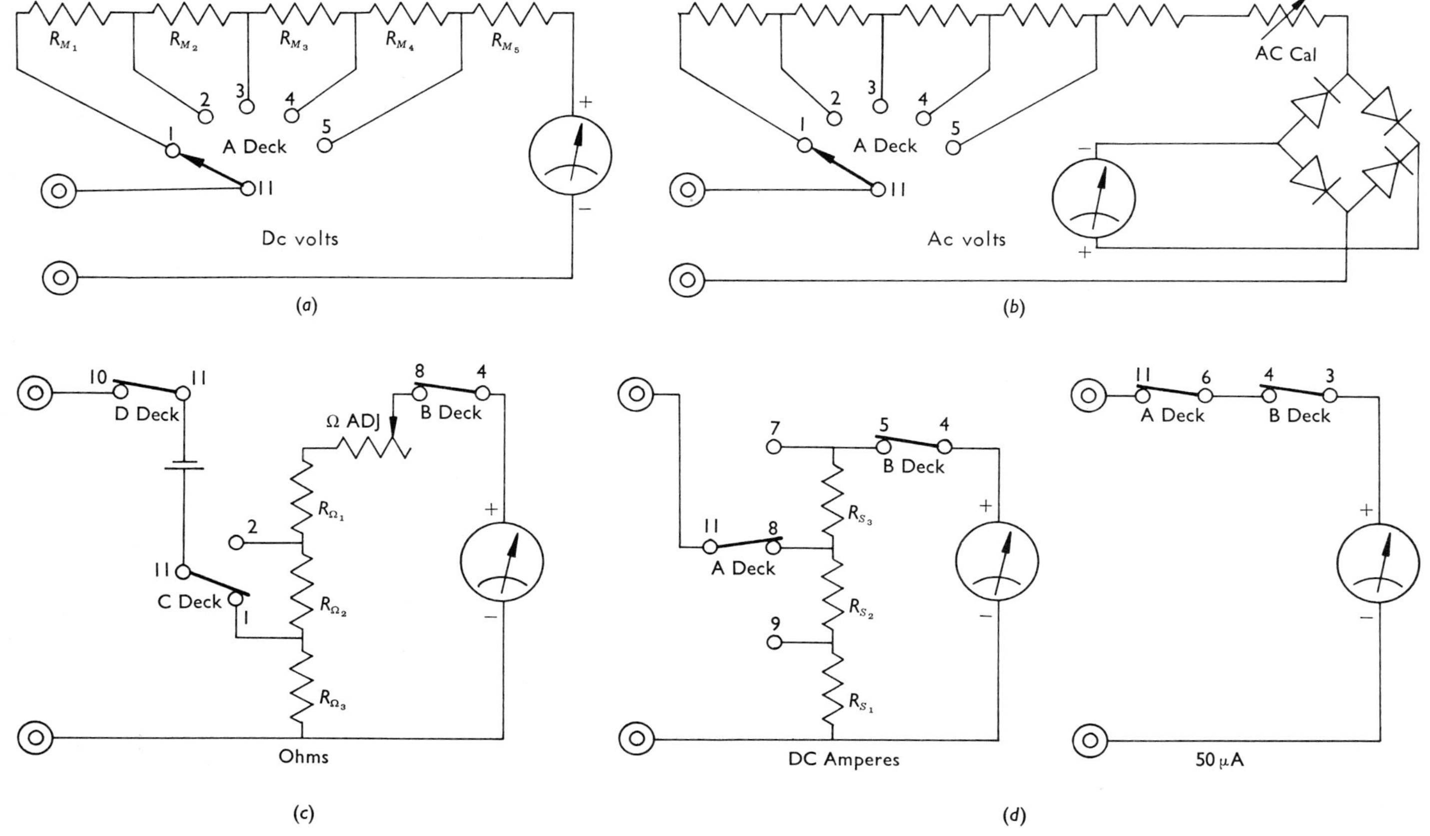

Fig. 15-8 Simplified diagrams of VOM functions: (*a*) DC volts; (*b*) AC volts; (*c*) ohms; (*d*) amperes.

15-9 AC VOLT FUNCTION

On ac voltage measurements a bridge rectifier is connected as shown in Fig. 15-8b. A calibration resistor is included in the circuit so that ac and dc readings do not correspond exactly at all needle deflections. A separate meter scale for ac volts is a feature of VOM instruments.

15-10 OHMS FUNCTION

When measuring ohms, the meter leads are first connected together to adjust for maximum meter current (0 Ω). Figure 15-8c shows how resistors R_{Ω_2} and R_{Ω_3} are used as shunt or series and shunt resistors to change ohms scales. The internal battery supplies current for resistance measurements.

15-11 CURRENT MEASUREMENTS

When measuring current the resistors R_{S_1}, R_{S_2}, and R_{S_3} are connected in shunt and series combinations to provide various ranges. On the highest range (pin 11 and 9 contact on A deck of the rotary selector switch) shunt resistor R_{S_1} detours most of the current, and very little reaches the meter (Fig. 15-8d) via R_{S_2} and R_{S_3}. On the lowest range (50 μA) the meter is connected directly to the input terminals. It is extremely vulnerable in this setting; most VOM circuits use some form of overload protection on the meter to safeguard it from accidental overloads.

15-12 DIGITAL VOLTMETER (DVM)

Digital measuring instruments are designed with circuits that produce numerical readouts using LEDs or Nixie tubes. This eliminates the need to interpret the position of a pointer employed on the usual meter movement.

Summary

Test equipment can be broadly subdivided into *signal-generating* equipment and *measurement* equipment.

Electronic measurements in order of difficulty are (1) *steady-state,* (2) *time-varying periodic,* and (3) *time-varying random,* especially short-duration and infrequent events.

The *cathode-ray oscilloscope* is useful because it can perform all types of measurement either directly or indirectly.

At high writing speeds an intense electron beam is required.

To achieve a high-energy beam *post-deflection anodes* are used.

A cathode-ray beam can be deflected *electrostatically* with plates or *electromagnetically* with coils. Electrostatic deflection is more suitable if speed is essential. Electromagnetic deflection is more suitable when the angle of deflection is relatively great and speed is not critical, as in television picture tubes.

A ramp-shaped voltage (sawtooth) is applied to the horizontal-deflection circuits of an oscilloscope. It is generated by a charging capacitor in the *sweep generator.*

A *volt-ohm-milliammeter (VOM)* is a portable multirange measuring instrument. By utilizing ingenious switching arrangements a single meter movement can be connected to provide a variety of measurements.

Questions and Exercises

1. What two broad classes does test equipment fall into?
2. List in order of difficulty three kinds of measurements which a technician might be required to perform.
3. Why is a ramp-shaped waveform used in most instances as the deflection signal on

the horizontal plates of an oscilloscope?

4. Sketch the figures you would expect to see on a CRT screen if the following signals were applied.
   a. A 60-Hz sine wave on both horizontal and vertical plates.
   b. Same as (*a*) but the connection to one set of plates is reversed.
   c. A 60-Hz sine wave on vertical plates. The same 60-Hz sine wave shifted 90° in phase on the horizontal plates.
   d. A 1200-Hz sine wave on the vertical plates and a 60-Hz square wave on the horizontal plates.
   e. A 200-Hz sine wave on the vertical and a 50-Hz ramp wave on the horizontal.

NOTE In these exercises it is wise to plot each set of waveforms on the same time axis. Then obtain relative amplitudes at a few time points to give you an idea of where the beam moves to trace out its figure on the screen.

5. How is an electron beam focused to a *point* on a CRT screen?
6. Why is post-deflection acceleration used on some but not all oscilloscopes?
7. Why is an attenuator included in an oscilloscope's vertical amplifier?
8. Why were vacuum tubes used in the inputs of early solid-state oscilloscopes?
9. If D_1 in Fig. 15-4 became open-circuited, would the oscilloscope cease functioning?
10. Why would it be unwise to use an oscilloscope if the -2-V source on D_2, Fig. 15-4, became accidentally disconnected?
11. What function does the holdoff flip-flop perform in Fig. 15-5?
12. How would oscilloscope performance be affected if in Fig. 15-5 transistor Q_1 were to become short-circuited between emitter and collector?
13. What is meant by voltmeter sensitivity?
14. A 1000-Ω/V voltmeter set to the 100-V range is being used to measure the voltage drop across a resistor. To the circuit under test the voltmeter appears as a resistor placed in parallel with the resistor being measured. What value of resistance does the meter appear to have at this range setting?
15. Why is an internal rectifier required when making ac volt measurements?
16. Why must the equipment under test be switched off before attempting to make resistance measurements?
17. How would VOM performance change if R_{S_1} (Fig. 15-7) were accidentally burned out (open-circuited)?
18. Obtain an electronics supply catalog and list model, make, and price for as many as possible of the items listed in Table 15-1.

16-1 THE OPTICAL-ELECTRICAL INTERFACE

For electronic information transmission an interface is needed between the real world where the information exists and the electronic circuits which process and transmit it. In audio and radio circuits, it is relatively easy to understand the interface. The original signal exists in the form of sound, i.e., air-pressure waves. The interfacing device is a microphone which converts the motion of air particles into an exact replica of electric current variations. Then electronic circuits can take over until it is desired to retrieve the signal in its original form. Again an interface is required. This time it is a speaker which when driven by the signal current produces a replica of the original sound-pressure waves.

In a video system the original signal exists in the form of light waves. An interface is needed which converts the light waves into electrical signals. This is done by the television camera. To reconstruct the original image an interface is again required. The device must emit light waves when driven by the camera signal.

There is a striking difference in what our senses find acceptable in the two systems. In an audio system, our ears require that both sound intensity information (loudness) and pitch (frequency) be transmitted. In a video system only intensity (lightness) need be transmitted to reconstruct an acceptable image. This will result in a *monochrome* or black and white picture. If color *hue* (frequency or wavelength) information is also transmitted and reconstructed, a color image results.

16-2 TELEVISION-CAMERA TUBE

A television camera reduces a scene from an optical image to an electrical signal. Mounted at the front of a television camera is a set of lenses which focus the image on a plane just as in a photographic camera. In place of the film there is a photosensitive mosaic screen.

On microscopic sections of the screen electric charges accumulate. Their density is proportional to the intensity of the illumination falling on the section. In a typical scene there will be many brightness levels. When these are focused onto the camera mosaic screen it will hold an image pattern in terms of varying charge density. It is the job of the camera to examine each section of the mosaic in sequence and to report the charge level at each segment. This process is known as *scanning* the image.

An image can be scanned in a variety of ways. It can be done vertically, horizontally, diagonally, or by a combination of all three. On the North American continent an image is scanned horizontally 525 times; i.e., a picture is broken up into *525 horizontal lines*.

Figure 16-1 shows a simplified diagram of the structure of an iconoscope camera tube. Scanning is accomplished with a thin beam of electrons. The electrons are given off by thermionic emission in the electron gun and

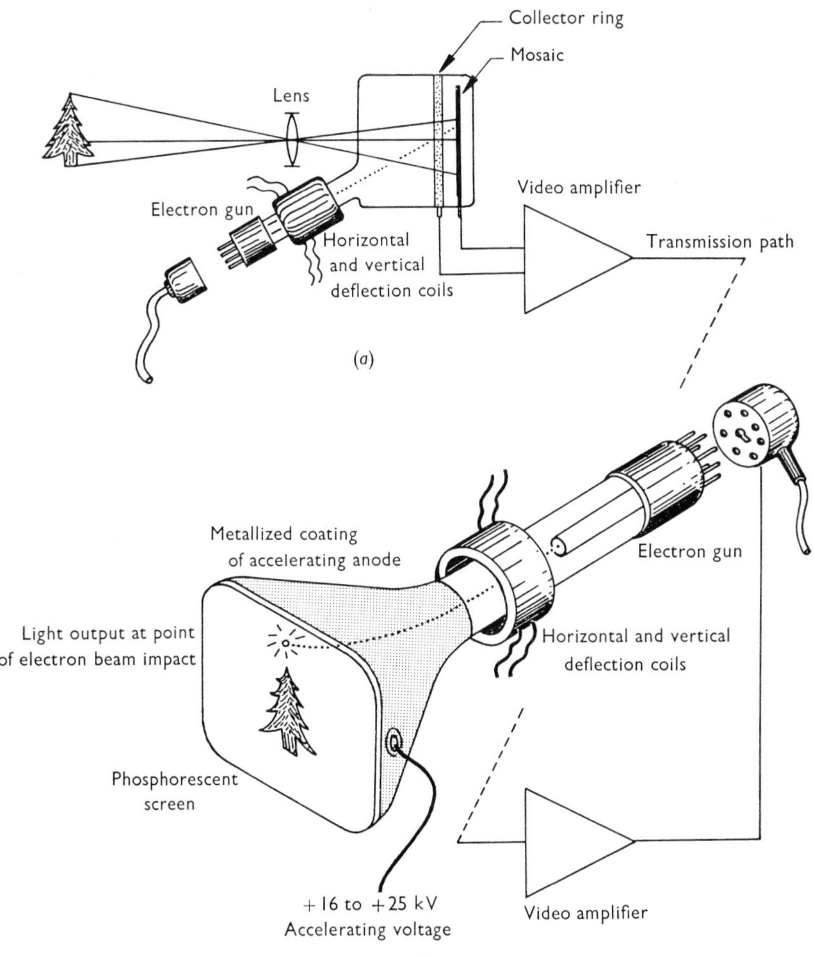

Fig. 16-1 The optical-electrical interface in a television system: (*a*) Iconoscope converts an optical image to an electrical signal; (*b*) Cathode-ray picture tube converts a signal into an optical image.

focused into a fine beam before they strike the mosaic screen. When the beam strikes the screen, the charge on the mosaic at the point of impact produces an output signal. It will be proportional to the light intensity falling on the segment. The electron beam is used only as a form of switch. Its job is to "connect" one segment of the screen after another to the input of the video amplifier. Be sure to understand that it is the light intensity which determines video-signal amplitude from instant to instant, not the intensity of the electron beam. Scanning-beam current is kept constant.

In order to scan the mosaic screen the beam must be constantly changing position. To "aim" the electron gun, two sets of deflection coils (or plates) are used on the neck of the tube. By changing the current through the horizontal and vertical deflection coils the electron beam can be fired at any point on the mosaic.

16-3 CATHODE-RAY PICTURE TUBES

To reconstruct an image from the video signal a cathode-ray picture tube (CRT) is used. Figure 16-1b shows a simplified diagram of the structure of such a tube. It too has an electron gun, but unlike the iconoscope gun which produces a beam of constant intensity, the CRT beam varies in intensity. Beam intensity is controlled by the video signal. When the electron beam strikes the screen it gives off light at that spot. The brightness is proportional to the beam current. As in the iconoscope, CRT beam current is "aimed" by means of horizontal and vertical deflection coils.

16-4 VIDEO TRANSMISSION

In its simplest form an image may be dissected by the camera and reconstructed by the CRT if the following conditions are met:

1. The aiming points of the electron guns in the iconoscope and CRT are "locked," so that at all times each is pointing to the same place on the respective screens.
2. The CRT beam intensity is modulated by the video signal from the iconoscope.
3. The iconoscope electron gun (master) and CRT electron gun (slave) fire repetitively in a given sequence at all points on the image; i.e., a scanning process takes place.

Figure 16-2 shows a simplified diagram of a minimum system. Both front views and side views are shown for each tube. A timing diagram is included to show how one scan line of video is produced. In a commercial broadcast system 525 lines of video are produced every $\frac{1}{30}$ of a second. A television station therefore transmits a total of $525 \times 30 = 15\,750$ lines of video every second. Since 525 lines make up a complete picture, there are $15\,750/525 = 30$ complete pictures or *frames* transmitted per second.

If all 525 lines were sent as a package every $\frac{1}{30}$ of a second, there would be a noticeable flicker because the eyes can distinguish events separated by $\frac{1}{30}$ of a second. A rather crafty trick is used to fool the eyes. Instead of transmitting a complete set of scan lines the odd-numbered scan lines are transmitted first. This takes half of $\frac{1}{30}$ or $\frac{1}{60}$ of a second. These are then followed by the even-numbered lines which take another $\frac{1}{60}$ of a second. In fact, only one complete picture is transmitted every $\frac{1}{30}$ of a second, but to the eyes it appears as though two have been sent, one every $\frac{1}{60}$ of a second. In this way flicker is reduced to the point where it is not noticeable.

16-5 HORIZONTAL AND VERTICAL SCANNING

Note that on the iconoscope screen the image is upside down and reversed as is any camera image produced by optical lenses. To present an upright image at the receiving CRT, camera scanning must take place from right to

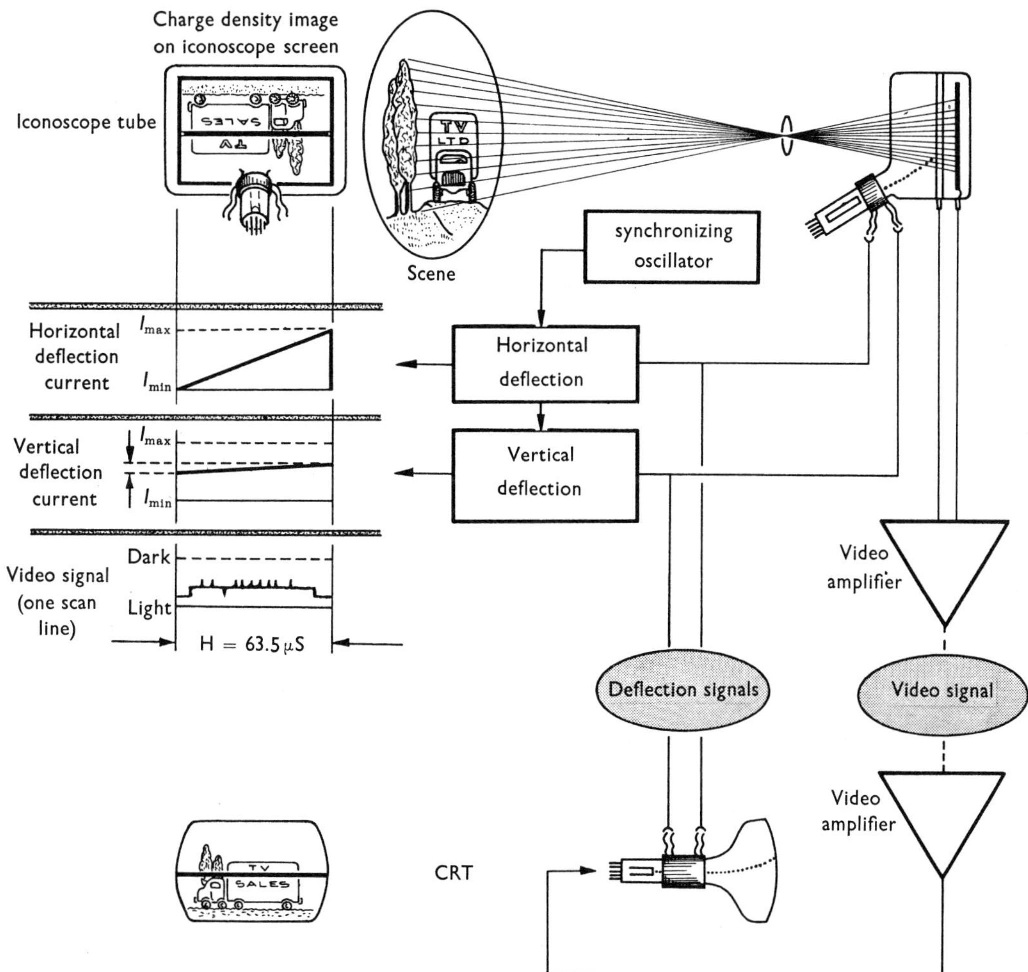

Fig. 16-2 Closed circuit television system.

left horizontally and from bottom to top vertically. The CRT screen on the other hand must be scanned from left to right and from top to bottom. Scanning direction is easily reversed by inverting the phase of the scanning voltage or current.

The waveshape of a scanning signal is a ramp of current (if coils are used) or voltage (if deflection plates are used). A ramp waveshape has the effect of applying a gradual, linearly increasing deflection force on the electron beam. The beam displacement is constant and uniform until a sweep is completed. Then there is a sudden drop in the ramp, forcing the beam to return quickly to the initial starting position of the sweep. This quick return is termed the "flyback." Horizontal flyback time in the North American television system must be less than 10 percent of the horizontal scan time. Vertical flyback time must be less than 7 percent of the vertical scan time. A CRT is blanked out during flyback so that the retrace is not normally seen.

16-6 HORIZONTAL AND VERTICAL SCANNING SIGNALS

For horizontal scanning a ramp signal of 63.5 μs duration is used. Of this time 57.15 μs is used for the linear sweep across the mosaic and 6.35 μs for the flyback. As the line scanning proceeds, the beam is also deflected (at a much slower rate) vertically. A vertical ramp lasts for 16 668.75 μs ($\frac{1}{60}$ s) including flyback time. This is long enough to produce 16 668.75/63.5 = 262.5 horizontal sweeps or one *field*, which is half a frame. The next 262.5 horizontal sweeps are interlaced between the first set and are completed during the second vertical sweep. Two vertical sweeps (i.e., two fields) are required to make one frame of 525 lines. Figure 16-3 shows the time relationship between the two sweep ramps for one complete frame of two fields. A diagram is also included to show how the interlaced scanning pattern is traced out on a CRT screen. Note that some scans are lost during vertical flyback time blankout. Even though they contribute nothing to the picture, they must be included to preserve ramp synchronization.

16-7 COMPOSITE VIDEO SIGNAL

In a television broadcasting system the deflection ramp voltages are never transmitted with the video signal. Instead, synchronizing signals which also drive the studio ramp generators are added to the video signal. When they arrive at the receiver they are used to lock or synchronize the local receiver ramp generators to those in the studio. The end result is as though the ramps themselves were transmitted. At every instant in every television receiver tuned to the same program each electron gun "points" at exactly the same spot on its screen. To achieve this remarkable synchronization horizontal and vertical sync pulses as well as flyback blanking signals are transmitted. The resulting video signal is known as the *composite video signal*. Composite video and its constituent components are shown in the diagram in Fig. 16-4.

When a horizontal line has been scanned, a blanking pulse is inserted to darken the retrace during horizontal flyback. Riding on top of the blanking pulse is a horizontal sync pulse. It is this pulse which is used to synchronize the horizontal ramp generator (horizontal oscillator) in the television receiver.

At the end of each field is a long blanking period. During this period the vertical synchronizing signal is transmitted. It consists of a pulse train from which the vertical sync pulse is recovered by an integrator circuit (Sec. 12-13). The equalizing pulses in this pulse train serve two purposes:

1. They adjust the scanning process so that interlacing occurs.
2. They continue to synchronize the horizontal oscillator during the transmission of the vertical sync pulse train.

The important specifications for the composite video signal are shown in Fig. 16-4. At the television station the video transmitter is amplitude-modulated with the composite video signal.

16-8 AUDIO AND VIDEO TRANSMISSION

The resolution of a television camera is not the limiting factor on the amount of picture detail which is transmitted. For high resolution the camera video signal will have high-frequency components because it is assumed that it will distinguish between light and dark areas closely spaced. This would result in a high modulating frequency at the transmitter and as a consequence the sidebands would occupy a wide bandwidth. Economic considerations dictate some sort of compromise between picture fidelity and frugal use of the radio-frequency spectrum. On the North American continent a television broadcast

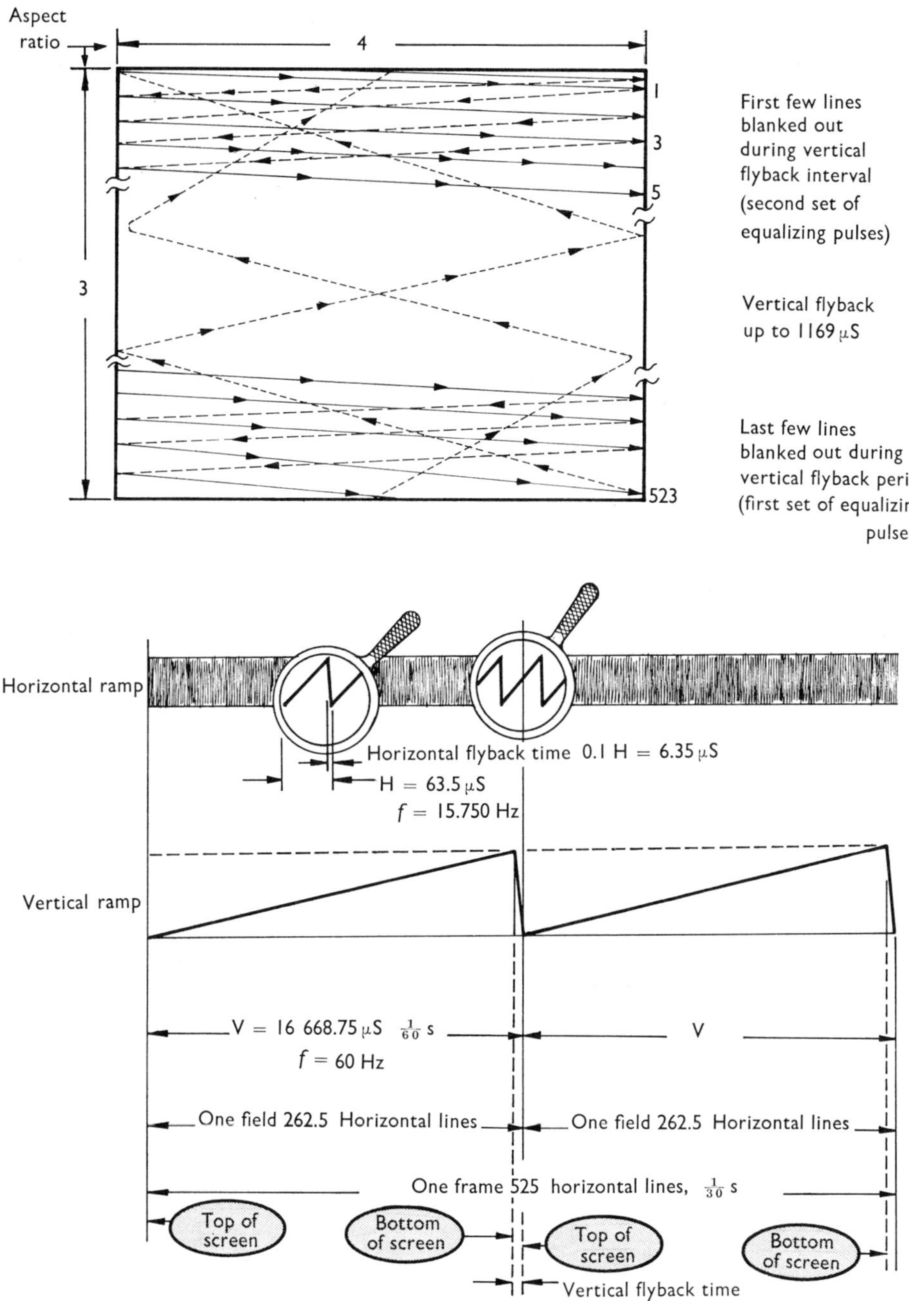

Fig. 16-3 Scanning pattern—time relationship between horizontal and vertical sweep ramps.

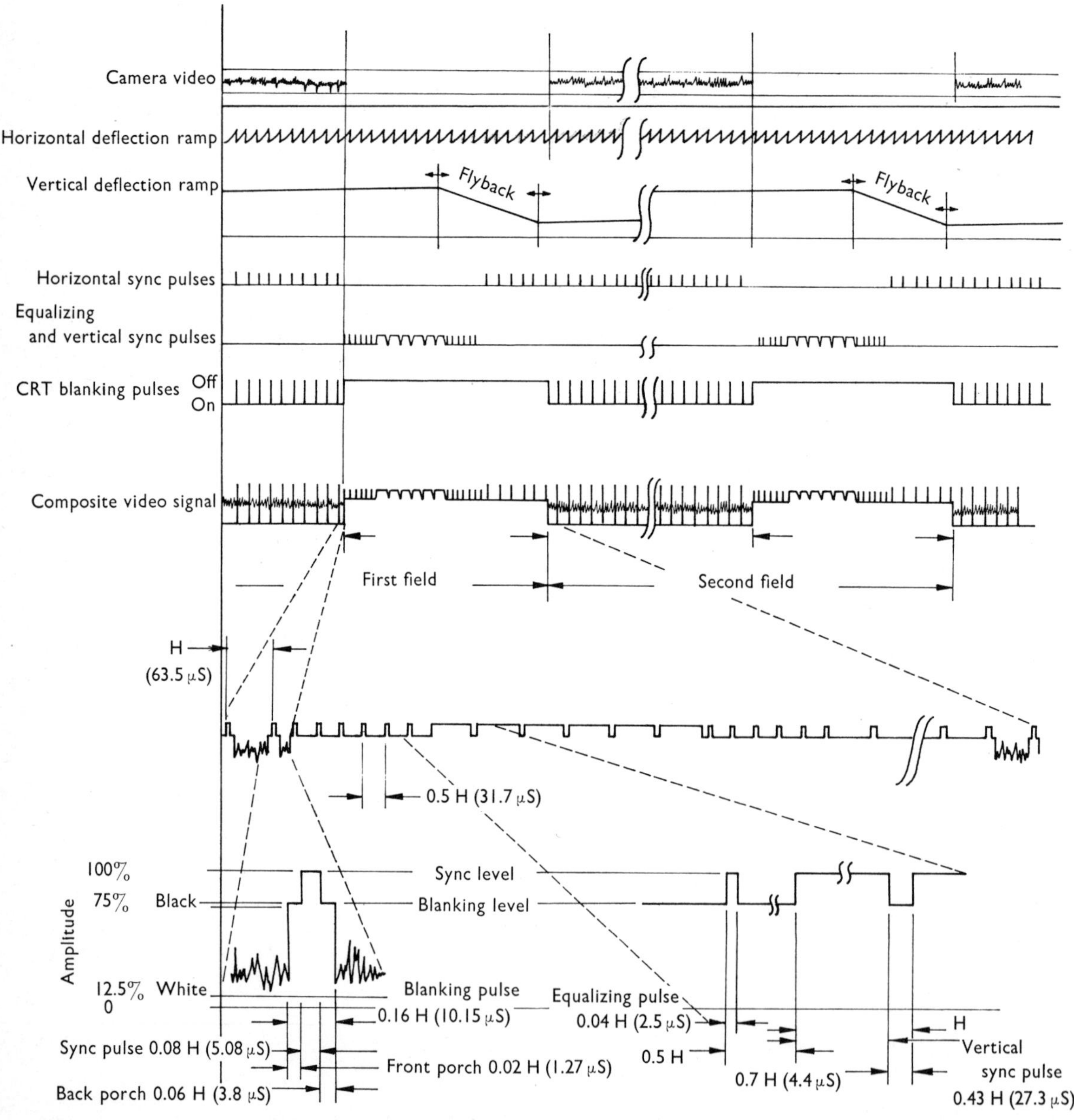

Fig. 16-4 Composite video signal and its components.

channel is 6 MHz wide. Within such a channel must be accommodated the video carrier, video sidebands, and the audio carrier and its sidebands. Figure 16-5a shows how the 6-MHz channel is utilized.

Only the upper video sideband is transmitted in full. Lower sideband frequencies which deviate from the carrier frequency by more than 0.5 MHz are suppressed at the transmitter. A television receiver passband is adjusted to com-

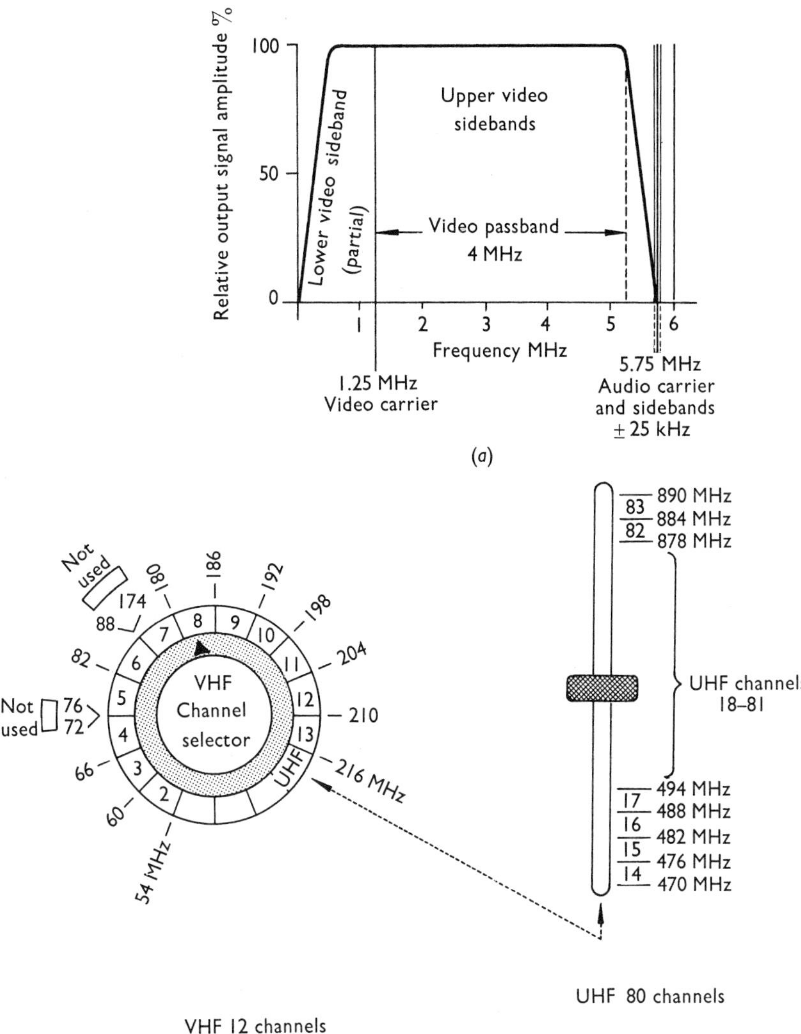

Fig. 16-5 Television channels: (*a*) Utilization of the 6-MHz television channel; (*b*) Relative location of television channels in the radio-frequency spectrum.

pensate for this loss of transmitted sideband power and is one of the factors which make receiver alignment a critical procedure.

Picture resolution is limited by the receiver passband. Assuming a full passband of 4 MHz is available, this represents a period of 0.25 μs per cycle. Assuming further that each cycle represents a period of transition from a light to a dark picture element (or vice versa) and since one scan line lasts for approximately 53 μs before blanking, there could be a maximum of $53/0.25 = 212$ picture-element pairs in one line. Because there are 525 such lines, it should be possible to transmit $2 \times 212 \times 525 = 222\,600$ picture elements in one frame. In round figures this means that the picture you see on the screen is composed of nearly a quarter of a million pieces.

The audio signal is transmitted by an FM transmitter broadcasting on a carrier frequency of 0.25 MHz from the upper channel limit. The audio and video carriers therefore differ by 4 MHz, and there is minimum interference.

16-9 TELEVISION-RECEIVER FUNCTIONAL BLOCKS

The essential parts of a black and white television receiver are shown in Fig. 16-6. A television receiver is basically a superheterodyne circuit (Sec. 11-11). The VHF tuner for channels 2 to 13 produces the intermediate frequency. It is not continuously tunable but varies in 6-MHz steps by switching in different coil segments for each channel. In addition it has one channel setting which is tuned to a signal produced by the UHF tuner. A UHF signal therefore undergoes double conversion. First it is converted to a VHF frequency which is fed to the VHF tuner. The VHF tuner then again converts the converted signal into the video IF. UHF tuners are continuously tunable because it would be impractical to provide an 80-channel selector switch.

After conversion the IF amplifier raises signal level to an amplitude suitable for detection. The IF passband must be at least 4.5 MHz because the IF amplifier must amplify both the video carrier, some of the lower video sidebands, all the upper video sidebands, and the sound carrier and sidebands. At the detector the sound carrier and its sidebands are removed by a filter for further amplification and FM detection to produce the audio signal. The video and blanking signals are applied to the CRT cathode where they modulate the CRT beam to produce the various brightness levels.

Some of video output is routed to a circuit which separates the horizontal and vertical sync pulses (sync separator). These are then passed to control circuits which control the output frequencies of the horizontal and vertical oscillators. These oscillators produce the horizontal and vertical ramp currents in the CRT deflection coils. The ramp currents must be locked or synchronized to those in the studio. A loss in synchronization would misplace the various picture elements on the screen. The effect is easily simulated on your own television receiver by rotating the horizontal or vertical "hold" controls.

The horizontal-output section not only produces the horizontal-deflection currents but the high electron-beam-accelerating voltage required by the CRT. A horizontal-output transformer (also known as a flyback transformer) steps up the pulses produced by horizontal-output power amplifier. These high-voltage pulses are then rectified and applied to the CRT anode. Because the circuit produces voltage in the neighborhood of 16 to 25 kV it represents a hazard and is therefore nearly always enclosed in a metallic cage.

16-10 REQUIREMENTS FOR COLOR TRANSMISSION

Only brightness or luminance information is required to reconstruct a monochrome image. A luminance signal can produce only light and dark sections on the screen and typically appears as a black and white picture. To produce a color picture two or more monochrome images of different color must be overlaid. In this way the colors mix and blend in various proportions to produce the illusion of multicolor. Color blending can easily be demonstrated by mixing paints. On a television screen the color image is composed of a blending of three primary colors. At any given instant there are actually three images on the screen, but the eye cannot distinguish one in the presence of the other two. The eye assimilates all three to give the illusion of full color.

A compatible color television system must be capable of transmitting a color picture and a black and white picture as well. It is this requirement plus the need to preserve a 6-MHz channel width that makes color television circuits relatively complex.

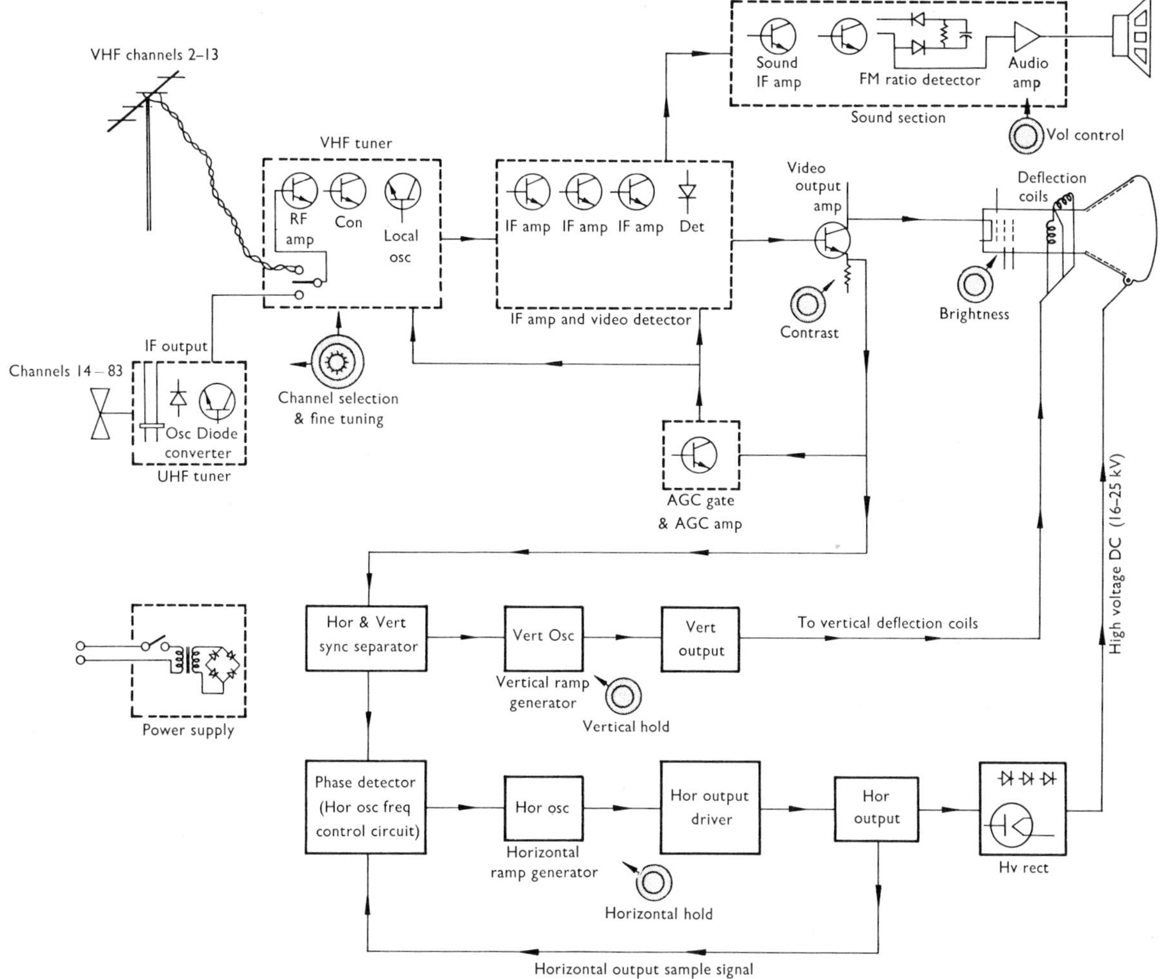

Fig. 16-6 Functional block of a monochrome television receiver.

16-11 COLOR VISION AND COLOR BLENDING

If white light is dispersed by a prism, the constituent colors are individually seen. A rainbow pattern results. The human eye has its greatest response at three primary colors, red, green, and blue, but the response is not equal for all three. Just as it is possible to scatter white light into constituent colors, it is also possible to blend constituent colors to produce white. Not all colors are required in the blending process. If the primary colors, red, green, and blue are combined in the correct proportions, white is obtained. Because the eye does not respond equally to all three colors, white on color television screens can only be produced if these colors are mixed with relative amplitudes. In a color telecast three camera tubes equipped with suitable color filters are used to produce the video signal. Each tube responds only to its primary color so that at the camera output there are red (R), green (G), and blue (B) video signals.

The *R, G, B* video voltages are added together in the proportions $0.59G$, $0.30R$, and $0.11B$ in a resistive summing network (Sec. 6-3) to produce a white or *luminance signal*. The luminance signal will produce a black and white image on either a color or black and white television screen.

It is now possible to state fairly concisely the principle of operation of color television. A tricolor camera produces the primary *R, G, B* video signals. These are electronically summed in a resistive network (also known as a matrix) to produce a luminance video signal. The luminance signal and components of the *R, G, B* signals are then transmitted to the color receiver. The receiver uses the luminance signal to produce what would be a black and white image; however, components of the *R, G, B* are also electronically added to the luminance video at the receiver. This upsets the instantaneous balance between the amplitudes of the *RGB* components in the luminance signal so that other than white (color) is produced.

16-12 COMPATIBLE COLOR TELEVISION

Even when a color camera is used, a black and white image can be produced from it. When the three primary colors are combined two at a time we obtain three additional colors, yellow, cyan, and magenta. If a picture element contains more than one primary color, more than one color camera tube responds. When white is picked up in the scene all three camera tubes respond, because white contains all three primary colors.

The video outputs from three camera tubes are applied to an electronic summing network. This circuit blends them in the correct proportions to produce a luminance signal which is very similar to a black and white video signal. The resistors in the summing network are chosen so that at the node point will appear 59 percent of the green video ($0.59G$), plus 30 percent of the red video ($0.30R$) and 11 percent of the blue ($0.11B$). The resultant voltage is then $Y = 0.59G + 0.30R + 0.11B$. This is known as the *luminance* or simply the *Y video* signal. By itself *it produces a black and white picture on either a monochrome or color television screen.*

To produce a color image, each of the *R, G,* and *B* signals can be applied to a color CRT. Here they each create the overlaid red, green, and blue images to produce a color picture. But what if the program originates from a black and white camera? Then there would be no *R, G,* and *B* signals as such, yet in this instance, for the sake of compatibility it is necessary to produce a black and white picture on a color CRT. One way to achieve compatibility is to use a more complex video drive arrangement. Instead of applying the *R, G, B* video directly, inverted *Y* video ($-Y$) is added to the *RGB* signal. In this way three color-difference signals are created. These are $R - Y$, $G - Y$, and $B - Y$. Since Y is also applied to the CRT the Y and $-Y$ cancel during color telecasts and the CRT receives only *R, G, B* video. During a black and white telecast *color-killer* switches are closed (electronically) so that only the *Y* video input acts on the CRT. A black and white image is then produced.

16-13 COLOR CRT

A black and white CRT generates only varying quantities of white light and therefore requires only one screen composed of a single phosphor coating. A color CRT requires a screen for each primary color so that in effect it has three screens in the same plane each made up of a different phosphor. To further complicate the construction, at no point must any two of the screens overlap, because then an electron beam would produce two colors, even if only one were required. This means that each screen must be deposited in the form of isolated microscopic dots of phosphor. Each point on the screen is composed of an arrangement of *R, G,* and *B* phosphor dots. A triangular arrangement is shown in Fig. 16-7a. Directly behind each cluster of three dots is a mask through which the illuminating electron beam must pass before it can strike any dot. This "shadow" mask is used to prevent the electrons bound for a given cluster of dots from illuminating neighboring clusters.

Each screen in a color CRT has its own electron gun. The relative brightness of each primary color in a three-dot cluster therefore depends on the beam current coming from its

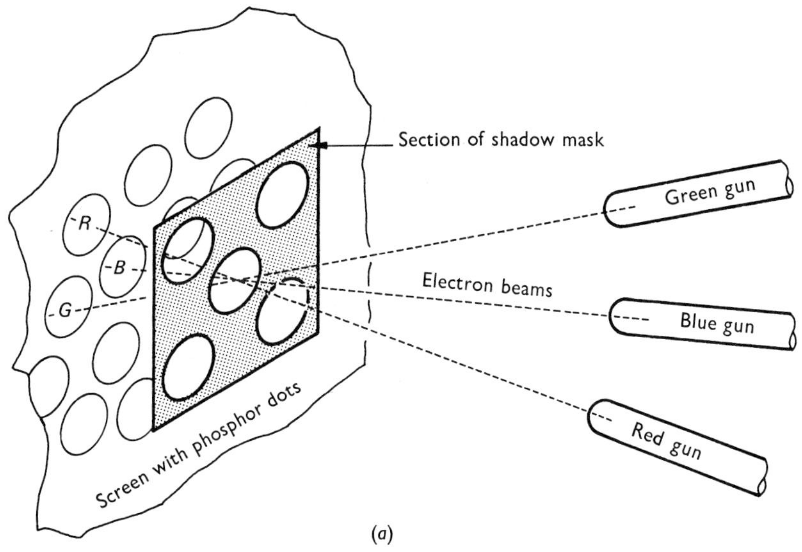

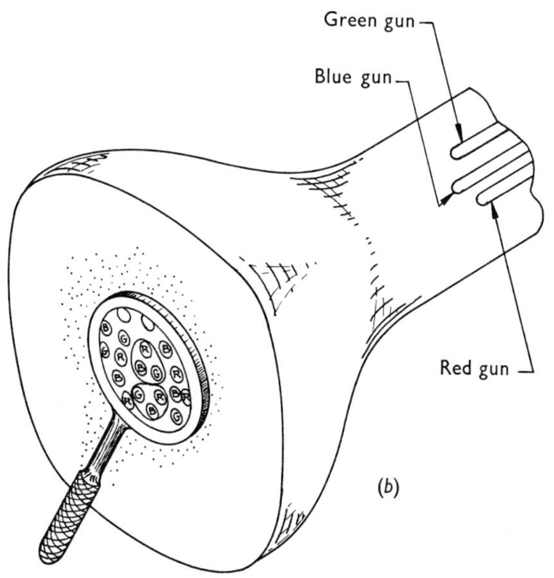

Fig. 16-7 Important features of a shadow mask type color CRT: (*a*) Structure of screen segment to show beam convergence through shadow mask; (*b*) CRT showing phosphor dot arrangement on screen.

electron gun. This in turn depends on the instantaneous video amplitude applied to the gun. Note that the guns are arranged in a triangular fashion so that each gun aims only at its own phosphor dot (Fig. 16-7b). All three beams are deflected simultaneously so that each passes through the same hole in the shadow mask.

It is an exacting procedure to adjust a color television receiver to ensure that all three beams converge on the same hole over the entire screen area. A *convergence adjustment* is required if color tints appear on black and white pictures. This indicates that the relative illumination from the phosphor dots is out of balance, probably because one or more of the dots is not receiving its fair share of beam current. The cause is likely due to the guns striking the shadow mask instead of the hole.

16-14 COLOR INFORMATION VISUALIZED IN VECTOR FORM

Suppose that the three primary colors, *R, G, B,* were arranged in the form of a blending triangle. Only at each apex would there be a saturated primary color. At all other points in the color triangle we find blends in a proportion corresponding to the distance from each of the apexes. The center of such a color triangle is white.

Now suppose that a vector free to rotate 360°, but pinned at the center, were affixed to the color triangle. The direction of this vector would tell us the color *hue* and the length would tell us the *saturation,* i.e., how pure the color is. This vector can be known as the *chrominance* (**C**) *vector* because it completely specifies the color content.

Next let us place the color triangle into a sphere. Through the center runs the black-white axis which is perpendicular to the plane in which the **C** vector rotates. If the *Y* video signal is also visualized as a vector, then it lies along the black-white axis. The bottom of the sphere at the black-white axis is maximum dark or black. Similarly, the top of the sphere represents maximum brightness. The center of the sphere then corresponds to medium bright (a very light gray).

Finally, visualize a **T** vector which is the resultant of **C** and *Y*. This imaginary **T** vector specifies how a spot on the screen will glow from instant to instant. It can point in any direction from the center of the sphere. As a line is scanned by the color camera the **T** vector is busily lengthening and shrinking and pointing in various directions at incredible speed. Yet, if we could stop it at each picture element and instruct an artist to put a dab of paint on a canvas as specified by the **T** vector, a complete color image would result. In a sense the color picture tube is a "high-speed artist" taking its directions from the **T** vector and from the horizontal and vertical deflection circuits.

In practical television circuits the **T** vector does not exist; i.e., you cannot measure a voltage representative of the **T** vector. You can, however, isolate the *Y* voltage and the signals which would be combined to form the equivalent of a **C** vector, i.e., the *chroma signal.* How such a chroma signal is generated from the *RGB* video signals and how it is transmitted is explained in the following sections.

16-15 COLOR TRANSMISSION

If all four video signals, the Y, $R - Y$, $G - Y$, and $B - Y$, were transmitted to produce the color image, four television channels would be required. Since only one 6-MHz channel is assigned to each television station some method must be found in which only the essential signal content is transmitted. In practice, it has been found that only two of the color videos out of *R, G, B,* need to be transmitted. The luminance (*Y*) signal contains all three, *R, G,* and *B*. This means that if two others are available at the receiver they can be used electronically to subtract or cancel their counterpart in the Y signal. All that remains of the Y signal after this has been done is the third color. For example, suppose the Y, R, and B signals are recovered in the television receiver. Then by inverting the R and B signals and using a suitable summing network so that their amplitudes are the same as in the Y signal, they can be canceled from the Y signal.

Adding in the summing network gives:

$$(Y) + (-0.30R) + (-0.11B)$$

Substituting for Y gives:

$$(0.59G + 0.30R + 0.11B) - 0.30R - 0.11B = 0.59G$$

Only green luminance video remains. Since only the luminance and two of the three color video signals need to be transmitted, a certain amount of channel width is saved. Further channel-width reductions are possible by restricting the two transmitted color video signals to a narrow band of 0.5 and 1.5 MHz. It has been found that when the bandwidth squeeze is applied in the right places there is little noticeable color degradation.

In actual practice, two chrominance signals are transmitted. Known as the I and Q signals, they are defined as

$$I = 0.60R - 0.28G - 0.32B$$

$$Q = 0.21R - 0.52G + 0.31B$$

Between them they contain all the color information needed to generate the chrominance vector.

Keep in mind that the chrominance vector is a mathematical concept whereas electronic signals are required to drive the red, green, and blue guns of the color CRT. These voltages can be generated from the Y, I, and Q signals by means of suitable resistive summing networks and inverters. It is not too difficult to give a mathematical proof that the I, Q, and Y signals can be used to generate the electron-gun signals. The expressions for I, Q, and Y are three equations in three unknowns:

$$0.60R - 0.28G - 0.32B = I$$

$$0.21R - 0.52G + 0.31B = Q$$

$$0.30R + 0.59G + 0.11B = Y$$

This system of equations can be solved to give R, G, or B in terms of I, Q, and Y. Once the RGB are available, it is only necessary to add $-Y$ to each to generate the difference signals $R - Y$, $G - Y$, and $B - Y$. These are the signals used to drive the guns of the color CRT.

16-16 COLOR TELEVISION TRANSMITTER

In the color television transmitter, Y signal is used to amplitude-modulate the carrier just as it would in a black and white transmitter. However, in a color transmitter, the Y signal is not the only signal which modulates the carrier, although it is the only signal detected by black and white receivers. Also, indirectly modulating the carrier are the I and Q signals. They first modulate a subcarrier (which is not transmitted). In the TV receiver the I and Q sidebands are mixed with a locally generated subcarrier so that the I and Q signals can be reconstituted as video voltages. From these are generated the gun voltages to drive the color picture tube.

A simplified functional block diagram of a color video transmitter system is shown in Fig. 16-8. A color camera provides the RGB video signals. The Y or luminance signal is generated in the resistive summing matrix, and all frequency components above 4.15 MHz are filtered out. It is then passed through a delay line to bring it into the same phase and time

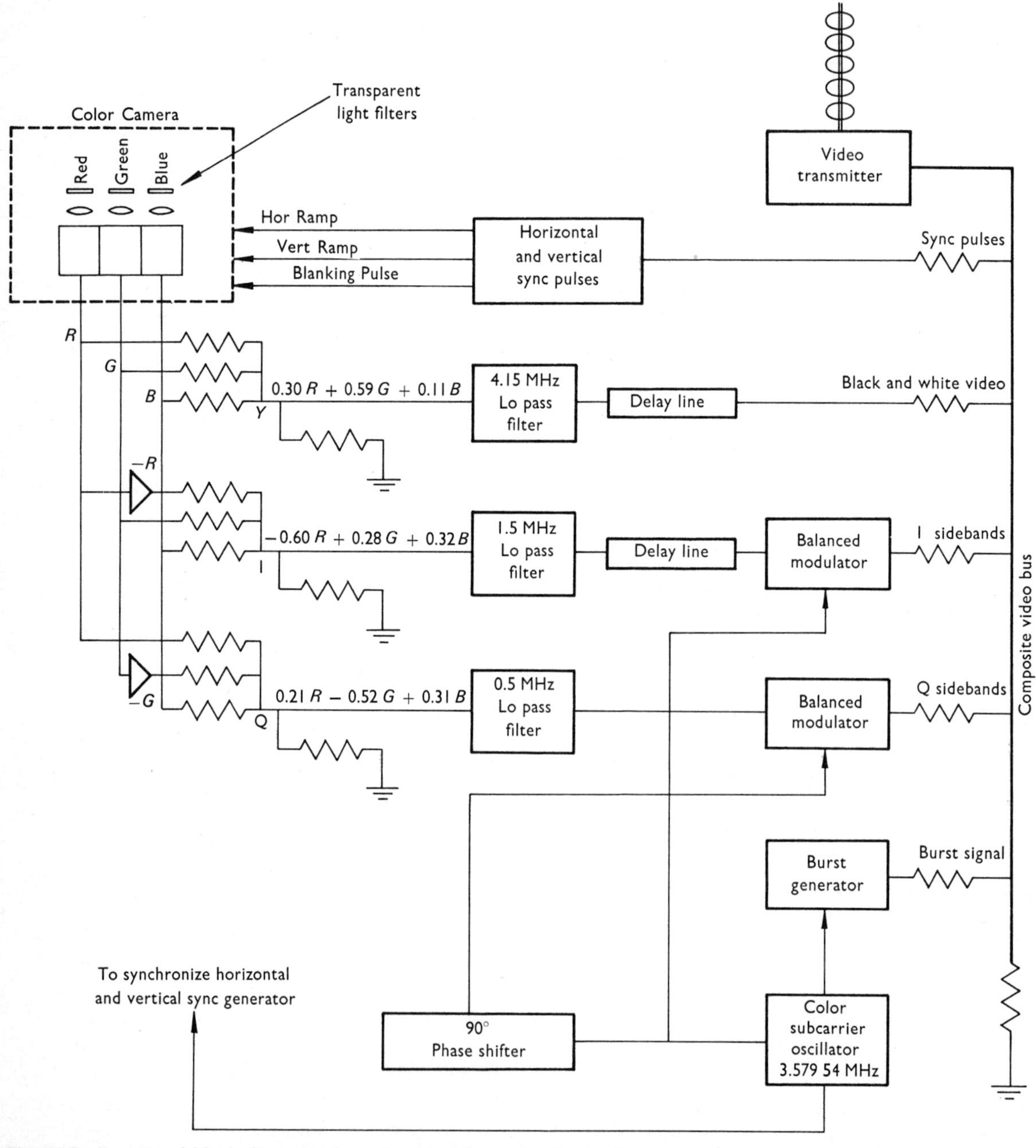

Fig. 16-8 Functional block diagram of a color television video transmission system.

relationship as the *I* and *Q* signals. The filtered and time-delayed *Y* signal is then ready for transmission and is added to the composite video bus.

The *I* and *Q* signals each go to a balanced modulator where they modulate the color subcarrier. The *I* signal modulates the subcarrier directly. (The *I* stands for In phase). The *Q* signal modulates the subcarrier too, but the subcarrier is first phase shifted 90° and is "in Quadrature" with the original subcarrier. (The *Q* stands for Quadrature phase.) Since balanced modulators are used, the subcarrier frequency is removed so that only *I* and *Q* sidebands remain and are passed to the composite video bus. The sidebands will not interfere with each other because of the 90° phase displacement between them. A 3.57954-MHz subcarrier frequency was chosen because it can be shown that it produces the least interference when mixed with harmonics of the horizontal-sweep frequency. Subcarrier oscillator frequency is divided down with binary counters to synchronize the horizontal and vertical sync and sweep generators.

When the *I* and *Q* sidebands arrive at the receiver they must first be demodulated in a detector. The subcarrier is required by the detector circuit, but since it is not transmitted it must be locally generated. To lock the local subcarrier oscillator to the same frequency (and phase) as in the transmitter a few cycles of the subcarrier are transmitted during horizontal blanking time. This is known as the *burst* signal. It appears on the back porch of the horizontal sync and blanking pulse, as shown in the composite color video signal in Fig. 16-9.

Except for the addition of the burst signal and the *I* and *Q* chroma sidebands, note the similarity between the color and monochrome composite video signals in Figs. 16-4 and 16-9. A black and white television receiver ignores the burst and chroma signals.

16-17 COLOR TELEVISION RECEIVER

A functional block diagram of a color television receiver is shown in Fig. 16-10. Many of the circuits are similar to those found in a monochrome receiver. The tuner and IF stages require critical alignment for good color reproduction. Poor high-frequency response will attenuate the color signals which lie in the 3- to 4-MHz region of the composite video signal.

After detection the composite television signal is distributed to various sections of the receiver for separation and further processing. The sound IF signal is removed at the detector, amplified, and passed to an FM discriminator for FM detection. The sound signal is recovered, amplified, and passed to the speaker. Horizontal and vertical sync pulses are recovered by the sync separator and are used to synchronize the deflection circuits.

Luminance video (*Y* signal) is amplified and delayed to compensate for the frequency filtering introduced in *I* and *Q* channels. It is then used to modulate the beam intensity of all three CRT electron guns simultaneously. After conversion to the correct polarity it is also applied to the three resistive matrix networks to take part in the recovery of the red, green, and blue video voltages at the CRT electron-gun cathodes.

Two additional outputs not found in a monochrome receiver are taken from the color-receiver detector. One of these is for the burst signal. During a colorcast the burst signal is amplified and used in a comparison circuit where the burst and local color carrier are matched for frequency and phase. Since the color carrier is removed at the transmitter, an exact replica of it is required in the *I* and *Q* detectors to demodulate the color signal. The burst signal ensures that an exact replica can be reconstructed in the viewer's home.

Lack of a burst signal signifies to a color receiver that a monochrome telecast is taking place. When this happens the color signal channels are disabled by the color-killer circuit, and only the *Y* video is applied to the CRT.

The second output at the detector which is unique to color receivers is the chroma output. Here the *I* and *Q* signals, which appear as the sidebands of the color subcarrier and its 90° quadrature signal, are removed. These sidebands are first amplified and then passed to the *I* and *Q* detectors. These are synchronous detectors and the 3.5796-MHz color carrier and its 90° quadrature twin can be regarded as switch-

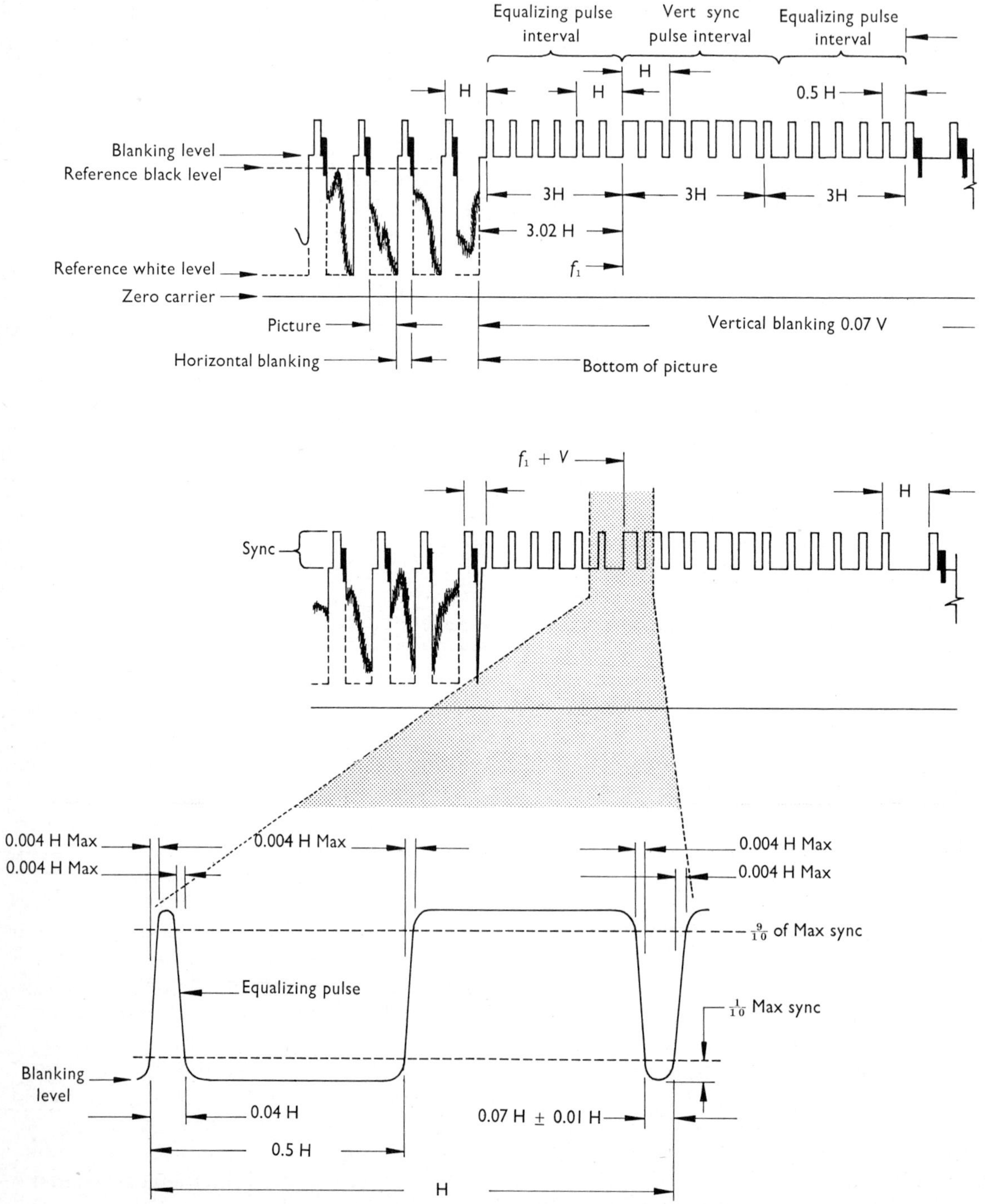

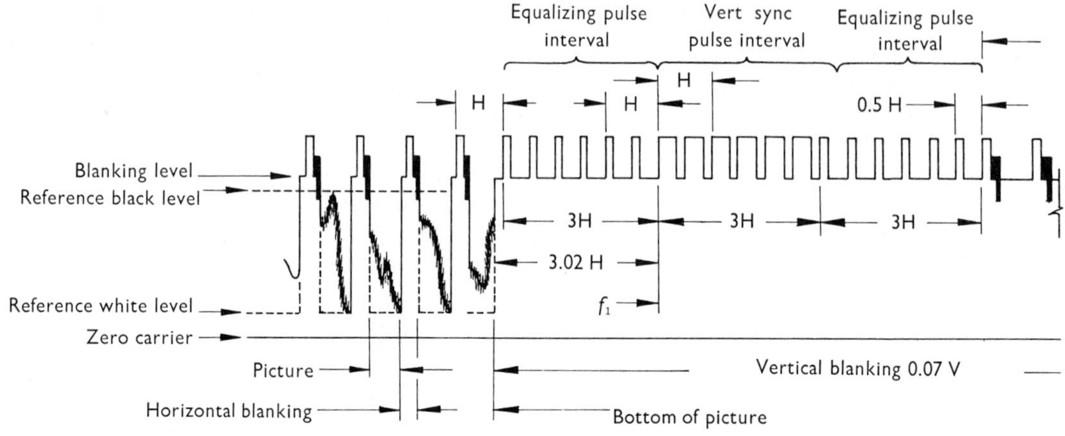

Equalizing pulse interval
Vert sync pulse interval
Equalizing pulse interval
H
H
H
0.5 H
Blanking level
Reference black level
3H
3H
3H
3.02 H
f_1
Reference white level
Zero carrier
Picture
Vertical blanking 0.07 V
Horizontal blanking
Bottom of picture
$f_1 + V$
H
Sync
0.004 H Max
0.004 H Max
0.004 H Max
0.004 H Max
0.004 H Max
$\frac{9}{10}$ of Max sync
Equalizing pulse
$\frac{1}{10}$ Max sync
Blanking level
0.04 H
0.07 H ± 0.01 H
0.5 H
H

Fig. 16-9 Composite color video signal.

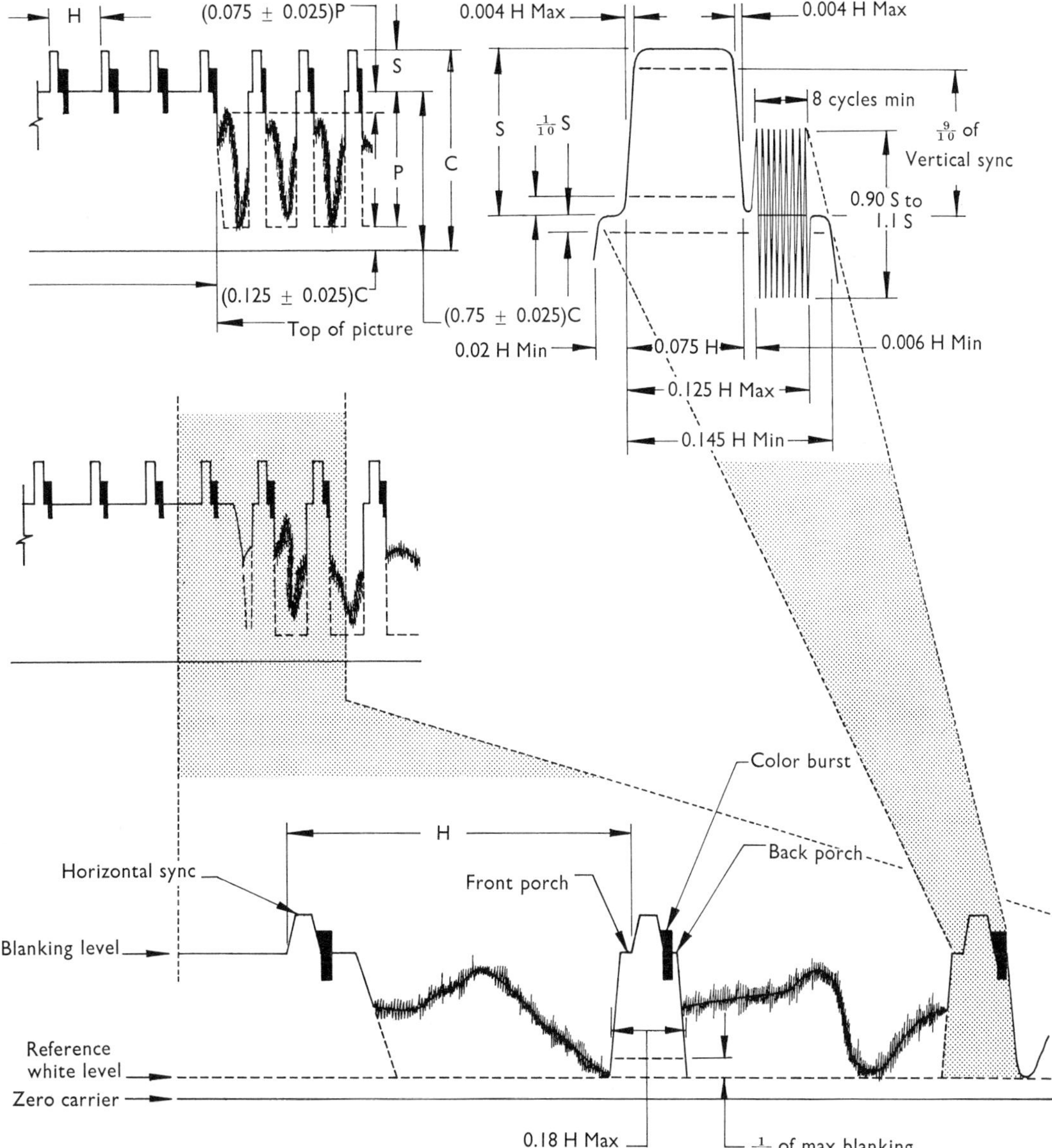

Fig. 16-9 *(Continued.)*

ing or gating signals. They in turn switch the *I* and *Q* detectors on and off at exactly the right moments to recover the *I* and *Q* color video voltage from the sideband signals. After recovery from the detectors, the *I* and *Q* signals pass to the resistive matrices which recover the $R - Y$, $G - Y$, and $B - Y$ signals, as shown in the block diagram, Fig. 16-10.

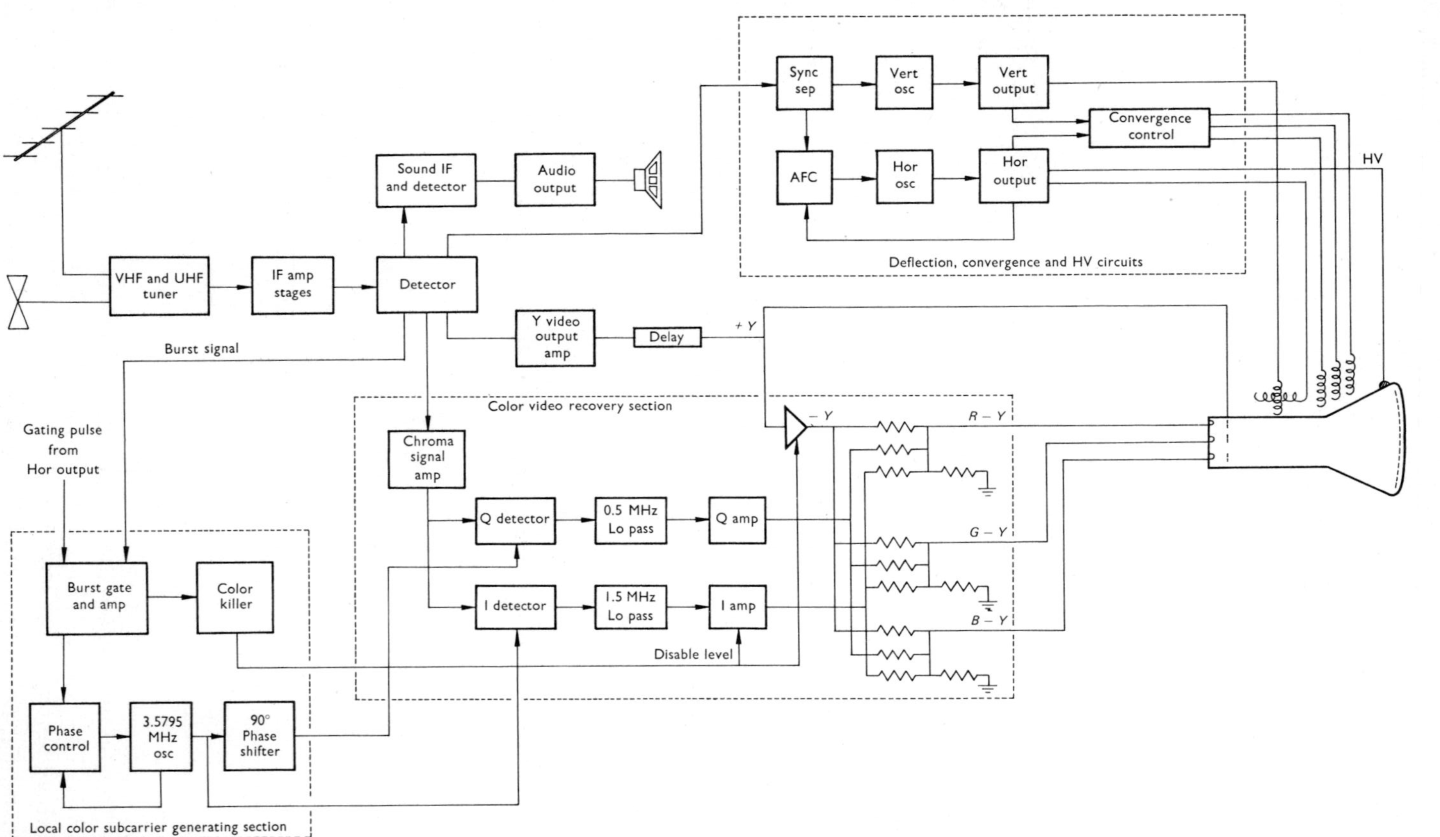

Fig. 16-10 Functional block diagram of a typical color television receiver. Enclosed sections are unique to color television receivers only. Other sections are common to both monochrome and color receivers.

Those circuits unique to a color receiver are shown in Fig. 16-10 enclosed in broken lines. For a more thorough discussion of actual circuits see Appendix L.

Summary

A television camera is an optical-electrical interface for converting scenic information into an electrical signal.

To produce a video signal the image on the camera screen is *scanned*.

The horizontal scanning rate is 525 lines (one frame) every $\frac{1}{30}$ of a second. This gives a horizontal scanning frequency of 15.75 kHz.

To decrease flicker, interlaced scanning is used. First the odd scanning lines are transmitted in $\frac{1}{60}$ of a second. This is one *field*. Then the even scanning lines, the second field, are transmitted in the next $\frac{1}{60}$ of a second.

The picture CRT is blanked out during both horizontal and vertical *flyback time*.

The vertical scanning frequency is much lower than the horizontal, i.e., 60 Hz to produce one field each $\frac{1}{60}$ of a second.

Composite video contains (1) the camera signal, (2) the blanking pulses, and (3) the horizontal and vertical sync pulses.

A television channel occupies 6 MHz of bandwidth.

A television receiver is a superheterodyne circuit with circuits that separate video, audio, and deflection-synchronizing signals.

Color and monochrome television systems must be compatible in all respects. This means that a monochrome receiver must produce a black and white image from a colorcast. Similarly, a color receiver must produce a monochrome (black and white) image from a monochrome telecast.

A color image is produced by the blending of three primary color images overlaid on a color television screen.

When mixed in the proportions $0.59G$, $0.30R$, and $0.11B$, the three color video signals will give a black and white picture. The signal $Y = 0.59G + 0.30R + 0.11B$ is known as the *luminance* signal.

During monochrome telecasts the lack of a *burst signal* activates the *color-killer* circuits in a receiver. Then only the luminance signal is applied to the CRT.

A color CRT screen is actually three screens in one. It is composed of clusters of three-dot phosphors, each producing its own color when excited by the electron beam.

A *shadow mask* prevents the electron beams from striking more than one cluster of dots at a time.

Convergence is the term used when all three guns strike only their own phosphor dots at any position on the screen.

Color information is transmitted by two video signals, (1) the *I signal:* $I = 0.60R - 0.28G - 0.32B$, and (2) the *Q signal:* $Q = 0.21R - 0.52G - 0.31B$.

The I and Q signals are used to modulate a *color subcarrier* and its 90° phase-shifted quadrature subcarrier. Only *sidebands* are produced in the process; the subcarriers are both suppressed. The sidebands so produced become part of the composite video signal.

When recovered at the receiver the Y, I, and Q signals are added in resistive networks to produce the red, green, and blue video signals.

Questions and Exercises

1. What name is given to the television process of dissecting a picture into small elements?
2. How many total horizontal sweeps are performed during the transmission of one complete picture in the North American television system?
3. How is the electron gun "aimed" at points

on the screen in a television camera tube?

4. To transmit a picture the aim of the camera and CRT guns must be "locked." Explain this statement.
5. Why is only half a picture (odd- or even-numbered scan lines) transmitted at a time?
6. What is the name given to one complete picture (all scan lines)?
7. What is the name given to half a picture (odd or even scan lines only)?
8. How often is a complete picture transmitted?
9. How many horizontal scan lines are transmitted each second?
10. How many vertical sweeps must be made each second?
11. What is flyback?
12. What is the total duration of (*a*) one horizontal scan line, (*b*) horizontal flyback time, (c) vertical scan (one sweep including flyback), (*d*) vertical flyback time, (e) horizontal sync pulse, and (*f*) horizontal blanking pulse?
13. What constitutes the composite video signal?
14. How is the vertical sync pulse recovered from the vertical sync pulse train?
15. What is the transmission channel width allotted to North American television stations?
16. How is sound transmitted along with video?
17. What purpose do the equalizing pulses serve?
18. Why is it impractical to use channel switching on a UHF tuner?
19. Explain briefly how the high accelerating voltage required by the picture tube is produced?
20. What is meant by the term compatible color transmission?
21. Explain briefly how color is produced on the screen of a color picture tube.
22. Why cannot color be produced on an ordinary monochrome picture CRT?
23. What proportions of the color video voltages from a color camera are required to produce a black and white image?
24. How are the exact proportions of color video produced from the outputs of the color camera?
25. What is a resistive matrix?
26. Define the term luminance signal.
27. Why is a color-killer circuit required in a color television receiver?
28. How is the color-killer circuit activated?
29. What adjustment is required if a black and white image has color tint on some parts of screen but not on others?
30. What two pieces of information are conveyed by the chrominance signal?
31. How is color video extracted from the *Y* and the chroma signal?
32. Why is it necessary only to transmit the *Y* signal and two of the three color video signals coming from the color camera?
33. Which three video signals (write their equations in terms of *RGB* content) are finally selected for transmission to the home receiver?
34. The *I* and *Q* signals are not used to modulate the transmitter carrier directly. What happens to these signals before they become part of the color composite video used to modulate the main carrier?
35. Why is a burst signal transmitted during a colorcast?
36. What is the frequency and duration of a burst signal?
37. What purpose does the reconstructed color subcarrier serve in the television receiver?
38. Obtain an equation for each of *R*, *G*, and *B*, by solving the simultaneous equations in Sec. 16-15.

appendix

- **a** Typical Power-supply Circuits
- **b** Vacuum-tube Amplifier Circuits
- **c** Coupled Vacuum-tube Amplifiers
- **d** Typical Amplifier Circuit
- **e** Vacuum-tube Power Amplifier
- **f** Transformerless Power Amplifier
- **g** Typical *CR* Test Oscillator
- **h** Vacuum-tube Oscillator Circuits
- **i** A Home-Entertainment Stereo System
- **j** Vacuum-tube Circuits
- **k** CMOS and I^2L
- **l** Color Television Receiver

A Typical Power-supply Circuits

A-1 LOW-VOLTAGE POWER-SUPPLY CIRCUIT

Figure A-1 illustrates the low-voltage power supply in the Philco model 17JT41 television receiver. A power transformer (PT) receives 110-V ac primary power via the receptacle, through

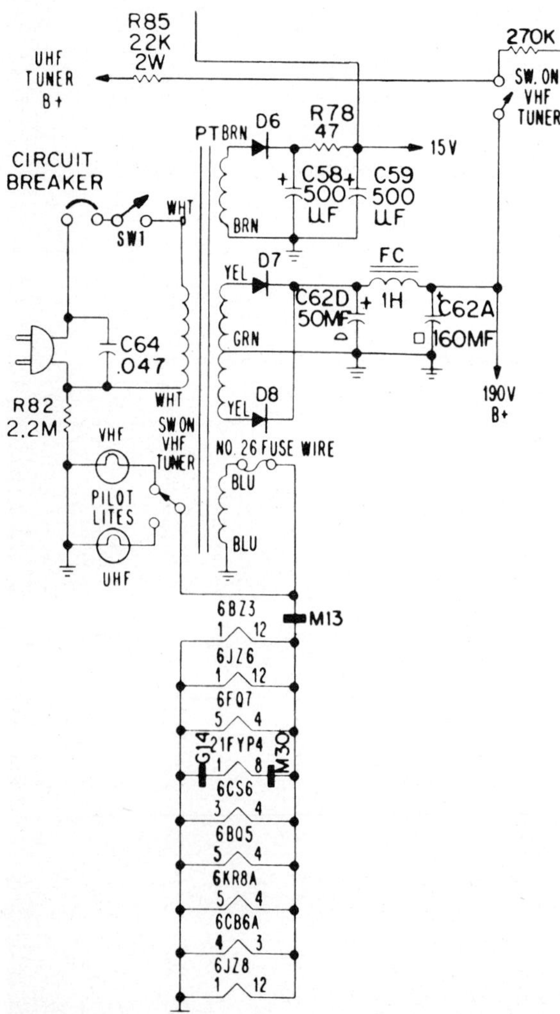

Fig. A-1 Low-voltage power-supply circuit in Philco model 17JT41 television receiver. (*Courtesy Philco-Ford of Canada Limited*)

the automatic overload circuit breaker and switch SW_1. Capacitor C_{64} does not interfere with the 110-V ac primary power because of its high capacitive reactance at 60 Hz. At very high frequencies, such as those present in bursts of noise or other interference, the capacitor has a low reactance and therefore acts as a short circuit. It is a low-pass filter which reduces high-frequency interference coming to or leaving the television receiver via the power line. Resistor R_{82} permits electric charge which may accumulate on the chassis to leak away into the grounded power line.

There are three secondaries on the transformer. One supplies power to the *half-wave rectifier* diode D_6. Capacitors C_{58}, C_{59}, and resistor R_{78} form a pi filter. The output of the half-wave power supply is 15 V dc and operates the transistors. A higher-voltage *full-wave* split phase secondary supplies power to rectifier diodes D_7 and D_8. The pulsating dc, filtered by pi filter C_{62D}, C_{62A}, and choke *FC*, operates the tubes in the receiver. A third secondary winding supplies 6.3 V ac to tube filaments and pilot lamps. The components designated as M_{13}, M_{30}, and G_{14} are mounting lugs and grounding lug, respectively.

A-2 BRIDGE RECTIFIER POWER-SUPPLY CIRCUIT

A *full-wave bridge*-type rectifier circuit is used on the RCA model CTC25A color television receiver. The power-supply schematic is shown in Fig. A-2. Primary power is applied to either the black-white or black-red lead of power transformer T_{105}, depending on the customer's line-voltage level. Capacitor-resistor networks CPR_{101}, CPR_{102}, and chokes L_{115A} and L_{115B} perform noise and interference filtering. Two secondary coils supply ac power to two sets of tube filaments. One set of filaments is held at approximately 190 V above ground potential by resistors R_{125} and R_{171} to minimize filament-to-cathode leakage current. The remainder of the

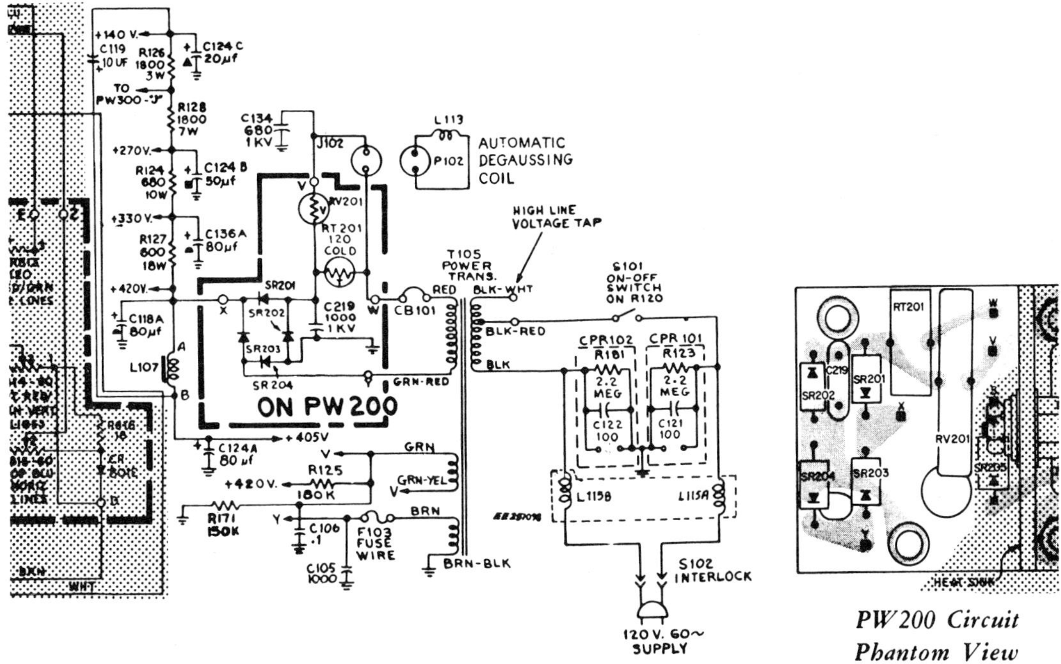

Fig. A-2 Power-supply circuit in RCA model CTC25A color television receiver. (*Courtesy RCA Limited*)

filaments are held at ground potential, and the brown-black lead of the filament coil is returned directly to chassis ground.

The high-voltage secondary (red and green-red leads) is connected through circuit breaker CB_{101}, degaussing coil L_{113}, and resistor RV_{201} to the diodes SR_{201}, SR_{202}, SR_{203}, and SR_{204} connected as a *bridge*. The load is connected between chassis ground and point X. Capacitors C_{118A}, C_{124A}, and choke L_{107} are a pi filter for the 405-V dc output. Other voltage levels are obtained at the *CR* filters consisting of R_{127}, C_{136A}, R_{124}, C_{124B}, R_{128}, R_{126}, and C_{124C}. Resistors RV_{201} and RT_{201} are temperature-sensitive. When the power is first switched on RT_{201} has a high resistance and RV_{201} a low resistance. Load current must therefore flow through the degaussing coil and RV_{201}, with only a small current through RT_{201}. Nevertheless, this small current heats RT_{201} and as the temperature rises its resistance drops, resulting in still larger current. In a few seconds the temperature of RT_{201} reaches and maintains a level such that RT_{201} is the much lower resistance path, whereas RV_{201} is a high resistance. There is negligible degaussing current after a few seconds of operation.

A-3 HALF-WAVE VOLTAGE DOUBLER

Figure A-3 shows a *half-wave voltage doubler* as used in the Canadian General Electric model M663 television receiver. Line voltage is applied between chassis ground and R_{401} through noise-filter chokes L_{402} and L_{403}. The charge path for C_{402} on the negative power-line voltage alternation (with respect to chassis ground) is through diode Y_{402}. On the positive alternation, the sum of line voltage and accumulated voltage across C_{402} is applied to C_{403A} through diode Y_{401}. Choke L_{401} and C_{403B} are ripple

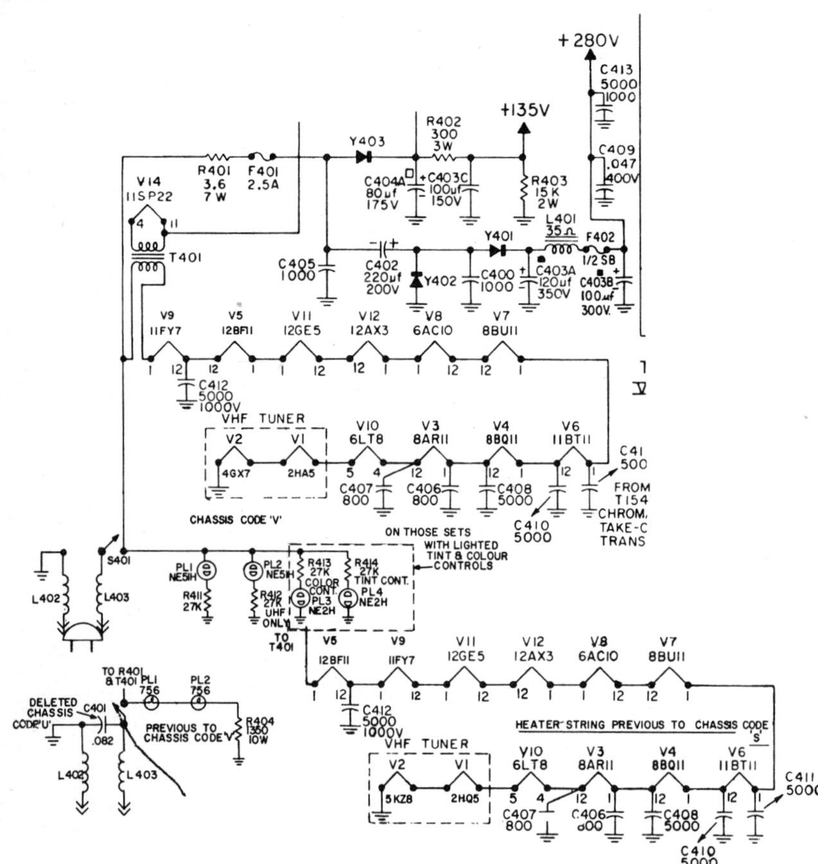

Fig. A-3 Power-supply circuit in Canadian General Electric model M663 color television receiver. *(Courtesy Canadian General Electric Co., Ltd.)*

filters. Since no power transformer is used, the filaments are connected in series such that their *IR drops* add up to line voltage.

A-4 FULL-WAVE VOLTAGE DOUBLER

In the power-supply schematic of the Zenith model 23XC38Z color television receiver shown in Fig. A-4, a *full-wave voltage doubler* is used. Voltage from the secondary of transformer T_{17} is applied through fuse F_1 and the degaussing coil network consisting of L_{41}, L_{42}, L_{43}, and resistor R_{103}. When the red lead of the secondary coil is negative with respect to the red-yellow lead, capacitor C_{156A} receives charge current through diode SE_3. On the next alternation the secondary-coil polarity reverses and C_{157} receives charge current through diode SE_4. The load is connected between ground (minus lead of C_{156A}) and the positive lead of C_{157} and therefore receives the sum of the voltages on the two capacitors. Choke L_{44} and capacitor C_{156B} are ripple filters for the 350-V output. Additional filtering is provided by resistor R_{104} and capacitor C_{32D} for the 250-V output.

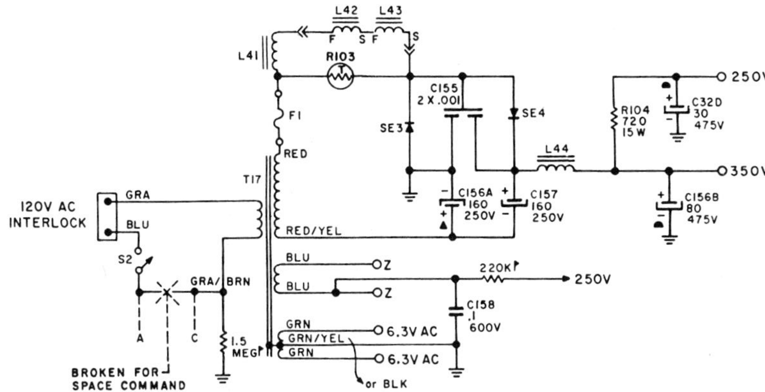

Fig. A-4 Power-supply circuit in Zenith model 23XC38Z color television receiver. (*Courtesy Zenith Radio Corporation*)

B Vacuum-tube Amplifier Circuits

B-1 ADVANTAGES OF VACUUM-TUBE AMPLIFIER CIRCUITS

A vacuum tube amplifier circuit is able to withstand momentary overloads better than either FET or bipolar transistor circuits. In applications such as the input amplifier in oscilloscopes and other test instruments, careless range-control settings often cause very large input signals to be applied to the amplifier. Vacuum tubes usually can withstand all but the most severe momentary overloads.

B-2 TRIODE AND PENTODE CIRCUITS

Triode and pentode amplifier circuits are shown in Fig. B-1. Except that much higher power-supply voltages are used, the circuits are quite similar to FET amplifier circuits. In operation, a signal voltage is impressed via R_1 onto the grid bias voltage developed by the cathode resistor R_K. Varying the grid-voltage controls, cathode-to-plate electron flow and a varying plate current passes through the load. To construct a load line, the two extreme points, namely cutoff and full conduction, are assumed. The two points $(V_{B+}, 0)$ and $(0, V_{B+/R_L})$ are plotted on the characteristic curves axes and are joined to make the load line.

A pentode amplifier circuit requires a positive voltage (with respect to cathode) on its screen grid. In nearly every instance this voltage is lower than the plate voltage. An *IR* drop across resistor R_S (Fig. B-1b) reduces the power-supply voltage to a lower value for the screen. Ohm's law is used to calculate the value of R_S, because the required *IR* drop voltage is known as is also the screen current. The following example illustrates how this is done.

EXAMPLE B-1 What value of screen-dropping resistor is required to drop screen voltage to 150 V from a power-supply voltage of 200 V? Quiescent-screen current is assumed to be 3.5 mA. The *IR* drop required is 200 − 150 = 50 V.

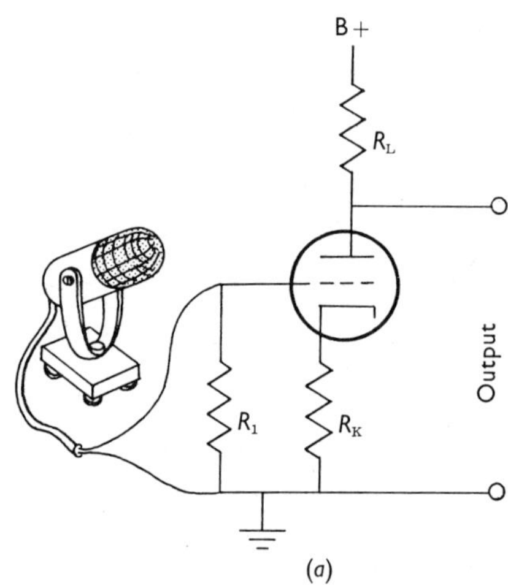

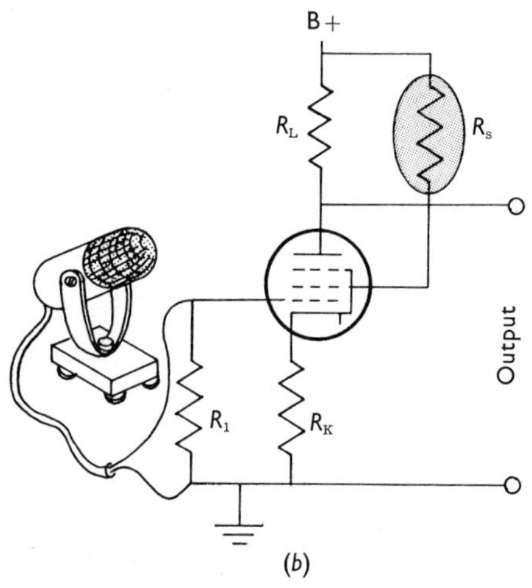

Fig. B-1 Vacuum-tube amplifier circuits: (*a*) Triode amplifier; (*b*) Pentode amplifier.

Applying Ohm's law gives

$$R_S = \frac{50}{3.5 \times 10^{-3}}$$

$$= \frac{50\ 000}{3.5}$$

ANSWER $= 14.3\ \text{k}\Omega$

B-3 VACUUM-TUBE AMPLIFIER CONNECTIONS

Vacuum-tube cathode, plate, and grid connections in amplifier circuits closely parallel those used in transistor amplifier circuits. Compare, for example, the connections of the emitter, collector, and base in the bipolar transistor circuits of Fig. 3-17 with those in Fig. B-2.

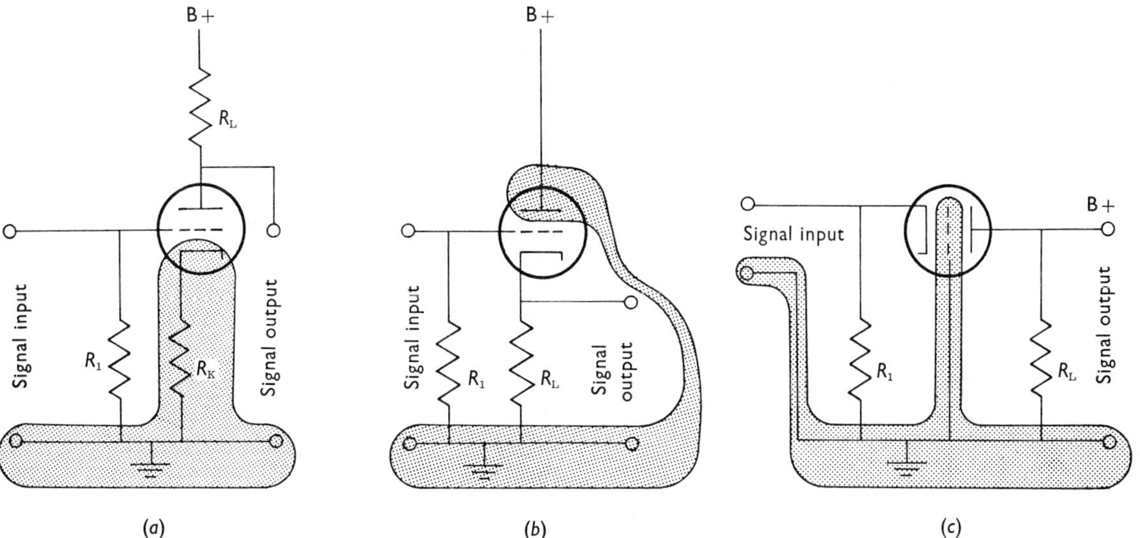

Fig. B-2 Fundamental vacuum-tube amplifier connections: (*a*) Common cathode; (*b*) Common plate; (*c*) Common or grounded grid.

C Coupled Vacuum-tube Amplifiers

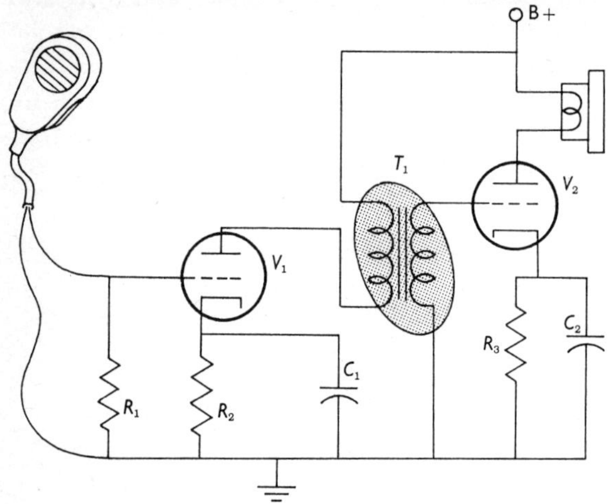

Fig. C-1 Transformer-coupled vacuum-tube amplifier.

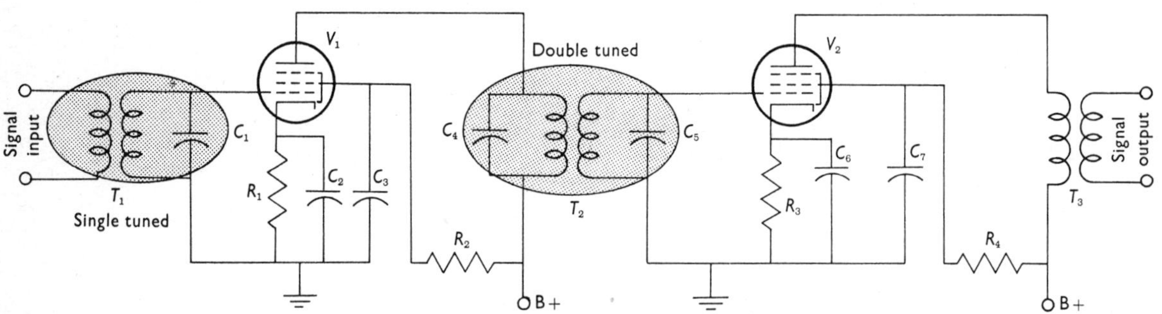

Fig. C-2 Single- and double-tuned transformer-coupled pentode vacuum-tube amplifier.

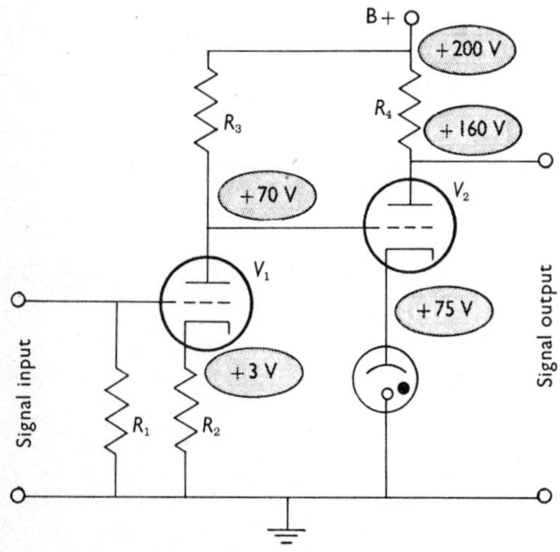

Fig. C-3 Direct-coupled vacuum-tube amplifier.

D Typical Amplifier Circuit

D-1 AUDIO AMPLIFIER CIRCUIT

Figure D-1 shows a schematic diagram of a portion of the audio amplifier section of radio chassis model 132.41501 (Simpson-Sears Limited). The first audio amplifier and driver stages (Q_5 and Q_6) have been isolated to illustrate how the principles discussed in Chaps. 3 to 6 are applied in practice.

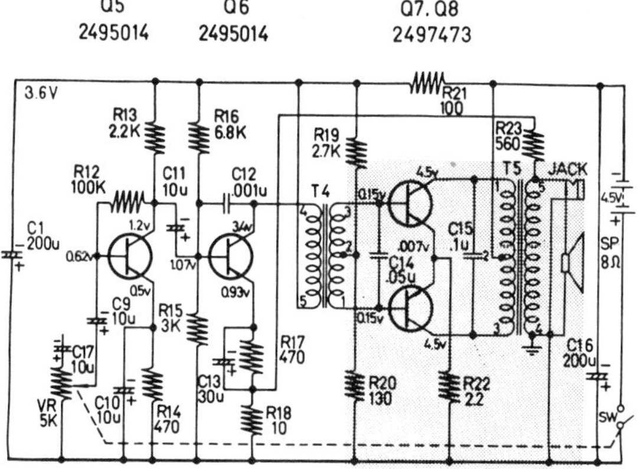

Fig. D-1 Audio amplifier section of a portable transistor radio receiver: schematic diagram. (*Courtesy Simpson-Sears Ltd.*)

D-2 *CR* DECOUPLING NETWORK

Resistor R_{21} and capacitor C_1 form a decoupling network between the power-output stage and the remainder of the radio circuit. The power-output stage operates class B. Although it draws negligible quiescent current, on large audio signal excursions it draws considerable momentary current. As the batteries approach exhaustion, they possess considerable internal resistance. Fluctuating current demand causes a fluctuating terminal voltage at the batteries. Power-source voltage fluctuations should not be passed on to the low-level amplifier stages. To minimize fluctuation, the decoupling network acts as a *CR* filter, the action of which is the same as that of a ripple filter as discussed in Sec. 2-6.

D-3 SIGNAL COUPLING

Audio signal voltage enters the first amplifier stage from the second detector via coupling capacitor C_{17}. It allows the ac component to pass but blocks dc levels developed in the second detector. The entire audio signal is dropped across *VR*, the volume control, but only a fraction (depending on the setting of the control shaft) is passed to the base of Q_5 via capacitor C_9. This coupling capacitor passes the signal ac component but prevents the dc base-bias level (0.62 V) from reaching the volume control.

The cutoff frequency for C_{17} and resistor *VR* is the frequency at which the reactance of C_{17} is such that only 70.7 percent of the signal voltage is being passed to *VR*. This occurs when $X_c = VR$, and (neglecting input-stage loading) can be calculated by using the capacitive-reactance equation,

$$X_c = \frac{1}{2\pi fc}$$

solving for f gives

$$f = \frac{1}{2\pi CX_c}$$

or

$$f = \frac{0.159}{CX_c}$$

Substituting $C = 10\ \mu\text{F}$ or 10×10^{-6} F, and $X_c = 5\ \text{k}\Omega$ or 5×10^3 gives

$$f = \frac{0.159}{10 \times 10^{-6} \times 5 \times 10^3}$$

$$= \frac{159}{50}$$

$$= 3.18\ \text{Hz}$$

The cutoff frequency lies well below the lowest audible frequency.

A similar calculation could be carried out with coupling capacitor C_9, but as it is working into a common-emitter transistor base, the effective resistance is likely to be at least as high or higher than that of the volume control *VR*. If the resistance is indeed higher, the cutoff frequency is even lower than 3.18 Hz, and good low-frequency response can be expected from the interstage coupling network.

D-4 THERMAL-RUNAWAY PROTECTION AND EMITTER-BYPASS CAPACITOR

Emitter resistor R_{14} protects against thermal runaway and gives a degree of temperature stabilization. It is bypassed with a 10-μF capacitor. When $X_{C_{10}} = 470\ \Omega$, half the signal current flows through R_{14} and half through C_{10}. This occurs at a frequency of

$$f = \frac{0.159}{10 \times 10^{-6} \times 470}$$

$$= \frac{1590}{47}$$

$$= 33.9 \text{ Hz}$$

Bypass action is effective at all but the lowest audio frequencies (considering the audio range to be 20 Hz to 20 kHz).

D-5 INPUT-STAGE BIASING

Base-bias current for the first stage is provided by R_{12}, which is returned to the collector rather than the supply voltage. Returning the resistor to the collector introduces some negative-feedback voltage from the output to the input of the stage. The dc bias can be calculated by applying Ohm's law. Reference to Fig. D-1 shows that the voltage difference across R_{12} is

$$1.2 \text{ V} - 0.62 \text{ V} = 0.58 \text{ V}$$

Applying Ohm's law gives a base current of

$$I = \frac{E}{R}$$

$$= \frac{0.58}{100\ 000}$$

$$= 5.8\ \mu\text{A}$$

The low quiescent base current results in better battery life, yet the output from this stage is sufficient to drive the next stage to full volume.

D-6 INTERSTAGE COUPLING

It is more difficult to assess the cutoff frequency for the interstage coupling network. Capacitor C_{11} is working into R_{16}, R_{15}, and Q_6 base resistance in parallel (insofar as signal voltage is concerned). Ignoring base resistance, the two resistors appear as an equivalent resistance of

$$R_T = \frac{R_{16} \times R_{15}}{R_{16} + R_{15}}$$

$$= \frac{6.8 \times 3}{6.8 + 3}\ \text{k}\Omega$$

$$= \frac{20.4}{9.8}\ \text{k}\Omega$$

$$= 2.04\ \text{k}\Omega$$

Assuming a base resistance of approximately the same magnitude gives a total parallel resistance in the neighborhood of 1000 Ω. A calculation will show that with this value and C_{11}, the cutoff frequency is well within the lower audio limit.

D-7 SECOND-STAGE BIASING

Quiescent base current for Q_6 can be calculated from data on the schematic diagram. Current through R_{15} is, by Ohm's law,

$$I_{R15} = \frac{E_{R15}}{R_{15}}$$

$$= \frac{1.07}{3 \times 10^3}\ \text{A}$$

$$I_{R_{15}} = \left(\frac{1.07}{3 \times 10^3}\right) \times 10^3 \text{ mA}$$
$$= 0.356 \text{ mA}$$

Also, current through R_{16} can be found by Ohm's law

$$I_{R_{16}} = \frac{IR \text{ drop across } R_{16}}{\text{resistance of } R_{16}}$$
$$= \frac{3.6 - 1.07}{6.8 \times 10^3} \text{ A}$$
$$= \left(\frac{2.53}{6.8 \times 10^3}\right) \times 10^3 \text{ mA}$$
$$= 0.372 \text{ mA}$$

Since 0.372 mA of current comes down from R_{16} and only 0.356 enters R_{15} by Kirchhoff's current law, the remainder must be base current. This is

$$0.372 - 0.356 = 0.016 \text{ mA or } 16\ \mu\text{A}$$

D-8 NEGATIVE FEEDBACK

Two sources of negative feedback are present in the second stage. Capacitor C_{12} gives voltage feedback from the collector, but only at higher frequencies when capacitive reactance is low. In this way some bass boost is achieved.

The second source of negative feedback comes from outside the stage. Feedback voltage is taken from the secondary of the output transformer; i.e., a portion of the signal at the speaker is fed back across R_{18}. In this way, distortion introduced by the driver stage, the driver transformer (T_4), the power-output amplifiers, and the output transformer is minimized.

Vacuum-tube circuits, referred to in Chap. 6, are shown in Figs. D-2 and D-3.

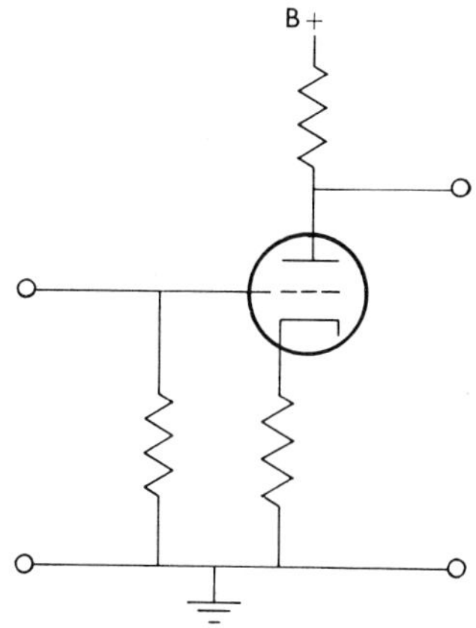

Fig. D-2 Negative feedback achieved with an unbypassed cathode resistor.

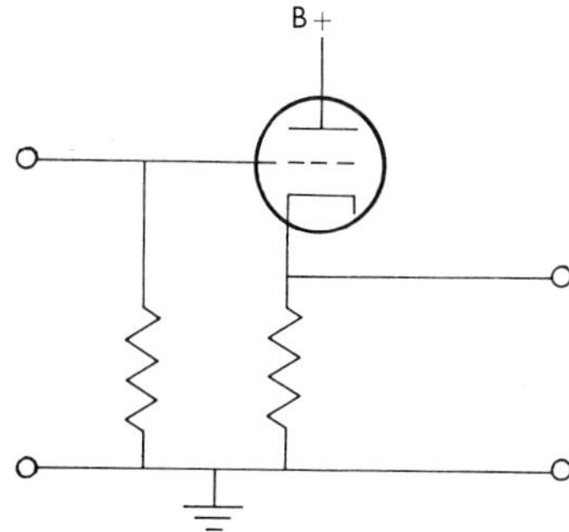

Fig. D-3 Cathode follower.

E Vacuum-tube Power Amplifier

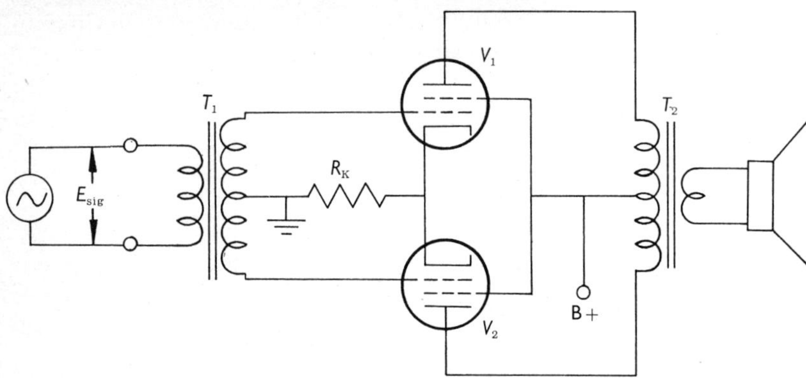

Fig. E-1 Tetrode push-pull power amplifier.

F Transformerless Power Amplifier

F-1 OUTPUT STAGE

Figure F-1 is a schematic diagram of the left-channel power amplifier of an Admiral model 22D5X stereo system (complete circuit shown in Fig. I-1, Appendix I). The right-channel amplifier is identical and the following discussion applies to both.

Transistors Q_{21} and Q_{22} form a totem-pole OTL circuit which drives a speaker through output capacitor C_{105}. Various points on the diagram are labeled with voltages as measured from ground. Reference to the diagram shows that both output transistors operate with approximately 0.7 V of forward bias.

F-2 DRIVER STAGE

The output transistors are driven by a complementary-symmetry-type phase splitter, Q_{17} and Q_{18}. Bias for the phase splitter is developed by transistor Q_{24} in a manner discussed in Sec. 8-14. Since the phase splitter is directly (dc) coupled to the output stage, both phase splitter and output-stage bias are adjusted by varying R_{95}. The manufacturer recommends a bias setting which results in 25 mA quiescent current in the output transistors to minimize cross-over distortion.

F-3 NEGATIVE FEEDBACK

Note that bias for the predriver transistor Q_{14} is obtained via R_{85}. This resistor is tied to the output connection to obtain negative feedback. Every stage in the amplifier is included in the negative-feedback loop because a sample of the signal being applied to the load (speaker) is fed back to the base of the input transistor. Resistor R_{85} supplies not only bias current to the base of Q_{14} but negative-feedback signal as well.

To show that negative feedback does in fact occur, it is necessary to follow signal polarity through the amplifier. Assume the base of Q_{14} is momentarily rising (positive-going signal). Its collector is then negative-going, and this signal is applied to Q_{18}. The output from Q_{18} is taken from the emitter and no phase inversion takes place. Output transistor Q_{22} then receives a negative-going signal. It operates as an emitter follower and develops a negative signal to the output. The negative-going signal developed at the output is applied via R_{85} to base of the input transistor to cancel some of the positive-going input signal.

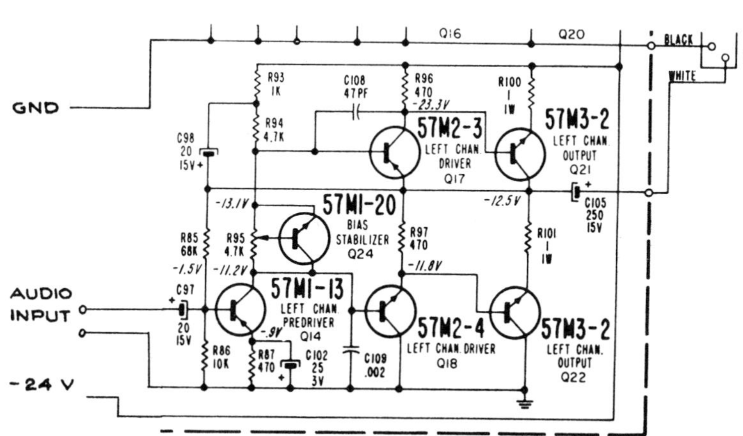

Fig. F-1 Left-channel output transformerless (OTL) power amplifier in Admiral 22D5X AM-FM stereo set. *(Courtesy Canadian Admiral Corp. Ltd.)*

G Typical *CR* Test Oscillator

G-1 GENERAL DESCRIPTION

Oscillators with *CR* frequency-determining networks are found in a variety of electronic devices. Probably electronic organs and test instruments are the most commonly encountered devices with *CR* oscillators.

Figure G-1 (page 293) shows a schematic diagram of a widely used *CR* test oscillator available in kit form. Its principle of operation is similar to that described in Sec. 9-16.

Transistors Q_2, Q_3, Q_4, and Q_5 constitute the amplifier section. Note that Q_3 drives a complementary-symmetry totem-pole-type output stage (see Sec. 8-4). This type of output stage has good drive capability on both positive and negative signal excursions, thereby maintaining the spectral purity of the output. It is also able to provide the necessary amplitude to drive three loads: (1) the square-wave generator (Q_6, Q_7, and Q_8), (2) the output-attenuation section and level meter connected via C_6, and (3) the feedback loop.

G-2 POSITIVE FEEDBACK

Positive feedback takes place as follows. Assume a positive-going signal at the output (junction of R_{13} and R_{14}). Follow it through lamp L_1 (where it receives some automatic amplitude control) to the slider on potentiometer R_7. One side of R_7 is connected to the base of the input transistor Q_2 via a 10-kΩ resistor (R_6). The other end of R_7 is in effect grounded for signals via capacitor C_6. The amount of positive feedback can be adjusted by the potentiometer. With the slider (labeled point 2) contacting point 1, maximum positive feedback occurs, whereas if it contacts point 3 all positive feedback is shunted to ground via C_5. Upon arrival at the base of Q_2 the positive-going feedback signal is inverted and amplified at the collector. It is then applied to Q_3 as a negative-going signal. Further, amplification and inversion occurs as it passes through Q_3, thus a positive-going signal reaches the input of the totem-pole output stage. No phase inversion takes place in the output stage as it is of the emitter-follower type. The original positive-going signal is reinforced, and positive feedback does indeed take place.

G-3 NEGATIVE FEEDBACK

Negative feedback occurs at all frequencies except the one rejected by the notch filter. It is applied to the base of Q_1, which introduces degeneration across the emitter resistor (R_4) shared by Q_1 and Q_2. Note that the notch filter consists of a variety of resistors and capacitors which can be switched in to obtain many output frequencies.

A typical vacuum-tube oscillator is shown in Fig. G-2.

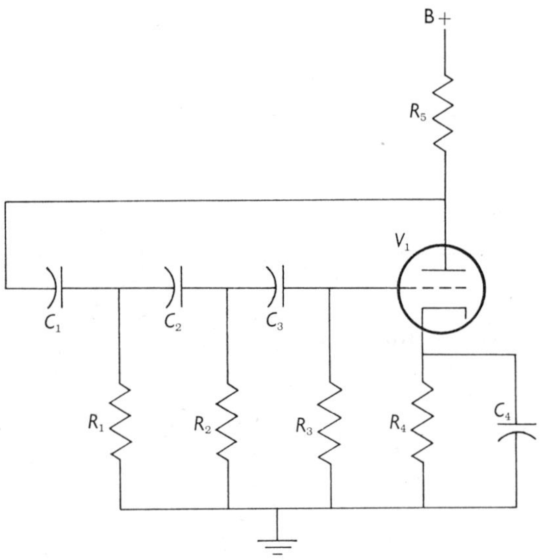

Fig. G-2 Vacuum-tube *CR* lag line oscillator.

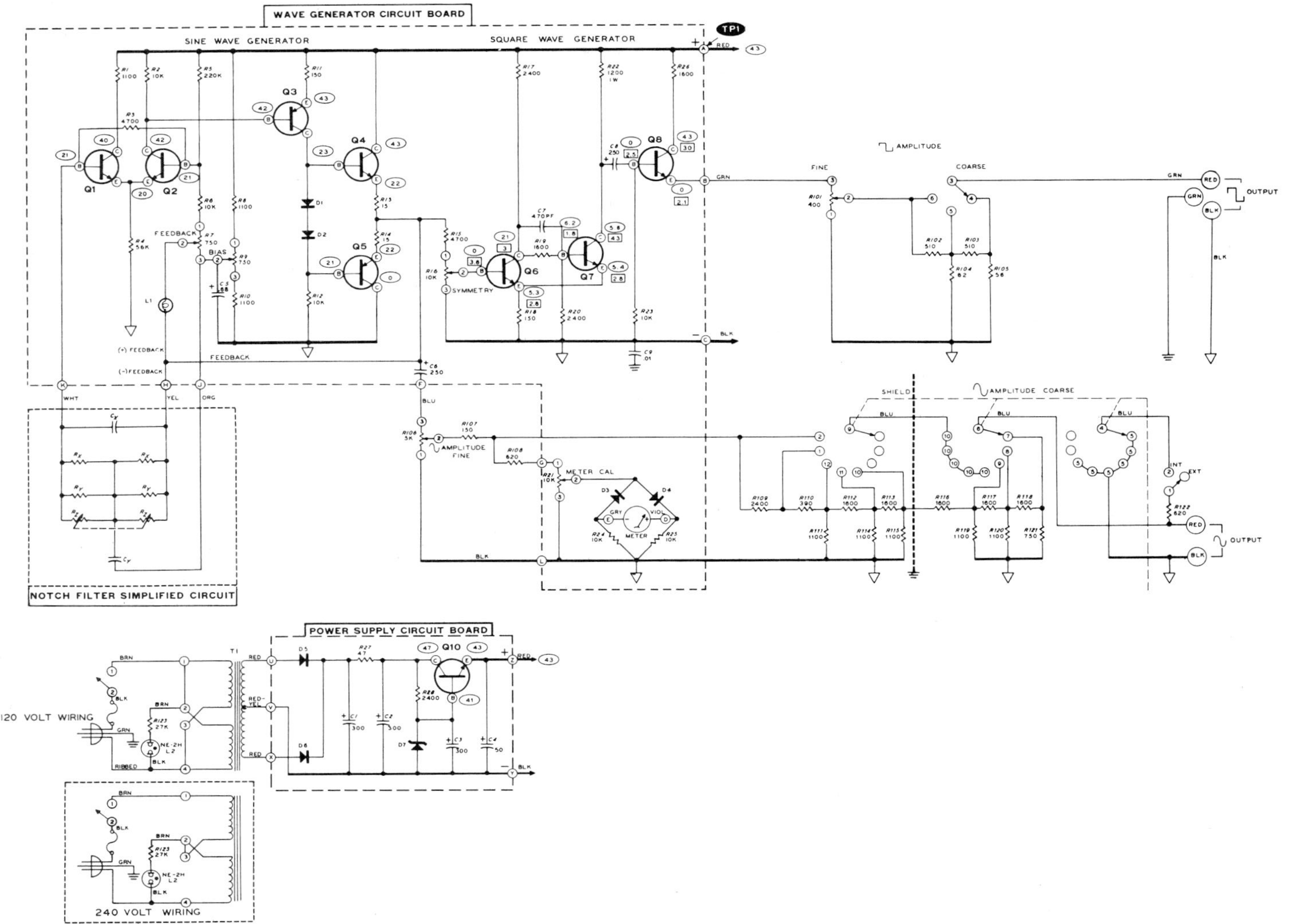

Fig. G-1 Heathkit 1G-18 test oscillator. *(Courtesy Heath Company, A Division of Schlumberger Canada Limited)*

H Vacuum-tube Oscillator Circuits

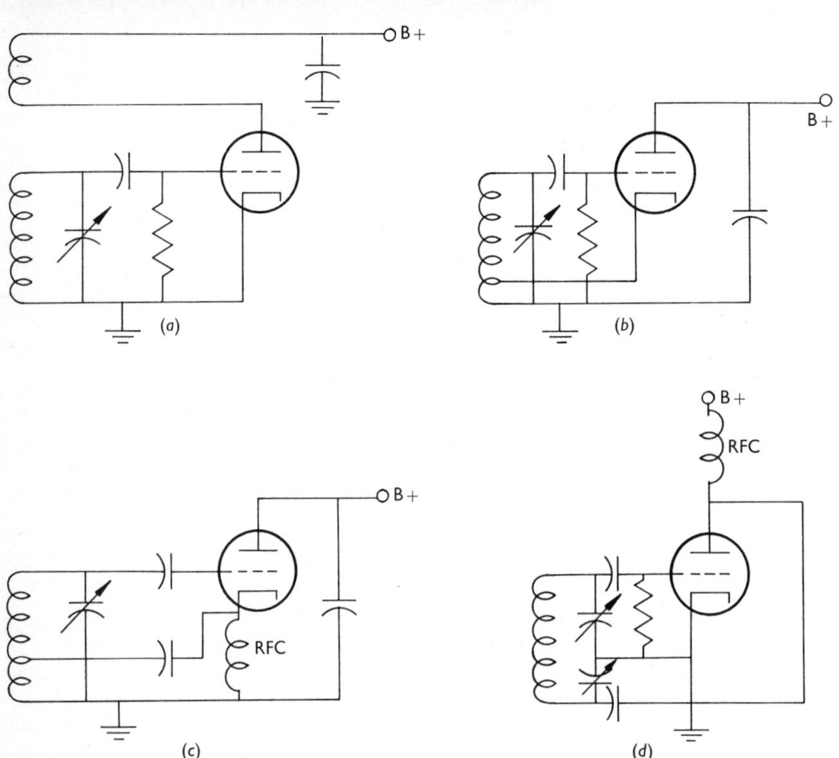

Fig. H-1 Vacuum-tube oscillator circuits: (*a*) Tickler; (*b*) Series-fed Hartley; (*c*) Shunt-fed Hartley; (*d*) Colpitts.

I A Home-Entertainment Stereo System

I-1 CIRCUIT DIAGRAM

Figure I-1 shows a schematic diagram of an AM-FM stereo unit manufactured by Canadian Admiral Corporation. The unit contains the receiver and the audio amplifier sections on two separate printed circuit boards.

I-2 AM CONVERTER SECTION

Transistor Q_7 functions as an autodyne converter for the AM broadcast band. Radio signals are picked up and applied to the base of Q_7 by the ferrite rod antenna, L_7. L_{10} is the local oscillator coil. The IF signal is applied via IF transformer T_5 and capacitor C_{30} to the base of Q_5 (through T_2 an FM IF transformer).

I-3 FM TUNER SECTION

Transistor Q_1 amplifies the FM signal induced in the FM antenna and passes it via C_6 and C_{12} to the base of Q_2, the FM mixer or converter. Local oscillations are produced by transistor Q_3. Its power-supply voltage is stabilized by zener diode CR_2 to minimize frequency deviation due to line-voltage variations. Further frequency control is provided if desired by a frequency-control voltage developed in the ratio detector and applied to varactor diode CR_1. The AFC switch makes it possible to disable this voltage if automatic frequency control is not desired, and is used mainly to disable AFC when tuning in stations. Once tuned in, AFC is switched on to maintain optimum reception. Local oscillator voltage is applied to the converter (Q_2) via capacitor C_{20}. The FM IF signal is applied to the base of transistor Q_4 via IF transformer T_1.

I-4 COMMON AM-FM IF AMPLIFIER SECTION

Three stages of FM IF amplification are provided by transistors Q_4, Q_5, and Q_6. IF transformers T_2 and T_3 provide interstage coupling for the FM IF signals. Only two stages are required for the AM IF section. Transistors Q_5 and Q_6 are the first and second AM IF amplifiers, and transformer T_6 provides the interstage coupling.

I-5 AM DETECTOR

The AM IF signal is applied to the AM diode detector (CR_5) via transformer T_7. AM AGC voltage is developed across capacitor C_{42}. Audio signal voltage is developed across R_{46} and is passed to pin 5 of the *FM-AM switch* S_4.

I-6 FM RATIO DETECTOR

Diodes CR_3 and CR_4 comprise the ratio detector. The ratio detector supplies AFC voltage (filtered by R_{41} and C_{48}) to the local oscillator via L_{16}. FM AGC voltage is not taken from the ratio detector but is developed by rectifying some of the FM IF signal with diode CR_{13}.

I-7 AM AUDIO SECTION

The *FM-AM switch* S_4 routes either the AM or FM audio signal to its own pin 3 from where it is applied to the base of transistor Q_8 via L_{13}, L_{12}, and C_{65}. Coils L_{13} and L_{12} perform no useful function for AM audio signals and are used only in connection with FM multiplex stereo detection. Transistor Q_8 acts as an emitter-follower amplifier for AM audio signals which are developed across its emitter resistor (R_{52}). Capacitor C_{67} couples the AM audio signal to pin 6 of the *FM stereo switch* S_3 where it is transferred to pin 3, then through R_{48} to pins 1 and 2 and internally to pins 4 and 5. Capacitors C_{59} and C_{60} next pass the AM audio signal to pins 11 and 12 of the *radio-phono switch* (S_2), where in the radio position it is applied internally to pins 7 and 8. These are connected directly to the inputs of the right- and left-channel audio amplifier loudness controls R_{66A} and R_{66B}.

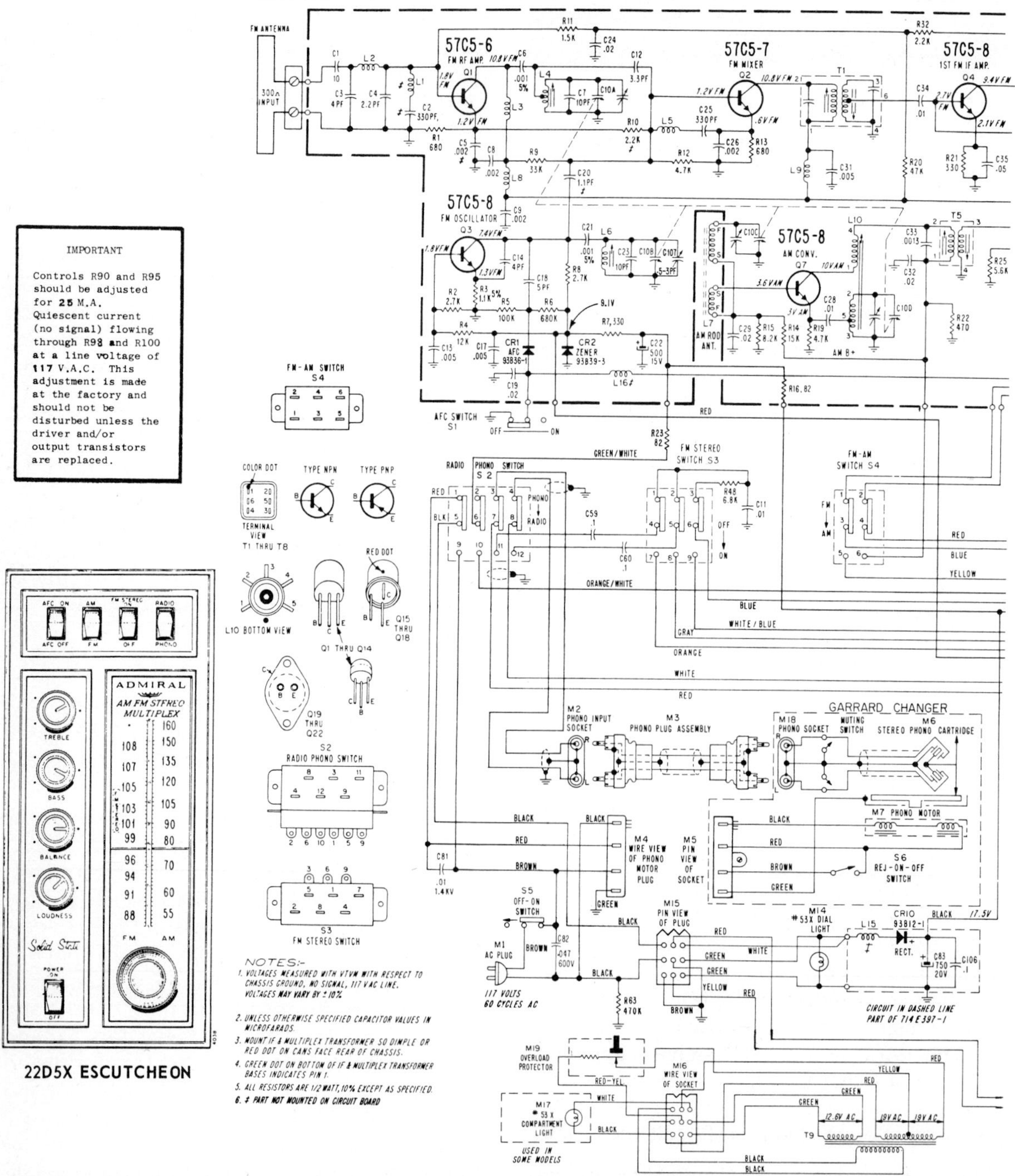

Fig. I-1 Schematic diagram of an AM-FM stereo radio and stereo phonograph. (*Courtesy Canadian Admiral Corp. Ltd.*)

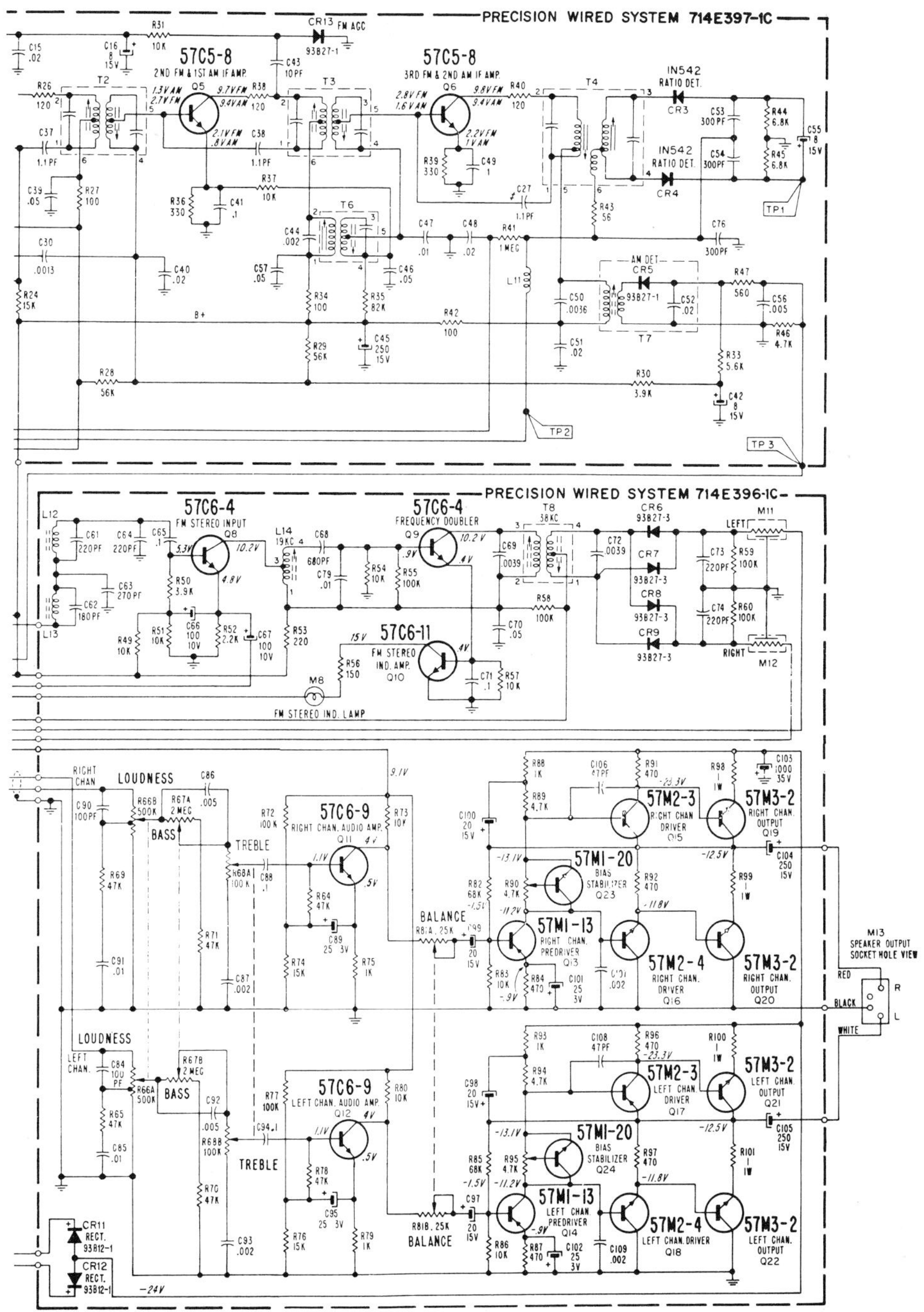

PRECISION WIRED SYSTEM 714E397-1C
CR13 FM AGC
57C5-8
2ND FM & 1ST AM IF AMP.
Q5
57C5-8
3RD FM & 2ND AM IF AMP.
Q6
T2
T3
T4
T6
T7
IN542 RATIO DET.
CR3
CR4
AM DET.
CR5
B+
TP1
TP2
TP3
PRECISION WIRED SYSTEM 714E396-1C
57C6-4
FM STEREO INPUT
Q8
57C6-4
FREQUENCY DOUBLER
Q9
T8
38KC
L14
19KC
57C6-11
FM STEREO IND. AMP.
Q10
M8
FM STEREO IND. LAMP
LEFT
RIGHT
M11
M12
RIGHT CHAN.
LOUDNESS
BASS
TREBLE
57C6-9
RIGHT CHAN. AUDIO AMP.
Q11
BALANCE
57M1-13
RIGHT CHAN. PREDRIVER
Q13
57M1-20
BIAS STABILIZER
Q23
57M2-3
RIGHT CHAN DRIVER
Q15
57M2-4
RIGHT CHAN. DRIVER
Q16
57M3-2
RIGHT CHAN. OUTPUT
Q19
57M3-2
RIGHT CHAN. OUTPUT
Q20
M13
SPEAKER OUTPUT SOCKET HOLE VIEW
RED
BLACK
WHITE
R
L
LOUDNESS
LEFT CHAN.
BASS
TREBLE
57C6-9
LEFT CHAN. AUDIO AMP.
Q12
BALANCE
57M1-13
LEFT CHAN. PREDRIVER
Q14
57M1-20
BIAS STABILIZER
Q24
57M2-3
LEFT CHAN. DRIVER
Q17
57M2-4
LEFT CHAN. DRIVER
Q18
57M3-2
LEFT CHAN. OUTPUT
Q21
57M3-2
LEFT CHAN. OUTPUT
Q22
CR11 RECT. 93B12-1
CR12 RECT. 93B12-1
-24V

I-8 FM AUDIO SIGNAL

When ordinary FM reception is taking place (not FM stereo multiplex), the FM audio signal from pin 3 of S_4 takes the same route as AM audio. The *FM-AM switch* (S_4) also disables either the AM or the FM converter by removing collector-supply voltage from one or the other.

I-9 PHONO SIGNAL

When using the stereo recordplayer the left- and right-channel signal from the cartridge is applied via the phono plug assembly (M_3) to pins 3 and 4 of the *radio-phono switch*. The remainder of the route (from pins 7 and 8) is the same as for AM or FM audio signals.

I-10 AUDIO OUTPUT SECTION

From the loudness control the audio signals for each channel are passed through bass and treble controls (R_{67A}, R_{68A} for the right channel and R_{67B}, R_{68B} for the left channel). Then the signals are applied to the respective channel audio amplifiers of fairly conventional design. Note that the emitter resistors are not bypassed to ground with capacitors C_{89} and C_{95}. This is done deliberately to increase the input impedance of the amplifiers to match the impedance of the phono cartridge.

The outputs of the audio amplifiers are connected to the inputs of their respective power amplifiers. The operation of the power amplifiers is described in Appendix D.

I-11 FM STEREO BROADCASTING

Figure I-2 shows a block diagram of an FM system designed for multiplex stereo transmission. The left (L) and right (R) audio signals are summed by resistive networks to produce two composite audio signals, the L + R signal and the L − R. At the receiving end these composite audio signals are combined to recover the individual L and R audio signals. The system is monophonic compatible, because a listener with only a monophonic receiver will hear only the L + R signal. This is essentially a monophonic signal produced by two microphones. The system can also transmit monophonic programs by feeding a monophonic signal to both the L and R channels. After summation L − R will always be zero (because L and R are identical during a monophonic transmission). The L − R signal modulates a 38-kHz subcarrier, but the subcarrier itself is suppressed and only sidebands leave the modulator. A zero L − R signal will result in no L − R sidebands being added to the transmitter composite audio bus.

The composite audio is a scrambled "mess" of L + R audio, L − R sidebands of a 38-kHz subcarrier, a 19-kHz pilot tone and possibly music on the auxiliary channel for use by shopping centers, etc., on a subscription basis. An FM receiver must sort out these signals and retain only those it requires.

Since only the sidebands of the L − R signal are transmitted, the 38-kHz subcarrier must be reconstructed at the receiver. For this reason the 19-kHz pilot tone is transmitted to synchronize the subcarrier oscillator in the multiplex receiver. A 19-kHz tone is beyond the range of hearing for most people, and in any case it is transmitted at reduced amplitude so that it causes no interference for monophonic listeners.

I-12 FM MULTIPLEX STEREO RECEPTION

Figure I-3 shows the major functional blocks in a multiplex stereo receiving system. An ordinary FM receiver is used to recover the composite audio signal. For good stereo separation the receiver must have a bandpass of at least 150 kHz so that the upper sidebands of the L − R signal are not attenuated. [The upper sideband could have a frequency as high as 38 kHz (subcarrier) + 15 kHz (audio) = 53 kHz.]

Composite audio frequencies above 58 kHz (the auxiliary music subcarrier at 67 kHz and its sidebands) are filtered out. Then the 19-kHz pilot tone is used to generate the 38-kHz subcarrier. In some circuits a local oscillator

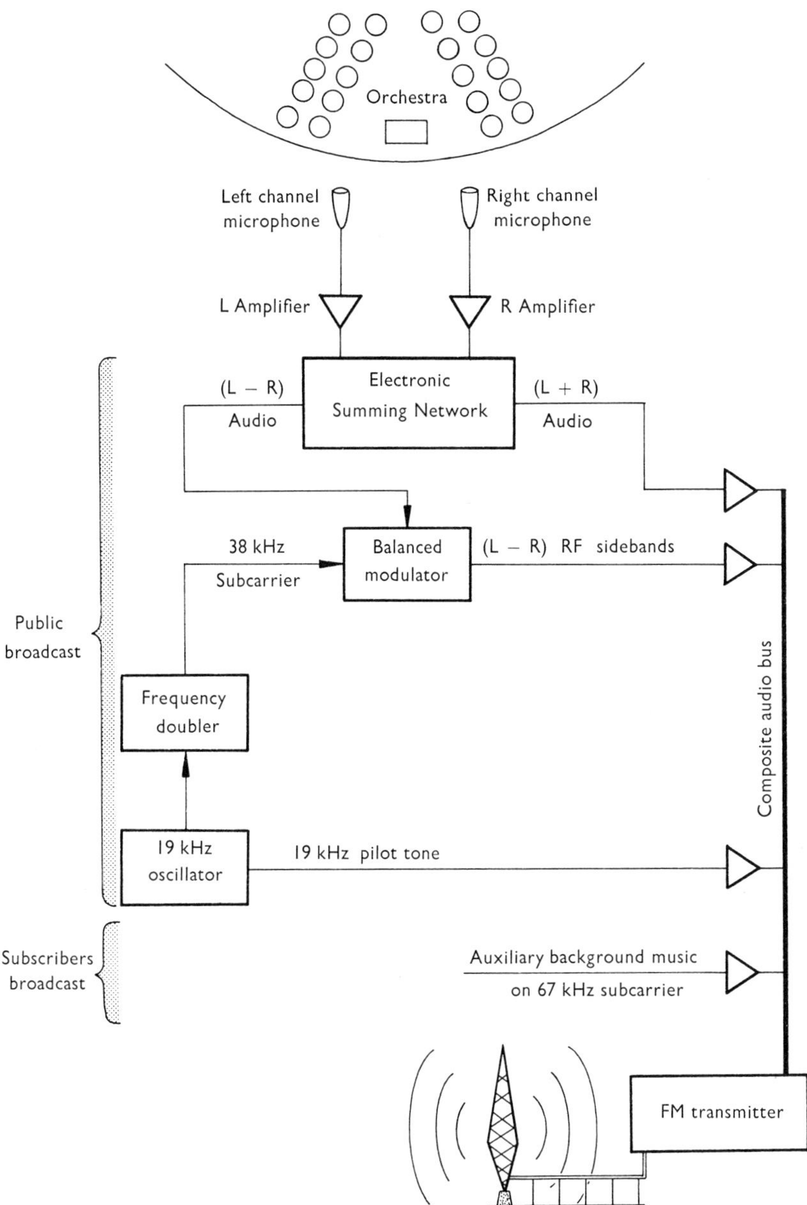

Fig. I-2 Functional blocks of an FM broadcast transmission system with multiplex stereo capability.

is phase-locked with the 19-kHz tone. In others the tone itself is amplified and passed to a frequency doubler to produce the 38-kHz subcarrier.

The 38-kHz subcarrier and the composite audio signal can be fed to a detector from which the L − R signal can be recovered and added to the L + R signal in correct proportions and phase to produce the L and R audio signals. A more commonly used recovery method is to employ the 38-kHz subcarrier as a switching signal to time-decode the L − R, L + R mixture of signals. The latter method is known as *synchronous detection*.

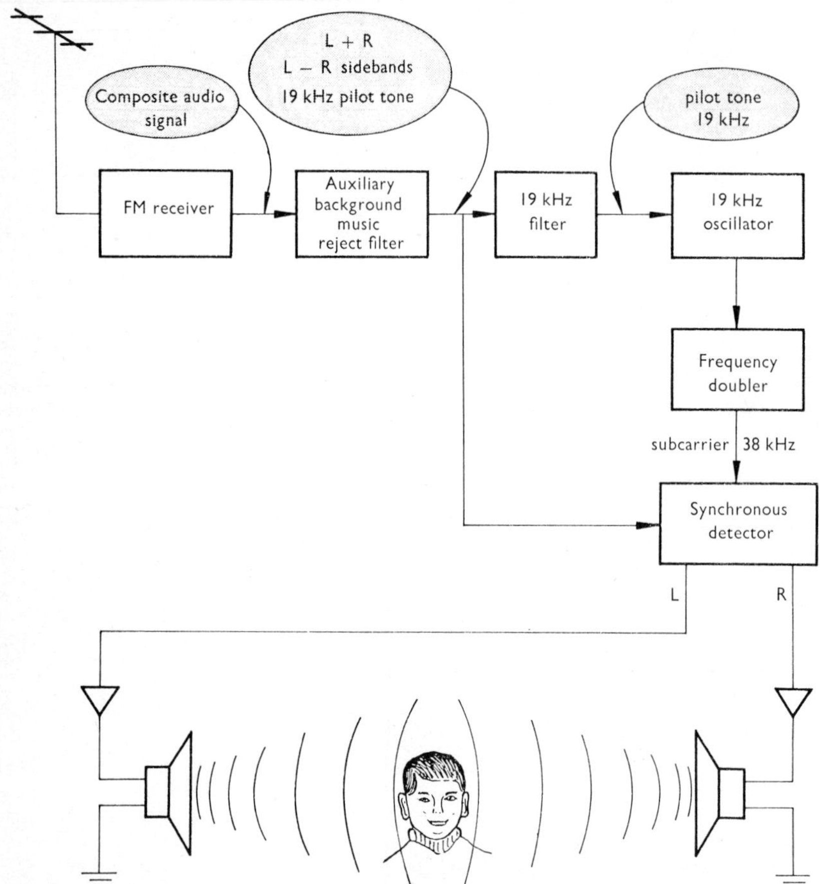

Fig. I-3 Functional blocks of an FM stereo multiplex receiving system.

I-13 SYNCHRONOUS DIODE DETECTOR

A circuit diagram of a popular FM stereo multiplex synchronous detector is shown in Fig. I-4. In this circuit the 38-kHz subcarrier is used as switching signal. Each time the 38-kHz signal changes polarity, one set of diodes, either CR_1 and CR_2 or CR_3 and CR_4, is biased into conduction. Whenever a pair of diodes is "switched" on in this manner, a circuit path is provided between the composite audio signal and either the left- or right-channel output load resistor (R_1 or R_2). This fact is shown in Fig. I-4b, c, d, and e. At precisely the time that CR_1 and CR_2 are switched on, left-channel audio is present in the composite audio signal. Left-channel audio is then impressed across R_1 as shown in Fig. I-4b and c. When the polarity of the 38-kHz subcarrier reverses, diodes CR_3 and CR_4 conduct and allow right-channel audio present in the composite audio signal to reach R_2.

I-14 FM MULTIPLEX DECODER

The FM stereo multiplex decoder in Fig. I-1 consists of transistors Q_8, Q_9, and diodes CR_6, CR_7, CR_8, and CR_9. Composite audio arrives from pin 3 of the *FM-AM switch* S_4 and is filtered by L_{13}, C_{62}, L_{12}, C_{61}, C_{63}, C_{64} to remove components above 53 kHz and is applied via C_{65} to the base of Q_8. The collector of Q_8 drives a 19-kHz tank circuit which provides the base of Q_9 with a 19-kHz pilot signal whenever a stereo broadcast is tuned in. Transistor Q_9 is heavily

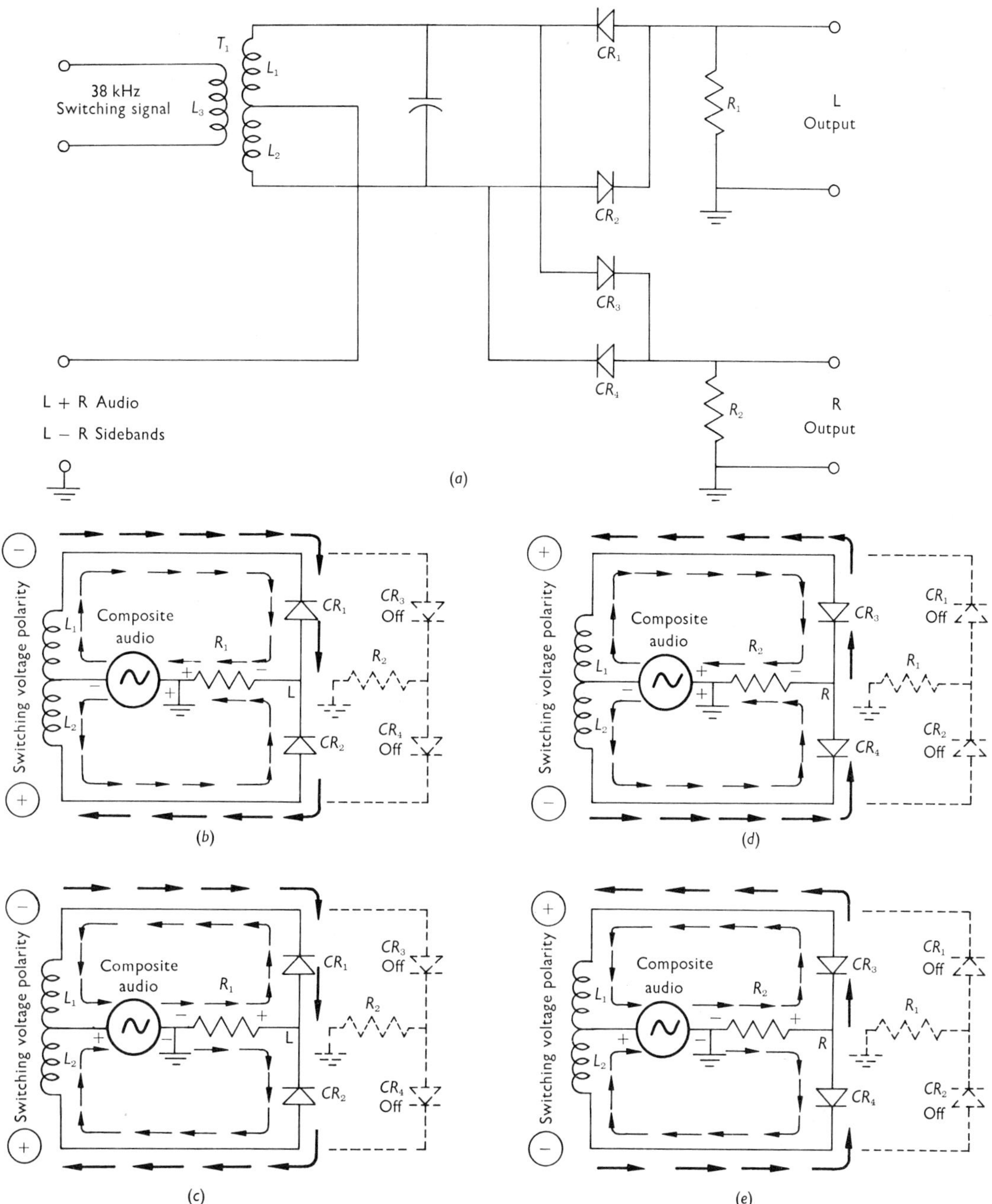

Fig. I-4 FM stereo multiplex synchronous diode detector: (*a*) Schematic diagram; (*b*) Simplified diagram, left channel audio recovery when composite audio signal is instantaneously negative; (*c*) Left channel audio recovery, composite audio signal is positive; (*d*) Right channel audio recovery, composite audio negative; (*e*) Right channel, composite audio positive.

driven and produces a strong second harmonic in its collector circuit. Transformer T_8 is tuned to the second harmonic of 19 kHz (38 kHz). In this way the 38-kHz subcarrier is produced directly from the transmitted pilot tone. At pins 1 and 4 of transformer T_8 a 38-kHz switching signal is available for driving the synchronous detector diodes CR_6, CR_7, CR_8, and CR_9.

Composite audio is taken from the emitter output of Q_8 via C_{67} and after passing through the *FM stereo switch* (pins 6 and 9) is applied to the secondary center tap of T_8. From the center tap of T_8 the audio is routed to R_{59} (left channel) or R_{60} (right channel) under the control of the switching diodes and the 38-kHz subcarrier. Switching transients are filtered out by M_{11} and M_{12}. The two audio signals then go to the *radio-phono switch* for distribution to the audio amplifiers.

Transistor Q_{10} is turned on by base current supplied from Q_9. This occurs only when a 19-kHz pilot tone is present, i.e., whenever a stereo broadcast is being received. When Q_{10} turns on, its collector current lights the FM stereo indicator lamp M_8 to show the operator that the FM station is broadcasting stereo.

J Vacuum-tube Circuits

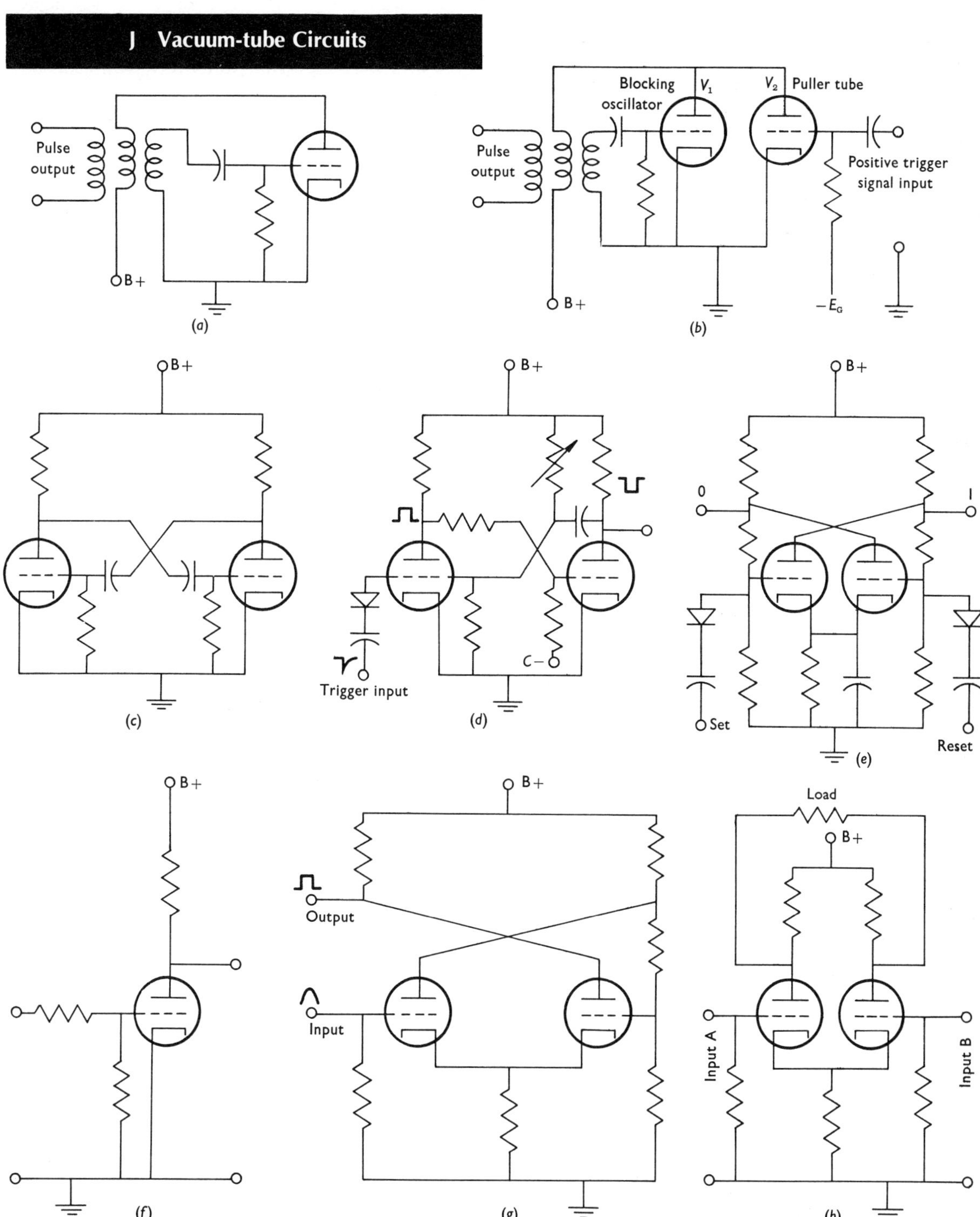

Fig. J-1 Vacuum-tube circuits: (*a*) Free-running blocking oscillator; (*b*) Triggered blocking oscillator; (*c*) Free-running multivibrator; (*d*) Monostable multivibrator; (*e*) Bistable multivibrator or flip-flop; (*f*) Limiter; (*g*) Schmitt trigger; (*h*) Differential amplifier.

K CMOS and I^2L

K-1 PROPERTIES

Complementary-symmetry metal-oxide semiconductor (CMOS) logic gates are noted for their low power requirement. This property makes them useful in battery-operated devices such as electronic wrist watches, electronic calculators, etc. Their low-power dissipation makes possible high-density packaging and large-scale integration (LSI circuits).

K-2 METAL-OXIDE FIELD-EFFECT TRANSISTOR (MOSFET)

Figure K-1 shows the structure and properties of both N- and P-channel enhancement-type MOSFET devices. In each device, a gate voltage applied with the correct polarity with respect to the source causes source-to-drain current flow in the channel beneath it.

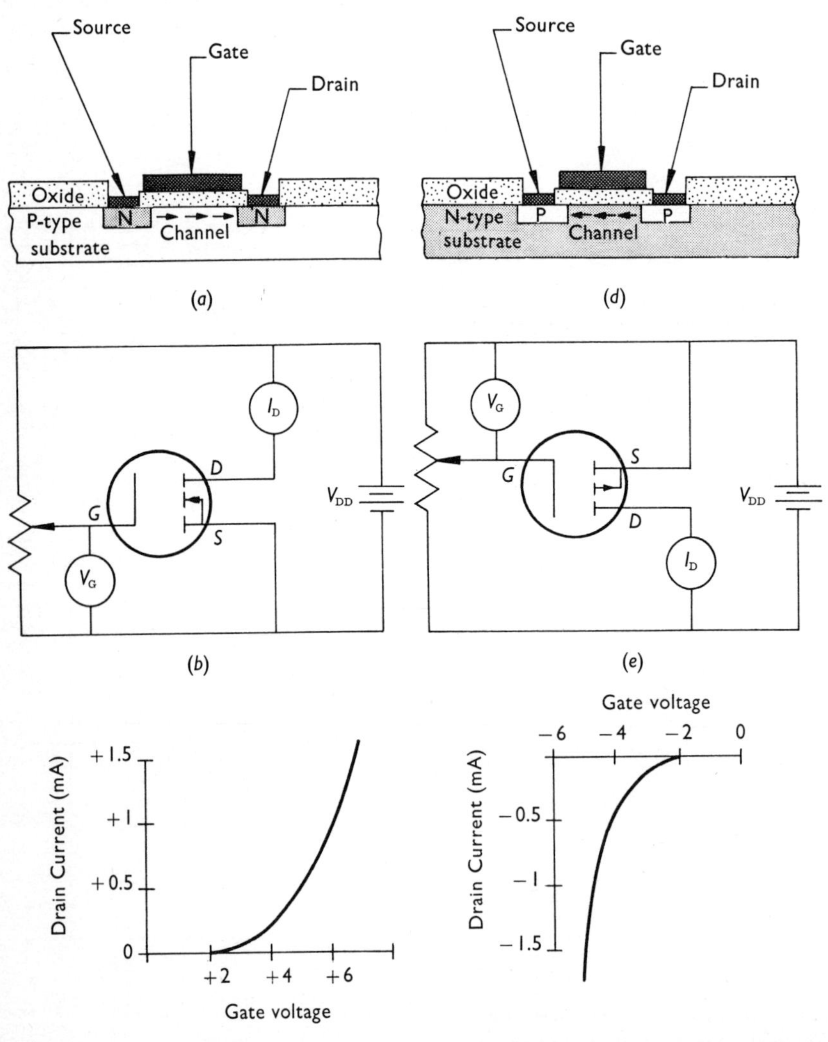

Fig. K-1 Structure and electrical properties of enhancement-type MOSFET: (*a*) N-channel device; (*b*) Test circuit to show that positive (with respect to source) gate voltage enhances drain current; (*c*) Drain current-gate voltage characteristic curve; (*d*) P-channel device; (*e*) Test circuit; (*f*) Characteristic curve.

K-3 CMOS INVERTER CIRCUIT

A basic complementary-symmetry inverter circuit is constructed as shown in Fig. K-2. CMOS inverters are fabricated as integrated circuits on a single chip.

If the gate electrodes (Fig. K-2) are connected to a high (H) the N-channel unit is turned on and the P-channel unit is turned off. A low (L) appears at the output.

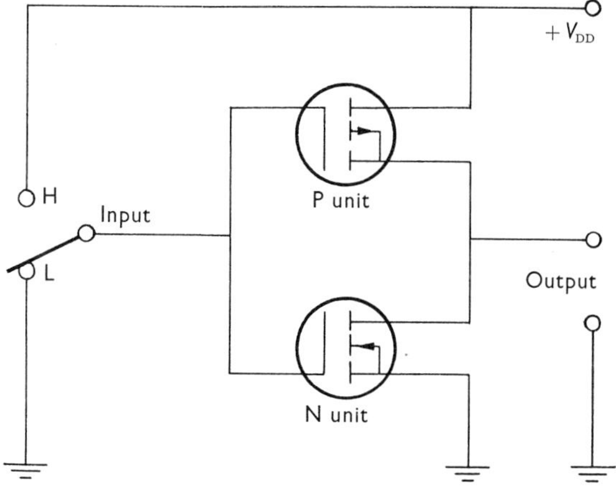

Fig. K-2 Basic CMOS inverter circuit.

If the gate electrodes are connected to a low (L) the N-channel unit is turned off and the P-channel unit is turned on. A high (H) appears at the output.

A CMOS circuit can operate satisfactorily with drain voltages (V_{DD}) ranging from approximately 2 to 15 V.

K-4 CMOS GATE CIRCUIT

Figure K-3 shows how CMOS gates can be constructed. By connecting the N-channel units in parallel and P units in series, a positive NOR gate results. A high (H) on any input causes a low output. All inputs low cause a high output as shown in the truth table. When the N-channel units are series connected and the P-channel units are in parallel, a positive NAND gate is formed.

By interconnecting CMOS logic gates, it is possible to construct flip-flops, various types of

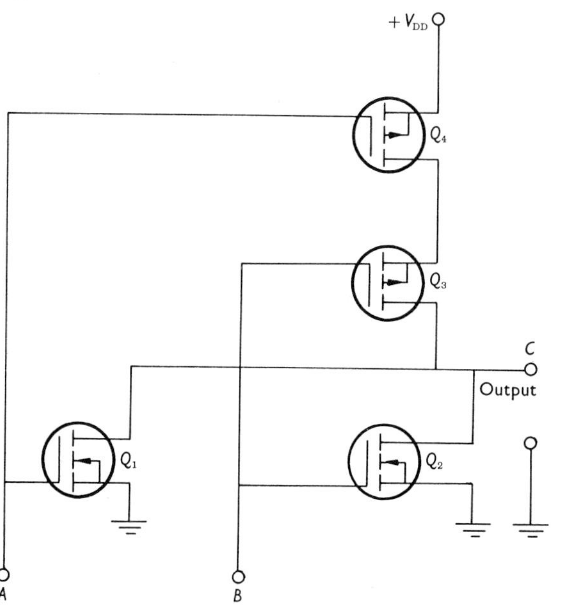

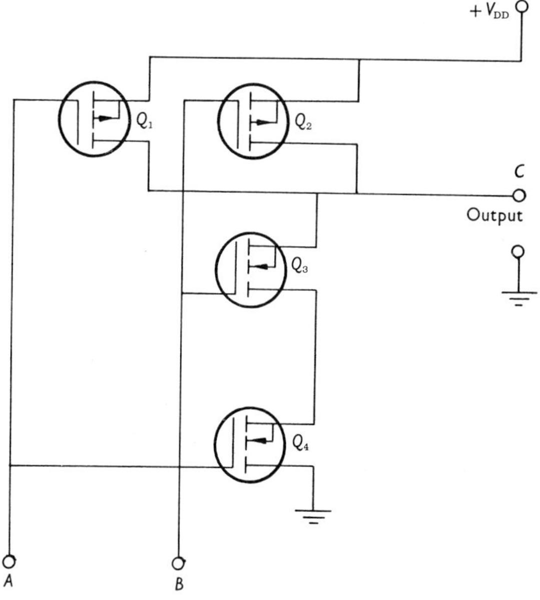

(*a*)

Fig. K-3 (pages 305–306) CMOS gate circuits: (*a*) Positive NOR and NAND schematic diagrams; (*b*) Logic symbols; (*c*) Truth tables; (*d*) Equivalent circuits using mechanical switch representation.

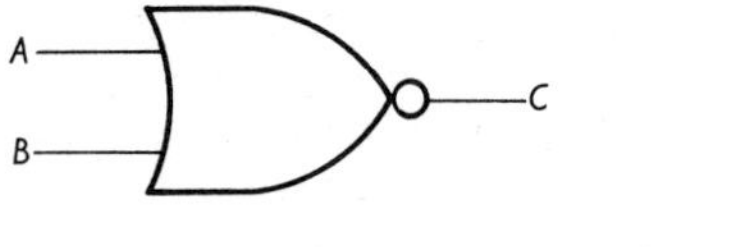

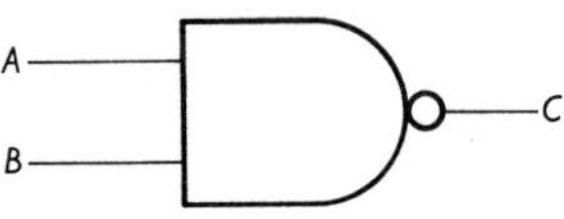

(*b*)

A	B	C
L	L	H
L	H	L
H	L	L
H	H	L

A	B	C
L	L	H
L	H	H
H	L	H
H	H	L

(*c*)

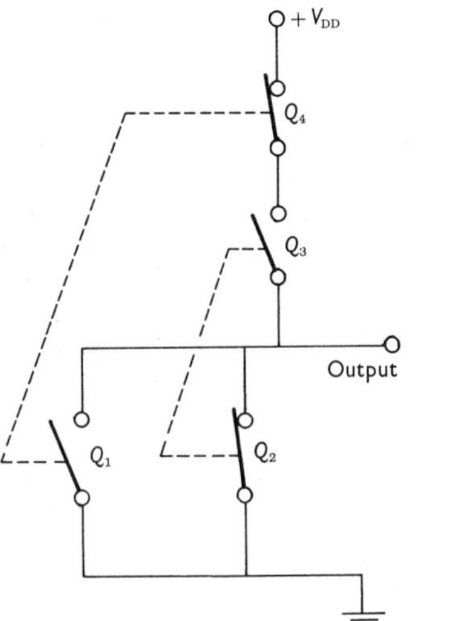

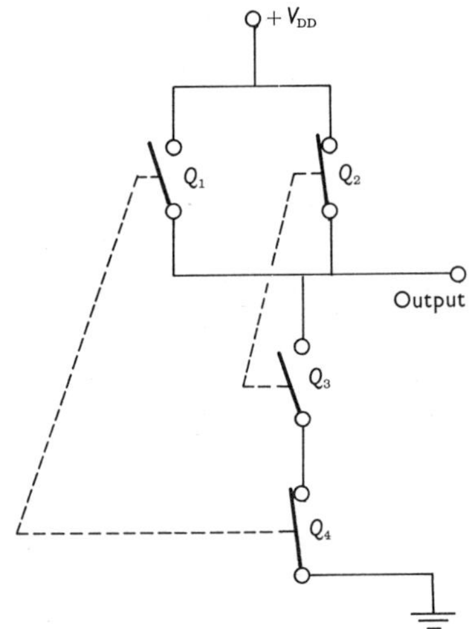

(*d*)

Fig. K-3 (*Continued.*)

registers, memory devices, and logic circuits. A CMOS circuit draws current only when it is changing level, i.e., when going from the H to L state or vice versa. This means that a single gate output can drive large numbers of inputs (high fan-out capability). These traits, high fan-out capability and low power dissipation, make it possible to construct large and complex subsystems on a single silicon chip.

K-5 FEATURES OF I^2L

Integrated injection logic (I^2L) has many of the desirable features of CMOS such as low supply voltage and high packing density, yet operates at bipolar speeds. Circuit simplicity permits relative ease of fabrication and allows large numbers of units to be placed on a single chip.

K-6 CIRCUIT CONSTRUCTION

Figure K-4 shows the circuit of a single I^2L unit. It consists of two transistors, a PNP (known as the injector) and an NPN switch. The injector supplies forward-bias current for the base of the switch. Switch transistors may be fabricated with several collectors which may be connected to other units on the same chip. The emitters of a large number of injectors can be put down as a single rail on the chip with the gates they serve arranged on either side. A cross-section diagram of such an arrangement is shown in Fig. K-5.

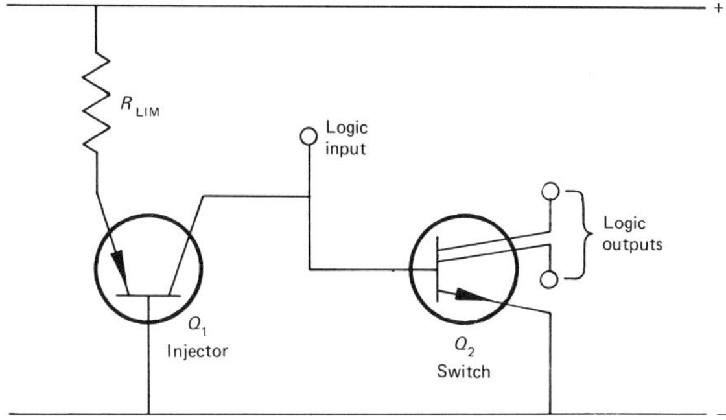

Fig. K-4 Schematic diagram of a single I^2L unit.

K-7 CIRCUIT OPERATION

Assume first that the logic input in Fig. K-4 is disconnected from any other circuits. The injector Q_1 is supplying base current to Q_2. Since Q_1 is always forward-biased, injector current flows from the positive-supply rail through R_{lim}, the emitter of Q_1, collector of Q_1 to the base of Q_2, to the emitter of Q_2 and the ground rail. (Actually electron flow follows this path in the reverse order.) Since the switch Q_2 is receiving base current, it is switched on.

To switch off Q_2, the logic input must be grounded, i.e., the base of the switch must be short-circuited or shunted. This can be achieved by connecting one of the collectors from another switch on the same chip.

A single I^2L unit can be regarded as an inverter in that a "shunt" input is changed to an "open" output, and an open to a shunt.

K-8 GATING ACTION

Figure K-6 shows one possible connection for logic gating. Two switch transistor collectors from two different units on the same chip are connected to the logic input of a third unit. To explain gate action assume the nomenclature L to signify a shunt condition and H to signify an open.

If either switch *A* or *B* receives an L at its logic input, it shunts the base of switch *C* which causes its collectors to open or to produce an H.

Listing all possible input combinations gives the truth table in Fig. K-7.

Since no resistors appear on the chip, time loss due to charging and discharging stray capacitance is minimal. Current switching is employed; therefore, voltage swings are also minimal. Supply voltages in the order of a single cell (1.5 V) may be used.

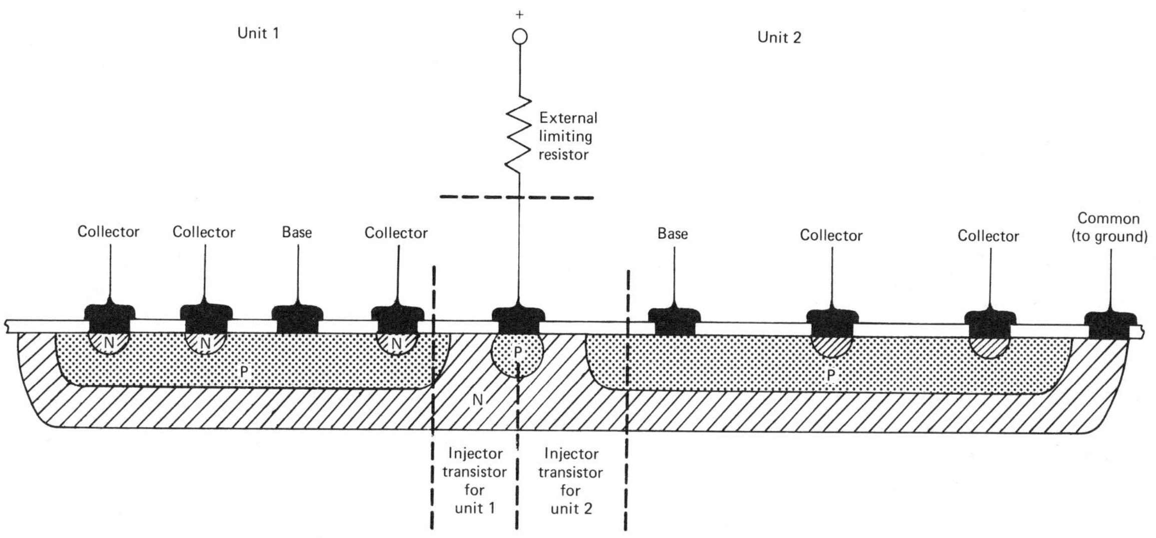

Fig. K-5 Cross-section diagram of two of many possible I^2L units on a single chip.

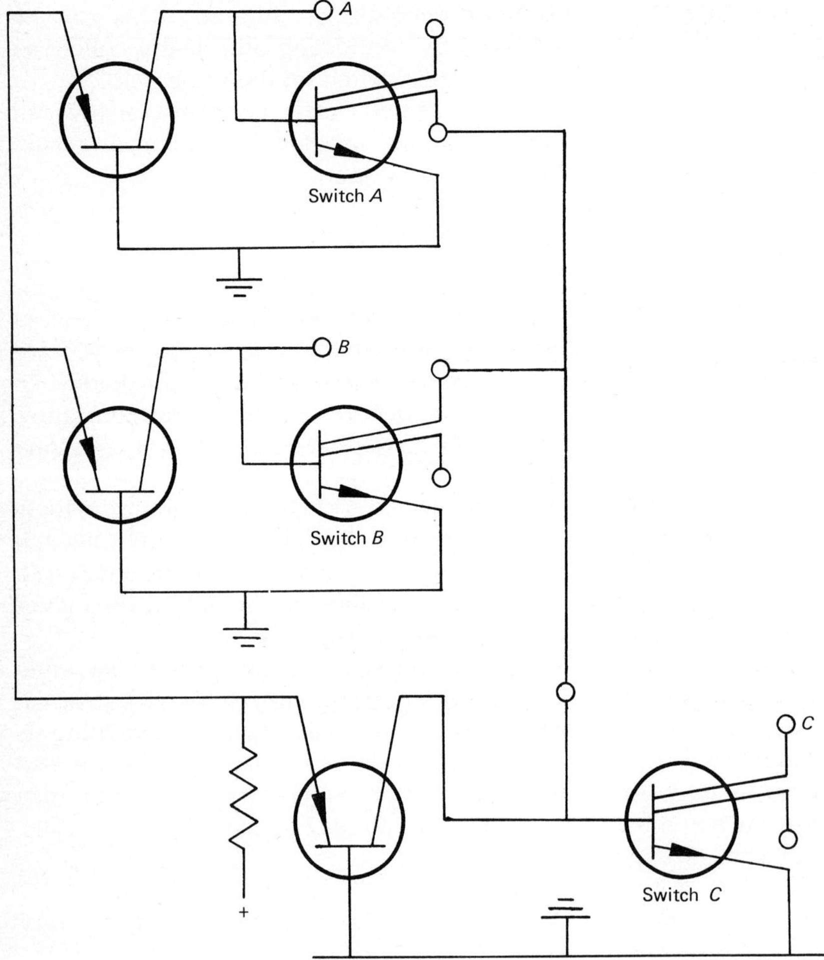

Fig. K-6 Three I^2L units connected to form a gate circuit.

Inputs A	Inputs B	Output C
L	L	H
L	H	H
H	L	H
H	H	H

Fig. K-7 Truth table and logic symbols for an I^2L gate.

L Color Television Receiver

L-1 COLOR TELEVISION CIRCUIT

A functional block diagram of a color television circuit (Fig. L-1) should be studied in conjunction with the schematic diagram for the circuit (Fig. L-2).

L-2 LOW-VOLTAGE POWER SUPPLY

Ac power is applied to the CRT filaments from transformer T_{902}. D_{602} and D_{603} rectify the applied ac which after initial ripple filtering by C_{902} emerges from series regulator Q_{903} as 115 V dc.

L-3 TUNER CIRCUITS

Figure L-3 shows the schematic diagrams of the UHF and VHF tuners. Transistor Q_{01} (Q_{104} in block diagram Fig. L-1) is the UHF local oscillator and diode D_{01} (D_{102}) is the converter. The IF is passed to the VHF tuner UHF input where it enters via C_{10}.

In the VHF tuner Q_1 (Q_{101}) is the RF amplifier, Q_2 (Q_{102}) the converter, and Q_3 (Q_{103}) the local oscillator. When switched to receive UHF (channel 1 position) Q_1 and Q_2 operate as UHF IF amplifiers, and oscillator Q_3 is disabled.

L-4 VIDEO IF AMPLIFIER AND DETECTOR

Transistors Q_{201} through Q_{203} are the video IF amplifiers. Note that gated AGC is used with the IF amplifiers, whereas the VHF tuner gain is regulated by continuous AGC produced by Q_{205} and Q_{206}. Diode D_{201} is the video detector and Q_{210} is a wideband video amplifier which passes both the Y and chroma signals.

L-5 SOUND SECTION

A separate detector (D_{203}) demodulates the composite IF signal taken from the collector of Q_{203} via C_{229}. Transistors Q_{207}, Q_{208}, and Q_{209} are sound IF amplifiers and diodes D_{204}, D_{205} are the ratio detectors. Note the limiting circuit (clipping by D_{207}, D_{208}) used to remove possible amplitude variations from the FM signal. The audio signal is applied via volume control VR_{901} to the audio amplifiers Q_{701} and Q_{905}.

L-6 *Y* AMPLIFIER SECTION

The luminance or Y signal is initially amplified by Q_{211} and Q_{212}, which operates as an emitter follower. Diode D_{206} clips noise pulses which exceed the nominal video level. Note the delay line DL used to compensate for the delay introduced by band limiting of the chroma signal.

From the emitter of Q_{212} the Y signal is applied to VR_{902}, a level adjustment. After removal of chroma components by the 3.58-MHz trap it is applied to Q_{409}, the Y signal-drive amplifier. This amplifier also operates as an emitter follower and supplies individually adjusted Y signal to the three color video output amplifiers. Blanking is accomplished by Q_{407} and Q_{408}.

L-7 CHROMA AND BURST SIGNAL AMPLIFIERS, COLOR SUBCARRIER RECOVERY

Composite video is applied from Q_{210} to chroma input transformer T_{301} via C_{301}. Transformers T_{301} and T_{302} are tuned to attenuate low-frequency signals and retain mainly the burst and chroma signals. Q_{301} amplifies these and applies them to VR_{904}, VR_{908} for use by the color video amplifier Q_{303}, and to the base of the burst amplifier Q_{305} via C_{308}. Note that the base of Q_{305} receives bias current only at horizontal sync pulse time via R_{331} from the sync separator.

The burst signal is filtered by crystal X_{301}. It provides the synchronizing signal (amplified by Q_{306} and Q_{310}) to lock the local color subcarrier oscillator Q_{311} into phase with the transmitter color subcarrier. VR_{905} permits a fine adjustment for phase to achieve optimum color demodulation.

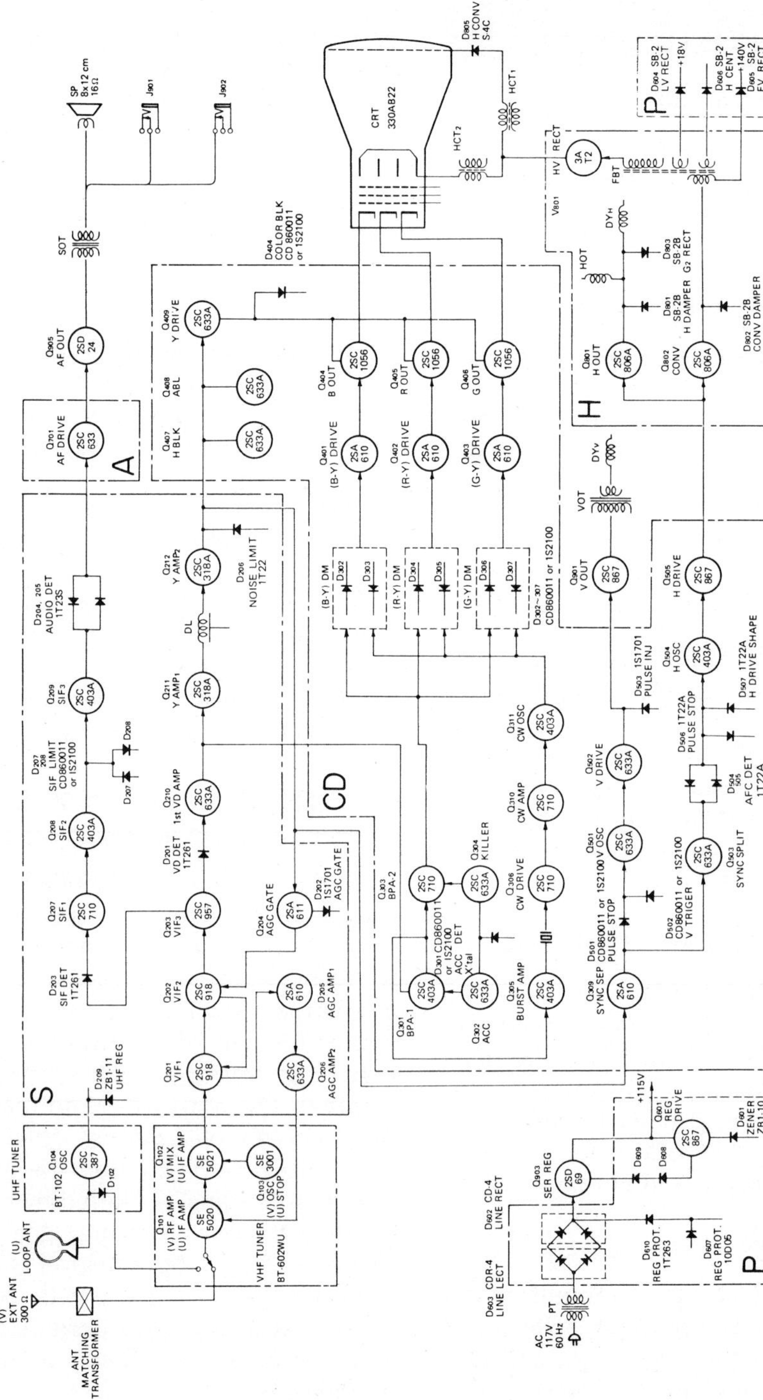

Fig. L-1 Block diagram of Sony KV-1200V color television receiver. *(Courtesy General Distributors Limited)*

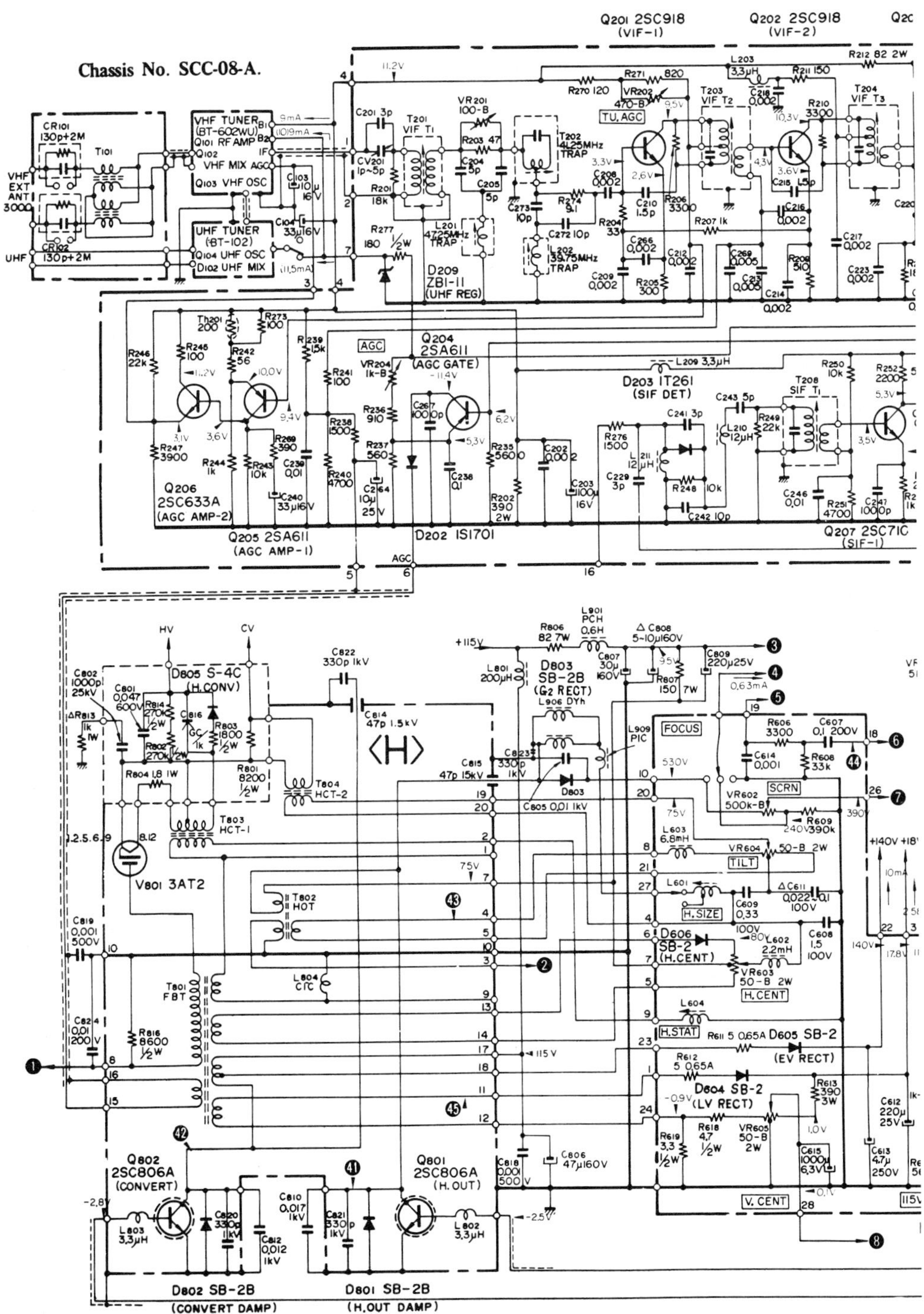

Fig. L-2 (pages 311–314) Sony KV-1200V color television receiver. (*Courtesy General Distributors Limited*)

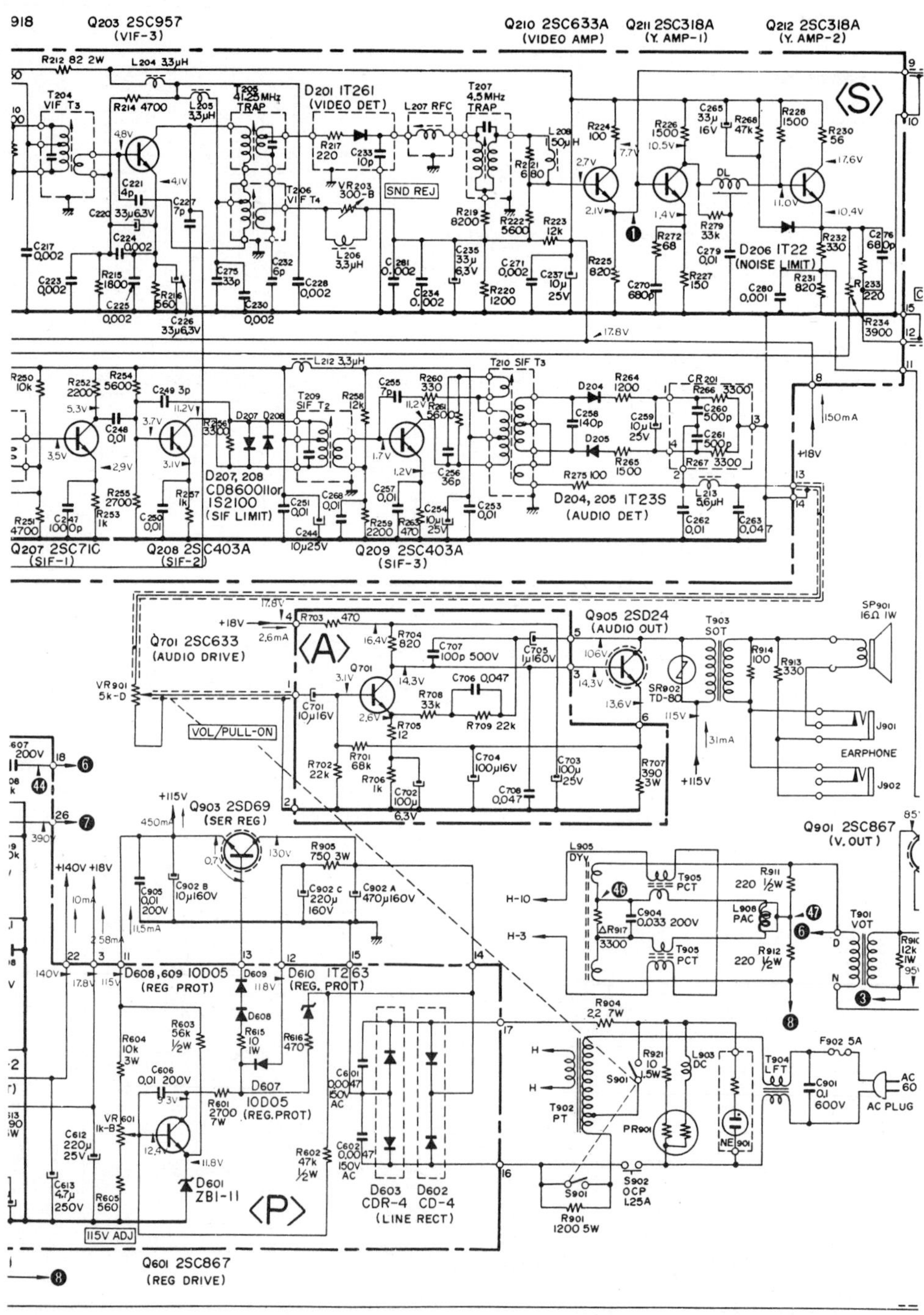

Fig. L-2 (*Continued.*)

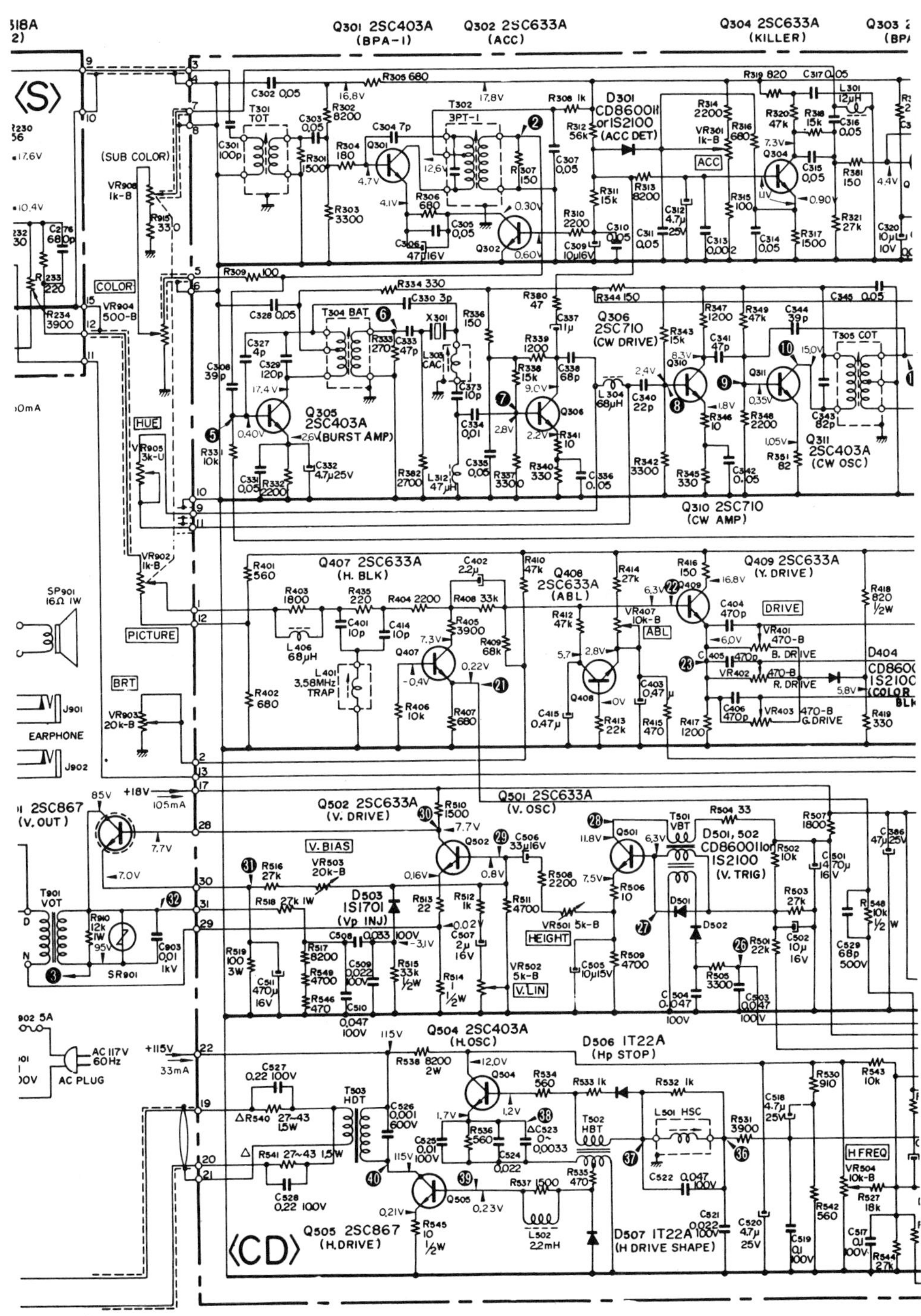
Q301 2SC403A (BPA-I)
Q302 2SC633A (ACC)
Q304 2SC633A (KILLER)
D301 CD86001I or IS2100 (ACC DET)
Q305 2SC403A (BURST AMP)
Q306 2SC710 (CW DRIVE)
Q310 2SC710 (CW AMP)
Q311 2SC403A (CW OSC)
Q407 2SC633A (H. BLK)
Q408 2SC633A (ABL)
Q409 2SC633A (Y. DRIVE)
Q501 2SC633A (V. OSC)
Q502 2SC633A (V. DRIVE)
D501, 502 CD86001I or IS2100 (V. TRIG)
D503 IS1701 (Vp INJ)
Q504 2SC403A (H. OSC)
Q505 2SC867 (H. DRIVE)
D506 IT22A (Hp STOP)
D507 IT22A (H DRIVE SHAPE)
2SC867 (V. OUT)
⟨S⟩
⟨CD⟩
SUB COLOR
COLOR
HUE
PICTURE
BRT
ACC
DRIVE
ABL
V. BIAS
HEIGHT
V. LIN
H FREQ
SP901 16Ω 1W
EARPHONE
AC 117V 60Hz
AC PLUG

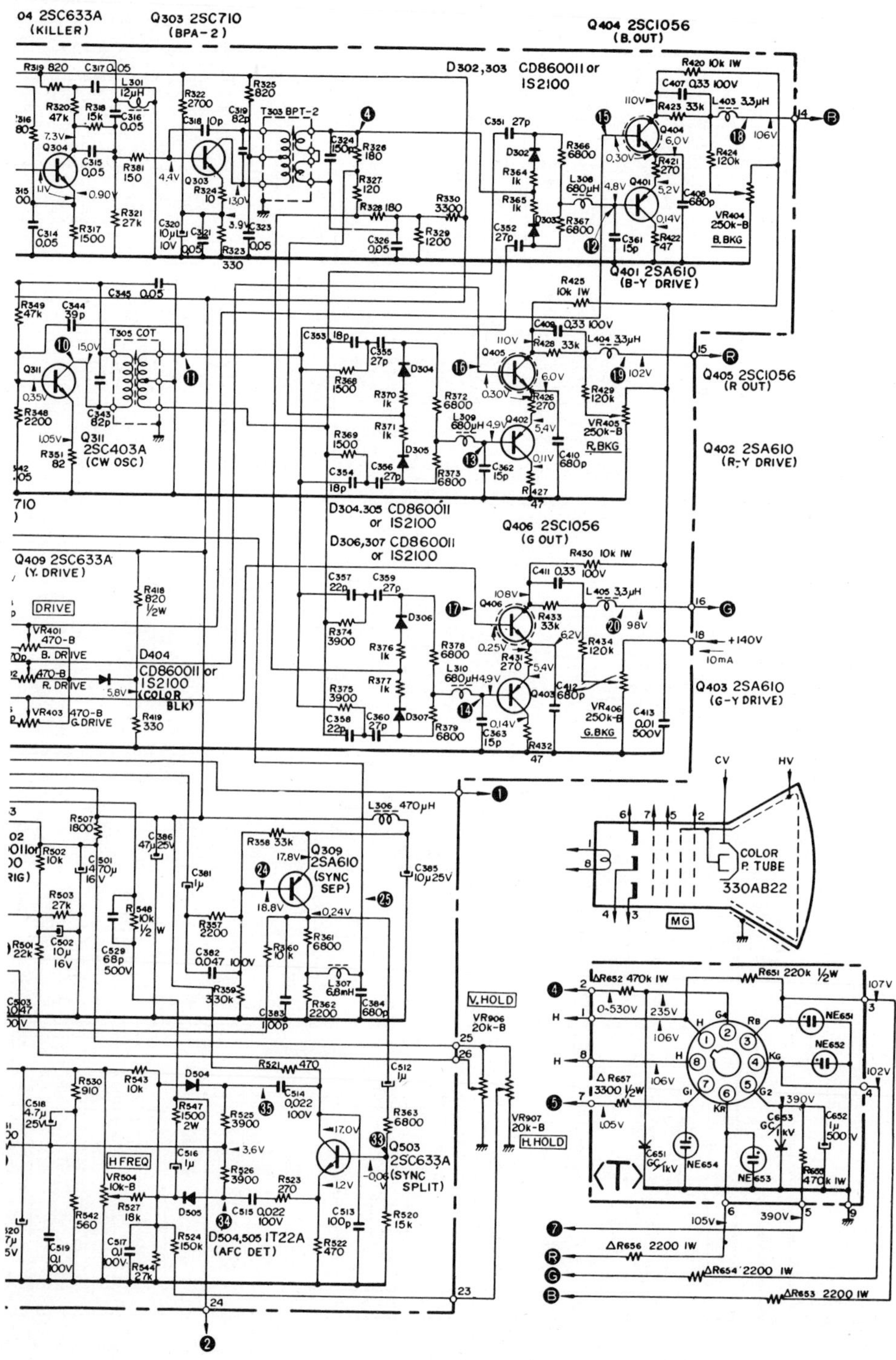

Fig. L-2 (*Continued.*)

UHF TUNER

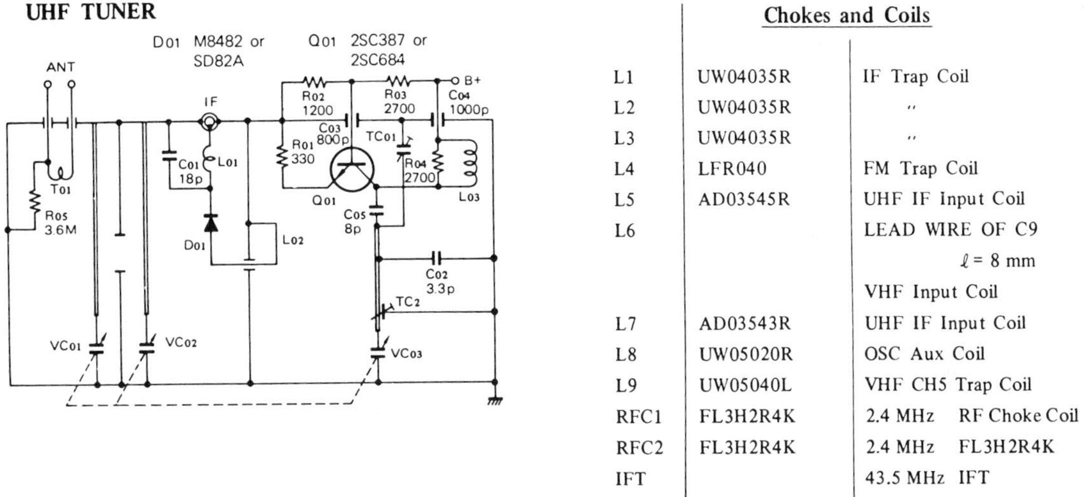

Chokes and Coils

L1	UW04035R	IF Trap Coil
L2	UW04035R	"
L3	UW04035R	"
L4	LFR040	FM Trap Coil
L5	AD03545R	UHF IF Input Coil
L6		LEAD WIRE OF C9 ℓ = 8 mm
		VHF Input Coil
L7	AD03543R	UHF IF Input Coil
L8	UW05020R	OSC Aux Coil
L9	UW05040L	VHF CH5 Trap Coil
RFC1	FL3H2R4K	2.4 MHz RF Choke Coil
RFC2	FL3H2R4K	2.4 MHz FL3H2R4K
IFT		43.5 MHz IFT

VHF TUNER

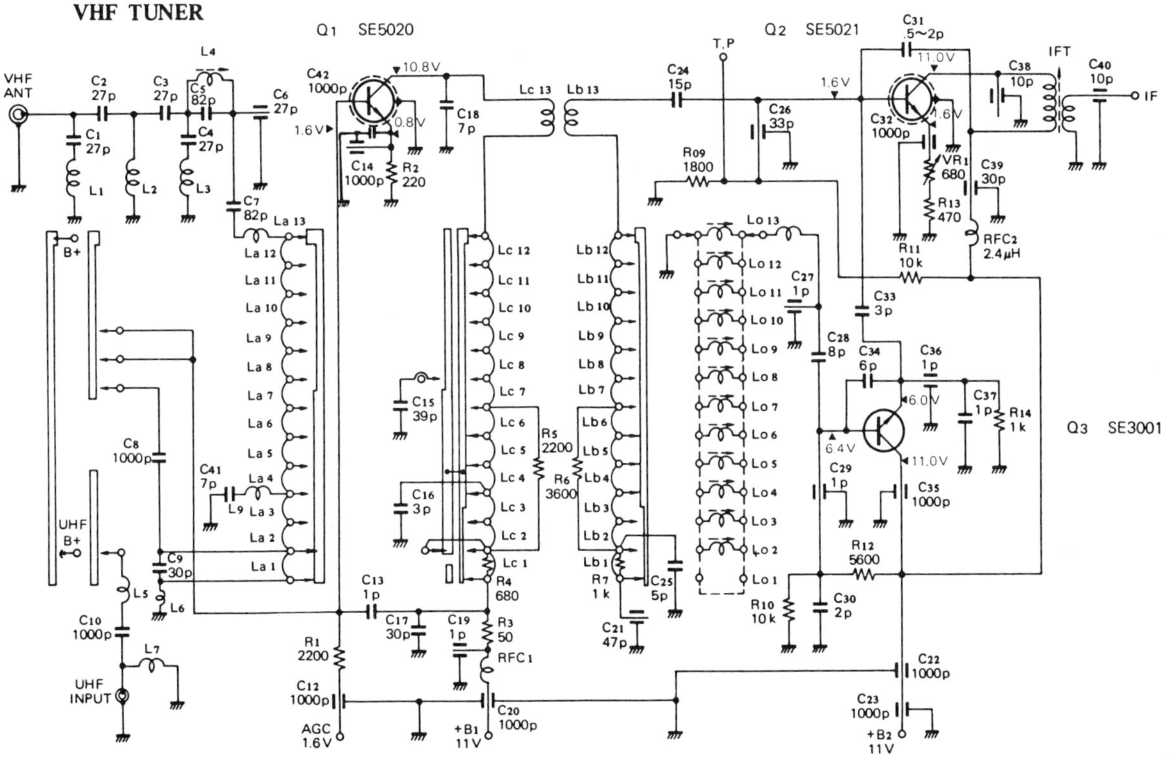

Fig. L-3 Tuner circuit Sony TV-1200V color television receiver. (*Courtesy General Distributors Limited*)

L-8 COLOR DEMODULATORS

Outputs from the color subcarrier cw oscillator Q_{311} and from the chroma amplifier Q_{303} are applied to the synchronous detectors D_{302} through D_{307}. Note that color signals are applied via a resistive matrixing network (R_{326}–R_{330}) to achieve the blending of correct proportions of color video.

L-9 *RGB* OUTPUT AMPLIFIERS AND COLOR KILLER

The three output video amplifiers Q_{401} to Q_{406} are identical. Y video is applied to the upper transistors (Q_{404}, Q_{405}, Q_{406}), but their emitter current is regulated by the color difference amplifiers (Q_{401}, Q_{402}, Q_{403}). In this way addition of $Y + (B - Y) = B$, $Y + (R - Y) = R$, and $Y + (G - Y) = G$ is performed. During a monochrome telecast there is no burst signal and transistor Q_{304} which is normally biased off by burst signal (from Q_{306} via C_{337}) turns on. It essentially shorts the input of the chroma amplifier Q_{303} and prevents color demodulation. Then transistors Q_{401}, Q_{402}, and Q_{403} behave as fixed emitter resistors for Q_{404}, Q_{405}, and Q_{406}. Only Y output then comes from the video output stages and the CRT produces a monochrome image.

L-10 SYNC SEPARATOR AND DEFLECTION CIRCUITS

Video from Q_{212} (from R_{231}) is applied via C_{381} to the sync separator Q_{309}. The vertical sync pulses integrated by R_{360}, R_{505}, C_{503}, and C_{504} are used to synchronize the vertical oscillator Q_{501}. Q_{502}, the vertical driver, and Q_{901}, the vertical output amplifier, complete the vertical-output section.

The horizontal sync pulse is applied to the sync phase splitter Q_{503} where the horizontal oscillator control voltage is developed by D_{504} and D_{505}. Horizontal drive is applied to both the horizontal output deflection amplifier (Q_{801}) and the convergence deflection amplifier (Q_{802}). High voltage for the CRT is provided by rectifier V_{801}.

L-11 DYNAMIC CONVERGENCE DEFLECTION CIRCUIT

Because the screen of the CRT is a relatively flat plane and because each of the electron beams converges at a fixed distance from the guns, they will not converge at all points on the screen. Convergence at the center is usually accomplished by means of small magnets mounted at the neck of the tube. This is termed *static* convergence.

Convergence off center is provided by a second set of deflection coils (convergence coils) which compensate for the loss of convergence due to normal deflection. The normal deflection coils carry ramp-shaped waveforms, but convergence coils operate with parabolic-shaped waveforms. The adjustment of the amplitude of the waveforms is termed *dynamic* convergence. In the circuit of Fig. L-2 the various adjustments are located on the power-supply board P. Note also that in this circuit the high voltage for the CRT is produced by the convergence deflection amplifier which generates only a corrective deflection. It need not expend as much deflection energy as the horizontal output amplifier and can generate the high voltage for the CRT.

index

D

P

Q

R

S

T

U